Meet the Faculty Candidate Poster Session – Sponsored by the Education Division 2018

AA009205

Core Programming Area at the 2018 AIChE Annual Meeting

Pittsburgh, Pennsylvania, USA
28 October - 2 November 2018

ISBN: 978-1-5108-7607-1

Printed from e-media with permission by:

Curran Associates, Inc.
57 Morehouse Lane
Red Hook, NY 12571

Some format issues inherent in the e-media version may also appear in this print version.

Printed by Curran Associates, Inc. (2019)

AIChE
120 Wall Street, FL 23
New York, NY 10005-4020

Phone: (800) 242-4363
Fax: (203) 775-5177

www.aiche.org

Additional copies of this publication are available from:

Curran Associates, Inc.
57 Morehouse Lane
Red Hook, NY 12571 USA
Phone: 845-758-0400
Fax: 845-758-2633
Email: curran@proceedings.com
Web: www.proceedings.com

Meet the Faculty Candidate Poster Session – Sponsored by the Education Division 2018

Core Programming Area at the 2018 AIChE Annual Meeting

Pittsburgh, Pennsylvania, USA
28 October - 2 November 2018

AIChE
The Global Home of Chemical Engineers

TABLE OF CONTENTS

Some abstracts in this book refer to images that were not made available by the publisher

541621 Presentation and Workshop on Career Planning

Abstract:

The transition to a faculty position can be a difficult process. This workshop is intended to help prepare prospective faculty prepare for the search process and successfully transition into an academic position. It will offer advice on identifying suitable career goals and specific objectives, and developing a plan to achieve them. Special attention will be devoted to establishing a funded research program. Elements of the career plan will include:

- Navigating the Job Search Process (research and teaching plans, interviewing, start-up packages)
- Establishing a Research Program (obtaining funding, Career Award preparation, problem identification & selection, contacting funding sources, network development)
- Understanding the University Environment (tenure & promotion, service activities, sabbaticals, reward structure, mentoring expectations)
- Managing Your Time (goal setting & priorities, stress reduction, good habits)

541756 Functional Biomaterials for Smart Delivery of Therapeutics

Abstract:

Research Interests: The rising prevalence of biological therapeutics has created the need to develop safe and effective delivery strategies. Functional biomaterials that interact with their environment can be tuned for precise spatial, temporal, and stimuli-responsive release of therapeutic agents. *My primary research interest is in the rational design and development of readily translatable responsive biomaterials.*

M.Phil. Thesis: "Synthesis and Characterisation of Protein Microgels for Drug Delivery Applications"
Advisor: Prof. Tuomas Knowles, Department of Chemistry, University of Cambridge, UK

My Master's research project focused on the self-assembly of proteins into hierarchically structured materials. During this time, I developed expertise in microfluidic techniques, biophysical characterization, and protein fibrillation. Using droplet microfluidics, I fabricated protein microgels with localized tunable structures and properties, including supramolecular core-shell microgels for the co-delivery of two different classes of therapeutic agents. These protein microgels are comprised of natural materials and formed under mild conditions in aqueous solution, making them promising biomaterials for a variety of biomedical applications.

Ph.D. Thesis: "Approaches for Creating Smart Insulin Delivery Systems"
Advisor: Prof. Daniel Anderson, Department of Chemical Engineering, MIT

The focus of my PhD research is the application of smart biomaterials as glucose-responsive insulin delivery systems. To this end, I have been synthesizing pH-sensitive polymers and formulating them as nanoparticles with glucose-sensing enzymes. I have developed biomaterial-based systems that rapidly respond to changes in blood sugar with changes in insulin release rate *in vivo* as well as a novel therapeutic formulation that affords over 3 weeks of glycemic control in a diabetic mouse model with only two doses. I have simultaneously been developing biomaterial-insulin formulations based on noncovalent interactions and nontraditional glucose sensing moieties. Throughout my PhD, I have gained a variety of skills in chemical synthesis, formulation, *in vitro* characterization and release, and *in vivo* analyses using multiple mouse models. The self-regulated treatments that I have developed provide a promising advancement toward better glycemic control for diabetic patients.

As a future faculty member, I aim to study the interaction of responsive biomaterials with their environment to better understand their role in complex biological systems and inform the rational design of next-generation therapies. Based on the techniques developed during my PhD, I aim to create platform delivery systems that respond to various stimuli to treat a wide range of diseases with the goals of improving patient quality of life and advancing the field of healthcare.

Teaching Interests: Throughout my undergraduate and graduate career, I have sought out teaching opportunities with the goals of increasing my students' intrinsic motivation and helping them develop critical thinking and problem solving skills. As an undergraduate, I taught a seminar course for freshman engineering students and developed a day-long workshop to introduce underrepresented minority middle school students to engineering. During my PhD, I TAed a polymer laboratory class in which I developed a novel laboratory module on glucose-responsive polymers and was invited to return as a guest lecturer the following year. To build upon my teaching experiences, I completed the Kauffman Teaching Certificate Program at MIT, which included 16 hours of formal instruction on course development and best teaching practices.

With these experiences in addition to my training and research in chemical engineering, I am prepared and excited to teach core courses at either the undergraduate or graduate level including Thermodynamics, Transport Phenomena, Fluid Mechanics, Separation Processes, and Chemical Kinetics and Reaction Engineering. I am also passionate about teaching bioengineering-related subjects and aim to develop specialized courses in the areas of Biomaterials and Drug Delivery with a focus on the rational design of systems based on material properties and transport phenomena.

For up-to-date information on my research projects and directions, visit my website: lisarvolpatti.com

542371 Molecular Recognition: From Polymer Science to Precision Medicine

Abstract:

Mission: To engage students in solving real-world healthcare problems through sound engineering fundamentals and rational design principles, within the laboratory and classroom.

Research Interests: Molecular recognition is of utmost importance in a variety of biomedical applications, ranging from targeted drug delivery, to the fabrication of scaffolds for tissue engineering, and diagnostic sensing. There is a need for further investigation in this area, and I plan to pursue a number of complementary research directions:

1. *Fabrication of Biohybrid Materials for Molecular Recognition:* My dissertation work in the laboratory of *Prof. Nicholas A. Peppas* has focused on the development of peptide-modified and molecularly imprinted polymers for targeted drug delivery applications. My work has highlighted the need for cost-effective, environmentally robust materials with natural and synthetic components, which act as *biological actuators*, coupling specific recognition of analytes to a mechanical output.

Relevant Publication: **Clegg, J. R.,** Zhong, J. X., Irani, A. S., Gu, J., Spencer, D. S., & Peppas, N. A. (2017). Characterization of protein interactions with molecularly imprinted hydrogels that possess engineered affinity for high isoelectric point biomarkers. *Journal of Biomedical Materials Research Part A, 105*(6), 1565-1574.

2. *Development of Predictive Models for Protein-Biomaterial Interactions:* Protein adsorption to injectable pharmaceuticals and biomedical implants drives important functions such as biocompatibility, bioactivity, and immune response. However, no existing tools predict the adsorption of proteins to biomaterials.

Relevant Publication: **Clegg, J. R.,** Gu, J., Sun, J., Roy, M., Venkataraman, A., & Peppas, N. A. Elucidating Independent and Combinatorial Effects of Gel Diameter and Hydrophobicity of Protein Adsorption. (in preparation)

3. *Biohybrid Platforms for Disease Modeling and Therapy:* I will use combinations of natural and synthetic components to reverse-engineer the relevant physiological environment of diseased states. Relevant aspects of this process will be mimicking the proper mechanical properties and maintaining the necessary chemical cues to mimic the native environment. This process will test the fields' understanding of a disease, as well as generate drug-screening platforms.

Relevant Publication: **Clegg, J. R**., Wechsler, M. E., & Peppas, N. A. (2017). Vision for Functionally Decorated and Molecularly Imprinted Polymers in Regenerative Engineering. *Regenerative Engineering and Translational Medicine, 3*(3), 166-175.

4. *Next-Generation Biosensors:* I will use combinations of natural and synthetic materials to generate new biosensors, with particular emphasis on using materials as biological actuators into mechanical outputs that are amenable to low resource settings

Relevant Publication: Culver, H. R., **Clegg, J. R.,** & Peppas, N. A. (2017). Analyte-responsive hydrogels: intelligent materials for biosensing and drug delivery. *Accounts of chemical research, 50*(2), 170-178.

Research Funding and Awards: NSF Graduate Research Fellowship, Student Award for Outstanding Research from the Society for Biomaterials (Awarded to one Ph.D. candidate annually).

Teaching Interests: The atmosphere of our classrooms, in parallel with course content, shapes students' perceptions of entire Chemical Engineering disciplines, which has an enduring impact on their life and career trajectory. My teaching mission, as a faculty member, will take *three major forms*:

The **first** is in graduate student education, in the areas of laboratory, technical, and communication-oriented training necessary for success as a professional engineer or educator. My approach will involve clinical practitioners in the design process, while simultaneously involving engineering students in generalizable mixed-methods research, developing transferrable skill to complement their technical prowess for industry or academia.

The **second**, in undergraduate education, will be the development and instruction of both core and elective undergraduate courses through *inquiry-based learning*. At UT Austin, I received formal training in teaching. I completed my Master's degree in STEM Education in May 2018, and was the first student to earn a graduate certificate in engineering education. Building upon this training as a faculty member, I am prepared to teach core (e.g. transport phenomena, thermodynamics, materials) and elective (e.g. polymer engineering, biomaterials, other graduate courses) chemical engineering courses. My inquiry-based approach will place substantial emphasis on applying chemical engineering analysis to understand case studies, solve real-world examples, construct design solutions to open-ended prompts, work in teams, and communicate to diverse audiences, while minimizing emphasis on memorization.

Third, I plan to develop an undergraduate research program, guided by student-centered outcomes, to simultaneously advance the biomaterials discipline and empower the next generation of engineering researchers. I have developed a four-point model for student-centered undergraduate research programs that includes the development of technical expertise, increasing familiarity with uncertainty and experimentation, participation in scientific discourse, and maturation of professional identity.

Relevant Publications to Education Mission: **[1] Clegg J.R.** & Diller K.R. Challenge Based Instruction Cultivates Transferable Skill and Confidence in Biomedical Problem Solving. *European Journal of Engineering Education.* (in revision). **[2] Clegg, J. R.,** & Diller, K. R. (2018). Evolution of Biomedical Engineering Students' Perceptions of Problem Solving and Instruction Stategies During a Challenge-Based Instruction Course. *Proc. American Society for Engineering Education, Annual Meeting.* Salt Lake City, UT, 6/27/18.

542417 Morphological Aspects in Materials for Biotechnological Applications

Abstract:

Research Interests:

My scientific research focuses on the use of both experimental and computational tools to study the effects of morphological aspects of structurally modified or complex heterogeneous materials on the mechanical, electrokinetic and transport behavior.

Research Experience

Transport and separation in structurally modified materials: Research involving polymer gels with embedded nanoparticles of varying properties is quite attractive because of the multitude of potential applications, including separation of biomacromolecules for environmental proteomics; more effective and less time consuming clinical diagnostics, growth of tissue in synthetic or natural scaffold etc. Tuning the nanometer-scale pore structure has been shown to improve mechanical properties and also alter the electrophoretic mobility of the biomolecules thus improving separation efficiency. Using analytical and computational based approaches the heterogeneous structure was modeled by considering the material as a network of small and large domains having idealized geometries connected to each other through which the solute particles transport under different driving forces such as 1D or 2D electric-fields, and pressure gradient. This research also introduced a novel approach to investigate the morphological effects of nanocomposite gel electrophoresis by integrating the numerical simulations based on finite element method and population-based search algorithm such as differential evolution. Execution of the efficient methodology outputs the final optimal values of the parameters and the separation resolution for the range of parameter values specified by the user.

Capturing mechanical behavior of filamentous network in complex biomaterials: From the nano/microscale filaments found in cells, tissue and gels to macroscale high-tensile ropes and cables, many structural components are designed as slender elements. Since slender elements bend more easily than stretch and quickly undergo nonlinear deflections several analytical, and numerical approaches have been proposed to tackle this geometric non-linearity but present different kinds of drawbacks, the most important being their high computation time. In this work, string-of-continuous beams (SOCB) methodology to capture the three-dimensional nonlinear bending dynamics is presented which is applicable for large deflection systems while considerably reducing the computational cost. The SOCB approach serves as a generalized platform for solving the nonlinear deformation of arbitrary 3D networks of slender filaments. The model has been able to successfully capture the unusual mechanics such as the reversible softening exhibited by compressed actin networks in dendritic morphology attributed to filament buckling.

Environmental bioremediation using heterogeneous biofilm matrix: Release of radioactive ^{129}I to the subsurface due to leaking tanks and direct disposal at the U.S. Department of Energy (DOE) Hanford Site has resulted in the generation of Iodine isotopes in groundwater. Speciation of radioiodine has shown that iodate comprises a major percent of iodine . One way to reduce iodate is by the action of biofilms which are predominantly found in natural environment. A biofilm is a surface or interface associated, sessile microbial community embedded in a matrix of self-produced extracellular polymeric substances (EPS). The polyionic nature, physical and chemical heterogeneity of biofilm matrices, combined with the reduced susceptibility of bacterial biofilms to toxicity by inorganic and organic pollutants, make them particularly suitable for bioremediation applications. This work develops a protocol for the proliferation of bacterial biofilms and investigates the potential of these biofilms to reduce iodate. These studies could lead to novel approaches and techniques for the bioremediation of radioiodine at contaminated sites.

Teaching Interests:

In addition to the research, I also have teaching experience, where I have designed and taught classes in Fluid Mechanics, Mass Transfer, Process dynamics and Control, Chemical Engineering Operations. During this time, I had the opportunity to develop the entire course from preparing syllabus to assessment. I was also involved in developing laboratory experiments for undergraduate engineering courses. I have also used various learning methods such as problem, team and project-based learning along with online tools which allow students to access course material and to collaborate on projects at any time. I am experienced in using the state-of-the-art computational engineering tools and would be interested in integrating some of the tools in my lectures. I have always been passionate about transfer of knowledge, and I bring that energy into the classroom to encourage and engage students.

542523 Transformation of Waste Biomass into Bioproducts (Bioenergy, Biomaterials, Biochemicals)

Abstract:

Research Interests:

Effectively partitioning waste biomass into its various fractions and converting these fractions into high value products is of high importance for the valorization and commercialization of bioproducts from bioresources from waste biomass. Waste biomass is released at over 2 billion tons per year in the U. S and has the potential to serve as a source of bioproducts with high commercial value. My research interests involve advancing green engineering and green materials, particularly with regard to their link to new production methods for a *sustainable energy infrastructure*. This typically involves the application of green chemistry/engineering principles, towards the design of sustainable products and processes that minimize the use and generation of hazardous substances. The core of my research strategies fall within the chemical and biomolecular engineering specialty, however, I often draw on key techniques in biology/microbiology, organic/analytical/computational chemistry and materials processing/characterization. In particular my expertise encompasses the development of *effective practices for the valorization of waste bioresources into commercially viable bioproducts (biofuels/bioenergy/biomaterials)*. My unique contribution is to apply the idea of sustainable development (i.e. the triple bottom line—economics, environment and society) towards establishing long lasting research innovations in my chosen field.

Proposed Research Thrusts

- Isolation, Purification and Characterization of Compounds of Significance from Waste Biomass (Biopolymers and Bioactives)
- Application Development (Chemical Functionalization/Deriviatization, Material Formation and Product Formulations)
- Process Development (Life Cycle Analyses, Techno-Economic Analyses and Process Modeling/Scale up)

Teaching Interests:

In addition to my research, I have developed my capacity for engineering principles through hands-on instruction. My major teaching experience is related to the Chemical Engineering curriculum, although, I have also taught Chemistry and Mathematics courses in the past. I have also acquired formal training/experience in teaching lectures through the "Certificate of Accomplishment in Teaching (CoAT)" program at North Carolina State University. As part of this program, I was required to lead at least forty-five hours of lecture, attend six training seminars on fundamentals in teaching and develop a teaching portfolio. During this program, I taught Transport Processes and notably, I co-developed a course titled: *Engineering Challenges at the Energy Frontier.*

Postdoctoral Research (**Fulbright Postdoctoral Fellowship**): Extraction and Recovery of Sinapic Acid from Oleaginous Biomass (Mustard Bran): A Sustainable Access to a Valuable Phenolic Platform Chemical

Supervisor. Professor FLorent Allais, Chaire ABI Agro-Biotechnologies Industrielles-AgroParisTech

Ph.D. Dissertation: Isolation and Recovery of Lignin from Lignocellulosic Biomass Using Recyclable Protic Ionic Liquids (PILs) for a Cost-Effective Biomass Processing Technique.

Ph.D. Advisor: Dr. Wesley Henderson, Chemical and Biomolecular Engineering; North Carolina State University

Selected Publications

1. **Achinivu E**. Protic Ionic Liquids for Lignin Extraction—A Lignin Characterization Study. *International Journal of Molecular Sciences*, **2018**, *19*, 428; DOI:3390/ijms19020428
2. **Achinivu E. C.**, Howard R. M., Li G., Gracz H., Henderson W. A. Lignin extraction from biomass with protic ionic liquids. *Green Chemistry*, **2014**, *16*, 1114-1119; DOI: 10.1039/C3GC42306A

542542 Bio-Inspired, Self-Organizing Soft Materials

Abstract:

Research Interests:

From adhesives, foams, shape-shifting materials, and paints, to the structure of biological cells, soft materials are ubiquitous. Because of their tendency to deform under thermal stresses, soft materials have intriguing characteristics such as self-healing, mechanical adaptivity, and self-organization. I am particularly fascinated by a class of soft materials that derives from biological polymers. Distinct from manufactured materials, these biological materials are active and dynamic, driven by accessory proteins that catalyze enzymatic reactions to create internal stresses and regulate polymer dynamics. Aided by advances in biochemistry and molecular biology that continue to reveal components of the biomolecular toolbox, these biological systems provide exciting frontiers in understanding and designing exotic soft materials. I propose to establish an interdisciplinary and collaborative research program that focuses on elucidating macromolecular mechanisms of controlling material properties and self-organization in soft, biopolymer-based materials.

As a postdoctoral researcher at the University of Chicago, my research focuses on active, biopolymer materials derived from the cytoskeletal proteins which regulate cell mechanics and shape control. In particular, my research focuses on a liquid condensed phase formed from cross-linked biopolymer filaments. These condensates are made of densely packed, aligned filaments, which cause the condensates to be anisotropic. Inspired by colloids and molecular liquid crystals, my research interests involve understanding these anisotropic condensates to create self-organized and shape changing composite assemblies, with structures and properties evocative of biological assemblies. As a graduate student at the University of California, Santa Barbara, my research focused on elucidating mechanisms of self-assembly and deformation in model biological membranes.

Teaching Interests:

I value inquiry-based interactive teaching with a focus on training independent scholars, positioned for interdisciplinary careers. With a Ph.D in an interdepartmental program, Biomolecular Science and Engineering, and undergraduate background in Physics, I have diverse training at the interface of physical and biological sciences and engineering. I look forward to applying my background to teach core courses and am excited to develop courses at intersection of chemical engineering and the physical and biological sciences. I am particularly interested in teaching experimental courses and am especially enthusiastic about introducing undergraduates to research. As a graduate student, I emphasized inquiry as teaching assistant for nine quarters in both upper and lower division, laboratory and lecture courses. Throughout my graduate and postdoctoral research, I have enjoyed mentoring several undergraduate and graduate students in interdisciplinary and collaborative research projects.

542159 Application of Ultrasound for Synthesis of Carbon Capture Microcapsules

Abstract:

Dr. Srinivas Mettu, Research Fellow at The University of Melbourne.

https://scholar.google.com.au/citations?hl=en&user=KtG2bY0AAAAJ&view_op=list_works&sortby=pubdate

Research Interests:

Colloids and Surface Science, Emulsions, Atomic Force Microscopy (AFM), Micro-Fluidics, Soft Matter, Wetting, Food Science.

Teaching Interests:

Colloids and Surface Science, Numerical Methods, Heat Transfer, Transport Phenomena, Thermodynamics, Reaction Engineering

Successful Proposals:

Subsequent to my PhD research, I have always been trying to diversify my research by developing new ideas either in the same area or in an inter-disciplinary area.

1. US $90,000 During my Post-Doc work at Xerox, my research grant was ranked as the top grant in all of Xerox that year and was funded for 3 years for a total of $90,000 by Xerox Foundation (University Affairs Committee) that supports fundamental research at national and international universities. This grant supported a Masters student in the research group of Assoc/Prof Susan Daniel in the Chemical and Biomolecular Engineering Department at Cornell University, Ithaca, USA.

2. AUS $10,000 At the University of Melbourne, I developed a new research project every year from 2014 to 2016 for ECR grant (Early Career Researcher) for funding from Melbourne Research Office on various fundamental aspects of forced contact line movement on soft surfaces. So far, I have received $10,000 from Faculty of Engineering through ECR Scheme.

Postdoctoral Research Experience and Projects:

1. Xerox Research Project: âFundamental interactions of liquids with solids in the context of developing next generation inkjet and offset printing technologiesâ Under the supervision of Dr. Kock-Yee Law.

2. The University of Melbourne, Melbourne, Australia. Worked on measuring the interaction forces between colliding emulsion drops using atomic force microscope (AFM) to correlate it to the stability of emulsions in presence of mixed surfactant systems in relation to food, cosmetics and pharmaceutical emulsions.

Developed emulsion formulations for Confectionery Industry Partner. Under the supervision of Prof. Raymond Dagastine and Prof. Muthupandian Ashokkumar.

3. The University of Melbourne, Melbourne, Australia. Project: âProbing microbial emulsions to break barriers to green oil productionâ Working to developing fundamental surface and interfacial chemistry needed to reduce cost of biodiesel production from Algae. Also working on using Ultrasound to improve extraction efficiency of lipids from Algae cells. Under the supervision of Prof. Muthupandian Ashokkumar and Dr. Greg Martin.

4. PhD Dissertation: Motion of Liquid Drops on Surfaces Induced by Vibration: Role of Contact Angle Hysteresis Under supervision of Prof. Manoj K Chaudhury, Department of Chemical Engineering, Lehigh University, Bethlehem, PA, USA. My PhD thesis was nominated for Stout Dissertation Award.

4. Masters Thesis: Momentum and heat transfer from an asymmetrically confined circular cylinder in a plane channel â Under supervision of Prof. Nishith Verma and Prof. Raj K Chhabra, IIT Kanpur, India.

Research Experience: Throughout my research career, I extensively worked in the area of colloid and surface science. For my PhD thesis, I studied the effects of contact angle hysteresis on movement of liquid drops on a solid surface. As a post-doctoral research scientist at the Xerox Research Centre, I studied the fundamental interactions of complex liquids (colloidal suspensions) with solids in the context of inkjet and offset printing technologies For the last three years (2014 to 2017) at The University of Melbourne, I worked as post-doctoral researcher on the ARC funded ITRP (Industrial Transformation Research Project) Hub on Food Industry Transformation for ASEAN markets (http://foodvaluechain.unimelb.edu.au). I worked closely with Mondelez Inc (Cadbury) on developing new emulsion based confectionery products which are healthy and cheap to produce. In the ITRP project, I also measured the strength of flavor encapsulated microcapsule to test their viability to be embedded into confectionery products. I worked with with Prof Ashok Kumar Muthupandian and Dr Meifang Zhou from Chemistry Department on the microcapsules work. As a post-doctoral research scientist at The University of Melbourne (from 2013 to 2014), I worked in the area of colloids, soft matter and surface science. My research focused on measuring the interaction forces between colliding emulsion drops using Atomic Force Microscope (AFM). This work was done to correlate the fundamental understanding to the stability of emulsions in the presence of mixed surfactant (ionic and non-ionic) systems in relation to food emulsion. This study has a very direct consequence in the design and formulation of detergents, cosmetics and food emulsions using mixed surfactants. At Xerox Research, I contributed extensively to Nanotechnology research group which had a goal of using the recently developed nanotechnology tools to improve and enhance the efficiency of print surfaces used in various Xerox printers.

Current:

July 2017 to current working as a research fellow on Probing microbial emulsions to break barriers to green oil production. With Professor Muthupandian Ashokkumar and Dr. Greg Martin.

- This project uses ultrasonic processing methods to improve the extraction efficiency of biofuels from marine microalgae cells.
- Studying the fundamental surface chemistry of stable microbial emulsions with the goal of breaking the emulsions for oil recovery

Teaching Experience:

1. During my doctoral study at Lehigh, I taught a Physical Chemistry Lab Course for two semesters and evaluated detailed lab reports of undergraduate students to assess and provide suggestions to improve their writing abilities. 2. I also worked as a Teaching Assistant for a Fluid Mechanics course for a semester.

Future Directions: As a faculty candidate I would like to apply the knowledge that I acquired in Colloids and Surface Science through my experience in Emulsion Formulations, Microcapsules fabrication using ultrasound, Extraction of bio fuels from Algae, and Wetting to various fundamental and industrial problems in Chemical Engineering Area. For example, Ultrasound cn be used to synthesize microcapsules for Carbon Capture nad Recovery.

Selected Publications:

1. Sam Q. K. Law, **Srinivas Mettu**, Peter J. Scales, Gregory J. O. Martin "Emulsifying properties of ruptured microalgae cells: Barriers to lipid extraction or promising biosurfactants?" **Published online in Journal** *Colloids and Surfaces B: Biointerfaces*, **2018.** DOI

2. (*Corresponding author) Srinivas Mettu*, Meifang Zhou, Qianyu Ye, Raymond Dagastine and Muthupandian Ashokkumar "Organic liquid-filled microcapsules: **Part 2**: Characterization using AFM and combined AFM-Confocal laser scanning fluorescence microscope" *Soft Matter*, **2018,** *14*, 3192-3201. **DOI:** 1039/C8SM00065D

3. (*Corresponding author) Qianyu Ye, **Srinivas Mettu***, Meifang Zhou, Raymond Dagastine and Muthupandian Ashokkumar "Synthesis of organic liquid-filled microcapsules: **Part 1**: Tuning physical & functional properties" *Soft Matter*, **2018,** *14*, 3202-3208. **DOI:** 1039/C8SM00064F.

4. Srinivas Mettu, Chu Wu, and Raymond Dagastine "Steric interaction forces between emulsified water drops coated with Poly-Glycerol-Poly-Ricinoleate (PGPR) in Canola oil" published online *Journal of Colloid and Interface Science*, **2018,** *517*, 166-175. https://doi.org/10.1016/j.jcis.2018.01.104

5. **(*Corresponding author) Srinivas Mettu**, Joe Berry and Raymond Dagastine "Charge and film drainage of colliding oil drops coated with the non-ionic surfactant C12E5" Published online, **Langmuir**, **2017**, 10.1021/acs.langmuir.6b04632.

6. Joe Berry, **Srinivas Mettu** and Raymond Dagastine "From Reissner to Hertz: Precise measurements of capsule and particle mechanical properties using indentationâ **Soft Matter, 2017,** 13, 1943-1947.

7. (***Corresponding author**) S Tan, **Srinivas Mettu**, MD Biviano, M Zhou, B Babgi, J White, RR Dagastine, and M Ashokkumar "Ultrasonic synthesis of stable oil filled microcapsules using thiolated chitosan and their characterization by AFM and numerical simulations" **Soft Matter, 2016**, 12, 7212-7222.

8. **Srinivas Mettu**, M Zhou, BL. Tardy, M Ashokkumar and RDagastine Temperature dependent mechanical properties of air, oil and water filled microcapsules studied by Atomic Force Microscopy (AFM) **Polymer, 2016**, 102, 333-341.

9. **Srinivas Mettu**, MKanungo, KYLaw, âAnomalous Thermal Induced Pinning of a Liquid Drop on a Solid Substrate **Langmuir, 2013**, 29, 10665-10673.

10. **Srinivas Mettu**, MKChaudhury, âVibration Spectroscopy of a Sessile Drop and Its Contact Line **Langmuir, 2012**, 28, 14100-14106.

11. **Srinivas Mettu**, MKChaudhury, âMotion of Liquid Drops on Surfaces Induced by Asymmetric Vibrations: Role of Contact Angle Hysteresis **Langmuir, 2011**, 27, 10327-10333.

12. **Srinivas Mettu**, MKChaudhury, âStochastic Relaxation of the Contact Line of a Water Drop on a Solid Substrate Subjected to White Noise Vibration: Roles of Hysteresis, **Langmuir, 2010**, 26, 8131-8140.

13. **Srinivas Mettu**, NVerma, and RPChhabra, "Momentum and Heat Transfer from an Asymmetrically Confined Circular Cylinder in a Plane Channel", **Heat Mass Transfer, 2006**, 42, 1037-1048.

542247 Leveraging Mechanistic Understanding of Triacylglycerides and Astaxanthin in Supercritical Carbon Dioxide for Selective Separation Processes from Microalgae

Abstract:

Research Interests: Sustainable technologies for biomass refinement and valorization. I have expertise in supercritical carbon dioxide and seek to investigate novel mixtures of supercritcal fluids as green solvents as well as leveraging molecular compression and expansion for advanced bio-materials, separations, and conversion technologies.

Teaching Interests: Developing students to develop as engineers with an entrepreneurial mindset is critical given the rapid adoption of technology in our society. I am interested in bridging the maker-movement with chemical engineering to prepare and expose students with entrepreneurship as well as provide relevant and modern context to our discipline. Secondly, educating students to be sustainably oriented is equally important and I seek integration of environmental, social, and economic aspects of chemical engineering in my teaching.

542256 Rational Fabrication of Biomaterial-Based Scaffolds, Devices and Films for Tissue Engineering, Drug/Gene Delivery, Biomedical Processes and Flexible Electronics

Abstract:

Research Interests:

Biodegradable Polymer-based Functional Scaffolds for Stem Cell Therapies and Tissue Regeneration:

Mimicking the complex 3D microenvironment of natural extracellular matrix is crucial to provide an available environment for cellular attachment, growth, proliferation, spreading, migration and differentiation.(1,2) Our research is concentrated on biomaterial design for stem cell-laden implantable scaffolds-based cellular therapies with a particular interest in stem cell differentiation targeting neuroregeneration. We uniquely integrated various physical and chemical cues on porous polymeric films, which possess longitudinal surface micropatterns and surface concentration gradient of neurotropic factors providing directed nerve growth as well as neurotropic factors loaded microparticles throughout the film matrix providing controlled release properties.(3) In addition, we demonstrated that the 3D microstructural and mechanical properties of the scaffolds has profound effect on stem cell differentiation.(4) Our results suggested that the scaffolds with larger pore size and higher porosity and possessing less stiff structure are providing the most favorable environment for differentiation of mesenchymal stem cells (MSCs) into Schwann cell-like phenotypes (SCs) via chemical induction. Recently, we have shown that the electrical stimuli applied through conductive graphene circuits developed on 2D substrates can be used as an alternative strategy for differentiation of MSCs into SCs.(5) This study selected as cover art for Advanced Healthcare Materials Journal, received **Star Award from the Society for Biomaterials and resulted in provisional patent filing.** Recently, we identified significantly regulated proteins and related cellular pathways upon transdifferentiation of MSCs into SCs via chemical stimuli.(6) This comparative proteomics analysis provided valuable information about specific proteins and pathways that play significant roles in myelination, axonal guidance, cytokine and growth factor secretion. Based on this background, we will focus on designing stem cell-laden, implantable, 3D bioprinted scaffolds to investigate material-cell and cell-cell interactions and *in vitro* and *in vivo* stem cell differentiation and neural regeneration. The future directions will be to serve the ultimate goal of making realistic 3D tissue models that mimic the actual cellular arrangement of a neural tissue using 3D bioprinting.

Nanoscale Drug/Gene Delivery System Design for Cancer Treatment and Theranostics:

Currently applied treatments for cancer, involving surgical cytoreduction followed by systemic cytotoxic agent chemotherapy, have not provided desired outcomes and resulted in chemo-resistance and tumor relapse.(7) As an alternative, RNA based therapies (microRNAs (miRNAs) and small interfering RNAs (siRNAs)) that influence major driver pathways in cancer development by directing sequence-specific degradation of target messenger RNAs (mRNA), have been proposed as a potential therapeutic for cancer treatment.(8) However, the potential clinical application of RNA therapy is limited by challenges related to delivery, stability, transport, diffusion, cellular entry, endosomal escape, efficient release and activity in the cytoplasm as well as the co-incorporation of drug molecules and RNAs in a single nano delivery device.(7,8) We have developed stimuli responsive polymer, peptide and inorganic nanoparticle based small interfering RNA (siRNA) delivery systems for cancer treatment. Our works have demonstrated that the pH and temperature responsive, cationic, amphiphilic and biocompatible multi-block copolymer-base systems are able to electrostatically condense and protect siRNA as well as providing enhanced cellular uptake and endosomal escape due to their temperature and pH responsive micellization enhancing the biological siRNA activity.(8) The cell penetrating and fusogenic peptide based peptideplex and conjugate systems are also alternative platforms for effective siRNA delivery.(9) We have also showed an efficient use of gold nanoparticles for theranostics and siRNA delivery applications.(8,10) The gold nanoparticle based multicomponent systems developed through subsequent layer-by-layer addition of siRNA and stimuli responsive polymer or cell penetrating and fusogenic peptide layers resulted in enhanced siRNA delivery and activity.(8,9,10) We have recently concentrated on developing nanoscale delivery platforms for co-delivery of siRNA and/or miRNA with chemotherapeutic drugs to target various intracellular pathways to reduce drug resistance and increase drug activity for ovarian and pancreatic cancer treatment.(11) As future direction, we will focus on the development of functional nanoscale delivery systems for efficient delivery of genome editing tools (Crispr-Cas9-sgRNA), along with chemotherapeutic agents for cancer treatment.

Design of Biomimetic, Polymer based Films for Biomedical Applications and Development of Biodegradable Flexible Electronic Films:

The objective of my research is to develop biomimetic synthetic membranes that are responsive to multiple external stimuli and exhibit selective transport capabilities, anti-fouling and self-cleaning features as in biological cell membranes. These bioinspired synthetic membranes are potentially promising for various biomedical applications including cell sorting, cell transfer, protein separation, hemodialysis and so on. We developed various polymer based membranes using dry or dry-wet phase inversion techniques and further carried out surface modification via layer-by-layer approach or surface chemistry. We developed nano/ultrafiltration membranes with enhanced surface properties to prevent chemical or biological fouling for waste water treatment applications.(12,13) We developed active agent incorporated polymeric membranes and developed different controlled release strategies (i.e. changing the membrane structure, incorporating nanoparticles in membrane matrix etc.) to provide sustained release, antibacterial activity and increase the shelf life of foods.(14,15) Recently, we have proposed bioinspired, biomimetic membranes functionalized through surface initiated atom transfer radical polymerization and protein localization to mimic selective separation, anti-fouling and self-cleaning properties of natural cell membranes for biomedical applications. This study received the Baxter Young Investigator Award. We will work on the development of functional polymeric films for biosensor and microfluidics applications. In alignment with these studies, another aspect of my current research is focused on developing biodegradable polymer based, 3D microstructured, flexible electronic devices/films using a novel approach coupled with 3D origami as an easy, fast, cost effective and green fabrication method. Our efforts for this study have resulted in submission of an invention disclosure and together with 3D bioprinting, it will be the main foundation of my future research endeavors. These flexible electronic devices hold promising potential to be used in various biomedical, tissue engineering and robotics applications. Thus, we investigate the biointerface of these devices for the purpose of microfluidics and organ/lab-on-a-chip applications for neural regeneration, which will be the main future directions.

Teaching Interests:

I am interested in teaching fundamental chemical and biological engineering principles and helping students understand how to use these principles in current technological problems. My teaching philosophy is based on embracing diversity/inclusion and using various learning methods such as, flipped classroom, problem, team and project-based learning, along with technological and online teaching tools in the class. I served as a teaching assistant for various chemical engineering undergraduate core courses and recently taught 3 credit chemical engineering undergraduate course "CHE 160: Chemical Engineering Problems with Computer Applications Laboratory" at Iowa State during 2015 Fall Semester. In addition to teaching core courses of chemical and biological engineering, I am also interested in curriculum development.

References:

1. **Uz, M,** and Mallapragada, S.K.; Smart Materials for Nerve Regeneration and Neural Tissue Engineering. Smart Materials for Tissue Engineering: Applications, Qun Wang (Ed.), Royal Society of Chemistry, 2017, pp.382-408. *(Print ISBN: 978-1-78262-484-4)*
2. J. Sandquist, **M. Uz,** A.D. Sharma, B.B. Patel, S.K. Mallapragada, D.S. Sakaguchi, Stem Cells, Bioengineering, and 3-D Scaffolds for Nervous System Repair and Regeneration: G.L. Zhang and L.D. Kaplan (Eds.), Neural Engineering: From Advanced Biomaterials to 3D Fabrication Techniques, Springer International Publishing 2016, pp. 25-81.
3. **Uz, M.,** Sharma, A., Adhikari, P., Sakaguchi, D., and Mallapragada, S.K., "Development of Multifunctional Films for Peripheral Nerve Regeneration Conduits, *Acta Biomat.* (2017). Available online. https://doi.org/10.1016/j.actbio.2016.09.039
4. **Uz, M.,** Buyukoz, M., Sharma, A., Sakaguchi, D., Altinkaya, S., and Mallapragada, S.K., "Gelatin-Based 3D Conduits for Transdifferentiation of Mesenchymal Stem Cells into Schwann-cell Like Phenotypes", *Acta Biomat.*, 53, 293-306 (2017).
5. Das, S.,# **Uz, M.,#** Ding, S., Lentner, M., Hondred, J., Cargill, A., Sakaguchi, D.S., Mallapragada, S., and Claussen, J., "Electrical Differentiation of Mesenchymal Stem Cells into Schwann Cell-Like Phenotypes using Inkjet Printed Graphene Circuits", *Healthcare Mater.*, 6(7), 1601087 (2017). Journal cover image. doi: 10.1002/adhm.201601087. *(# These authors put equal contribution)*
6. Anup D. Sharma, Jayme Horning, **Metin Uz,** Pawel Ciborowski, Surya K. Mallapragada, Howard E. Gendelman, and Donald S. Sakaguchi; "Proteomics Analysis of Mesenchymal to Schwann Cell Transdifferentiation.", *Journal of Proteomics, June 2017, In Press,* https://doi.org/10.1016/j.jprot.2017.06.011
7. **Uz, M.;** Altinkaya, S. A.; and Mallapragada, S. K.; Stimuli responsive polymer-based strategies for polynucleotide delivery. Journal of Materials Research, April 2017. (DOI: https://doi.org/10.1557/jmr.2017.116)
8. **Uz, M.;** Mallapragada, S. K.; Altinkaya, S. A., Responsive Pentablock Copolymers for siRNA Delivery. RSC Advances 2015, 5 (54), 43515-43527.
9. **Metin Uz,** Volga Bulmus and Sacide Alsoy Altinkaya; Evaluation of Cell Penetrating and Fusogenic TAT-HA2 Peptide Performance in Peptideplex, Multicomponent and Conjugate siRNA Delivery Systems. *(Under review, draft manuscript available upon request)*
10. **Uz, M.;** Bulmus, V.; Alsoy Altinkaya, S., Effect of PEG Grafting Density and Hydrodynamic Volume on Gold Nanoparticle–Cell Interactions: An Investigation on Cell Cycle, Apoptosis, and DNA Damage. Langmuir 2016, 32 (23), 5997-6009.
11. **Metin Uz,** Satyanarayana Rachagani, Surinder K. Batra and Surya K. Mallapragada; Dual Delivery Nanoscale Device for miR-345 and Gemcitabine Co-Delivery to Treat Pancreatic Cancer. *(Under review, draft manuscript available upon request)*
12. **Uz, M.;** Mahlicli, F. Y.; Polat, M.; Altinkaya, S. A., Characterization of Polysulfone Based Hemodialysis Membranes by AFM. *Procedia Engineering* 2012, 44, 1166-1167.
13. **Metin Uz,** Filiz Yasar Mahlicli, Erol Seker, Sacide Alsoy Altinkaya; Silver Release from TiO_2-nAgCl Xerogels Incorporated Mixed Matrix Ultrafiltration PAN Membranes in the Presence of Excess Chloride Ions: Long-term Silver Retaining, Biofouling Resistance and Antibacterial Activity. *(Under review, draft manuscript available upon request)*
14. Ozer P.B.B. **‡; Uz M. ‡;** Oymaci P.; Altinkaya S.A, Development of a Novel Strategy for Controlled Release of Lysozyme from Whey Protein Isolate Based Active Food Packaging Films. *Food Hydrocolloids* 2016, 61, 877–886. (*‡These authors contributed equally to this work.*)
15. **Uz, M.;** Altınkaya, S. A., Development of Mono and Multilayer Antimicrobial Food Packaging Materials for Controlled Release of Potassium Sorbate. LWT - *Food Science and Technology* 2011, 44 (10), 2302-2309.

542261 Biomolecular Engineering and Ultrasound-Enhanced Transport in Neuroscience

Abstract:

Research Interests:

The Szablowski laboratory will develop technologies for perturbing neural circuits to study brain function and treat neurological and neuropsychiatric disease. We will achieve these goals through a combination of unique engineering strategies aimed at overcoming transport limitations in the brain using focused ultrasound, and allowing specific and efficient control of neuronal cells through biomolecular engineering.

Ultrasound can be used to deliver molecules locally to tissues with millimeter precision, a level of accuracy required to selectively modulate brain circuits. Ultrasound-mediated gene delivery will allow us to achieve spatially selective long-term perturbation of brain circuits in a fully non-invasive manner. We will combine enhanced transport across tissue boundaries with biomolecular engineering. This work will utilize genetic engineering, protein receptor engineering, and next-generation sequencing to develop tools to noninvasively modulate neurons and investigate circuit function. Neuroscience is central to human health and due the complexity of the brain, controlling neural circuits is a perfect testing-ground for creative biomolecular engineering strategies. This research program will establish new engineering principles and yield discoveries in both basic science and translational medicine.

Research experience

Postdoctoral: Ultrasound-Modulated Transport of Genetic Vectors Into the Brain for Non-Invasive Control of Neural Circuits (Advisor: Mikhail G. Shapiro, Chemical Engineering, Caltech, 06/2015-current)

The treatment of neuropsychiatric and neurological diseases, such as mood disorders, or epilepsy, requires control of the right cells within the right brain region at the right time. However, currently, this cannot be achieved noninvasively. For example small molecule therapeutics can act on selected protein receptors, but are delivered non-selectively throughout the brain. On the other hand deep brain stimulation can target a specific region within the brain but requires a surgery and a permanent intracranial placement of an electrode.

During my postdoctoral training I initiated a new area of research to develop a methodology that can be both noninvasive and have regional, cell-type, and temporal specificity of neuromodulation. This technology, called Acoustically Targeted Chemogenetics[1] (in press, Nature Biomedical Engineering), or ATAC, combines three cutting-edge techniques with clinical potential. First, ultrasound is used for gene delivery across blood-brain barrier (BBB). When microscopic vesicles containing gas are infused intravenously focused ultrasound (FUS) stimulates their oscillations which exert weak pressure on the brain vasculature temporarily loosening junctions between endothelial cells thus opening the BBB (BBBO). Consequently, viral vectors that cannot pass through an intact BBB, but readily do so after BBBO, can be selectively delivered to a sonicated brain region. Second, by using those viral vectors to deliver engineered protein receptors that respond to BBB-permeable bio-inert drugs (chemogenetic receptor-ligand pairs), selected neurons can be controlled pharmacologically, which results in noninvasive neuromodulation. And third, by using cell specific promoters a subpopulation of neurons can be targeted with spatial accuracy beyond what is offered by FUS. In this project, we pharmacologically activated and inhibited neurons specifically targeted by ATAC. We then used this technique to inhibit formation of traumatic memories in mice and demonstrated a new method of sub-millimeter accuracy of targeting. Current efforts include further improvements of safety and costs, and translation of ATAC into non-human primates.

In addition to this main project, I contributed to another effort focused on engineering new biomolecular reporters for magnetic resonance imaging[2] (Nature Materials, 2018), and co-authored two comprehensive reviews on the combination of biomolecules and ultrasound.

Graduate: Molecular Engineering of programmable therapeutics for oncology: drug delivery and mechanism of action (Peter Dervan, Bioengineering, Caltech, 09/2009-05/2015).

Altering gene expression is one of the most attractive therapeutic strategies. Multiple diseases, such as cancer, are a result of improper gene regulation. Additionally, DNA has a simple and predictable structure that is independent of its sequence, which allows similar compounds to target multiple DNA sequences and thus multiple diseases. Consequently, a number of therapeutic approaches focused on DNA have emerged such as CRISPR gene editing, RNA interference, or gene therapy. However all of these methods require gene delivery which is more expensive and less established than use of small molecule therapeutics. The Dervan lab has developed small molecules that are cell-permeable, have pharmacokinetics and biodistribution of small molecules, but can bind to DNA sequence-specifically to regulate gene expression similarly to genetic approaches.

In my graduate work I focused on understanding *in vivo* selectivity, mechanisms of action, and clinical applicability of DNA-sequence specific small molecule drugs. We evaluated gene expression effects and biodistribution following polyamide treatment in a mouse model of breast cancer and glioma. Using intravital bioluminescence imaging and next generation mRNA sequencing studies, we discovered that one compound targeted to the Estrogen-Response-Element was capable of downregulating estrogenic gene expression in tumors[5]. We also evaluated pharmacokinetics and biodistribution of Py-Im polyamides in both tissues and tumors[4].

For my next project[3], I focused on one of the most critical pathways in cancer – the hypoxic response. Resistance to hypoxia is one of the hallmarks of tumor cells and is one of the major sources of resistance to radiation and chemotherapy. Existing anti-angiogenic therapies, which attempt to starve tumors of their blood supply, can actually make tumors more resistant to hypoxia, thereby promoting tumor progression and metastasis. I attempted to develop an antiangiogenic therapy that does not have this confounding effect. Using a Py-Im polyamide compound targeted to Hypoxia Inducible Factor-1 Response Elements (HRE), I discovered that systemic administration of this compound reduced tumor growth, inhibited blood vessel formation, and sensitized the tumor cells to hypoxia. The Py-Im polyamides are being actively translated to the clinic as cancer therapeutics by Dr. Dervan (through his startup Gene Sciences Inc. in San Diego).

Teaching Interests: My teaching interests include the fundamentals of chemical and biomolecular engineering and an elective focus on therapeutics. Thanks to my undergraduate and graduate training in highly quantitative Bioengineering programs, I have experience with core thermodynamics, transport and kinetics concepts that are critical for Chemical Engineering students. I will leverage my graduate and postdoctoral training and provide students with an ability to intuitively understand how molecules interact and distribute in cells, tissues and organisms. This knowledge will allow them to apply what they learned in core courses to design and engineer new materials and therapeutics. In the future I would like to develop a course on molecular recognition and selectivity *in vivo*.

Teaching experience:

Mentoring: 2 undegraduate, 1 graduate student (*Caltech, 2016-present)*

- Guest lecturer, ChemE 188, Molecular Imaging (*Caltech, 2017-present*)
- Biology Tutor for Biology Undergraduate Education and Office of Minority Education (*MIT, 2007-2008*)
- Teaching a seminar course in introductory biology, for Office of Minority Education (*MIT, 2008*)

Selected publications (out of 12 total)

First-author contributions are underlined

1. **Szablowski JO**, Lue B, Lee-Gosselin A, Malounda D, Shapiro MG, Acoustically Targeted Chemogenetics for Noninvasive Control of Neural Circuits, *in press, Nature Biomed. Eng.* (2018)
2. Lu GJ, Farhadi A, **Szablowski JO**, Barnes SR, Lakshmanan A, Bourdeau RW, Shapiro MG, Acoustomagnetic imaging with gas-filled protein nanostructures, *Nature materials 17 (5), 456* (2018)
3. **Szablowski JO,** Raskatov JA, Dervan PB. An HRE-binding Py-Im polyamide impairs hypoxic signaling in tumors. *Mol. Cancer Ther.* 15 (4), 608-617 (2016)
4. Raskatov JA, **Szablowski JO**, Dervan PB, "Tumor Xenograft Uptake of a Py Im Polyamide Varies as a Function of Cell Line Grafted", *J. Med. Chem.*, **57**:8471-8476 (2014)
5. Nickols NG**#**, **Szablowski JO**#, Hargrove AE, Li BC, Raskatov JA, Dervan PB. "Activity of a Py-Im Polyamide Targeted to the Estrogen response Element," *Mol. Cancer Ther.*, **12**:675-684, (2013). # co-first authors, (Article selected as one of the 'AACR hot topics, 2013').
6. Shapiro MG**#**, Westmeyer GG**#**, Romero P, **Szablowski JO**, Küster B, Shah A, Otey CR, Langer R, Arnold FH, & Jasanoff AP, "Directed evolution of an MRI contrast agent for noninvasive imaging of dopamine". *Nature Biotech.*, 28:264–270 (2010), # co-first authors

Patents and patent applications

1. Szablowski JO, Shapiro MG, Acoustically Targeted Chemogenetics, CIT File No.: CIT-7921-P, Provisional filed: 12/7/2017
2. Lu G, Farhadi A, Szablowski JO, Shapiro MG, Gas Filled Structures and related compositions, methods and systems for magnetic resonance imaging, CIT File No.: CIT-7580-P, Provisional filed: 7/28/2016, Patent application filed: 7/28/2017
3. Szablowski JO, User-adjustable knee orthosis for patellar instability and related disorders, CIT File No.: CIT 12-216, Provisional Filed: 8/7/2012

542274 Protein Engineering for Cell- and Ligand-Based Immunotherapy

Abstract:

Research Interests: Immunotherapy is a treatment method that harnesses a patient's own immune system to fight disease. One such technology, chimeric antigen receptor (CAR) engineered T cells, has reinvigorated cancer immunotherapy with outstanding clinical success treating subsets of leukemia and lymphoma. A CAR is a synthetic molecule that directs a T cell to bind to a cancer-associated marker that it wouldn't ordinarily recognize, causing the T cell to respond in a disease-targeted manner. Development of these receptors is often empirical with minimal optimization; previously established domains are fused to mediate cancer recognition and T cell response. Currently, the biggest challenge in this field is lack of potency when applying these therapies to diseases outside of hematologic malignancy. This stems from two disparate areas: 1) lack of optimization of new CARs can yield undesirable characteristics such as low expression, tonic signaling, or the inability to control tumors *in vivo*, and 2) the microenvironment of many tumors is suppressive to T cells and other immune cells, leading to a lack of therapeutic impact. My laboratory will address these challenges 1) by building a fundamental understanding of the impacts of domain-level alteration on CAR function via protein engineering directed by natural homology, structure-guided design, and macromolecular modeling and 2) by studying novel combination therapies – via hypothesis-driven experiments guided by protein and immune cell profiling in disease models and patient samples – for their ability to condition the tumor microenvironment to CAR T cell response.

The lab's long-term goal is to apply our experience in protein engineering, immunotherapy development, and translational research to design new CARs that are optimized for therapeutic impact and pair these with engineered "companion" molecules to increase their impact on difficult-to-treat cancers. To meet these ends, we will employ protein engineering and high-throughput screening techniques to understand the impact of structure-function relationships in individual domains of CARs on therapeutic potential, leading to development of techniques for efficient optimization of these domains. Our experience in protein design and expression will be used to engineer and study companion molecules and their effects on the tumor microenvironment. Collaboration will be sought for clinical translation in the hope of turning our fundamental scientific advances into technologies with a positive impact on patients and their families.

Research in my group builds from the fundamentals of biology, chemistry, engineering, and immunology. This broad knowledge will allow us to adapt our approaches to diseases other than cancer, with potential applications in autoimmune disease and infectious disease.

Our initial investigations will focus on three primary aims:

1. **Engineering CARs for unaddressed cancer biomarkers:**

Although numerous CARs have been developed preclinically, substantial limitations remain and many diseases remain unaddressed. With our expertise in protein selection, we are uniquely positioned to develop new CARs from *de novo* approaches. We will engineer candidate antigen recognition domains with systematic variation in multiple design elements to study the interplay of affinity, antigen density, and bound epitope geometry on downstream signaling and anti-tumor effect in a variety of diseases. These studies will yield new therapeutic options, while also providing a blueprint for the field to break its reliance on previously discovered antigen recognition sequences.

2. **Understanding and optimizing domains of CARs for potent, safe responses:**

Previous CAR development has been largely empirical, relying on fusing a binding domain to known natural domains for membrane anchoring, signaling, and co-stimulation. We aim to understand how engineering the transmembrane, co-stimulatory, and stimulatory domains can lead to changes in CAR expression, potency and selectivity of response, and downstream signaling. Although initial studies will focus on engineering natural domains, future work will involve developing panels of synthetic domains capable of tuning CAR response for the unique challenges of individual diseases.

3. **Engineering companion molecules for modification of the tumor microenvironment:**

Suppression of CAR T cells and other immune populations in the tumor microenvironment poses a major challenge to immunotherapy. We aim to understand - via proteomic and immune cell profiling techniques - the combinations of suppressive molecules present in individual diseases and build proteins capable of antagonizing these axes. We will study these molecules alone and in combination with CAR T cells in appropriate mouse models to elicit effects of these treatments on immune cell recruitment and anti-tumor response.

Successful Proposals: Postdoctoral Fellowship Award, Tobacco-Related Disease Research Program (2018); Parker Institute for Cancer Immunotherapy Grant (Co-PI, PI: Xiuli Wang) (2017); NIGMS T32 Predoctoral Fellowship (2012)

Postdoctoral Project: *Yeast Surface Display Techniques Enhance Development of Chimeric Antigen Receptors*, Beckman Research Institute of the City of Hope (Advisor: Dr. Stephen J. Forman)

PhD Dissertation: *Cellular Selections Aid Translational Binding in Ligand Discovery*, Department of Chemical Engineering and Materials Science, University of Minnesota – Twin Cities (Advisor: Associate Professor Benjamin J. Hackel)

Research Experience: My research focused on the optimal development of proteins to meet diagnostic and therapeutic needs. These projects were driven by my education in chemical engineering and my training in protein engineering, molecular biology, and high-throughput screening. Much of this work was collaborative, allowing me the privilege of working with engineers, biologists, chemists, and clinicians to meet our goals. I believe these experiences position me well to work both as an independent investigator and as a member of a thriving research community.

My graduate work focused on a prevalent but rarely published problem in engineering protein-protein interactions: standard methods for *in vitro* development of binding ligands using recombinant target proteins often yield interactions that are not recapitulated with cell-expressed target proteins. To meet this challenge, I optimized and developed yeast surface display methods for selecting binding ligands directly against mammalian cell monolayers, reducing the prevalence of reagent binders. In addition to methodological contributions, I engineered ligands that are currently under study for molecular imaging and cancer therapeutics.

My postdoctoral research leverages my knowledge of protein engineering for the development and optimization of CARs with translational focus. Many CARs in development face challenges related to potential toxicities. One such toxicity is the formation of an immune response against the binding domain of the CAR, a molecule which is often of non-human origin. A second toxicity is off-tumor killing of healthy tissue that expresses molecules considered cancer biomarkers. To meet these challenges, I am currently demonstrating high-throughput methods for reduction in immunogenicity and optimization of CAR selectivity to spare healthy tissue while mediating potent anti-tumor response. In addition to developing candidates with translational intention, the parameters that I am studying will help develop a mechanistic understanding of CAR function that the field currently lacks.

Teaching Experience: During my PhD, I served as a teaching assistant for a core undergraduate Chemical Engineering course (Introduction to Biomolecular Engineering) and a core graduate course (Physical and Chemical Thermodynamics). My duties included grading assignments and exams, and holding weekly office hours to supplement lecture. As a senior graduate student, I became a recitation instructor for Introduction to Biomolecular Engineering. I led a weekly discussion section in supplementary lectures and problem sets. I also wrote and contributed to weekly problem sets and gave lectures to the full class in the instructor's absence. In addition to my formal teaching experiences, I participated in education outreach opportunities including giving lectures on clinical diagnostics to Minnesota's Profoundly Gifted Youth and to Honors Chemistry classes at a local high school, and leading laboratory tours for high school students. As a graduate student, I also had the honor of mentoring six excellent undergraduate students to success in the laboratory. As a postdoctoral fellow, I continue to mentor young scientists including one research associate and one summer undergraduate student.

Teaching Interests: My teaching interests encompass core areas of chemical engineering including kinetics, transport, and separations. I am also very interested in developing and teaching courses in biomolecular engineering and bioprocesses in manufacturing. In addition to teaching in the classroom, I will welcome undergraduate researchers into my laboratory with mentorship on independent research projects. Through these projects, I will aim to teach them the fundamentals of laboratory research while cultivating their curiosity and problem-solving skills.

Selected Publications:

1. **Stern LA**, Schrack IA, Johnson SM, Deshpande A, Bennett NR, Harasymiw LA, Gardner MK, Hackel BJ. "Geometry and Expression Enhance Enrichment of Functional Yeast-Displayed Ligands Via Cell Panning." *Biotechnology and Bioengineering*, 2016, 113(11):2328-2341 (Highlighted as a "Spotlight").
2. **Stern LA**, Csizmar CM, Woldring DR, Wagner CR, Hackel BJ. "Titratable Avidity Reduction Enhances Affinity Discrimination in Mammalian Cellular Selection of Yeast-Displayed Ligands." *ACS Combinatorial Science*, 2017, 19(5):315-323.
3. **Stern LA**, Case BA, Hackel BJ. "Alternative Non-Antibody Protein Scaffolds for Molecular Imaging of Cancer." *Current Opinion in Chemical Engineering*, 2013, 2:425-432.
4. Csizmar CM, Petersburg J, Hendricks A, **Stern LA**, Hackel BJ, Wagner CR. "Engineering Reversible Cell-Cell Interactions with Lipid Anchored Prosthetic Receptors." *Bioconjugate Chemistry*, 2018, 29(4):1291-1301.
5. Woldring DR, Holec PV, **Stern LA**, Du Y, Hackel BJ. "Sitewise Diversity to Optimize Fitness of a Three-Helix Bundle Combinatorial Library." *Biochemistry*, 2017, 56(11):1656-1671.
6. Case BA, Kruziki MA, **Stern LA**, Hackel BJ. "Evaluation of Affibody Charge Modification Identified by Synthetic Consensus Design in Molecular PET Imaging of Epidermal Growth Factor Receptor." *Molecular Systems Design & Engineering*, 2018, 3:171-182.

542406 Life Science Systems Engineering

Abstract:

Research Interests:

Throughout my academic journey so far, I have been focusing on the development and application of Process Systems Engineering (PSE) methods to various topics in the Life Sciences.

Namely, during my MSc research project I developed a high-fidelity mathematical model to describe the events that took place during the clinical trial of TGN1214 mAb. Although, due to confidentiality hurdles, this work has not been published yet, as part of my MSc thesis, I developed a full scale mathematical model describing the functions of human cell and cytokine subsets both in homeostasis, as well as under sudden events. The model was validated against patient data and can be used to assist the development of novel immunotherapies and test their effect in silico prior to administration. Following that, my PhD research focused on the design of in-silico tools for the design and testing of continuous processing in monoclonal antibody production. Within this study, a mammalian cell culture in a bioreactor and a multi-column, chromatographic separation process were modelled, validated and simulated, aiming to explore the potential for process intensification. In addition, novel control strategies were designed for both systems and tested against experimentally optimal profiles. Lastly, I am currently investigating the development of novel supply chain and manufacturing concepts/routes in cancer immunotherapies. The project brings together knowledge stemming from product manufacturing, regulatory environment and techno-economic optimization, with end-goal to provide novel solutions for affordable therapies with decreased manufacturing and delivery times.

Based on my research background and scientific interests, I am envisioning the initiation of a research group that incorporates three research strands: (1) computational tools for life science engineering, (2) supply chain optimization in life sciences and (3) design, testing and distribution of mathematical tools that will assist the development of novel, sustainable solutions. I am suggesting the development, testing and distribution of mathematical tools and scientific frameworks targeted to: (1) the development & delivery of frameworks for robust model development, (2) models & algorithms testing on "real-world" case studies and (3) the creation of a model library focused on life-science engineering. For the "real-world" cases studies, priority will be given to sectors such as: (1) pharmaceutical product and process development and (2) food engineering and manufacturing. The models and/or procedures will be looking into process dynamics, as well as the underlying supply chain models of the relevant industry. During these years, I have built a strong and international network in both academia and industry that will allow me to establish meaningful cross-departmental/-organizational collaborations to support and advance my research objectives.

Teaching Interests:

My teaching experience spans a wide range of populations both in the European and the American academic environment. I have spent significant time, teaching at final year undergraduate and MSc students at Imperial College London, MSc & PhD students at Texas A&M University, as well as PhD students at the Technical University of Denmark. These experiences have trained me to efficiently deliver lectures to a highly competitive and demanding audience, of exceptional technical background. Moreover, given the diversity in the student level (from final year undergraduate to PhD), I have learned how to adapt course material based on the class needs and expectations. I managed to successfully design course and tutorial plans and assign tasks to fellow colleagues who were also assisting with the courses. I have a reputation for being especially organized and well-planned. Moreover, I have been in charge for the development of lecture notes, tutorials, exercise sets and course projects, as well as the coordination of office hours and student assistance.

I can teach subjects related to Process Modelling & Optimization of Biological, Biopharmaceutical and Pharmaceutical Systems, Pharmaceutical Process Development.

542407 Optical Imaging of the Brain at Nanoscopic Resolution

Abstract:

Research Interests:

Conventional fluorescence microscopy suffers from tradeoffs in resolution, imaging depth and photobleaching. In this work, we implement a novel scalable 3d imaging strategy to overcome these limitations by combining the physical expansion and clearing capabilities of hydrogel-based expansion microscopy (ExM) and the fast and high resolution imaging volumetric capabilities of advanced light microscopy. We show that (a) rational design of hydrogel monomers can lead to isotropic expansion of biological samples down to 10-100 nm length scale; (b) integration of expanded tissue-hydrogel composites with ultra-fast, low-photobleaching light-sheet microscopy can realize nanoscopic imaging of brain tissues up to 1000 times faster than electron microscopy (EM) and super-resolution light microscopy. Finally, we discuss tissue expansion/clearing technologyâ€™s broad implications for neuronal network tracing and functional omics.

Teaching Interests:

Broadly in the fields of material science, chemistry, nanoscience and nanotechnology, brain science, biomaterials and bioengineering. Five years of teaching experience at Harvard University as a Teaching Fellow.

541862 Engineering Protein Specificity: New Tools and Biologics to Remediate Human Diseases

Abstract:

Research Interests:

Recombinant protein drugs and protein-based diagnostics have opened up new ways by which we understand, diagnose, and treat diseases. Antibodies are now prescribed for the treatment of cardiovascular and infectious diseases, cancers, autoimmune and inflammation. Enzyme replacement therapies (ERTs) have well-established clinical examples and enzymatic amino acid depletion strategies hold great promise in fighting several types of cancers. Nevertheless, in order to expand the pharmacopeia of protein drugs and overcome our toughest health challenges, existing protein and cell therapies need to be further improved or introduced and novel protein-based therapeutic approaches need to be developed.

Owing to recent advances in protein engineering, cellular engineering and synthetic biology, protease therapeutics have started to gain more attention for the treatment of diseases associated with protein dysregulation and misfolding. Unlike antibodies and small molecule drugs which bind their targets stoichiometrically, protease therapeutics offer a distinct mechanistic and therapeutic approach because they can catalytically degrade their targets. Another major advancement in protein therapeutics has been the emergence of targeted therapies enabled by protein-drug conjugates (PDCs) whereby a payload is conjugated to a targeted protein scaffold via a site-selective chemical or enzymatic post-translational modification (PTM). As of now, these two technologies are not yet fully established. Potential therapeutic proteases often fail in the clinic due to toxicity stemming from their off-target activities or inactivation by serum inhibitors. PDC scaffolds are currently mostly limited to antibodies and cell surface engineering techniques to produce cell-based therapies and diagnostics are only starting to be developed. This is in part because many enzymes that enable protein-drug conjugations (e.g. transpeptidases) suffer from low activity and a restricted substrate specificity which limits their general applicability.

Work in my laboratory aims to tackle the challenge of engineering proteases and protein-modifying enzymes to bring these two technologies to the foreground of biotechnology and biomedicine. I aim to **redesign the specificity of proteases to (1) degrade proteins involved in neurodegenerative and autoimmune diseases and (2) to develop ERTs that treat rare monogenic disorders and thrombotic diseases.** I will further **evolve protein-modifying enzymes for site-specific protein and cell surface manipulations**. As I seek these two broad objectives, I aim to develop novel high-throughput technologies that facilitate protein engineering. My work in the **Huimin Zhao** and **Brent Iverson** labs have provided with the necessary foundation to successfully pursue these goals. In the Huimin Zhao lab, I gathered expertise in directed protein evolution and synthetic biology, leveraged enzyme specificity to enable creative chemo-enzymatic processes for asymmetric synthesis and gained expertise in cellular engineering. In the Brent Iverson lab, I further enriched my experience in protein engineering and cellular engineering by building *in vitro* and *in vivo* tools that facilitate protease engineering, in particular through the development of high-throughput compartmentalization platforms. With this accumulated experience, my independent career will be dedicated to engineering enzymes for therapeutic and diagnostic applications, as well as fundamental biomedical research.

In the future of biomedicine, I envision targeted protein and cell-based therapies that are highly functionalized such as stimulus-responsive prodrugs with biochemical logic capable of both effector and biosensor functions. I also envision completely novel engineered therapeutic protein scaffolds entering the drug market, beyond antibodies.

Research experience

PhD research

My graduate work centered on the development of chemo-enzymatic one-pot reactions that combine chemical and biocatalysis in ways that capitalized on their underutilized cooperativity, to convert readily available starting materials into enantiopure products. Specifically, I developed catalytic routes to convert mixtures of alkenes into a single chiral product, a fundamental building block in synthetic chemistry. In a set of investigations, I ***combined for the first time*** an olefin cross- metathesis reaction that establishes an equilibrium of alkenes of different lengths and functional groups with a P450-catalyzed epoxidation that is selective for the cross-metathesis product. By engineering the enzyme and reaction conditions, **up to 90% yield of the final epoxide** product could be obtained, which is twice as high as the yields of two sequential reactions. The off-equilibrium chemo-enzymatic reactions I developed were an important addition to the bio-catalysis field, as it showed that these combinations of chemical and biocatalysts could not only be made possible, but also be made better than the sum of their parts. Throughout my PhD work, I applied a similar strategy to synthesize chiral 2-succinate derivatives in high yields and enantioselectivity. Lastly, using an immobilized glucose isomerase and a supported acid catalyst, I prepared hyroxymethylfurfural (HMF) selectively starting from glucose in a two-step cascade reaction. Importantly, at the center of my graduate work were my ability to work in multidisciplinary teams, as well as to introduce novel and creative approaches to tackle synthetic challenges. In addition, I gained a synthetic biology toolkit by engineering translational control of gene expression in mammalian cells. **Through this set of investigations, I developed in-depth analytical, experimental, and problem-solving abilities, garnered expertise in a multitude of disciplines including chemical biology, bioengineering, enzyme engineering and immobilization, and firmly enriched my synthetic biology background**.

Post-doctoral research

To complete the foundation of my research program, I wanted to couple my experience in reaction engineering and synthetic biology with more in-depth experience in protein evolution and high-throughput screening. I joined the Iverson Lab to develop state-of-the-art high-throughput screening platforms for protease engineering, in an effort to produce recombinant human protease drugs as treatments for autoimmune diseases.

I developed and optimized several tools for protease engineering. First, I improved an *in vivo* yeast endoplasmic reticulum sequestration system (YESS), an approach previously developed in our lab that enables both protease engineering and substrate specificity profiling using fluorescence-activated cell sorting (FACS). In particular, I restructured this platform to allow a fast, modular, and efficient assembly of genetic elements that control transcription, translation, spatial sequestration and reporter tags. This seamless design-build-test cycle greatly speeds up protease engineering and screening efforts. Second, I used YESS to narrow the substrate specificity of human neutrophil elastase (*hne*) towards the selective cleavage of soluble tumor necrosis factor-alpha, a target in inflammatory bowel disease treatment. Third, using Next-Generation sequencing (NGS), I have analyzed the degradome of proteases and their variants, including tissue kallikreins which are biomarkers in cancers and inflammation. This high-throughput substrate specificity profiling allows a truly comprehensive analysis of sequence space that is not provided by positional scanning and proteomics-based methods. This in-depth analysis will help gain further insight into the roles of proteases in diseases and aid in the design of more specific peptide inhibitors, fluorogenic substrates, and protease-activatable prodrugs. Lastly, I independently developed an *in vitro* compartmentalization directed evolution platform for the engineering protease substrate specificity. **This tool offers the advantages of 1) bypassing protease-induced cytotoxicity, 2) enabling engineering on full proteins in the context of their 3D structures rather than peptide substrates and 3) offering a level of control over screening conditions not achievable *in vivo***. Using this platform, I am engineering human matrix metalloproteases (MMPs) to cleave IgG hinge regions as a treatment for IgG-mediated pathogenic conditions (rheumatoid arthritis, systemic lupus erythematosus, and multiple myeloma). Recent evidence has showed that ablating IgG levels through cleavage of IgG hinges by the bacterial enzyme IdeS (immunoglobulins degrading enzyme from *Streptococcus pyogenes*) alleviates autoimmune disease symptoms in a number of animal models of antibody-mediated disorders and reduces transplant rejection in humans. Although effective, we hypothesize that an engineered human protease would be preferred for repeated injections since IdeS inevitably triggers an immune reaction that diminishes or abrogates its efficacy.

My post-doctoral work allowed me to gather a new set of skills, namely in the in-depth analysis of DNA libraries using NGS, in creating large DNA libraries and manipulating the genome of *S. cerevisiae*. In particular, I learned how to create emulsion-based *in vitro* engineering platform, which will form the basis of my early independent career. In summary, my accumulated experience during both by PhD and post-doctoral work puts me in a unique position to establish a successful research program 1) building high-throughput directed evolution platforms to 2) engineer new and improved enzymes for therapeutic applications.

Teaching Interests:

Teaching experience. During my graduate studies, I was a teaching assistant (TA) for three chemical engineering classes. In Biomolecular Engineering (ChBE 473), I designed homework and quiz questions, formulated exam questions and developed grading rubrics. In particular, I was tasked to give three guest lectures, which provided me the opportunity to teach a class of about 50 students. I learned how to effectively prepare class material. I implemented many active learning methods during my guest lectures. In particular, I found that asking students to explain to a classmate how they understood the material was a valuable teaching tool. In Momentum and Heat Transfer (ChBE 421), I led weekly discussion sessions to a class of about 30 students. In these sessions, students and I went over homework and coursework. These discussion sessions became a more intimate setting that allowed students to obtain personalized help, which they wouldnâ€™t have gotten in a 300-student class.

As a postdoctoral fellow, I attended a class on evidence-based teaching in order to sharpen my teaching skills, namely through the introduction of various teaching and assessment strategies. Through this class, I learned how to effectively plan lessons using backwards design, delineate learning objectives and design assessments that specifically address these objectives. I also learned a variety of learning strategies. Most importantly, evidence-based teaching guided and cemented my teaching philosophy.

Teaching philosophy. Throughout my undergraduate, graduate and postdoctoral studies, I have had the opportunity to observe, analyze and practice many different teaching styles. These formative years have provided me with a strong foundation to become an effective teacher at many educational levels. **My core teaching philosophy is to build and foster an environment that promotes: (1) accountability and clear expectations from both teacher and students, (2) collaborative learning, (3) connection to the real world, (4) organized study, (5) an inclusive learning environment and lastly, (6) enthusiasm.**

Teaching interests and goals. I am prepared to teach all core chemical engineering classes, as well as introductory classes such as biochemistry, molecular biology, and higher-level graduate-level applied courses in protein and cellular engineering and biocatalysis. Throughout my career, I would also like to develop postgraduate applied courses centered on the needs of the department and students while taking inspiration from my research interests and background. In undergraduate level classes, students will be expected to 1) demonstrate understanding of major concepts related to First Principles, 2) apply their knowledge to formulate solutions to real-life problems and 3) develop a strong foundation to tackle more advanced classes. In graduate-level classes, my goals are to instruct students to 1) grasp more advanced concepts, 2) competently evaluate the current literature of their respective fields, 3) build a set of tools which will help them become experts in their fields and 4) apply knowledge acquired in my classes to their research. To achieve these goals, I will apply several effective teaching methods.

First and foremost, the learning objectives and expectations will be clearly delineated at the beginning of each lecture or set of lectures. Second, as an effective teacher, I will structure my classes in a way that intersperses short lecturing times with student activities or short assessment questions. Activities can include discussion questions (open-ended, multiple-choice) that can be revisited at different points during a lecture to not only introduce concepts in a layered and organized manner, but also to instantly evaluate understanding. Third, I plan to pursue many additional formats to promote student engagement. To this end, I will illustrate text-book problems and provide historical or current problems related to the lecture content. In addition, I plan to engage students using problem-based learning (PBL). In the classroom, PBL will encourage peer-teaching through the formation of study groups and other active learning tools, including think-pair share activities, multiple choice questions, jigsaw discussions, literature reviews or semester-long projects. In particular, for introductory classes, I will implement two-stage exams where appropriate, whereby students first complete the exam individually, and then working in small groups, answer the exam questions again. During the group exam, students will receive immediate, targeted feedback on their solutions from their fellow classmates and learn about alternative solutions. This makes the exam itself a valuable learning experience. I will continue to encourage student learning outside the classroom, by posting videos on the class website that expands on concepts covered in class and set up a message board where I can have discussions with students on more elaborate problems and concepts. This will help me disseminate class material in different formats and provide students with many options to remain engaged. Lastly, I will ask students to evaluate my teaching, not at the end of the semester, but once during the semester. That way, I can identify knowledge gaps and misconceptions and quickly adjust my teaching methods.

Creating an inclusive learning environment. For efficient learning, creating an inclusive environment is very important. As part of my overall teaching philosophy, I guarantee that students in my classes will feel free to express themselves and ask questions, as I will treat them equally and respectfully. Students, especially freshmen and sophomores, may come in to challenging classes with many misconceptions, knowledge gaps, as well as different learning styles. Encouraging students, especially minority students who may otherwise hesitate before actively participating, is one of my primary goals in that respect.

Teaching through research mentorship. I have been fortunate to mentor six undergraduate students, as well as several graduate students. It has always pleased me to hear from them of the great things they have gone to accomplish and the awards they have won. Through these experiences and observing my own mentors, I learned that there is no one-size-fits-all mentoring style and each individual may require a different mentoring approach. During graduate studies, I learned various management skills that I will build on.

First, encouraging students in my lab to read literature and understand their field and the larger implications of their work is critical to their success. To maintain the awareness of my trainees beyond the scope of their project, they will be asked to lead literature review sessions and asked to write a review paper early in their graduate career. I will also encourage my trainees to think independently, to feel free to advance their own ideas, establish collaborations with their colleagues as well as colleagues from other labs. Second, to further improve communication and writing skills, graduate students and post-docs will be encouraged to apply for fellowships, participate in grant writing, and attend conferences yearly. Third, while my door will always be open to my mentees, I will hold a formal weekly meeting to discuss their progress, and I will ask graduate students and postdocs to turn in a yearly performance review that will include a self-assessment, an advisor evaluation and a refinement of their career goals. Such a process I believe is critical to their professional development.

541905 Engineering 3D Models of Cancer through Application of Biomaterials and Systems Biology

Abstract:

Research Interests:

My lab will use biomaterials and statistical modeling to generate 3D models of cancer in order to guide treatment selection, discover novel targets, and design potential treatment strategies in a high throughput manner. My graduate studies provided me with a strong foundation in using design of experiments multivariate analysis (DOE) to tailor biomaterials in order to instruct cell fate and engineer the immune response. Specifically, I used DOE to engineer the 3D microenvironment in order to optimize the survival, pro-angiogenic, and/or anti-inflammatory potential of mesenchymal stem cells for tissue repair. I carried this expertise into my postdoctoral work developing *in vitro* models of ovarian cancer, and have added complementary systems biology modeling tools such as partial least squares regression (PLSR) that allow me to analyze paracrine signaling in the metastatic niche. I have recently identified critical components of the pro-tumor macrophage secretome that drive ovarian cancer progression as well as the central signaling pathway that dictates this response. Additionally, by using RNA-Seq in conjunction with an *in vitro* model of the monocyte-ovarian cancer tumor microenvironment, I have elucidated a non-canonical pathway by which ovarian cancer cells corrupt naïve monocytes towards a pro-tumor phenotype.

My research program will consist of three major focus areas. First, I will develop a personalized predictive model of cancer for pre-clinical evaluation of combinatorial therapies, ultimately incorporating patient cells and accounting for their specific tumor microenvironment. DOE will enable high-throughput screening of multiple bio-inspired input variables such as extracellular matrix components, viscoelastic properties, and growth factor presentation and their combinatorial effects on multiple output variables such as cell proliferation and invasion with the ultimate goal of identifying a hydrogel formulation that elicits the same response *in vitro* as observed *in vivo*. This platform can then be combined with patient derived cancer cells and DOE can be used to evaluate the effectiveness of combinatorial therapies. Second, I will identify cross-talk signaling networks between cancer cells, pro-tumor macrophages, and endothelial cells. To accomplish this goal I will combine my postdoctoral work in designing *in vitro* models of cancer with my doctoral work in vasculogenesis and macrophage polarization in order to construct a tri-culture system of cancer cells, macrophages, and endothelial cells. This will then be used for rapid analysis of cross-talk signaling networks in order to identify treatment strategies to prevent or delay critical processes in cancer metastasis. Finally, I will engineer a delivery vehicle for patient-derived anti-tumor macrophages. Both my doctoral and postdoctoral research projects have focused on modulating macrophage phenotype through presentation of soluble and physical cues; I will build upon this work to design a delivery vehicle for anti-tumor macrophages that maintains the desired phenotype as well as counteracts the pro-tumor capacity of the macrophages already present in the tumor microenvironment.

My lab will focus these tools on the study of cervical cancer, the second leading cause of cancer-related deaths in women aged 20-39. While surgery, chemotherapy, and radiation therapy are effective treatment strategies for early stage cervical cancer, 20% of patients will have recurrent and/or metastatic disease. Immunotherapy is a promising avenue for these patients, as it can disrupt the immune systemâ€™s contribution to cancer progression. In tandem, precision medicine enables clinicians to tailor the treatment regimen to the specific characteristics of a tumor. However, a patientâ€™s response to these therapies relies on complex interactions within the tumor microenvironment including the extracellular matrix and surrounding cells. Thus, in order to guide treatment selection and design new treatment strategies the tumor microenvironment must be accounted for in a way that enables high-throughput screening.

Successful Proposals:

Rivkin Center for Ovarian Cancer Scientific Scholar Award; 2017 - Present

American Heart Association Predoctoral Fellowship; 2015 - 2017

Postdoctoral Project: Systems biology approach to interrogate the influence of macrophages on the progression of ovarian cancer

Under supervision of Dr. Pam Kreeger, Biomedical Engineering, University of Wisconsin - Madison

Ph.D. Dissertation: Engineering hydrogel biophysical properties and mesenchymal stem cell microenvironment for tissue repair

Under supervision of Dr. Kent Leach, Biomedical Engineering, University of California, Davis

Education:

Ph.D., Biomedical Engineering, University of California, Davis, 2016

B.S., Chemical and Biological Engineering, University of Wisconsin - Madison, 2010

Publication Highlights: 9 first author papers, 7 co-authored papers, 1 book chapter, 1 conference proceeding. Average impact factor = 5.

Selected Publications (of 16):

1. Carroll MJ, **Fogg KC**, Patel HA, Krause HB, Mancha AS, Patankar MS, Weisman PS, Barroilhet L, Kreeger PK. Alternatively activated macrophages upregulate mesothelial expression of P-selectin to enhance adhesion of ovarian cancer cells. *Cancer Res*. 2018 May 8. [Epub ahead of print]
2. **Murphy KC**, Whitehead J, Zhou D, Ho SS, Leach JK. Engineering fibrin hydrogels to promote the wound healing potential of mesenchymal stem cell spheroids. *Acta Biomater.* 2017 Dec;64:176-186.
3. **Murphy KC**, Whitehead J, Falahee P, Zhou D, Simon SI, Leach JK. Multifactorial experimental design to optimize the anti-inflammatory and proangiogenic potential of mesenchymal stem cell spheroids. *Stem Cells*. 2017 Jun;35(6):1493-1504.
4. **Murphy KC**, StilhanoRS, Mitra D, Zhou D, Batarni S, Silva EA, Leach JK. Hydrogel biophysical properties instruct co-culture-mediated osteogenic potential. *FASEB J.*2016 Jan;30(1):477-86.

Teaching Interests:

In college I was heavily involved with Camp Badger, an outreach program that brings in students from underprivileged communities for a one-week summer camp in order to teach them fundamental engineering principles. As a graduate student I served as a teaching assistant as well as a teaching associate for 10 different courses in four departments, including chemical engineering. As a teaching associate for transport phenomena as well as chemical kinetics and reactor design I prepared and gave lectures as well as lead discussions and held office hours.

From my B.S. in chemical engineering at UW-Madison, I am comfortable with a full chemical engineering curriculum; however, based on my research and teaching experiences I would most enjoy teaching transport phenomena, chemical kinetics and reactor design, and process synthesis at the undergraduate level. Additionally, I would enjoy teaching or developing graduate coursework in biomaterials, design of experiments for material and process design, and stem cell engineering.

541962 Integrated Gene Circuit Design and Cellular Engineering: Probing and Reshaping the Genome to Control Cell Fate

Abstract:

The ultimate goal of my research is to leverage synthetic biology to transform how we understand cellular transitions and engineer cellular therapies.

I. Summary. Synthetic biology aims to harness the power of biological systems to perform tasks such as tumor surveillance, pathogen identification, and metabolite homeostasis. Native biological circuits (e.g. connected networks of genes) query data within the cell to coordinate cellular behaviors in space and time. Within mammalian systems, there exists an enormous opportunity to use synthetic circuitry to dynamically access information in the cell. As a chemical engineer working in molecular systems biology, my research focuses on integrating synthetic circuitry to interrogate and drive cellular behaviors. As a postdoc, I identified topological stress across the genome as a primary barrier to cellular reprogramming. This finding opens completely new questions for how the structure of the human genome stabilizes cellular identity and buffers cells against transitions to pathological states. Additionally, my findings explain why particular circuit designs fail and suggest principles for improving the performance of circuits integrated into the genome. I will establish a leading research program focused on designing and constructing integrated synthetic circuits to probe and actuate changes in DNA topology that drive changes in cell fate.

II. Research Accomplishments. Spanning a range of model organisms from yeast to human cells, I have engineered systems for dynamic behaviors across multiple scales from the molecular design of noncoding RNA devices to optimization of large transcriptional networks.

Graduate: Working with Dr. Christina Smolke at the California Institute of Technology, I constructed synthetic gene circuits and developed a class of genetic control systems called molecular network diverters. By interfacing these molecular network diverters with the native MAPK signaling pathway, I could temporally and spatially control cellular decision-making events (Galloway, K.E. et al. *Science*. 2013,). Beyond controlling cell fate, my work highlighted that integrated negative regulators can buffer a system against noise amplification mediated through positive-feedback loops by providing a resistance to amplification (Galloway, K.E. and Franco, E. Computational Methods in Synthetic Biology, 2015). By constructing synthetic circuitry, I will continue to illuminate paradigms in biological control and apply those principles to enhance cellular engineering.

Postdoc: As a postdoc at USC in Dr. Justin Ichida's laboratory, my work has focused on elucidating and overcoming the reprogramming roadblocks to the robust generation of mature cell types. Accurately modeling neurological disorders with *in vitro* cellular models relies on reliable methods to generate the distinct neural subpopulations affected by the disease. When I began my project, the conversion process was extremely inefficient, requiring large-scale efforts to generate only a few hundred cells. Moreover, the central mechanistic rules for direct lineage conversion were undefined. Today, as a direct result of my work to improve the reprogramming process, I can robustly generate thousands of cells with signatures of enhanced maturity (Babos, K.N.*, Galloway, K.E.*, et al. Balancing hyperproliferation, transcription to drive cellular reprogramming. *In Preparation*.,). In addition to improving the reprogramming process, my work uncovers a previously unrecognized explanation for why only rare cells undergo transcription factor-mediated reprogramming successfully. I found that cellular reprogramming is limited to a small population of cells equipped to process the dual, competing demands of hyperproliferation and hypertranscription. High rates of transcription and replication accelerate the rate of DNA tangling (e.g. supercoiling). Only cells equipped with high expression of topoisomerases, enzymes that relax DNA supercoils, are capable of mediating the massive genomic and transcriptional realignment to convert from one state to another. My findings suggest that topological stress profoundly impacts the function of gene networks (e.g. native or synthetic circuits). By more precisely defining the impact of topological stress on gene circuits, I will elucidate principles for buffering (or alternatively, harnessing) topological stress to enhance the performance and capabilities of genome-integrated synthetic circuits.

III. Research Interests: Prokaryotic systems and single-celled eukaryotic organisms have dominated the field of synthetic biology and revealed important paradigms in cellular biology including the role of feedback, noise, and cooperativity. While translation of synthetic biology to mammalian systems has been slower, with the advent of improved genetic tools for mammalian cells (e.g. gene therapy approaches including CRISPR technologies, AAVs), synthetic circuits are positioned to massively reshape how we study and treat diseases. Elucidating the principles of mammalian circuit design offers the opportunity to engineer cellular behaviors and to identify and target diseased states. Further, understanding the how cell types differentially process classes of circuits will improve our ability to predict the response of native transcriptional networks and design systems that are optimally wired for their function and context.

Focus 1. Elucidating the systems-level principles of cell fate transitions.

>Mapping transcriptional states: Characterizing the influence of hyperproliferation on the transcriptional trajectories between cell states.

>Proliferation as a motif: Defining the molecular impact of proliferation on cellular transitions

>Measuring transition speed: Characterizing the dynamics of proliferation-mediated transitions between cell states.

Focus 2. Integrated gene circuit design

>Elucidating principles of three-dimensional circuit design.

>T.A.N.G.L.E.S. (Topologically-Affected Network of Genes Linking Expression to State) as probes.

> Identifying robust designs for genome-integrated circuits: Mining the mammalian genome for structural designs.

>Funding

Continuous independent funding as postdoctoral fellow with acquisition of additional mini-grants

NIH Ruth L. Kirschstein NRSA Postdoctoral Fellowship (Fall 2015 – Fall 2018)

California Institute of Regenerative Medicine Postdoctoral Fellowship (Fall 2013 – Fall 2015)

Doerr USC Stem Cell Challenge Award $10,000 project (2017)

Fluidigm USC Single Cell Project Grant $9,000 in reagents and materials (June 2016)

>Publications

Galloway, K.E.*, Babos, K.*, and Ichida, J.I. Balancing hyperproliferation, transcription to drive cellular reprogramming. (In preparation). *These authors contributed equally to this work.

Galloway, K.E., Babos, K., and Ichida, J.I. Enhancing *in vitro* disease models of ALS through precise motor neuron subtype engineering. (In preparation).

Galloway, K.E*., Yu, H. *., Segil, N. I., and Ichida, J.I. Building a motor neuron enhancer map using ATM-ChIPseq. (In preparation). *These authors contributed equally to this work.

Ichida,J.I. Staats, K., Davis-Dusenbery, B.N., Clement, K., **Galloway, K.E.**, Babos, K.N. Son, E.Y., Kiskinis, E., Nicholas Atwater, N. , Gu ,H, Gnirke, A., Alexander Meissner, Kevin Eggan. Comparative genomic analysis of embryonic, lineage-converted, and stem cell-derived motor neurons. Development. (In resubmission).

Galloway, K.E. and Ichida, J.I. Modeling neurodegenerative diseases and neurodevelopmental disorders with reprogrammed cells. *Stem Cells, Tissue Engineering and Regenerative Medicine. D.A. Warburton, Ed. (World Scientific, New Jersey, 2015).*

Franco, E., and **Galloway, K.E.** Feedback loops in biological networks. *Computational Methods in Synthetic Biology.* M. A. Marchisio, Ed. (Springer New York, 2015), vol. 1244, pp. 193-214.

Galloway K.E., Franco, E., and Smolke, C.D. Dynamically reshaping signaling networks to program cell fate via genetic controllers. *Science*. 2013. 341:1235005.

Chen, Y.Y*, **Galloway, K.E.***, and Smolke, C.D. Synthetic biology: advancing biological frontiers by building synthetic systems. *Genome Biology*. 2012. 13:240. * These authors contributed equally to this work.

Kostal, J., Mulchandani, A., **Gropp, K.E.**, and Chen, W. A. Temperature Responsive Biopolymer for Mercury Remediation. *Environmental Science & Technology*. 2003. 37, 4457-4462.

Teaching Interests:

My comprehensive Chemical Engineering education from UC Berkeley and the California Institute of Technology has equipped me to teach classes in the core Chemical Engineering curriculum (kinetics, transport, and thermodynamics) at both undergraduate and graduate levels. I'm also very familiar with control theory and molecular and cellular biology. The problem-solving skills I acquired as a chemical engineer have set me up for success in elucidating the systems principles of biological systems. Thus, I am enthusiastic to teach chemical engineering principles to the next generation of students. Over my four years as a postdoc, I have directly mentored 15 students (5 grad, 8 undergrad, 2 high school) which has motivated me to develop strategies for training my students 1) to be effective in lab and 2) to develop their research and presentation skills.

Teaching experience: In addition to serving as a teaching assistant for two classes at Caltech, I taught the freshman chemistry lab at Harvey Mudd College (HMC) in spring 2013. HMC is highly regarded for its focus on teaching and is currently ranked as the #2 engineering school without a doctorate program. Since all HMC students are obligated to take the freshman chemistry course, many students who were uncomfortable in a laboratory setting approached the class with dread. In order to capture my students' interest, I began each class by engaging them with a bigger picture view of the purpose of the lab (e.g. why do we care about error propagation, synthesis yields, carbonate chemistry, etc.) as well as a general overview of what to expect. One student noted in the teaching reviews, "I thought the way Prof. Galloway started each lab with a little PowerPoint overview (where she went over the prelab if it was confusing, and just generally went over the process (was good))." Helping my students understand the bigger picture of the material made the labs more meaningful and empowering. With this level of engagement, I could challenge the students to think deeper when troubleshooting their experiments and analyzing their results. Whenever a student asks a question, I try to refrain from giving direct answers, instead offering another question that leads them toward answering their initial question. Letting the students discover the answers for themselves helps them to build a better understanding and confidence in their deductive skills. Long-term, I want students to recognize that asking good questions is fundamental to science, whether in theory or in practice. One student reflected that I was "always available to answer questions about lab procedure and write-up, and provides ample help without any hand-holding." Overall, my students were engaged and left with a positive view of chemistry. I received excellent evaluations with scores of either 6 or 7 (out of 7) in all categories as well as this gratifying comment: "Prof Galloway is extremely supportive of her students and she clearly wants us all to succeed, not only in lab, but in our careers as scientists."

Philosophy: While at Caltech, I completed the Caltech Project in Effective Teaching (CPET) certificate program, which introduces teaching pedagogy and practices through six seminars. From CPET, I learned that effective communication is the key to being an excellent teacher. I have become an extremely effective communicator as evidenced by the awards I have won for scientific communication, including the Everhart Lecture Award and First Place at the Annual USC Postdoctoral Symposium. In teaching, I have used techniques introduced to me through CPET training including the Socratic Method, inductive teaching, and story-telling. My goal is to help my students engage with concepts and approach a deeper understanding of scientific principles. In particular, I find that requiring students to draw out systems improves mechanistic understanding, develops intuition for the way processes work, and reduces experimental errors.

542310 Imitating Nature's Approach: Molecular Engineering of Organic Materials for Energy and Sensing

Abstract:

Research Interests:

My research interests and area of expertise sit at the interface of soft materials science and device fabrication for applications in energy, human health, and environmental sustainability. While conventional electronic devices are composed of hard materials, the pliability and chemical reactivity of organic materials may enable game-changing solutions to pressing scientific challenges. Here, I present two examples of molecular engineering of organic materials for (1) mechanically robust organic photovoltaics (OPVs) and (2) bio-inspired selective chemical sensors.

First, I will discuss the relationships between mechanical deformability and charge transport in polymeric systems, and the rational design principles that afford intrinsically stretchable OPVs. Organic photovoltaics hold promises to produce devices with performance approaching that of silicon-based electronics, but with the mechanical stability of conventional plastics. However, obtaining this "plastic" deformability and high photovoltaic performance has proven challenging. In addition, the mechanical fragility of many high performing conjugated polymers limits their applicability for large-area, low-cost, and printable devices. Using mechanical, spectroscopic, photovoltaic device-based measurement, I demonstrate the impact of molecular structure and solid-state packing on bulk mechanical properties, allowing for the co-optimization of OPVs towards the "best of both worlds."

Second, I will also discuss the designs of two novel chemical sensors—based on single-walled carbon nanotubes (SWCNTs) or complex liquid colloids—that translate molecular and biological recognition into sensory systems for environmental and health monitoring. Although selectivity underpins the utility of any sensing platform, obtaining a selective and practical sensor is often elusive. To address this challenge, I integrate bio-inspired molecular receptors into the sensing platforms to impart high selectivity and responsiveness. Specifically, a heme-inspired iron porphyrin provides controllable sensitivity towards carbon monoxide in SWCNT-based gas sensors, and carbohydrate-lectin binding allows for the detection of *Escherichia coli* bacteria using complex emulsions. These nature-inspired examples serve as an important step in demonstrating the possibility of translating chemical principles to practical devices.

Teaching Interests:

I believe that the success of the students is the most important accomplishment of the teacher. Throughout my studies, I have had impactful mentors whose goals were to guide me toward the best possible path. I am dedicated to pay this debt forward to my students and future mentees.

Between my undergraduate and graduate trainings in chemical engineering at UC Berkeley and UC San Diego, my research in polymeric soft materials, and my postdoctoral training at MIT, I am confident that I can teach anything within the chemical engineering curriculum. I have a particular passion for the core courses (fluid dynamics, heat and mass transfer, and thermodynamics) as well as courses related to polymeric materials and chemical synthesis. I have taught a graduate level course in intermolecular and surface forces that sits at the interface of my research interest and my formal training in chemical engineering. In addition, I served as a TA for the undergraduate introductory class to polymer science.

Mentorship is a critical facet of my educational mission. As a graduate student at UC San Diego, I have mentored 9 undergraduate students, all of whom became co-authors on at least one papers. At MIT, I am currently the primary mentor of 3 graduate students and have mentored 3 other who have graduated with their Ph.D. I aim to continue mentoring students as a professor.

Selected Publications

Full list of publications: https://scholar.google.com/citations?user=_-Zx0l4AAAAJ&hl=en

S. Savagatrup,† V. Schroeder,† X. He, S. Lin, M. He, O. Yassine, K. N. Salama, X. Zhang, T. M. Swager. "Bio-Inspired Carbon Monoxide Sensors with Voltage-Activated Sensitivity" Angewandte Chemie International Edition, 2017, 56, 14066. († Equal contribution)

V. Schroeder, **S. Savagatrup**, M. He, S. Lin, T. M. Swager. "Carbon Nanotube-based Chemical Sensors" Chemical Reviews, 2018. (invited, submitted).

Q. Zhang, **S. Savagatrup**, P. Kaplonek, P. H. Seeberger, T. M. Swager. "Janus Emulsions for the Detection of Bacteria" ACS Central Science, 2017, 3, 309.

S. Savagatrup, A. S. Makaram, D. J. Burke, and D. J. Lipomi. "Mechanical Properties of Conjugated Polymers and Polymer-Fullerene Composites as a Function of Molecular Structure." Advanced Functional Materials, 2014, 24, 1169.

S. Savagatrup,† A. D. Printz,† D. Rodriquez, and D. J. Lipomi. "Best of Both Worlds: Conjugated Polymers Exhibiting Good Photovoltaic Properties and High Tensile Elasticity." Macromolecules, 2014, 47, 1981. († Equal contribution)

S. Savagatrup, A. D. Printz, T. F. O'Connor, A. V. Zaretski, and D. J. Lipomi. "Molecularly Stretchable Electronics." Chemistry of Materials, 2014, 26, 3028.

S. E. Root, **S. Savagatrup**, A. D. Printz, D. Rodriquez, D. J. Lipomi. "Mechanical Properties of Organic Semiconductors for Stretchable, Highly-Flexible, and Mechanically Robust Electronics" Chemical Reviews, 2017, 117, 6467.

S. Savagatrup, E. Chan, S. M. Renteria-Garcia, A. D. Printz, A. V. Zaretski, T. F. O'Connor, D. Rodriquez, E. Valle, and D. J. Lipomi. "Plasticization of PEDOT:PSS by Common Additives for Mechanically Robust Devices and Wearable Sensors." Advanced Functional Materials, 2015, 25, 427.

541977 Advancing Technologies for Protein Engineering, Metabolic Engineering, and High-Throughput Technologies

Abstract:

Research Interests:

Microbial production of natural product drugs from plants offers numerous advantages over traditional synthetic chemistry, plant cell culture, or recovery from the plant. However, to realize this potential, we must tackle problems including incomplete biosynthetic pathways, poor enzyme properties, imbalance of metabolic flux, and uncontrolled cellular physiology. To address these challenges, my research aims to advance technologies that enable efficient biotransformation at the enzyme, pathway, and system level. My research will integrate computational biology, metabolic engineering, synthetic chromosome biology, and droplet microfluidics to reprogram metabolic networks and cellular processes, to guide design of protein engineering and pathway configurations, and to parse out outperformed enzyme and strain variants for maximal biosynthesis. Specifically, by consolidating statistical methods and machine learning of large-scale evolutionary data, I will establish an open-source algorithm for directed evolution of proteins. To improve pathway performance, I will create genetically encoded synthetic enzyme complexes with precise control of active site orientations and stoichiometries. I will also develop synthetic biology tools to perturb global gene expression, which ensures normal physiology for cell growth and high product yields. New modalities of biosensors capable of detecting both intracellular and extracellular metabolite concentrations will be developed to screen millions of enzymes and strain variants. In the long term, I will leverage the power of bioinformatics and large datasets of genome sequences to decipher missing enzymes in key natural product pathways, thereby promising complete biosynthesis of drugs in the microbial host.

Past research

In working with Prof. Philip Romero for postdoctoral training, I have expanded my skills of computation and statistical analysis while developing new algorithms for recombination-based protein engineering. The involvement of a deep mutational scanning project has broadened my capability to apply droplet microfluidics for high throughput biochemical analysis. My expertise rewiring cellular metabolism and reprogramming gene expression for effective microbial production has been cultivated during my postdoctoral work with Prof. Hal Alper. Under his mentorship, I have developed a CRISPR dCas9-based tool for chromosome organization, and employed CRISPR dCas9 gene regulation system for pathway engineering. As a PhD student of Prof. Ian Wheeldon, by developing synthetic protein scaffolds for organizing metabolic enzymes in *S. cerevisiae*, I demonstrated that the pathway efficiency could be improved via co-localization effects. In addition, my enthusiasm toward high throughput technologies has led me to develop three fluorogenic or chromogenic assays to overcome shortage of traditional analytic methods.

Teaching Interests:

Teaching interests

My training as a chemical engineer and geneticist from undergraduate and graduate studies has been supplemented with formal studies of chemical engineering and life science courses. I am enthusiastic about the opportunity to teach undergraduate courses including Transport Phenomena, Kinetics and Reaction Engineering, and Process Dynamics and Control. In addition, I look forward to the opportunity teaching upper level courses including Chemical Engineering Kinetics and Transport Phenomena.

Besides the aforementioned courses, I plan to develop and instruct an upper level course in synthetic biology for chemical engineers and bioengineers. This course will cover topics including protein engineering, metabolic engineering, gene expression, genome editing, molecular and cellular reprogramming, microbiome engineering, and high throughput technologies. My expertise in molecular and cell biology, protein and metabolic engineering, synthetic biology, and microfluidics has provided me with a unique vantage to lecture this course. Research and current challenges in the fields under the scope of these topics will be incorporated into the course structure and discussed.

Mentoring experience

My graduate and postdoctoral studies have exposed me to a wide range of teaching environments. As a teaching assistant, I have given lectures and led discussions in the classrooms of Chemical Process Analysis, Biological Unit Processes, Process Dynamics and Control, and Biochemical Engineering Principles. These teaching opportunities have taught me how to prepare course materials, cultivate student engagement, and assess the learning process. As a graduate and postdoctoral researcher, I have mentored five undergraduate and eight graduate students for their senior design or research projects. Through close interactions with my mentees, I learned to tailor my mentorship for their needs as students come with different interests and backgrounds

Selected Publications:

Lin, J.L., Bremer, B., Duan, J., Greenhalgh, J., & Romero P. Evolution-guided recombination of proteins. (*in preparation*)

Roychowdhury, H., **Lin, J.L.**, & Romero, P. Sequence-function mapping for allosteric sites of caspases with droplet microfluidics. (*in preparation*)

Lin, J.L., Ekas, H., Deaner, M., & Alper, H. CRISPR-PIN: controlling gene position in the nucleus using dCas9-mediated tethering. (*under revision*).

Lin, J. L. †, Wagner, M. W. †, & Alper, H. (2017) Enabling tools for high-throughput detection of metabolites: metabolic engineering and directed evolution applications. *Biotechnol. Adv.* 35(8), 950-970. †*Co-first author*

Abatemarco, J. †, Sarhan, M. F. †, Wagner, J. M. †, **Lin, J. L.,** Liu, L., Hassouneh, W., Yuan, S. F., Alper, H., & Abate, A. (2017) RNA-aptamers-in-droplets (RAPID) high throughput screening for secretory phenotypes. *Nat. Commun.* 8:332. †*Co-first author*

Lin, J. L. †, Zhu, J. †, & Wheeldon, I. (2017) Synthetic protein scaffolds for biosynthetic pathway co-localization on lipid droplet membranes. *Acs Synth. Biol.* 6, 1534-1544. †*Co-first author*

Lin, J. L., Palomec, P., & Wheeldon, I. (2014) Design and analyis of enhanced catalysis in scaffolded multi-enzyme cascade reactions. *Acs Catal.* 4 (2), 505-511.

Lin, J. L. & Wheeldon, I. (2013) Kinetic enhancement in DNA-enzyme nanostructures mimic the Sabatier Principle. *Acs Catal.* 3, 560-564.

542091 New Routes Toward Biomass-Derived Carbohydrates Upgrading

Abstract:

Research Interests:

Over the past 4 years, my PhD research has been focused on the utilization of lignocellulosic biomass as a renewable carbon source for the production of high-value chemicals through two approaches: (i) organosolv biomass pretreatment at mild conditions and (ii) the catalytic upgrading of stabilized carbohydrates into building-block molecules and secondary chemicals. Critical challenges linked to biomass processing such as product separation, solvent recovery, kinetic and mass-transfer limitations were addressed. This was done by employing liquid-CO_2, protection group chemistry, flow-through reactor setups and heterogeneous catalysts.

Organosolv biomass pretreatment

The aim of this project was to use mild temperatures and low acid content in solvent-aided integrated biomass depolymerization. The utilization of aprotic polar solvents such as Î³-valerolactone (GVL), 1,4-dioxane or tetrahydrofuran (THF) allowed us to extract up to 80% lignin with almost full xylan recovery and cellulose retention in the pretreated substrate. When GVL was used, over 99.5% of the solvent was recovered by liquid-CO_2 extraction, which could significantly improve process economy. This approach generated a carbohydrates stream with a concentration of 185 g L^{-1} after enzymatic hydrolysis. Moreover, we found that the addition of formaldehyde during biomass pretreatment prevented the formation of inter-unit C-C linkages in lignin during its extraction, increasing yields of lignin monomers by 6-fold based on native lignin. Following this approach, we used a similar reversible stabilization of xylose and glucose by acetal formation with aldehydes, preventing sugar dehydration to furans and their subsequent degradation. This stabilization strategy pushes back the longstanding kinetic limits of polysaccharide depolymerization vs. dehydration and enabled the recovery of biomass-derived carbohydrates at high yields and concentrations.

Stabilized carbohydrates upgrading

Stabilized carbohydrates are versatile molecules produced almost quantitatively from lignocellulosic biomass, that contrary to unmodified carbohydrates, are volatile and soluble in alkanes. The utilization of stabilized carbohydrates as platform molecules presents an opportunity to produce building block molecules through new reaction pathways. For example, when diformylxylose (formaldehyde stabilized xylose) was used as a starting reagent, similar furfural yields to those obtained with xylose *via* hydrogen transfer were achieved without Lewis acid addition. Surprisingly, no difference in the reaction kinetics was observed in the presence of a Lewis acid, which suggested that diformylxylose proceeded to furfural through a different dehydration mechanism. In light of these results, we proposed a tandem hydrolysis-dehydration reaction mechanism by a new intermediate identified during *operando* ^{13}C NMR studies.

Currently, my research is centered in the upgrading of aldehyde-stabilized carbohydrates into furans and polyols using heterogeneous catalysts. The goal of this approach is to produce targeted molecules through new, potentially simpler reaction routes compared to those currently established. The formation of novel products through these routes is also considered.

In my future research, I would like to explore the production of hydrogen and liquid fuels from aldehyde-stabilized carbohydrates using bifunctional heterogeneous catalysts, particularly zeolites. The projected reaction pathways include aqueous-phase reforming and cascade processes involving hydrogenolysis and aldol-condensation. Given that aldehyde-stabilized carbohydrates exhibit different reactivity compared to unmodified carbohydrates, they present an opportunity to explore novel upgrading routes.

Teaching Interests:

My teaching interest has been molded by chemical reaction engineering and thermodynamics. Based on my research and teaching experience, I would be well suited for courses in material and energy balances, kinetics and reactor design, and thermodynamics. Additionally, I have particular interests in specialized courses in heterogeneous catalysis and micro-reactors. I also visualize the possibility of developing new courses related to biorefineries.

Website/Contact:

- https://people.epfl.ch/cgi-bin/people?id=244963&op=bio&lang=en&cvlang=en
- ydna.questell@epfl.ch

529064 Engineering Multienzyme Systems for the Next Generation of Biomanufacturing

Abstract:

Research Interests:

Rational design of nanoscale biochemical factories is highly appealing for the next generation of biomanufacturing. Enzymes are the primary building blocks to assemble such molecular factories. My research aims to understand the reaction kinetics and system dynamics of multienzyme systems, and to integrate the enzymes in a synthetic way to realize defined, programmable and life-like functions *in vitro*. Focusing on dynamic behaviors of a synthetic biological system will provide new insights on non-steady or non-equilibrium states, and will eventually lead to evolutionary developments in biocatalysis, diagnostics, and adaptive biochemical devices.

I participated in the rapid development of biocatalysis in the past decade. My doctoral research focused on the development of stimuli-responsive enzyme catalysts and the application of enzymes in organic synthesis. Then I started research on multienzyme catalysis in the last phase of my doctoral study. As a postdoctoral researcher in the lab of Prof. Henry Hess, I am mainly working towards the understanding of the activity enhancements of enzyme cascades and the complex dynamics of enzymatic reaction networks. I experimentally demonstrated that the microenvironment rather than proximity channeling causes the activity enhancements of enzyme cascades on DNA scaffolds. Later on, I discovered the smallest enzymatic reaction network which can generate complex dynamics and Rayleigh-Bernard convective patterns.

My future work will focus on the development and application of multienzyme systems, and the construction of synthetic intelligent devices with autonomous and adaptive behaviors. I plan to pursue the following research thrusts as a faculty member: (1) Application of non-aqueous multienzyme systems for the degradation of plastics and biosynthesis of fine chemicals; (2) Process engineering of non-ribosomal peptide synthetases for the sequential synthesis of non-ribosomal peptides; (3) Bottom-up assembly of enzymatic reaction networks for intelligent devices based on self-regulating biochemistry.

Teaching Interests:

I genuinely enjoy teaching and mentoring students. I have extensive experience as a teaching assistant in four courses at both undergraduate and graduate levels at Tsinghua University and Columbia University. I have mentored twelve students in nine research projects and eight of them published their research in peer-reviewed journals. Now most of these students have become productive researchers pursuing Ph.D. degrees at institutions such as the Massachusetts Institute of Technology, Stanford University, the University of Minnesota, the University of Washington and Tsinghua University.

As I obtained both my bachelor's and doctoral degree in Chemical Engineering at Tsinghua University, an ABET accredited program, I am comfortable to teach core courses such as thermodynamics, transport processes, chemical reaction engineering, and separation processes as well as specialized courses like biochemical engineering and bioseparation processes. I am also eager to develop a course, *Nanoengineering in Biocatalysis*, to introduce the fundamentals and recent developments of biocatalysis and the influence of nanotechnology on the field.

Selected Publications (28 total, 11 first author, 600+ citations):

1. **Yifei Zhang**, Stanislav Tsitkov, Henry Hess. Complex dynamics in a two-enzyme reaction network with substrate competition, ***Nature Catalysis*** 2018, 1, 276−281 (Featured on the cover).

2. **Yifei Zhang**, and Henry Hess. Toward rational design of high-efficiency enzyme cascade. ***ACS Catalysis*** 2017, 7, 6018−6027.

3. **Yifei Zhang**, Qin Wang, Henry Hess. Increasing enzyme cascade throughput by pH-engineering the microenvironment of individual enzymes. ***ACS Catalysis*** 2017, 7, 2047−2051.

4. **Yifei Zhang**, Stanislav Tsitkov, Henry Hess. Proximity does not contribute to activity enhancement in the glucose oxidase-horseradish peroxidase cascade. ***Nature Communications*** 2016, 7, 13982.

5. **Yifei Zhang**, You Yong, Jun Ge, Zheng Liu. Lectin agglutinated multienzyme catalyst with enhanced substrate affinity and activity. ***ACS Catalysis*** 2016, 6, 3789−3795.

6. **Yifei Zhang**, Jun Ge and Zheng Liu. Enhanced activity of immobilized or chemically modified enzymes. ***ACS Catalysis*** 2015, 5, 4503−4513.

7. **Yifei Zhang**, Fengjiao Lyu, Jun Ge, Zheng Liu. Ink-jet printing an optimal multi-enzyme system. ***Chemical Communications*** 2014, 50, 12919−12922.

External links:

1. Homepage at https://sites.google.com/view/yifeizhang

2. Google scholar at https://scholar.google.com/citations?user=sqc421QAAAAJ

532953 Unifying Engineering and Synthesis to Create Platform Biomaterials

Abstract:

Research Interests:

Throughout my Ph.D. and postdoctoral studies, my overarching goal has been to help answer a difficult question: when we create new biomaterials, how do we ensure that they will work for our desired application? Indeed, predicting a priori if a biomaterial will function properly is a major challenge that the biomaterials community still faces today. For example, when developing drug-eluting reservoirs, how do we ensure that our system is not only biocompatible, but is also capable of maintaining the appropriate material properties throughout the course of our study? Alternatively, when studying specific disease states in vitro, how do we predict which biomaterials can be molded into better therapeutic models than those that already exist? Moreover, how do we ensure that our materials are both generally accessible and reproducible so that they can be studied broadly in academic and professional settings?

Leveraging principles from engineering and fundamental chemical synthesis, our research aims to answer these questions by employing a three-step process. First, we *identify* chemical space with untapped potential for biomaterials applications. Second, we *design* new biomaterials within these spaces that address unmet challenges. Finally, we *synthesize* these materials, ensuring that our systems are both scalable and reproducible. In following this process, our goal is to create platform biomaterials that not only can be used to treat and study specific diseases, but can also be broadly implemented within the biomaterials community.

Inspired by these questions, I focused my Ph.D. studies at MIT on integrating concepts from drug delivery, materials science, and synthetic chemistry to better understand how structure impacts function within non-viral delivery vectors for messenger RNA (mRNA) therapeutics. Therapeutic mRNAs could be used for a range of biomedical applications including genomic engineering, cancer immunotherapy, and protein replacement strategies. However, the serum instability of mRNAs and their limited ability to passively transfect cellular membranes limits their clinical potential. Lipid nanoparticles (LNPs) have emerged as an attractive non-viral delivery platform for improving the safety and potency of RNA therapeutics. Nevertheless, their complex molecular composition makes it difficult to understand which parameters ultimately impact the function of the LNP, thereby making it challenging to improve the efficacy and tolerability profiles of next-generation RNA delivery vectors. Towards this end, we have *i.* synthesized and purified rationally designed lipid materials of precise chemical structure, *ii.* formulated these materials into mRNA LNPs using microfluidic approaches, and *iii.* evaluated LNP potency and safety in vitro and in vivo as a function of the ionizable lipid chemical structure. Our lipids, which are actively being pursued as lead materials with pharmaceutical partners, ultimately represent robust RNA delivery vectors with high potency, low toxicity, and tunable biodistribution profiles without the need for targeting ligands or sequence modifications.

In my Postdoctoral research with Professor Bob Langer at MIT, we have developed scalable and modular hydrogels for biomedical applications. Specifically, we have created a class of mechanically- and kinetically-tunable hydrogels whose gelation occurs at physiologically relevant temperatures without the need for initiators, specialized laboratory equipment, or complex monomer synthesis. Our approach, which involves the self-assembly of a commercially available small molecule and a decagram-scalable polyethylene glycol derivative, ultimately affords hydrogels that are easy to fabricate on mutligram-scale. We are actively exploring several applications for our hydrogels including electrospraying strategies for stem cell recapitulation on implantable scaffolds, 3-dimensional printing models of cardiac valve disease, and islet entrapment for diabetes management. Additional efforts to explore the controlled release of therapeutics to the entrapped cell populations are also underway.

Teaching Interests:

1. **RNA delivery from concept to clinic: This course will cover the historical background, delivery principles, and cutting edge technologies associated with local and systemic delivery of nucleic acid therapeutics. Proposed mechanisms of cellular uptake as well as the formulation and characterization of polymeric and nanoparticle based delivery vectors will be discussed.**
2. **Common Organic Transformations for Biomaterials Science: This course will provide an overview of select organic transformations that are commonly employed in the fabrication of biomaterials. The mechanistic underpinnings and recent applications of each reaction in the biomaterials field will be discussed.**
3. **Controlled Polymer Synthesis and Its Application in Engineering: This course will begin with the basic principles of polymer science, including how the physical behavior of polymers arises from their topology and chemical structure. Post-polymerization modification strategies will also be explored, particularly in the context of creating next generation biomaterials.**

537595 On-Demand Therapeutics: From Externally-Triggerable Drug Delivery Systems to Bioelectronics

Abstract:

Research Interests:

Current treatment of pain heavily relies on opioids, with significant side effects such as addiction, clouded sensorium, tolerance and respiratory depression. Conventional local anesthetics may be effective with limited side effects, yet their duration of analgesia after a single injection is often insufficient for many pain states, including acute pain. As a result, pain management is in need of an effective and safe pain relief therapeutic system with limited side effects. My goal in research is to bridge this gap and develop platforms that may provide the opportunity for patients to have a better quality of life.

Externally-triggerable drug delivery systems provide a promising strategy for the treatment of pain. Such systems enable personalize and flexible control on the release of therapeutics, at the desired dosage, timing, and location according to the changing pain states of the patient. During my Ph.D. studies, near infrared light-activated pain relief systems using light-triggerable liposomes loaded with local anesthetics were developed. The incorporation of mechanisms such as photosensitization and photothermal effects made light-triggerable pain relief achievable in vivo at the sciatic nerve. It was shown that by changing the activating parameters of the near infrared light (e.g. the intensity and duration of activation), the dosage of drug release (and thus the degree of nerve block) was adjustable for multiple events within a single injection. Such repeatable and adjustable drug delivery systems demonstrated the promising potential of on-demand pain relief. Furthermore, the activating trigger was not limited to near infrared light. Ultrasound penetrated more effectively through tissue than near infrared light, and was used as the energy source of an ultrasound-triggerable pain relief system that demonstrated strong effectiveness in vivo. In summary, on-demand pain relief systems can be versatile and provides an effective strategy for pain management.

Pain management may benefit not only from an effective therapeutic system, but also from a powerful diagnostic system to better determine the degree of pain for clinical evaluation. Bioelectronics that can monitor pain responses, such as changes in heart rate, respiration rate, temperature and blood pressure, may provide a quantitative measurement of pain. In my postdoctoral research, I am developing bioelectronic systems for the monitoring of critical tissue properties, such as brain oxygenation and hydration. I am also performing clinical measurements of vital signs on neonatal patients. My training in translational research and bioelectronics development equips me with the toolsets to contribute to pain management.

I have a strong interest in the development of novel diagnostic and drug delivery systems for pain treatment. To merge such interest with my current training in translational research, I aim to invent and develop systems that benefit those who suffer from pain.

Fellowships

TL1 Postdoctoral Training Award, National Institutes of Health (NIH), 2017-2019

Chyn Duog Shiah Memorial Fellowship, MIT Office of the Dean for Graduate Education, 2014-2015

Selected Publications

1. Enhanced triggering of local anesthetic particles by photosensitization and photothermal effect using a common wavelength

Rwei A, Wang BY, Ji T, Zhan C, Kohane DS, *Nano Letters* 2017; 17 (11): 7138–7145

2. Ultrasound-triggered local anaesthesia

Rwei A, Paris JL, Wang B, Wang W, Axon CD, Vallet-Regí M, Langer R, Kohane DS,

Nature Biomedical Engineering 2017; 1, 644–653

3. Multiply repeatable and adjustable on-demand phototriggered local anesthesia

Rwei A, Zhan C, Wang B, Kohane DS, *Journal of Controlled Release* 2017; 251: 68-74

4. Repeatable and adjustable on-demand sciatic nerve block with phototriggerable liposomes

Rwei A, Lee JJ, Zhan C, Liu Q, Ok M, Shankarappa SA, Langer R, Kohane DS, *PNAS* 2015; 112 (51): 15719-15724

5. Photoresponsive nanoparticles for drug delivery

Rwei A, Wang W, Kohane DS, *Nano Today* 2015; 10(4): 451–467.

Teaching Interests:

Education and mentorship have been important aspects of my career. During my engineering training, my mentors have led me to the world of science, engineering and research, equipping me with enthusiasm and passing on their profound knowledge. I am grateful to my mentors, and I aim to pass down this experience to my students. During my TA experience in graduate school, I have found that patience, enthusiasm, and empathy are some of the most important aspects in teaching. As a mentor, seeing the students understanding a new concept, solving a new problem, and demonstrating confidence and enthusiasm in the subject, is the most rewarding experience. I received the "John Wulff Award for Excellence in Teaching an Undergraduate Subject" award from the MIT Department of Material Science when I graduated from MIT, as well as the historically highest TA evaluation score from the students' end-of-semester evaluations of the subject. My goal and interest in mentorship is to equip our students with enthusiasm in engineering and science, and guide them to understand the importance of the subject and knowledge they are trained upon. My hope is that together, we can make this world a better place through the advancement of engineering and science.

540162 Stability of Recombinant Protein-Based Bio-Pharmaceuticals: Stability in the Glassy Lyophilized State, at Various Interfaces and in Bulk Bio-Manufacturing Flows

Abstract:

Research Interests:

"Stability of Protein-Based Bio-Pharmaceuticals: Stability in the Glassy Lyophilized State, At Various Interfaces and in Bulk Bio-Manufacturing Flows"

Peer Recognition, Key Strengths, Originality and Differentiation: The PI has (co)-authored 32 peer-reviewed publications in respected journals. The impact of the PI's recent research, published with his former post-doctoral fellow, Dr. Prasad Sarangapani, [1] at MedImmune/AstraZeneca, was recognized by Prof. John Prausnitz (Dept. of Chemical Engineering, UC Berkeley), who reviewed the paper and was invited by the editors of *Biophysical Journal* to write a "New & Notable" Comment. Prof Prausnitz wrote about [2] the significance and impact of the work of Sarangapani *et al.:*

"Many like me will be secretly unhappy about the demise of the colloidal theory of globular protein solutions. Although we knew that this theory was "sick," we hoped that it might "recover." But now after the report of Sarangapani *et al.*, the colloid-like theory is dead. We can take comfort in the remark of Sarangapani *et al.*, that while scientifically erroneous, the colloid-like theory may nevertheless be useful for some purposes in biotechnology. While we mourn with sadness, we also owe much thanks to Sarangapani *et al.*, that when describing nature, yes, by all means seek simplicity, but, with respect for complexity, don't forget to mistrust it."

The PI is a recognized expert in the solution rheology and bio-physics of proteins: he delivered an invited plenary lecture at the XVII International Conference of Rheology (Kyoto, Japan, Aug. 2016). He has nearly 10 years of experience in bio-physics. The PI has identified these research problems during his career, and has trained a post-doctoral fellow and graduate students while in industry. The PI has 11 publications in biophysics, and leverages his knowledge of macromolecular physics, fluid mechanics and rheology extensively. The PI has an extensive network of domestic and international collaborators in academia, industry and in federal labs (NIST, NIH etc.), which he will rely upon to get the appropriate subject matter experts involved in the inter-disciplinary research program proposed here and to secure funding from federal agencies. This network will also be leveraged for funding. The PI can perform fundamental scientific inquiry in problems of relevance to industry, which will enable rigorous training of Chemical Engineers as Ph.D. scientists. The PI will also develop and teach new courses to ChE graduate and undergraduate students to mentor them for jobs in the bio-tech. industry, and will involve undergraduates in research, with special encouragement to minorities and female students. The PI is the co-chair of the first Gordon Research Conference GRC on Development of Bio-therapeutics and Vaccines with Prof. Christopher Roberts (Univ. of Delaware). The PI was the lead author of the successful team-proposal, whose funding led to the creation of this new GRC; website is available below.

https://www.grc.org/biotherapeutics-and-vaccines-development-conference/2019/

The 21st century poses global challenges to mankind, especially in harnessing alternate sources of clean energy, responding to climate change, ensuring abundant food and water supply and revolution healthcare/medicine. In healthcare/medicine, bio-technology has unleashed a revolution since the discovery of recombinant DNA technology, which has unleashed the discovery, development and commercialization/regulatory approval of numerous biologics (bio-pharmaceuticals): large molecule drugs expressed from cells, microbes etc., for use as medicines (therapeutics) and vaccines (prophylactics).

Chemical engineers have made prolific contributions in bio-technology to the science of upstream technology (protein expression) and downstream technology (protein purification). Yet, numerous scientific/technical challenges abound. Protein stability towards aggregation and fragmentation degradation pathways during flow encountered ubiquitously in bio-manufacturing: flow during cell culture in bio-reactors, flow in chromatography columns for purification, flow in formulation vessels and filling piston/peristaltic/time-pressure pumps and fill nozzles during aseptic (sterile) fill/finish of drug products, flow during shipment of finished liquid drug product in vials/syringes and flow in syringes and other novel devices for self-administration or clinical administration. It is impossible to expresses, purify, process, ship and deliver protein drug products to patients without subjecting them to various flow fields: shear flow, extensional flow etc., wherein protein molecules in concentrated solution encounter a wide range of shear and extensional rates. **Protein aggregation and other instability pathways during these unit operations pose serious challenges to biologics development and bio-pharmaceutical manufacturing.**

1) Protein Stability in Flow Fields and at Interfaces: The proposed research theme is distinct from quiescent protein aggregation (e.g. during storage), which has been studied extensively by Prof. Christopher Roberts (Univ. of Delaware) and his research group; **the work proposed by the PI aims to study the interplay of thermodynamics, flow fields, protein solution rheology and surfaces/interfaces on protein stability during flow.** The highly non-Newtonian rheology of concentrated protein solutions stems from hydrodynamic and thermodynamic interactions, and is also determined by the presence of solid/solution and air/solution interfaces. The effects of flow fields on thermodynamic phase behavior, which are well documented in synthetic polymer solutions, have not been scrutinized for protein solutions. This fundamental phenomenon has practical consequences for bio-manufacturing. **This research will bridge the molecular, mesoscopic and** *macroscopic/continuum domains.* Bulk rheology (macro and micro) and interfacial rheology (interfacial shear and dilation) will be applied in conjunction with scattering (light and neutrons), *in-situ* circular dichroism, intrinsic/extrinsic fluorescence, optical microscopy (flow visualization) etc. to pinpoint **how the unfolding of protein molecules air/solution and solid/solution interfaces during flow leads to formation of microscopic aggregates (~ 10 nm – ~ 1 mm), followed by larger microscopic "sub-visible particles", as the FDA refers to them, (~1 – ~ 100 mm) and finally visible particles**.

Impacts, Stakeholders and Funding Agencies: The formation of aggregates and particles poses a roadblock to therapeutic/vaccine development by increasing development cycle time and added costs. For ensuring patient safety against adverse immunogenic reactions, regulatory agencies such as the FDA, EMEA etc. have strict regulations for the maximum aggregate content and proteinaceous particle size/count that can be injected into patients. Statutory compliance with these regulatory requirements ensures that the bio-pharmaceutical industry will be a key stakeholder in this basic research, as development scientists do not have the time, resources and expertise to tackle these challenges at a fundamental scientific level; strict development timeline requirements only allow for engineering quick-fix solutions.

This work will proceed on two parallel tracks: with biophysically well-characterized model proteins for proof of concept development and rapid publication, and with proprietary proteins in collaboration with bio-pharma companies. A collaboration with protein expression and purification experts in academia will be initiated to leverage the upstream/downstream expertise required for supply of proteins, antibodies etc. The National Institute for Innovation in Biopharmaceuticals (NIIMBL), a public-private consortium dedicated to advancing Biopharmaceutical Manufacturing Innovation (NIST/US Dept. of Commerce), the Biomolecular Interaction Technology Center (Univ. of Delaware) and major bio-pharmaceutical companies will be approached for funding. This research will be disseminated in the following journals: *Journal of Pharmaceutical Sciences*, *Biotechnology Bioengineering*, *Journal of Rheology*, *Soft Matter* etc.

2) The Physics of Protein-Protein Interactions in Concentrated Solutions: Research dedicated exclusively to the *molecular/microscopic domain* **will probe the delicate interplay of hydrodynamics and thermodynamics in determining the stability of protein molecules, protein-protein interactions and viscosity/mutual diffusion/sedimentation behavior of protein solutions.** Solution physical hydrodynamic and thermodynamic properties will be measured and interrogated against available literature theoretical models, which propose inter-relationships among them. Static light scattering (SLS) will be used to measure the osmotic second virial coefficient (B_{22}), dynamic light scattering (DLS) for quantifying the diffusion interaction parameter (k_D), sedimentation coefficient (k_S) measurements using Analytical Ultracentrifugation (AUC) and viscometry to determine protein intrinsic viscosity $[\eta]$ and the Huggins coefficient k_H. The PI is a rheology expert, who routinely performs dilute solution viscometry to determine intrinsic viscosity and Huggins coefficient, and he is is also an experienced practitioner of SLS and DLS.

Impacts, Stakeholders and Funding Agencies: While theoretical predictions from colloidal models, generalized Stokes-Einstein models and mode coupling theory are available for the inter-relationship amongst these hydrodynamic and thermodynamic quantities, they have not been interrogated in sufficient depth to elucidate the correct physics. This work will be done strictly in the hydrodynamic/rheology and thermodynamics realm, with multi-domain Immunoglobulins and with globular proteins in the public domain. The funding agencies for this work will be NIH - NIBIB (primary) and NSF - CBET (secondary). While this investigation will be at a fundamental level, its significance and impacts will have useful consequences for industry formulations scientists who perform these measurements routinely to determine protein-protein interactions, stability and high solution viscosity during the development of high concentration liquid protein formulations. Biophysical and thermodynamic measurements are done in dilute solution conditions, while HCLFs are far from dilute: inter-molecular hydrodynamic and thermodynamic interactions must be accounted for, and this work will bridge the gap from dilute to concentrated solutions, so no extrapolations are required in understanding the thermodynamic and hydrodynamic contributions to the viscosity of concentrated protein solutions. A wealth of knowledge can be extracted by performing these measurements on purified well-characterized proteins. The learning from this research will be disseminated in journals like *Biophysical Journal*, *Journal of Chemical Physics*, *Journal of Physical Chemistry*, *Physical Review Letters,* etc.

3) **Rheology and Non-Newtonian Fluid Dynamics of Protein Solutions: Inter-Relationship with Protein Stability in the Glassy State (near the glass transition temperature T_g') and FLow Instabilities above T_g':** Since protein molecules are often prone to aggregation and are hence unstable in concentrated solutions, their development as bio-therapeutics becomes challenging. A typical short-term fix for Phase I clinical trials, where it is urgent to get the molecule into the clinic for First Time in Human (FTIH) trials to generate human tolerability and safety profiles, is to lyophilize (freeze dry) the protein formulation. Lyophilization requires the development of a formulation with appropriate bulking agents and cryo-stabilizing preservatives, and development of a freeze-drying cycle. After freezing, nucleation and primary/secondary drying. the protein and other excipients are left in the form of an elegant cake inside the glass vial. The cake is reconstituted by addition of specified quantity of sterile Water for Injection in the clinic. While the molecule is in Phase I trials in the freeze-dried solid dosage form, a stable liquid dosage form formulation is developed in the interim for Phase II clinical trials.

While lyophilization removes water, and enables rapid translation of the molecule into the clinic for FTIH milestone, trace quantities of moisture act as a plasticizer that reduces the T_g' of the lyophilized formulation. These trace quantities of water fasten the β-relaxations (secondary relaxations associated with side chains of protein molecules), which are related to the shelf-life of the lyophilized formulation, thanks to the pioneering work of Marcus Cicerone *et al.* at NIST. While Cicerone *et al.* have used elegant inelastic neutron scattering techniques to quantify the *beta*-relaxations, most academic and industrial researchers cannot easily access beam lines that are only available on the beam line at a cold neutron reactor source.

The PI proposes to use high frequency shear rheology and possibly also dielectric/fluorescence correlation spectroscopy to quantify the temperature and frequency dependence of these β-relaxations, as is done for polymeric glasses. High frequencies can be easily accessed with oscillatory shear rheology, and making these measurements of the β-relaxations and α-relaxation (glass transition) over a broad temperature range will enable us to determine the temperature dependence of their relaxation timescales. The temperature dependence of this *a*-relaxation timescale will also help elucidate the storage shelf life of protein therapeutics. Typically, these liquid formulation therapeutics are stored in cold chain at 5 ^oC, while accelerated stability tests are typically performed at 40 ^oC, to kinetically accelerate the instability pathway, which is most commonly aggregation. The PI and his team will probe universality in the temperature dependence of the *a*-relaxation, which is a reasonable expectation from proteins, since they form fragile glasses.

Finally, the behavior of therapeutic protein solutions as they flow out filling nozzles will be visualized using high speed microscopy to observe the different regimes that are associated with non-Newtonian fluids: dripping, jetting, etc. The delicate balance between a plethora of physical forces, such as surface tension, gravity, inertia, elasticity etc. demarcates these various regimes observed in high speed video microscopy. This investigation has not been done in the context of concentrated protein solutions of pharmaceutical relevance. This research will be conducted on a fundamental basis, and explained in terms of the flow field and protein solution rheology and physical characteristics. This work has tremendous technological importance too, as improper filling before lyophilization can lead to splashing etc., which leads to rejection of costly drug product during mandatory regulatory visual inspection. Consequently, bio-pharma companies suffer yield hits, and sometimes entire batches are rejected.

Impacts, Stakeholders and Funding Agencies: The PI will try to fundamentally study the physics of protein glasses, and translate the learning to interested bio-pharma companies that are prospective collaborators. The learning from this research will be disseminated in journals such as *Molecular Pharmaceutics, Pharmaceutical Research etc.* The agencies that will be approached for funding the glass transition work include the NIH, the Burroughs-Wellcome Fund and the Bill and Melinda Gates Foundation, because this work can also be translated to improve the stability of vaccines in developing countries and the development of room temperature stable vaccines by addition of excipients that raise the T_g' above room temperature, thus kinetically retarding the degradation pathway. The funding agencies for flow visualization of protein solutions include the NSF-CBET and bio-pharmaceutical companies. That research will be published in *Journal of Non-Newtonian Fluid Mechanics* and *Journal of Pharmaceutical Sciences*.

The PI and his team will present at AIChE Annual Meeting, ACS- American Chemical Society Biotech Meeting, Biophysical Society Annual Meeting, American Physical Society March Meeting and the Society of Rheology Annual Meeting.

[1] "Critical Examination of the Colloidal Particle Model of Globular Models" Prasad S. Sarangapani, Steven D. Hudson, Ronald L. Jones, Jack F. Douglas and Jai A. Pathak. *Biophysical Journal*, 108, 724 – 737 (**2015**).

[2] "The Fallacy of Misplaced Concreteness" John Prausnitz, *Biophysical Journal*, 108, 453-454 (**2015**).

Teaching Interests:

Philosophy: I am a strong believer in teaching our undergraduates and graduate students well: they are our next generation of engineers and scientists, and deserve the same investment in teaching that my professors made in me. My pedagogical philosophy is outlined in detail below. I served as a TA for 6 semesters, and enjoy teaching.

For undergraduates, my main objective is to teach them to grasp the essential concepts of Chemical Engineering in ways that they can readily apply in their careers. My emphasis will be on teaching them how to make sense of the output from their calculations, done using whatever software package/code, and determine whether it is physically sensible and correct. It is great of they can learn how to write code. They will prove their mettle as chemical engineers if they can make sense of the inputs to and outputs of their code.

I will tailor elective courses in such a way to advanced undergraduates so that those students who have an aptitude for graduate school will be exposed to reading the literature and the process of building critical thinking skills. **Having worked in the bio-pharmaceutical industry, I will build a program that prepares students for careers in this vibrant industry, which is predicted to be a key growth sector in the US in this century. Revenue from bio-pharmaceuticals (biologics) already far outstrips the revenue from small molecule drugs. I propose to develop a new-course in bio-pharma manufacturing, wherein I will teach them the basics of upstream cell culture, downstream purification and formulation/aseptic fill-finish in a cGMP (current Good Manufacturing practices) environment.**

In addition to teaching them engineering, **I will also start a new course for graduating seniors to teach them skills that today's employers expect: communication (technical/regulatory writing and presentations), time management and prioritization, working in a matrixed environment, interviewing skills, workplace etiquette and communication, and soft skills (e.g. body language awareness and control): these are all critical for landing job offers after interviews and for successful career development and progression.**

For graduate students, my approach is radically different. With them, my over-riding emphasis will be on development of critical thinking skills and learning how to think fast on their feet to respond effectively and sensibly to tough questions. In addition to lectures, I will hand out papers and ask them to find logical flaws in them after reading them carefully. I will also require in-class presentations, and I will develop their ability to think quickly on their feet and handle challenging questions from the audience effectively. Students will also be required to learn how to decipher the physics underlying mathematical equations. Moreover, I will work with faculty to train the graduate students in this critical skill during graduate student colloquia. I will work with faculty to encourage graduate students to spend at least one summer in industry as an intern. Most graduate students pursue careers in industry, and need hands-on early preparation for this transition, so that the industry environment is not a mystery to them when they start their careers,

The list of courses I am interested in teaching is below.

- Undergraduates
- Mass and Energy Balances
- Transport Phenomena (Momentum, Mass and Heat Transfer)
- **Transition to the Workplace for ChEs (seniors only – new course to be developed)**
- Process Control
- Graduate Students[1]
- Core - Fluid Mechanics; Thermodynamics/Statistical Mechanics
- **Elective - Bio-pharmaceutical Manufacturing (new course – to be developed)**
- **Elective - Protein Solution Biophysics (new course – to be developed)**
- **Elective - Rheology of Complex Fluids (new course – to be developed)**

[1] These courses will also be open to advanced undergrads who can satisfy appropriate prerequisites.

541210 Multiscale Multiphysics Modeling of Blood Clotting and Thrombus Bio-Chemomechanics in the Vasculature

Abstract:

Research Interests:

- Multiphysics and Multiscale Modeling of Biological Systems and Complex Fluids (e.g., Blood)
- Coarse-grained Particle-based Methods (e.g., DPD) in Cellular and sub-cellular Modeling
- High-order Spectral Element Methods in Fluid Mechanics, Elastodynamics and Fluid-solid Interaction (FSI)
- Patient-specific and Probabilistic Modeling of Blood Flow in Arterial Network such as Coronary Arteries

Teaching Interests:

- Fluid Mechanics and Transport Phenomena
- Thermodynamics and Heat Transfer
- Introductory and Advanced Numerical Methods for Chemical Engineers
- Multiscale Modeling in Biophysics and Biomedical Engineering

ABSTRACT:

The process of clot formation and growth at the site of injury on a blood vessel wall involves a number of multiscale simultaneous processes including: multiple chemical reactions in the coagulation cascade, species transport, platelet adhesion as well as the hemodynamics and blood rheology. We model these processes at different levels of description (e.g. macro- vs. micro-scales). At continuum level, we solve Navier-Stokes equations for blood flow as well as the PDEs associated with advection-diffusion-reaction (ADR) of multiple species in the coagulation cascade using a spectral element solver. Depending on the required degrees of fidelity, platelets could be modeled either as a continuum concentration field or their transport and adhesion could be addressed by looking at individual platelets in a Lagrangian framework. At cellular and sub-cellular levels, however, particle-based methods such as Dissipative Particle Dynamics (DPD) have proven themselves to be more effective in addressing complex biological systems. I will discuss the recent developments in DPD framework that extends its applicability to transport problems, and discrete models of soft biological tissues. The major drawback of modeling such fully-resolved systems at the cellular scale, however, is the long-term computation of their whole course of evolutions. There are numerous biological and materials systems that evolve at slow rates (e.g. clotting), which make the simulations prohibitively long. I propose a multidelity technique based on co-krigging information fusion approach in statistical learning in order to facilitate acceleration in time. The multiscale framework's implications and future directions will also be discussed.

541836 Preparing of a Composite Nano Disperse Dye Using a Hydroxypropyl Sulfonated Lignin Dispersant and the Interaction of Dispersant and Dye Surface

Abstract:

Research Interests: Bio-energy and Biomass-mertinals, appalicaation of lignin, cellulose and cellulose nano fiber.

Teaching Interests: Communication lignin and cellulose, Conversion process of biomass to biofuel.

Lignin is the second or third most abundant organic polymer in terrestrial plants. It consists of chains of partly crosslinked phenylpropanoid units, mainly of guaiacyl and syringylpropans, but grasses also contain up to 20% of 4-alkoxy-phenylpropanes. Traditional technical lignins, mainly lignosulfonates, alkali lignins and organosolv lignin, are byproducts of pulp and paper industry. Enzymatic hydrolysis lignin (EHL) is a new kind of lignin, which derived from biofuel production possesses many differences compared to kraft lignin, lignosulfonates obtained by pulping processes. It is closer to the native lignin with weak chemical activity. In addition, an abundant source will be obtained with the fast development of the biofuel industry. So, developing a way to utilize EHL is a potential strategy to prosper biofuel industry.

Here we prepared a light-colored hydroxypropyl sulfonated lignin (HS-EHL) based on enzymatic hydrolysis lignin (EHL) by grafted-sulfonation and crosslinking reaction processes. The sulfonic group content and Mw of the product is elevated by this reaction. The color of HS-EHL turned to light yellow compared to the dark brown color of EAL. HS-EHL has also a low staining effect on fiber. Moreover, it has the best dispersion ability and temperature stability and dye uptake.

HS-EHL was further used as dye dispersant to prepare composite nano disperse dye (CL-NDD) by self-assembly in ethanol or acetone solution. Particles sizes of disperse dye in dye bath great affect its dyeing quality. CL-NDD with average particles size of 94.25 nm exhibited good dispersion, stability at high temperature compared with two milling dye (one commercial disperse dye and one is milling dye with HS-EHL analyzed by particle analyzer, SEM, TEM and AFM. The reducing effect of CL-NDD as low as 5.39% due to HS-EHL had low phenolic hydroxyl group content can prevent to reduce azo disperse dye. The uptake of CL-NDD reached to amazing of 94.27%. The adsorption characters of dispersants onto dye surface were tested by employing QCM-D and AFM to reveal the interaction of dispersant and dye surface. A higher adsorption amount of HS-EHL exhibited a more viscoelastic adsorption layer than sodium lignosulphonate (NaLS) due to HS-EHL had the long alkyl chain with higher sulfonic group content can touch with water phase to form strong electrostatic repulsion. Whatâ€™s more, higher molecular weight of HS-EHL can provided stronger hydrophobic effect and higher adsorption amount to keep dye particles stable in dye bath.

542356 Elastohydrodynamics and Soft Matter Mechanics to Understand Biological Adhesion, Human Touch, and Optics-Free Cytometry

Abstract:

Research Interests:

Research Experience: My experience and abilities lie at the interface between the mechanics of soft matter and applications in cell biology and artificial touch. In particular, I have exploited elastohydrodynamic phenomena—effects arising from the coupling of soft structures and fluid flow—to understand various observations in nature. My work has elucidated how tree frogs stick to surfaces while submerged, how fingerprints influence tactile sensation, and how the sidewalls of microfluidic channels deform during transit of biological analytes as a means of mechanical biosensing.

My graduate work with Prof. Joelle Frechette at Johns Hopkins University found how drainage channels in the toe pads of tree frogs reduced hydrodynamic resistance and promoted contact with flooded surfaces. Drainage channels, however, also reduce the adhesion of tree frogs once in contact. The improvements in making contact offset the reduced adhesion. During my postdoc in Prof. Darren Lipomi's group at UC San Diego, I continuously measured particles and fluid velocity in a microfluidic channel by monitoring the fluid-induced (elastohydrodynamic) deformation in the sidewalls using only a voltage signal. This new microfluidic sensor circumvents the need for costly optical measurements. This was done with an ultra-sensitive palladium nanoislands on graphene strain sensor embedded in the elastic sidewalls of the microfluidic channel. I also applied soft matter mechanics to the sense of touch and found that human subjects can be sensitive to monolayer differences in surfaces, due to cues from friction, and that human subjects determine how soft a surface is by using indentation depth and contact area. We are using these findings to design devices that can generate tactile sensations. For example, we have found that deformable, patterned interfaces amplify differences in friction, which would help make two surfaces feel more different.

Selected Publications:

Dhong, C., Lipomi D., *et al.* "Role of fingerprint-inspired relief structures in elastomeric slabs for detecting frictional differences arising from surface monolayers." (Submitted)

Dhong, C., Lipomi D., *et al.* "Optics-free, non-contact measurements of fluids, bubbles and particles in microchannels using metallic nanoislands on graphene." (Submitted)

Carpenter, CW.*, **Dhong, C.*,** Ramachandran, V.S., Lipomi, D., *et al.*, "Human ability to discriminate surface chemistry by touch." *Materials Horizon* (2018) *Authors contributed equally to the work

Dhong, C., Frechette, J. "Peeling flexible beams in viscous fluids: Rigidity and extensional compliance." *Journal of Applied Physics* (2017)

Wang, Y., **Dhong, C.,** Frechette, J., "Out-of-contact elastohydrodynamic deformation due to lubrication forces." *Physical Review Letters* (2015)

Dhong, C., Frechette, J. "Coupled effects of applied load and surface structure on the viscous forces during peeling." *Soft Matter* (2015)

Future Directions: My future research will create devices that are designed using soft matter mechanics with functional materials. This enables devices with unique functionalities and allows me to probe fundamental phenomenon in soft matter and biology. For example, it is known that circulating tumor cells are three times softer than regular healthy cells and there are no current diagnostic methods that take advantage of this fact. I propose identifying circulating tumor cells without the use of antibodies or prohibitively expensive equipment by measuring the cell stiffness. Cell stiffness as a biomarker is not limited to cancer; malarial cells are also mechanically distinct from their healthy counterparts. Using the nanoisland-on-graphene strain measurements of elastohydrodynamic phenomenon, the continuous, high-throughput measurements of cell stiffness can transform medical diagnostics. Detecting diseased cells is just one aspect, fundamental biomechanical studies will investigate drug efficacy have on the endothelial tissue by monitoring the viscoelastic response of arteries in real time.

In addition, current technology has not effectively addressed the sense of touch. State-of-the-art haptic devices can still only generate a simple vibration. I look forward to combining soft matter phenomenon with functional materials to create devices that generate haptic sensations that go beyond this limitation.

Teaching Interests:

I would enjoy teaching any core chemical engineering course, especially transport. At the graduate level, I would enjoy teaching any transport phenomenon course, which would incorporate numerical methods. I would also like to teach electives in continuum mechanics topics such as adhesion, contact and fracture mechanics. I have taught fluid/transport and helped to develop and taught a master's level lab course at Johns Hopkins University. At UC San Diego, I have also guest lectured two classes in a polymeric materials course.

543291 Using Microrobotic Tools for Probing Cellular Pattern Generation and Morphogenesis

Abstract:

Research Interests: Fate specification and differentiation of embryos during development is controlled by concentration gradients of morphogens. Most of these gradients are formed through diffusion of the signaling molecule from a localized source and can be described by a reaction diffusion equation with production and degradation terms. It is less clear, however, how subcellular processes lead to the observed gradient shapes. It has been shown that morphogen interactions with surface receptors or inhibitors with different diffusion rates affect gradient shape. As well, temporal variations in gradient amplitudes during pattern formation have been observed. Current methods in literature rely on static one-and-done patterns to study tissue-level gradient shapes, which are not sufficient to map the kinetics of pattern formation. Thus, there are significant theoretical and technical gaps to overcome if we are to determine how dynamic gradient behavior governs cell fate downstream. My goal is to use microrobotic systems to delivers signaling molecules to a network of engineered cells to generate user defined patterns with spatial and temporal control. This will help us to understand how single cells communicate in space and time to generate organs and to harness this knowledge to develop organoids. Using genetic technology and synthetic biology, and armed with the better control over cell differentiation, I will endeavor to create high fidelity 3D tissues from engineered stem cells.

An exciting avenue for further study is cell polarization. Asymmetric cell division generates two functionally different daughter cells with distinct fates because of the unequal distribution of cell fate determinants or signaling pathway components. The molecular mechanisms driving and maintaining asymmetric divisions are poorly understood, but signaling pathways such as Notch and Wnt are known to be key factors. Due to sub micrometer precision in the placement of microparticles using the robotic systems, I would be able to place a morphogen loaded microparticle at a particular position on the cell surface and then monitor cell division and the asymmetric distribution of the proteins in response to different directions of the morphogen source.

1. S. Das, E.B. Steager, M. A. Hsieh, K. Stebe, V. Kumar. 'Modelling and Ensemble Control of Multiple Catalytic Microrobots', Journal of Micro-Bio Robotics, In Press

2. S. Das, E. E. Hunter, 6. E. B. Steager, V. Kumar. 'Controlled Delivery of Signaling Molecules using Magnetic Microrobots', in 2018 International Conference on Manipula- tion, Automation and Robotics at Small Scales (MARSS), IEEE, July 2018, Nagoya, Japan

*3. S. Das, E. B. Steager, K. J. Stebe, V. Kumar. 'Simultaneous control of Spherical Microrobots using Catalytic and Magnetic Actuation', in 2017 International Conference
on Manipulation, Automation and Robotics at Small Scales (MARSS), IEEE, July 2017, Montreal, Canada*

4. S. Das, O. Shklyaev, A. Altemose, H. Shum, I. Ortiz-Rivera, L.Valdez, T. E. Mallouk, A. Balazs, A. Sen. 'Harnessing Catalytic Pumps for Directional Delivery of Microparticles

in Microchambers', Nature Communications, 8, 14384 (2017)

*5. S. Das, A. Garg, A. Campbell, J. Howse, D. Velegol, A. Sen, R. Golestanian, S. Ebbens. 'Boundaries can steer active Janus spheres', Nature Communications, 6, 8999
(2015)*

*6. W.Duan, W.Wang, S. Das, V. Yadav, T.E. Mallouk, A. Sen. 'Synthetic Nano- and Micromachines in Analytical Chemistry: Sensing, Migration, Capture, Delivery, and Sep-
aration', Annu. Rev. Anal. Chem., 8, 311 (2015)*

7. K.K.Dey, S. Das, M. Poyton, S. Sengupta, P. Cremer, P. Butler, A. Sen. 'Chemotactic Separation of Enzymes', ACS Nano, 8, 11941 (2014). Editors Choice

Teaching Interests:

As a child I delighted in competing to learn new things before my peers and in sharing my knowledge with others. I recall being scolded in primary school for teaching my classmates long division! This thirst for knowledge and desire to teach are still major parts of my life, motivating my pursuit of an advanced degree in chemistry in order to become an educator in the sciences. As an adult, I have come to appreciate the influence teachers have on their students. My goal as a teacher would be to challenge my students and inspire them to grow and expand their horizons. I want my students to learn from and mentor each other. I am a big believer in the power of collaboration and collective learning and would like my students to leave my classroom with the collaborative spirit in mind. I also hope to instill in my students the confidence to not know all the answers but the ability to go and find the answers. My desire to challenge and engage my students is not limited to just the classroom. I was a mentor for two first year female graduate students in my fourth year of graduate school. Also in my fourth and fifth year I participated in Research Snapshots; a program where groups of minority and female high school students spend a day under the guidance of a senior graduate student and learn about their research and life as a PhD student. I was a mentor for a group of 7 freshman and sophomore students and my consistent focus during the program was to encourage them to pursue an advanced STEM degree. I also mentored a minority REU (Research Experiences for Undergraduates) student on a daily basis in the summer of 2015. Moreover, I have been mentoring two junior graduate students in my lab on a weekly basis for the past two years on different projects.

543285 3D Bio-Printed Models of Vascularized Tissues

Abstract:

Research Interests:

My overarching research goals are to develop biofabrication technologies for creating full-size vascularized tissues/organs and to understand the interactions between vasculatures and surrounding cells during vessel formation and maturation process within dense tissues/organs. I have a wide variety of diverse experiences that have prepared me for this path. My research experiences include the design/development of bioprinting equipment and bioractors, vascular engineering, vascularized 3D tissue engineering, and cancer study model development. During my PhD, I developed 3D bioprinting systems, designed bioreactors, and established protocols to create various-sized functional vessels within thick tissue models. My postdoctoral work has focused on creating 3D *in vitro* models of brain tumor-vascular niche and utilizing the models for drug test and personalized therapy applications. I have also developed numerous professional skills including how to set up a new laboratory, craft grant proposals, establish and maintain collaboration projects, and manage research and personnel budgets. I believe I will be uniquely suited to pursue my research goals to solve challenges for large tissue engineering and study vascular biology within full-size tissues/organs.

Proposed Research Overview: Engineering full-size functional tissues and organs have a promise for patients on the organ transplant waiting list and scientist who want to test their theories on *in vitro* model systems. One of the major issues in creating full-size large tissues and organs is the lack of a vascular system that can provide a sufficient blood circulation throughout the engineered tissue/organ structures. A number of technologies have been developed to fabricate a vascular system within engineered tissues and organs, however, providing a physiologically-relevant vascular system that contains various-sized branching blood vessels with efficient circulation and maintaining the structural and functional integrity during the maturation of large and dense tissues/ organs still remain as major challenges. I will lead a multidisciplinary research group to understand vascularization processes in tissue-specific conditions and to exploit the knowledge and develop technologies for full-size vascularized tissue/organ engineering. Success in these aims may ultimately transform our understanding of vascular formation processes in tissue/organ-specific conditions, will provide customizable *in vitro* tissue/organ models to reveal new therapeutic solutions, and will broadly enable the scientific community with powerful tissue/organ engineering methodologies.

Teaching Interests:

My goal is to help students developing themselves as curious problem solvers who are capable of working with diverse teams. My teaching philosophy has been built on my experience as a scientist, as an engineer, as a colleague, and as a member of collaborative interdisciplinary teams. "Be curious and critical for science" and "Be positive and active for people" are two most critical values that I wish to cultivate in my classroom and mentoring sessions.

543999 From Macromolecular Science to the Skin Barrier: Engineering Novel Platforms for Transdermal Drug Delivery

Abstract:

Research Interests: In the coming years, I intend to extend my research on the development of different transdermal drug delivery methods in order to more precisely gain a comprehensive view of an individualâ€™s health state at molecular levels and provide deep and high-level interfaces for creating biocompatible and biodegradable materials onto the skin. Particular challenge in this field is that human skin is intrinsically complicated and acts as a strong barrier for external transportation of any molecules. In overcoming skin's barrier, we need to explore a variety of engineering challenges to increase skin permeability in order to develop more methods designed to enhance and embellish it. Starting with my background in transdermal drug delivery, the following methods are my future plans:

(a) The study of hot melt extrusion processes for transdermal patch development

Recently, there has been a growing interest towards the use of transdermal drug delivery systems (TDDS). My future plan is to overcome formidable barrier that can limit permeation of therapeutic agents as a great challenge by the skin. Therefore, my plan is to use my background studies in my PhD to develop hot melt extrusion technology (HME) to prepare these TDDS which has the potential to generate considerable interest in the pharmaceutical field. Briefly, this work combines these interests through preparation of several hot melt extruded formulations screened for their potential for transdermal development. So far, we have determined processing parameters (temperature, heating duration and mixing speed) for preparing hot melt extruded compositions comprising model drug such as clotirmazole and various carrier excipients can be established with minimal thermal degradation.

(b) The study of electrospinning processes for transdermal patch development

Electrospinning is the method for preparing drug-loaded nanofibers with ultrafine structure, a large surface area to volume ratio, and a high porosity with a small pore size. My second plan is to develop electrospinning method as a most cost effective one with simple and easy use operation tools which is applicable to produce ultrafine fibers with a simple step-up production for transdermal drug delivery applications. During my PhD and postdoc, I worked on the fabrication, advantages and limitations for topical/transdermal drug delivery applications. The combination of my background with the deep understanding of polymeric materials for drug delivery application can open up a new avenue for the transportation of drugs, vaccine and proteins into the skin.

(c) The study of diagnostic haptic patches that continuously monitor physiological markers from sweat

Innovations in mobile and electronic healthcare are revolutionizing the involvement of both doctors and patients in the modern healthcare system by extending the capabilities of physiological monitoring devices. My third goal is to provide flexible and comfortable wearable patches for patients, which is one of the keys for patient acceptance. There are many devices already on the market for fitness and wellness that use consumer-facing applications which can be

easily incorporated into clinical practice. However, the chemical detection of diseases from sweat and not only from blood or interstitial fluids can create more interest for future skin sensors.

Teaching Interests: Much of my teaching experience was gained from the Chemical Engineering Department of Urmia University of Technology as a faculty member. I taught Fluid Mechanics and Thermodynamics courses for undergraduate students. While my teaching focuses on the chemical engineering field, my teaching interests are much broader. At Case Western Reserve University, I transitioned into Macromolecular Science and Engineering field and gained interdisciplinary experience teaching courses in polymer physics, polymer chemistry and polymer processing topics in addition to chemical engineering core courses such as transport phenomena, and advanced thermodynamics that were new to me and outside of the field of my primary education. Beside my background experiences, my teaching philosophy was influenced by my participation in future faculty programs at both Georgia Institute of Technology and Case Western Reserve University. My teaching philosophy, which focuses on creating an inclusive learning environment, preparing students to serve diverse communities, and developing evidence-based teaching, enables me to meet your needs by taking a) an interdisciplinary approach in my work, b) maintaining an ongoing commitment to teaching, and c) deeply involving students in research.

The following abstracts have been accepted for an oral presentation at the 2018 AIChE Annual Meeting in Pittsburgh, PA.

Session: **Drug Delivery II: Small Molecules**
Date: **Wednesday, October 31, 2018**
Session Time: 12:30 PM - 3:00 PM

Presentation Title: Electrospun Patch for Transdermal Delivery of Contraceptive Hormone
Presentation Time: **2:18 PM - 2:36 PM**
Location: Westin Convention Center, Cambria

Session: **Drug Delivery III: Systems for Administration**
Date: **Wednesday, October 31, 2018**
Session Time: 3:30 PM - 6:00 PM

Presentation Title: Design, Structure, Material Strength of Dissolvable Microneedle Patch Vaccine Delivery Systems: From Fabrication to Characterization of Microscale Transdermal Patches
Presentation Time: **4:24 PM - 4:42 PM**
Location: Westin Convention Center, Cambria

Publications:

- **Mofidfar M.**, Prausnitz M. R. (2018) Electrospun Transdermal Patches for Contraceptive Hormone Delivery, under review.
- **Mofidfar M.**, Wang J., Long L., Hager L. C., Vareechon C., Pearlman E., Baer E., Ghannoum M., Wnek G. E (2016) Polymeric Nanofiber/Antifungal Formulations Using a Novel Co-extrusion Approach, AAPS PharmSciTech.

Patents:

- Erol U. L., Akkus O., Weidenbecher M., **Mofidfar M.**, (2018) A device for the construction and coating of tubular structures by spraying on a rotating mandrel, Patent Application No. US20180133950 A1.
- Baer E., Wnek G. E., **Mofidfar M.**, Wang J., (2017) Polymer fiber scaffolds and uses thereof, Patent Application No. WO2017041109 A1.

545939 Soft, Stretchable Wearable Platforms for Sensing and Energy Harvesting Applications

Abstract:

The human sweat represents an interesting biofluid due to its ease of collection and presence of biomarkers related to dehydration, physical fatigue, mental stress, and various diseases. Recently reported technologies for analyzing sweat rely either on active, battery-powered electronics for electrochemical detection or passive, colorimetric chemistries for visual readout. Complex construction and large size/weight represent disadvantages of the former; semi-quantitative operation and limited scope of measurable biomarkers are limitations of the latter. I introduce the first example of a unique, unconventional class of battery-free, wireless, flexible, lightweight, skin interfaced platform that perform sensing via schemes inspired by electrochemical biofuel cells and colorimetric assays embedded in chrono-sampling microfluidic networks with broad functionality in sweat analysis. NFC electronics, biofuel cells and microfluidics results in a wearable chemical sensor which is several orders lighter, cheaper and smaller than reported alternatives with no apparent effect on its performance. A demonstration device allows simultaneous monitoring of sweat rate/loss, along with quantitative measurements of pH and of lactate, glucose and chloride concentrations using biofuel cell and colorimetric approaches. Systematic studies of the electronics designs, the microfluidic systems and the integration schemes establish the key design considerations and performance attributes. Capabilities in measurements with short- and long range wireless reader systems in the context of personalized fitness monitoring suggest a broad range of application possibilities. Human trials involving measurements of sweat glucose and lactate concentrations and comparisons to levels in the blood over a period of two days highlight the ability for long-term monitoring of sweat analytes as a potentially non-invasive means to track corresponding blood concentrations. These successful studies represent the first examples of skin interfaced sweat sensors with multi-day operational use.

In addition to realizing new sensing platforms, developing viable wearable energy sources remain another key challenge hampering the widespread growth of the wearable sensors field. Most prior efforts rely on bulky coin batteries which severely compromise wearability. Efforts focused on developing thin, stretchable batteries and super capacitors address this issue but at the cost of limited energy capacity and frequent recharging. Various forms of wearable energy harvesters that harness energy from body heat, movement and sun demonstrate potential for powering wearable electronics. However, many of the energy harvesters and storage systems rely on toxic chemicals which pose serious health concerns. I describe the fabrication, characterization, and real-life application of a highly biocompatible, soft, stretchable electronic-skin-based biofuel cell (E-BFC) that scavenges energy from lactate present in human sweat. The E-BFC exhibits a power density of ~1.2 mWcm-2, representing the highest power density recorded by a wearable biofuel cell to date. Such performance is achieved via a unique combination of lithographically-patterned stretchable electronic framework and screen-printed, densely-packed three-dimensional carbonnanotube-based bioanode and cathode array arranged in a stretchable "island-bridge" configuration. The E-BFC maintains its performance even under repeated strains of 50% and can power a Bluetooth Low Energy (BLE) radio. This is the first example of a wearable biofuel cell powering a BLE radio. Such systematic studies and demonstrations mark a significant breakthrough in the field of wearable biofuel cells.

Selected Publications: (Total papers: 38; H-index: 26; Total citations: 2583)

Full List:
https://scholar.google.com/citations?hl=en&user=FYgDmoAAAAAJ&view_op=list_works&sortby=pubdate

1. "Battery-free, skin-interfaced microfluidic/electronic system for simultaneous electrochemical, colorimetric & volumetric sweat analysis" A. J. Bandodkar et al, Sci. Adv. (under review).
2. "Soft, stretchable, high power density electronic skin-based biofuel cells for scavenging energy from human sweat", A. J. Bandodkar, J.-M. You, N.-H. Kim, Y. Gu, R. Kumar, A. M. V. Mohan, J. Kurniawan, S. Imani, T. Nakagawa, B. Parish, M. Parthasarathy, P. P. Mercier, S. Xu and J. Wang, Energy Environ. Sci., 2017,10, 1581.
3. "All-printed magnetically self-healing electrochemical devices", A. J. Bandodkar, C. S. López, A. M. V. Mohan, L. Yin, R. Kumar and J Wang, Sci. Adv. 2016, 2, e1601465.
4. "A wearable chemical–electrophysiological hybrid biosensing system for real-time health and fitness monitoring", S. Imani, A. J. Bandodkar, V. Mohan, R. Kumar, S. Yu, J. Wang, P.P. Mercier, Nat. Commun. 2016, 7, 11650.
5. "Highly Stretchable Fully-Printed CNT-based Electrochemical Sensors and Biofuel Cells: Combining Intrinsic and Design-induced Stretchability", A. J. Bandodkar, I. Jeerapan, J. M. You, R. N. Flores and J. Wang, Nano Lett. 2016, 16, 721.
6. "All-Printed Stretchable Electrochemical Devices", A. J. Bandodkar, R. Nuñez-Flores, W. Jia and J. Wang, Adv. Mat. 2015, 27, 3060.

Keywords: Wearable chemical sensors, biofuel cells, microfluidics, colorimetric assays, wireless electronics, printed devices.

Research Interests: Wearable chemical sensors, implantable chemical sensors, biofuel cells, batteries, energy harvesting, stretchable devices, self-healing systems.

Teaching Interests: Teachers and mentors are important pillars in one's success and I owe my accomplishments to the teachers in my past and present who instilled scientific curiosity in me, taught me technological skills, helped me overcome my weaknesses, and guided me in tackling challenging tasks. A major motivation for me to become a faculty is to partake into the noble gesture of teaching and guide young minds towards a successful, fruitful and content life. During my PhD and now as a postdoc, I have mentored 14 undergraduate students of which 4 are currently PhD students in top tier schools; 3 work in industry and 7 are currently pursing undergraduate degrees. I have also been a TA for two graduate level courses (~50 students in each course). Apart from helping students understand the concepts during office hours, I also taught the class when the concerned professor was travelling. Considering my ability to explain complex scientific concepts in a simple manner to people without STEM background and my passion to motivate people towards research, I was invited as a guest lecturer for an undergraduate course (~150 students) on "Designing Information" at the UC San Diego's Fine Arts Department. During this lecture, I illustrated the different ways in which wearable devices offer an exciting avenue for achieving personalized Internet of Things and how these devices are having an impact in today's day and age. The overwhelming feedback and interest and curiosity ignited in the Fine Arts students, who before the lecture had given little thought to wearable devices and their impact, was quite satisfying to me. In fact, one of the students even collaborated with me on a project that combined chemical sensors with fine arts – a truly unique learning for both of us. In addition to these, I have also been invited as a speaker at the University's undergraduate outreach events to motivate and encourage young minds to take up careers in the STEM fields.

As a faculty I would be thrilled to teach courses on mass and heat transfer, thermodynamics, electrochemistry, chemical sensors, nanobiotechnology and surface science. Additionally, I would be excited to be involved in committees focused on increasing student participation in STEM fields.

547302 Soft and Biological Materials: Simulation and Theory (SoftBiM)

Abstract:

Collective motion in dense packings of cells occurs in wound healing, embryonic development, and cancerous tumor growth. Most current computational models of dense cell packings either treat the system as collections of spherical particles or assume that the system is confluent, with no extracellular space. We have developed a new model for dense cell packings in two spatial dimensions, where the cells are modeled as deformable particles that have a preferred area and perimeter. We measure the packing fraction, φj at jamming onset as a function of the asphericity, a, which is the ratio of the perimeter square to the area of the particle. We find that the jammed packing fraction increases monotonically with a and that the system becomes confluent with $\varphi_j = 1$ for $a > 1.16$. Using surface Voronoi analysis, we show that this value for a corresponds to the case when the cells completely fill their Voronoi-tesselated regions. We also demonstrate that the free area per cell obeys a k-gamma distribution, which has been found for jammed packings of non-deformable particles. Finally, we will describe results from our model concerning the mobility of deformable particles subjected to applied forces, as well as diffusion of deformable particles subjected to active forces to discuss the effect of geometry in active jamming of deformable particles.

Keywords: Soft materials, Granular materials, Complex fluids, Jamming of Cell monolayers.

Research Interests:

Soft Materials such as complex fluids, granular materials, and soft solids are ubiquitous in industrial applications and frequently used in the food, consumer, oil, and biotechnology industries. Their processability and stability are mainly affected by their flow behavior and rheological properties. To explain their complex rheological behaviors one needs to understand their microstructure and its evolution under applied deformation. In addition, there is a spectrum of length and time scales that must be resolved to explain their macroscopic flow behavior such as shear-thinning in polymer solutions and biological fluids like blood, shear-thickening in colloidal suspensions, and jamming and yielding in epithelial cell monolayers. Particulate systems composed of soft particles, such as emulsions, foams, bubbles, hydrogels, grafted core-shell particles, dendrimers, and star polymers can be used as the building blocks novel multifunctional materials, with optimized energy absorption, self-healing behavior, high mechanical strength, and other desirable properties. In addition, many biological systems such as biofilms, cell colonies, and tissues can also be considered as collections of soft and deformable particles. *In my future research, I will develop a theoretical framework that will enable the design of new multiscale soft materials with optimized and multifunctional properties.*

Postdoctoral Project: *"Jamming of soft deformable particles: From bubbles to tissues"*. This research was conducted under the mentorship of Prof. Corey S. O'Hern in the Department of Mechanical Engineering & Material Science, Yale University, New Haven, CT.

Ph.D. Dissertation: *"Computational studies on multi-phasic multi-component complex fluids"*. This research was conducted under the mentorship of Prof. Joao M. Maia from the Department of Macromolecular Science and Engineering, Case Western Reserve University, Cleveland, OH.

Research Experience:

My academic training in mechanical engineering, chemical engineering, rheology, and macromolecular science and engineering enabled me to study a wide range of materials using experimental, computational, and theoretical techniques. During my postgraduate studies in Europe (KU Leuven, Belgium, University of Ljubljana, Slovenia, and University of Minho, Portugal), I focused on the rheological properties of polymeric materials in both solid and liquid states with an emphasis on designing and fabricating novel polymeric nanofiltration membranes with antifouling properties. During my PhD training, I developed simulation techniques to investigate the rheological properties of a variety of soft materials such as colloidal suspensions, colloidal gels, emulsions, and polymeric blends. I have expertise in using micro- and mesoscale simulation techniques, such as molecular dynamics, dissipative particle dynamics, smooth particle hydrodynamics, and Lattice Boltzmann methods and macro-scale techniques such as finite element and discrete element method simulations. I have also developed multi-threaded, parallel, and GPU algorithms for high performance computing clusters.

Teaching Interests:

While at Yale, I served as a research mentor for several high school and undergraduate students during the summer through the Science, Technology, and Research Scholars (STARS) program and Research & Engineering Apprenticeship program (REAP), which is supported by the Army Educational Outreach Program (AEOP). These programs seek to encourage and prepare disadvantaged students and students from underrepresented groups to pursue careers in science, engineering, technology and mathematics (STEM) fields. I have also served as a teaching assistant for undergraduate and graduate courses in the rheology of complex fluids, theory of viscoelasticity, and computational methods in chemical and mechanical engineering. As a faculty member, I am interested in teaching courses at both the undergraduate and graduate levels in fluid mechanics, transport phenomena, microhydrodynamics, and mechanics of materials.

Future Directions in SoftBiM:

The main direction of my future research will be the development of highly efficient hierarchal computational methods to understand the mechanical and rheological properties of soft materials. My computational studies will be complemented by theoretical calculations to investigate the flow behavior of soft particulate systems with tunable interactions relevant to a wide range of industrial applications such as flow assurance, mixing, and separations. These studies will accelerate the development of novel multifunctional materials by tuning particle interactions, geometries, and tribological properties such as surface friction.

Fluid mechanics and rheology of complex fluids: As a faculty member, I will use mecoscopic simulation methods to study flows of multiphasic systems such as suspensions and emulsions in bulk and under confinement. These systems are of immediate importance in a wide range of industries such as oil, food, consumer products, and biotechnology. Flows in micro-channels are used to model biological phenomena, such as blood flow and cell migration where the suspended particles are soft and deformable. For example, in patients with sickle cell anemia, some red blood cells (RBC) stiffen into irregular shapes as a result of oxygen depletion (through mutant hemoglobin (HbS) nucleation and polymerization), which blocks small blood vessels. The blockages further deplete oxygen, which promotes vasoocclusion. To model sickle cell anemia, a multiscale approach is required since the problem involves a wide range of length scales (from nanoscale nucleation and growth to microscale shape changes and macroscopic clogging of cells). I will investigate the role of particle geometry, surface properties, interactions, and mechanical properties in determining the flow mechanics of a wide range of multiphasic systems ranging from colloidal suspensions and gels to emulsions and bubble flows to cell monolayers and tissues.

Granular materials: From sand and staples to animal swarms: While a pile of dry sand can flow freely like a fluid upon removing the container walls, the same is not true for highly nonspherical particles.

In fact, piles of long rods, flexible fibers, and staples can behave like cohesive solids and form a free-standing stable column with finite tensile and compressive strengths. The cohesive behavior of the pile is purely geometrical and is not currently understood. Geometric cohesion is used by swarms of insects as a survival mechanism to build stable structures such as rafts to survive floods or adaptable clusters of insects to protect themselves during their reproductive cycle against environmental, mechanical (gravity, wind), and thermal perturbations. My research group will investigate the physical mechanisms that give rise to geometrical cohesion and how it can be used to create and develop new materials that utilize geometric cohesion.

Selected Publications:

1. A. Boromand, A. Signoriello, F. Ye, C.S. O'Hern, M.D. Shattuck, "*Jamming of Deformable Polygons*", Phys. Rev. Lett. Under review, (2018).
2. A. Boromand, S. Jamali, B. Grove, J. Maia, "*A generalized frictional and hydrodynamic model of the dynamics and structure of dense colloidal suspensions*", J. Rheol. Cover Image 62, 905 (2018).
3. A. Boromand, S. Jamali, J. Maia, "*Structural fingerprints of yielding in short-range attractive colloidal gels*", Soft Matter. 13, 458 (2017).
4. A. Boromand, S. Jamali, J. Maia, "*Gaussian-inspired auxiliary non-equilibrium thermostat (GIANT) for Dissipative Particle Dynamics simulations*", Comput. Phys. Commun. 197, 27 (2015).
5. A. Boromand, S. Jamali, J. Maia, "*Viscosity measurement techniques in Dissipative Particle Dynamics*", Comput. Phys. Commun. 196, 149 (2015).
6. S. Jamali, A. Boromand, N. Wagner, J. Maia, "*Microstructure and rheology of soft to rigid shear-thickening colloidal suspensions*", J. Rheol. 59, 1377 (2015).
7. 7. S. Jamali, A. Boromand, S. Khani, J. Wagner, M. Yamanoi, J. Maia, "*Generalized mapping of multi-body dissipative particle dynamics onto fluid compressibility and the Flory-Huggins theory*", J. Chem. Phys. 142, 164902 (2015).

547355 Engineering biomaterial-based sensors for point-of-care diagnosis

Abstract:

Research Interests: Improving disease detection saves lives: in 2015 more than 4.8 million children died of preventable infections and 5 million people annually succumb to death by highly treatable cancers if identified early. Such challenges inspire and motivate my endeavors to develop forward-thinking biosensors for more accurate, sensitive and affordable diagnostics.
I am interested in combining my chemical and biomedical engineering expertise toward creating sophisticated, miniaturized diagnostic systems to detect disease much earlier than currently possible today. My vision is to apply novel sensing platforms for new biomarker discovery/ validation and effective preclinical/clinical exploration of disease pathogenesis and trajectory.
The emphasis of my laboratory will be on developing translational sensing technologies based on engineered biomaterials to address important biomedical and clinical problems. Specifically, I will study self-assembled plasmonic polymer nano-composites for the following purposes: 1) onsite diagnosis/high-throughput screening of infectious diseases, 2) single extracellular vesicles (EV) analysis via multiplexed profiling of proteins and RNAs for cancer diagnosis, and 3) smart coating as a doctor-on-a-chip for early diagnosis and prevention of bacterial infections on medical implant. All the proposed studies will involve development of new sensing platforms with the aim of diagnosing disease in a faster, more cost-effective manner. Most will involve collaboration with researchers at universities, national laboratories, and hospitals with diverse research backgrounds in chemistry, material science, oncology, pathology, and computer science. My research has the potential to provide insights into pathophysiology and immunology, ultimately accelerating development of optimized and personalized therapeutics.
This proposed research is made possible by my unique, interdisciplinary background. My undergraduate and graduate studies provide me the foundation in nanofabrication of biomaterials via chemical synthesis, extensive in vivo studies, and mathematical modeling of transport for drug delivery, and I will build on my strengths to design novel biomaterials in my independent research laboratory. My expertise has grown to encompass optic engineering, molecular imaging, deep-learning-assisted image analysis, and exosome isolation and analysis through my postdoctoral work, which uniquely positions me to proceed research at the interface of engineering and medicine.
I believe this proposal takes an important step in bridging imaging/sensing technologies with bio/nano-materials engineering and will give way to a long-term research program dedicated to the development of smart and affordable diagnostic and, if possible, integrated theranostics to win the battle against fatal diseases.

Teaching Interests: I am interested and qualified in teaching both Chemical Engineering and Bioengineering courses (undergraduate and graduate level) in the areas of biomaterials, reaction kinetics, transport, thermodynamics, and polymer physics. Additionally, I am excited to create a graduate course in the emerging fields of Biotechnology (Biosensors and Medical Devices), Biomaterials, and Drug Delivery.

548156 Rewritable Multi-event Analog Recording in Bacteria and Mammalian Cells

Abstract:

Stable recording of cellular events can significantly advance our understanding of a cell's history and its responses to stimuli. However, the construction of synthetic memory devices, especially ones with large storage capacities, remains challenging. We recently developed two platforms that use base editors and CRISPR-Cas9 nucleases to stably record molecular events of interest, including exposure to antibiotics, nutrients, viruses, and light, in living cells. Reliable readout can be obtained by sequencing as few as dozens of cells and the formed synthetic memories can be erased for repeated recording. Moreover, the base editor-mediated system functions in human cells, recording the presence of exogenous small molecules as well as changes in endogenous Wnt signaling in a safe-harbor locus of the human genome. We expect these CRISPR-mediated analog multi-event recording apparatus (CAMERA) systems to serve as "cell data recorders" that write a history of endogenous or exogenous signals into permanent DNA sequence modifications in living cells.

Postdoctoral Project: *engineering the CRISPR system for broadened applications*

Postdoctoral Fellowship: *the Jane Coffin Childs Memorial Fund*

Supervised by David R. Liu, Broad Institute of MIT and Harvard; Department of Chemistry and Chemical Biology, University of Harvard

PhD Dissertation: *mechanistic studies of lanthipeptide biosynthesis*

Supervised by Wilfred A. van der Donk, Chemical, Department of Chemistry, University of Illinois, Urbana-Champaign

Research Interests:

As a graduate student in Prof. van der Donk's lab, I focused on uncovering the biosynthetic mechanisms of lanthipeptides, a family of ribosomal natural products with extensive post-translational modifications. Using recombinant expression systems, I successfully prepared six lanthipeptides that could not be obtained from native producers and characterized their structures by NMR. During my study of cytolysin, an important virulence factor produced by *Enterococcus faecalis*, I discovered an unusual lanthionine stereochemistry that was different from all documented lanthionine structures in lanthipeptides. I further confirmed that, surprisingly, the formation of the unusual stereochemistry was controlled by a substrate motif rather than the synthetase. Following my discovery, more than a dozen other lanthipeptides have been reported to contain this substrate-determined LL stereochemistry.

After graduation, I moved to a different research area and worked on programmable genome-editing agents focusing on the CRISPR system in Prof. Liu's lab. In collaboration with graduate students in the Liu lab, I engineered Cas9 variants for expanded protospacer adjacent motif (PAM) specificities, enabling Cas9-mediated activities to be directed to a significantly expanded variety of genomic loci. In a second project, I took a synthetic biologist's perspective and developed a set of memory devices using the CRISPR technology to facilitate faithful and durable recording of stimuli and cell state changes in bacteria including exposure to antibiotics, nutrients, viruses, and light. These devices can also function in mammalian cells to record the activity fluctuation of critical endogenous signaling pathways.

Working on various research topics and interacting with scientists from different technical backgrounds have made me appreciate the advantages of a well-diversified research program, and have inspired me to pursue a wide variety of research projects in my own group.

Teaching Interests:

My experiences in science have made me passionate about teaching. I served as a teaching assistant in general chemistry classes at the University of Illinois, Urbana-Champaign and have mentored several Master's and junior PhD students both in graduate school and as a postdoctoral researcher. To better prepare myself for the upcoming teaching responsibility, I attended the Kaufman Teaching Certificate Program at MIT during the spring semester of 2018. In this 12-week program, I improved a variety of my teaching skills, including designing a course syllabus, interactive teaching and active learning, designing assignments and problem sets, and how to teaching inclusively.

Given my dual background in chemistry and biology, I can teach a variety of undergraduate courses such as *General Chemistry*, *Biochemistry*, and their lab sessions as well as graduate-level courses such as *Recent Advances in Chemical Biology*, *Synthetic Biology*, *Mechanistic Enzymology*, and *Modern Therapeutics in the Genomic Era*.

Future Direction

As an assistant professor, I aim to utilize my expertise in chemical biology and bioengineering to develop approaches that enable the investigation of three critical life science questions: 1) are novel DNA base modifications awaiting discovery? 2) can highly functionalized cyclic peptides provide improved pharmacokinetic properties and realize the ultimate therapeutic potential of peptide natural products? 3) can cell state dynamics be recorded by taking full advantage of the programmable nature of DNA?

More than 10 types of nucleotide thiolation have been reported in RNA, yet no such modification has been found in DNA, likely owing to the lack of characterization approaches. In my first project, I aim to develop sequencing techniques to read out sulfur-modifications in DNA at base resolution. These sequencing techniques will not only facilitate the investigation of the native presence of these thionucleosides in DNA, but also enable the application of sulfur-modified deoxyribonucleotide as DNA labeling probes to study genome dynamics at a heretofore inaccessible resolution and depth.

Cyclic peptides are better drug candidates than linear peptides because they offer better target binding affinity and specificity as well as improved pharmacokinetic properties. However, current strategies of synthesizing cyclic peptide libraries often provide limited complexity and are not compatible with high throughput screenings due to the lack of genotype-phenotype correlation. In my second project, I aim to construct highly functionalized cyclic peptide libraries by deriving a family of peptide natural products that are genetically encoded and post-translationally modified. A microfluidics-enabled selection platform will be engineered to screen for cyclic peptides that bind and regulate target proteins in live cells.

In my third project, I plan to engineer a targeted *in situ* DNA diversification tool that has broad applications ranging from construction of complex synthetic memory that facilitates single-cell lineage tracing to focused randomization of genes of interest in live cells for forward genetic studies and continuous directed evolution.

The proposed projects will not only use chemical biology principles to push the boundaries of science in areas of epigenetics, peptide drug design, and cell state dynamic tracking, but will also facilitate many future studies by offering valuable tools to efficiently label DNA with probes that can be mapped at base resolution, synthesize cyclic peptide libraries of high diversity and complexity, screen large compound libraries carried by live cells in therapeutic-enabling environments, and diversify the target DNA without affecting other sequences during directed evolution.

Selected Publications

Tang W; Liu DR*. Rewritable multi-event analog recording in bacterial and mammalian cells. *Science* **2018**, *360*, eaap8992

Tang W; Hu JH; Liu DR*. Aptazyme-embedded guide RNAs enable ligand-responsive genome engineering. *Nat. Commun.* **2017**, *8*, 15939

Tang W; Thibodeaux GN; van der Donk WA*. The enterococcal cytolysin synthetase coevolves with substrate for stereoselective lanthionine synthesis. *ACS Chem. Biol.* **2016**, *11*, 2438-2446.

Dong SH[#], **Tang W[#]**, Lukk T, Nair SK*, van der Donk WA*. The enterococcal cytolysin synthetase has an unanticipated lipid kinase fold. *eLife* **2015**, *4*, 07607 (# equal contribution)

Tang W, Jiménez-Osés G, Houk KN*, van der Donk WA*. Substrate control in stereoselective lanthionine biosynthesis. *Nat. Chem.* **2015**, *7*, 57-64

Tang W[#], Dong SH[#], Repka LM, He C, Nair SK*, van der Donk WA*. Applications of the class II lanthipeptide protease LicP for sequence-specific, traceless peptide bond cleavage. *Chem. Sci.* **2015**, *6*, 6270-6279 (# equal contribution)

Tang W, van der Donk WA*. The sequence of the enterococcal cytolysin imparts unusual lanthionine stereochemistry. *Nat. Chem. Biol.* **2013**, *9*, 157-159

Tang W, van der Donk WA*. Structural characterization of four prochlorosins: a novel class of lantipeptides produced by planktonic marine cyanobacteria. *Biochemistry* **2012**, *51*, 4271-4279

Patents

van der Donk WA, Garg N, Goto Y, **Tang W**. Class I and II lantibiotics from Geobacillus thermodenitrificans. U.S. Patent 9,326,523, issued on May 3, 2016; WIPO Pat. Appl. WO/2013/119821 A1, filed on August 15, 2013.

van der Donk WA, **Tang W**. Higher performance proteases for scarless tag removal. US-2017-0240878, published on August 24, 2017; WIPO Pat. Appl. WO/2015/175576 A3, filed on March 17, 2016.

Liu DR, **Tang W**, Hu JH. Aptazyme-embedded guide RNAs for use with CRISPR-Cas9 in genome editing and transcriptional activation. U.S. 62/505,175, provision filed on May 12, 2017.

Liu DR, **Tang W**. Cell data recorders and uses thereof. U.S. 62/631,455, provision filed on February 15, 2018.

542868 Genetically Engineered Probiotics to Target and Eliminate Colorectal Cancer

Abstract:

Research Interests:

I have always believed that a multidisciplinary approach to tackle complex scientific questions is critical for success, particularly in the cross-disciplinary field of biomedical science. To become a well-rounded scientist, I have made a point throughout my academic career to gain expertise in a broad array of fields. My path began as a chemical engineer modeling polymerization reactions, to a bioengineer investigating bacterial communication, to now a metabolic engineer manipulating pathways and enzymatic domains. As a principal investigator, I plan to unify my research skills towards the application of genetically engineered probiotics to treat various gastrointestinal diseases.

After my MS and time in industry, I decided to pursue a PhD in biomedical science and began investigating dynamic regulation and cell signaling under the guidance of Dr. William Bentley. Specifically, I elucidated interkingdom communication from quorum sensing (QS) molecules, clarified the mechanics of QS processing, developed quorum-quenching bacteria that could manipulate biofilms and other QS phenotypes, and established a metabolic engineering platform coordinated through QS molecules. By learning to design and develop genetic circuits towards bioengineering and biomedical applications, I established a foundation in synthetic biology, a rapidly advancing discipline that already impacts many aspects of society.

During my postdoc in Dr. Jay Keasling's laboratory, I have sought to match my knowledge of cellular coordination and genetic circuits by learning to design and manipulate the enzymes and pathways that lead to high-value products. I was awarded the NIH National Research Service Award (NRSA) to develop an in-depth expertise in polyketide synthases (PKSs), a modular assembly-like enzyme that is responsible for the production of many pharmaceutical drugs. My work has helped to develop design principles for these enzymatic pathways, and to programmatically produce high-value chemicals in *Streptomyces albus*.

By marrying my graduate school expertise of dynamic regulation with my newfound knowledge in metabolic pathways, I am developing a generalizable platform to produce disease-fighting molecules, specifically cancer, in the gastrointestinal system. As a principal investigator, I seek to establish *Bacillus subtilis*, a probiotic and GRAS organism, as a host for the production and release of cytotoxic drugs. Through a just-in-time release mechanism, these genetically engineered probiotics will release a cancer therapeutic at the cancer site and only when sufficiently accumulated. This project not only represents a vertical advance in cancer probiotic therapy, but also provides a basis for establishing *B. subtilis* as a metabolic engineering chassis for the heterologous production of novel polyketides.

Teaching Interests:

Fundamentally, my motivation to teaching is to train the next generation of people to contribute to society, hopefully in science, but most importantly in the manner they choose. The philosophy behind that desire is that I am devoted to evidence-based science, and I would apply this practice to my teaching.

I believe it is important to provide context for the lesson by relating new ideas to familiar concepts. Therefore, a hallmark of my lectures would be an interactive environment, and sometimes even a 'flipped' classroom. As a student, I enjoyed classes where we broke up into small teams in the middle of the lecture to answer questions. It not only helped me understand better by teaching and being taught simultaneously, I found that it helped build camaraderie among fellow students, who at a large university, I would otherwise not know. As a professor, I would also plan to provide a short quiz at the end of every lecture, covering what I discussed in the prior lecture. As a teaching assistant, a professor provided these short quizzes and it enabled the lecturers and the students to identify gaps in knowledge quicker, and any broad-based gaps were identified in the following lecture. The goal is to ensure an active classroom, where questions are not treated as interruptions but as opportunities to more firmly root scientific learning.

Personally, I am fascinated at the power of synthetic and systems biology to manipulate and optimize metabolic pathways. However, a laboratory course in synthetic biology is not in most curricula of institutions, not least because of the complexity and cost. I would hope to do an *in silico* version of a project to create a molecule of interest. In groups, the students would select plasmid backbones, design vectors to produce gene products, manipulate promoters and ribosome binding sites, and analyze secondary regulation to identify targets to increase metabolic flux. Fundamentally, the objective of this project would be to show the cross-disciplinary nature of bioengineering to create natural products.

My diverse background in chemical and biomedical engineering enables me to explain concepts and provide context in a multidisciplinary manner to retain the interest of diverse audiences. I am comfortable teaching undergraduate courses in courses in bioengineering such as biofluids and biocomputational methods along with traditional chemical engineering topics such as chemical kinetics and thermodynamics. My foundation in these broad topics will allow me to provide a unique perspective in this interdisciplinary field.

Selected Publications (17 total, 10 first-author)

Amin Zargar, JF Barajas, R Lal, JD Keasling. "Polyketide synthases as a platform for chemical product design", *AICHE Journal*, (2018)

Jesus Barajas, **A Zargar**, B Pang, V Benites, J Gin, E Baidoo, C Petzhold, N Hillson, JD Keasling, "Biochemical Characterization of b-Amino Acid Incorporation in Fluvirucin B2 Biosynthesis", *ChemBioChem* (2018).

Amin Zargar, CB Bailey, R Haushalter, CB Eiben, L Katz, JD Keasling, "Leveraging microbial biosynthetic pathways for the generation of 'drop-in' biofuels", *Current Opinion in Biotechnology* (2017)

Amin Zargar, DN Quan, N Abutaleb, E Choi, WE Bentley, "Biocompatible capsules to control quorum sensing", *Biotechnology & Bioengineering* (2016)

Amin Zargar, DN Quan, WE Bentley, "Reducing stochasticity in cell populations: rewiring quorum sensing to engineer dynamic gene expression with population-level control", *ACS Synthetic Biology* (2016)

Amin Zargar, GF Payne, WE Bentley, "A 'bioproduction breadboard': programming, assembling and actuating cellular networks", *Current Opinion in Biotechnology* (2015, journal cover) 36: p. 154-160

Amin Zargar, DN Quan, M Emamian, CY Tsao, HC Wu, CR Virgile, WE Bentley, "Rational design of 'controller cells' to manipulate protein and phenotype expression", *Metabolic Engineering* (2015). 30: p. 61-68.

Amin Zargar, DN Quan, KK Carter, M Guo, HO Sintim, GF Payne, WE Bentley, 'Bacterial secretions of nonpathogenic *E. coli* elicit inflammatory pathways: a closer investigation of interkingdom signaling", *mBio* (2015). 6:2 e00025-15.

542450 Engineered Hydrogel Biomaterials for Mimicking Tumor Microenvironments and Controlling Cancer Cell Fate

Abstract:

Research Interests: Biomaterials, Cancer, Metastasis, Tumor Dormancy, Angiogenesis, Vascularization, Microfluidics, Biofabrication

Teaching Interests: Cell & Tissue Engineering, Cancer Tissue Engineering, Biomaterials, Biomimetics & Biofabrication, Polymer Synthesis & Characterization, Transport Phenomena

Metastatic breast cancer is one of the most predominant causes of cancer-associated mortality in USA and throughout the world.[1,2] Current challenges toward developing effective treatment strategies for metastatic breast cancer include, amongst others, targeting of dormant chemoâ€‘resistant tumor cell subpopulations[3] and preventing crucial events in the metastatic cascade (e.g. epithelial-mesenchymal transition, invasion, extravasation).[4] These efforts are aimed at preventing incidences of secondary tumor recurrence and improving relapse-free periods in cancer patients.

My research focuses on developing engineered biomaterial-based models of the tumor microenvironment for investigation of disease mechanisms and anti-cancer drug development processes. The tumor microenvironment plays a crucial role in the malignant and metastatic progression of the disease, the importance of which is gradually being realized over the past few years.[5,6] The complex, heterogenous and altered microenvironment of the malignant primary site significantly affect progression of the disease and treatment outcomes. Hence, investigating this complex milieu of tumor cells, supporting cell types and cytokine signaling within extracellular matrices (ECM) is important in this context.

In previous studies, I employed biosynthetic hydrogels, poly(ethylene glycol diacrylate)-fibrinogen (PEG-Fb), as a tissue mimic for three-dimensional (3D) culture of various cancer cell types.[7,8] The biochemical and mechanical characteristics of the PEG-Fb matrix were modulated to varying degrees to modulate breast cancer cell behavior. Softer PEG-Fb matrices facilitated higher spreading, higher colony formation and higher proliferation of cancer cells when compared to stiffer PEG-Fb matrices.[8] PEG-Fb hydrogels were also fabricated into microspheres encapsulating cancer cells to realize a high-throughput system for anti-cancer drug testing applications.[7] The PEG-Fb hydrogel model was eventually implemented in a microfluidic-chip device where breast cancer and stromal cells were cocultured long-term with tumor-mimetic endothelial cell networks and therapeutic efficacy of doxorubicin and paclitaxel was evaluated *in vitro*.[9] Both drugs demonstrated cancer cell type-dependent and vascular network-dependent variations in drug efficacy, which could potentially be useful toward developing a more physiologically relevant drug-testing model for cancer treatment. Overall, we developed various biofabrication strategies to establishing 3D models of the primary breast cancer and microfluidic models to study tumor-stromal-endothelial interactions for drug efficacy testing.

In our current study, we focus specifically on recapitulation of metastatic breast tumor dormancy in secondary organotropic niches. Disseminated tumor cells (DTCs) from the primary tumor often lie dormant in secondary organs (bone marrow, brain, lung liver etc.) for extended periods of time and can revert to an actively proliferating state upon appropriate stimulation by inflammatory or proangiogenic signaling. Targeting DTCs as an approach towards reducing metastatic outbreaks is difficult owing to their quiescent behavior, poorly understood cellular characteristics, and limits of clinical detection.[10,11] Tumor dormancy is manifested in one of two ways: cellular dormancy (G0-G1 arrest) or dormant micrometastasis (angiogenic dormancy). In our studies, we find that metastatic breast cancer cells cultured within threeâ€‘dimensional (3D) hydrogel matrices can be made to remain dormant or become invasive by tuning the matrix adhesivity, crosslinking density and overall matrix compliance. Our developed hydrogel platform can be used to specifically identify and target dormant tumor subpopulations which would otherwise have been difficult to isolate and study in currently established models.

In the tumor dormancy model, MDA-MB-231 metastatic breast cancer cells were cultured within tunable PEG-based hydrogels containing the proteolytic-degradable peptide sequence, GGGPQGIWGQGK (PQ), and the integrin-ligating RGDS peptide. The RGDS concentration and co-monomer N-vinyl pyrrolidone (NVP) concentration were varied to modulate matrix adhesivity and crosslinking density respectively. Cells encapsulated within hydrogels with 1-10 mM RGDS displayed high viability and low apoptosis over 15 days, though cells in non-adhesive hydrogels (0 mM RGDS) underwent high cell death. Viable cell density and proliferation varied as a function of both NVP and RGDS incorporation. Soft (2.5 kPa, no NVP) hydrogels containing 5-10 mM RGDS displayed higher proliferation, higher cell density, higher metabolic activity, and the formation of large, invasive colonies. However, stiff (5.7-8.6 kPa, with NVP) hydrogels induced low proliferation, higher metabolic quiescence and a constant viable cell density over time irrespective of matrix adhesivity. Cells within dense matrices either remained solitary or formed small multicellular foci. Overall, matrix crosslinking density and adhesivity were found to jointly regulate four different fates for metastatic cancer cells: restricted survival, single cell dormancy, balanced proliferation/death and invasive growth. By tuning the matrix characteristics, DTCs were maintained in a dormant, state with high viability, low proliferation, low apoptosis, and reduced metabolic activity over 15 days in culture. Eventually, we aim to combine our established techniques of generating organ-mimetic vasculature within microfluidic devices[12-14] with the developed tumor dormancy model to investigate mechanisms of extravasation and dormancy within secondary organotropic niches (bone marrow, lung, brain etc.).

Overall, our research paves the way for designing improved engineered models to study cancer metastasis and associated pathological conditions *in vitro*. These models can be used to study cancer-specific mechanisms, cell-microenvironment crosstalk and increased chemo-resistance of cells. Eventually, these models could be used for predicting efficacy of new drug candidates and for improved translation towards *in vivo* and pre-clinical studies. In the future, I plan to extend these models and findings towards investigating specific vascular diseases, obesity, systemic inflammation and their association with cancer metastasis through microenvironmental crosstalk.[15] Specifically, I am interested in investigating the role of obesity-induced inflammation in disrupting normal vasculogenic development and in fueling tumor growth (breast, colorectal, liver etc.) through activated cytokine signaling mechanisms.[16] Ultimately, these initiatives would help establish causal relationships between rising obesity trends and cancer incidences, and thereby facilitate the development of novel therapeutic interventions.

References:

1. Gomis et. al., Molecular Oncology 2017;11:62-78.
2. Giancotti, Cell 2013;155:750-764.
3. Aguirre-Ghiso et. al., Nature Medicine 2013; 19(3): 276â€"277.
4. Brabletz et. al., Nature Reviews Cancer 2018; 18: 128-134.
5. Whiteside, Oncogene 2008; 27:5904-5912.
6. Place et. al., BMC Breast Cancer Research 2011; 13:227
7. Pradhan et. al., Biomaterials 2017; 115:141-154.
8. Pradhan et. al., Journal of Biomedical Materials Research Part A 2017; 105:236-252.
9. Pradhan et. al., Scientific Reports 2018; 8:3171.
10. Aguirre-Ghiso et. al., Nature Medicine 2013;19:276-277.

11. Almog, Cancer Letters 2010;294:139-146
12. Heintz et al., Advanced Healthcare Materials 2016, 5(17):2153-2160.
13. Heintz et. al., Journal of Visualized Experiments. 2017, 119.
14. Pradhan et. al., Advanced Healthcare Materials 2017, 6(24):1700681
15. Deng et. al., Annual Review of Pathology 2016; 11:421-449.
16. Picon-Ruiz et. al., CA Cancer Journal for Clinicians 2017; 67:378-397.

542455 Interstitial Fluid Flow and Transport in Neural Trauma and Disease

Abstract:

Research Interests:

Nearly 100 million Americans suffer from at least one of over 1,000 described neurological diseases. Many of these diseases are associated with alterations in cerebral blood flow, the blood-brain barrier, the extracellular matrix, and/or drainage pathways such as the meningeal lymphatics. Hence, regulators of fluid flow within neural tissue are emerging as an underlying connection between seemingly disparate neuropathologies. While bulk flow around the brain is well studied, researchers are only starting to characterize the effects of fluid transport within the interstitial space between neural cells, coined interstitial fluid flow. The magnitudes of interstitial fluid flow and the resulting shear forces are typically low in the brain under normal physiological conditions, so even small perturbations can affect neural cell function. One prime example is in glioblastoma, the most common and deadly brain tumor. My postdoctoral research has focused on studying effects of interstitial fluid flow and shear stress on tumor-adjacent neural cells called the tumor microenvironment. We have shown that interstitial fluid flow significantly increases at tumor-brain interfaces, and these convective forces stimulate neural cells to enhance cancer cell invasion. We can effectively reduce glioma cell dissemination by preventing this flow stimulation, thereby demonstrating the therapeutic potential of altering responses to flow in the brain.

In my own research program, **I will combine my expertise in interstitial fluid flow with my training in neural tissue engineering and biomaterial development towards creating novel strategies for neural regeneration**. My initial approach will be to 1) characterize interstitial fluid dynamics and the contributions to neurodegeneration after spinal cord injury, 2) engineer hydrogel systems for controlling interstitial flow *in vitro* and *in vivo*, and 3) develop polymer-based strategies for modulating neural cell phenotype after injury and in disease. My overarching goal is to understand fluid flow dynamics in the nervous system and the resulting impacts on the neural cell microenvironment such that we can engineer better therapies for these debilitating disorders.

Teaching Interests:

My passion for teaching started in high school when I first tutored a friend three grades my senior in her Advanced Algebra class. This experience motivated me to develop my one-on-one teaching skills in college by tutoring student athletes, and I eventually became a primary tutor of the center's "Math Lab." In graduate school, I led small groups as a teaching assistant for the Fundamentals of Chemical Engineering lab course at the University of Texas and gave a guest lecturer for the University of Florida's Student Science Training Program. As a postdoctoral researcher, I co-instructed the Integrative Design and Experimental Analysis lab course at the University of Virginia where I both delivered lectures and directed lab classes focused on mass transport in tissue engineering. I am currently helping develop an international cancer course titled "Frontiers in Cancer Engineering" in which my postdoctoral advisor and I will introduce a small class of graduate students at Virginia Tech to cancer research in Europe through telecommunicated lectures and site visits to research laboratories across Switzerland. I want to continue to develop my skills as an educator and look forward to the challenge of continually tailoring lectures to the needs of the class, both in established courses and courses I develop.

Teaching Philosophy

My goal as an educator is to share with students my fascination of the physical world and the human body, highlighting methods that scientists and engineers are leveraging to treat real-world problems. I will focus my teaching philosophy on three main objectives: 1) provide resources to enable students to enhance their educational experience as needed to fit their learning needs, 2) demonstrate connections across the range of course topics to promote an integrative understanding, and 3) provide context to the real world by incorporating examples based on cutting-edge research. I am most interested and qualified to teach introductions to mass and energy balances and transport phenomena. For more advanced courses, my interests and research expertise make well suited to teach courses on mass separations, polymeric biomaterials, and engineering physiology. I am also particularly interested in developing an advanced course on neuro-immune engineering.

Mentorship

I have been fortunate enough to work in both a well-established, medium-sized program (12+ trainees) and a newer, smaller program (3-6 trainees). These experiences have taught me that my mentor style will need to adapt to both the requirements of individual mentees as they progress throughout their training as well as my career stage. Nonetheless, my guiding principle for mentoring students will always be to cultivate their intellectual independence while highlighting the importance of collaboration. I will use group and individual meetings to develop my mentees' abilities to design and implement impactful, innovative research projects, critically examine the literature and reflect on their own methods and goals, effectively communicate research results to the scientific and general communities, and hone their professional development skills for realizing their career objectives.

Education:

Ph.D., University of Texas at Austin, Chemical Engineering, 2015

B.S., University of Tennessee at Knoxville, Chemical and Biomolecular Engineering, 2011

Postdoctoral Project: "Defining and targeting the effects of interstitial fluid flow on the brain tumor microenvironment" under supervision of Jennifer M. Munson, originally at University of Virginia and currently at Virginia Polytechnic Institute and State University.

Ph.D. Dissertation: "An injectable acellular nerve graft as a platform for treating spinal cord injury" under supervision of Christine E. Schmidt, originally at University of Texas and currently at the University of Florida.

Selected Peer-Reviewed Publications:

1. **Cornelison RC**, Brennan CE, Munson JM. "Convective forces increase CXCR4-dependent glioblastoma cell invasion in GL261 murine model." (*under revision at Scientific Reports*)
2. **Cornelison RC**, Wellman SW, Park JH, Porvasnik SL, Song YH, Wachs RA, Schmidt CE. "Development of an apoptosis-assisted decellularization method for maximal preservation of nerve tissue structure." (*under review at Acta Biomaterialia*)
3. Da Mesquita S, Louveau A, Vaccari A, Smirnov I, **Cornelison RC**, Kingsmore KM, Contarino C, Onengut-Gumuscu S, Farber E, Raper D, Viar KE, Baker W, Dabhi N, Oliver G, Rich S, Munson JM, Overall CC, Acton ST, Kipnis J. "Functional aspects of meningeal lymphatics in aging and Alzheimer's diseas" Nature, (in press).
4. **Cornelison RC** and Munson JM. "Perspective on translating biomaterials into glioma therapy: Lessons from in vitro models." Frontiers in Materials, 5 (2018): 27.

5. **Cornelison RC**, Gonzalez-Rothi EJ, Porvasnik SL, Wellman SM, Park JH, Fuller DD, Schmidt CE. "Injectable hydrogels of optimized acellular nerve for injection into the injured spinal cord." Biomedical Materials, 13.3 (2018): 034110.

Patents:

Schmidt CE, Wachs RA, **Cornelison RC.** "Tissue Decellularization Methods" (2017). International Publication No.: WO/2017/011653.

542465 Quantitative Label-Free Dynamic Phenotyping of Highly Metastatic Cancer Cells for Emerging Liquid Biopsy Applications

Abstract:

Research Interests:

My Ph.D. work at Texas A&M University (TAMU) focused on developing optical methods to probe surface interactions of microspheres and biomimetic systems such as polymer vesicles. The combination of theoretical, modeling, computational, and experimental approaches produced a significant breakthrough using an interference-based technique to enable fast and nanometer-scale reconstruction of the geometry of curved microscopic objects near surfaces. During my post-docs at the University of California, Los Angeles (UCLA), and TAMU, I have grown as a researcher in the development of next-generation imaging and sensing platforms based on lens-free computational microscopy, label-free quantitative imaging tools, smartphones, and microfluidics, applied to emerging liquid biopsy analysis in my most recent work.

To pursue research as a faculty, I am uniquely equipped to tackle interdisciplinary problems at the interface between engineering, optics, biology, and medicine. In particular, personalized medicine enabled by liquid biopsy of bodily fluids for early disease diagnosis, treatment monitoring, and patient prognosis, faces important technological challenges depending on the targeted biomarker (e.g., cells, exosomes, and cell-free nucleic acids). For instance, circulating tumor cells (CTCs), diseased cells in circulation that could be responsible for metastasis at distant sites, are present in extremely low numbers relative to blood cells. Conventional antibody-based approaches for CTC isolation and detection involve complex workflows, limiting throughput and potentially missing metastatic cells not targeted by the antibodies used. Alternatively, physical differences between CTCs and blood cells (e.g., size and deformability) can be exploited for high-throughput CTC isolation and enrichment using microfluidics. Then, label-free interference-based quantitative imaging is performed at the single-cell level for identification of highly metastatic cells using label-free fingerprints such as adhesion and filopodia dynamics. This approach overcomes limitations of conventional analysis methods while offering great potential for label-free phenotyping of cells in a broader context. Similarly, liquid biopsy is not limited to cancer, with advancements in exosome-based analysis enabling minimally invasive access to a wide range of pathologies, including diabetes, inflammation, acute kidney injury, concussion, cardiovascular disease, alcoholic liver disease, and central nervous system diseases.

Therefore, building on my expertise and skills, I aim to integrate novel and powerful imaging and sensing tools and approaches into portable and low-cost liquid biopsy analysis platforms to facilitate personalized medicine in resource-limited settings. This will be achieved through fundamental research, a combination of theoretical, modeling, computational, and experimental approaches, and emphasis on the development of platforms for high-throughput analysis of bio-colloidal systems. Recent advances in lens-free computational microscopy and smartphone-based microscopy and sensing aided with artificial intelligence techniques represent extremely valuable tools for telemedicine and point-of-care applications and are expected to play a key role in my future research.

Teaching Interests:

Teaching in chemical engineering offers a unique opportunity for motivating students, especially at the undergraduate level, by providing a variety of real-life scenarios where new concepts can be applied, even to understand natural phenomena around us. This is fundamental in my teaching philosophy, where inquisitive questions (e.g., how do tall trees transport water from their roots to the top?) are presented to the students before a new topic (e.g., capillary phenomena) is introduced. In addition, interactive in-classroom experiences that help measure/visualize/illustrate content being learned are integral part of my approach. For this purpose, smartphones can be readily exploited as powerful scientific/educational tools, capturing images, videos, and incorporating virtual and augmented reality technologies. For instance, I worked with Honors students in the Fluid Operations course to develop a portable smartphone-enabled capillary-based viscometer. It is also important to recognize that knowledge is not complete, and concepts learned in classrooms could lead to more questions and technological developments. Thus, I finish course materials by making students wonder, highlighting new challenges ahead in the ever-evolving chemical engineering field.

I feel confident in my ability to teach most of the traditional subjects in the chemical engineering curriculum at the undergraduate level, and my preferences are Transport Phenomena and Process Dynamics and Control. I am also interested in developing an elective course on colloidal science and quantitative imaging methods that can be taken by both graduate and undergraduate students.

542499 Biofilm Engineering for Human Health and Environmental Sustainability

Abstract:

Biofilm Engineering for human health and environmental sustainability

Research Statement:

My research focuses on the interface of electrochemistry and biofilm processes in medical and environment applications. The medical research focused on understanding and developing new strategies for controlling pathogenic biofilms, such as wound biofilms. We developed a novel technology for combatting wound biofilm infection by controlled production of biocides on flexible electrodes embedded in a wound bandage. The work included the development of primary wound scaffold containing flexible electrodes and the design of control electronics. The controlled continuous production of biocides on electrodes allowed the elimination of biofilm infections on wound surface, without damaging the wound tissue. In addition, microsensors were used to study the microscale gradients in wound biofilms and understand their response to antibiotic treatment. This understanding allowed the development of enhanced antibiotic delivery in biofilms based on diffusion-reaction theory. Environmental applications focused on the utilization of biofilms that grow and exchange electrons with inert electrodes, called electrochemically-active biofilms (EABs). EABs can be used to speed up wastewater treatment, recover energy from organic waste streams, or for the production biocommodities. At fundamental level, my research involved in developing novel methods to measure depth-resolved biofilm conductivity using innovative microelectrode array assembly, which was used to investigate fundamental electron transfer mechanisms in EABs. In addition, we investigated factors controlling the scaleup of EAB electrodes, and operated a large pilot-scale reactor for wastewater system. The innovation her is using solid electrodes as electron acceptors, rather than the conventional aeration method (currently accounting for 3-4% of US electrical energy consumption). In addition, we developed a novel strategy improving the long-term stability of EABs based on potentiostatic control. We developed two autonomous electronic platforms for monitoring bioelectrochemical systems and for in-situ enrichment of EABs in remote locations. For microorganisms that do not directly interact with electrodes, we utilize electron shuttles to alter the fermentation profiles of microorganism for the electrosynthesis of higher value reduced metabolic end products. Overall, my PhD research demonstrated the use of electrochemistry and engineering tools to control biofilm process to 1) prevent, delay, or treat pathogenic biofilm infections for medical applications, 2) speed up wastewater treatment using an innovative EAB-based system, and 3) shift the metabolic profile of microorganisms to produce higher value products.

Research Interests:

Biofilm engineering

Electrochemically active biofilms

Electrochemical Engineering

Microsensors (electrochemical) and applications in biology

Environmental microbiology and biotechnology

Bioinstrumentation

Teaching Interests:

Electrochemistry and electrochemical engineering

Transport phenomena

Numerical methods for chemical engineers

Bioinstrumentation

Bioprocess engineering

542502 Multi-Scale Biomolecular Modeling and Design for Engineering and Medicine

Abstract:

Research Faculty

Research Interests:

My research focuses on multi-scale modeling of proteins and leveraging these models to gain insights into disease, as well as to design potential therapeutics. Biophysical and physicochemical understanding of protein function is a central element of my work, but one theme of my research has been developing models across physiological scales by using data mining and machine learning to integrate multiple types of experimental and computational data. A primary application area of my work has been computational immunology, where I have been involved in projects focused on topics such as the complement immune system and HIV. My recent research has focused on a family of proteases known to be upregulated in various types of cancer, but the development of tools for general protein structure prediction and protein design is also of great interest. My recent work, has started to combine molecular modeling with systems-level models to better understand the contribution of molecular changes in complex protein networks. One major interest of mine is the development of computational tools for research and teaching, including the development of multiple R packages and more recently the development of intuitive online interfaces to allow nonexperts to interact with models using the R Shiny platform.

Research Experience:

The core of my training and experience has been in computational modeling, starting with an emphasis biophysical models of proteins when I was a graduate student, including continuum electrostatics and molecular dynamics simulations. My graduate work also included computational drug design based on computational chemistry methods, such as pharmacophore searches and virtual screening. As a postdoctoral researcher, my work remained computational, but transitioned to an approach more centered on data science, utilizing machine learning and global optimization. My current project still involves biophysical modeling for understanding protein interactions, but also includes a systems modeling component for which I have also been developing an algorithm for parameter estimation. In addition to having the opportunity work on a diverse collection of computational projects, throughout my academic career I have also had an interest in coupling computational modeling with experiments. This interest began as a graduate student where I used NMR spectroscopy to characterize peptide structure, as well as, gaining experience in measuring the potency of designed inhibitors using ELISAs. As a postdoctoral researcher, one of my primary projects involved performing solid-phase peptide synthesis for over 40 peptides, that where then used in high-throughput peptide solubility measurements. More recently I have gained experience with basic cell and molecular biology techniques including gel electrophoresis, Western blots, and tissue culture.

Successful Proposals: UCOP Tobacco Related Disease Research Program Dissertation Fellowship, NSF/DAAD Central Europe Summer Research Institute Research Fellowship, NSF/JSPS East Asia and the Pacific Summer Institute Research Fellowship.

Current Project: "Computational modeling of cysteine cathepsin proteases and the proteolytic network."

Under the supervision of Manu O. Platt, Coulter Department of Biomedical Engineering, Georgia Institute of Technology

Postdoctoral Project: "Application of global optimization and machine learning to predict protein structure and function."

Under the supervision of Christodoulos A. Floudas, Department of Chemical and Biological Engineering, Princeton University (2012-2014) and Texas A&M Energy Institute (2014-2016)

PhD Dissertation: "Design and Applications of a Computational Framework for Protein and Drug Design."

Under supervision of Dimitrios Morikis, Department of Bioengineering, University of California, Riverside

Teaching Interests:

For my current appointment, in addition to my research I have also been teaching an undergraduate course for junior and senior biomedical engineering students, Biomedical Systems and Modeling, that introduces the fundamentals of modeling biomedical systems. The course covers a wide variety of topics from system biology, and success in the class requires a grasp of modern biology, analyzing graphs and networks, models consisting of nonlinear differential equations, several types of simulations, and a variety of auxiliary mathematical methods. The course has a pseudo-flipped structure with weekly overview lectures and problem solve sessions that give the students hands-on modeling experience. To complement the problem solving sessions, I have also initiated and been leading an effort to develop new online teaching tools that cover various modeling topics. The students have found these web-based applications very useful in learning difficult concepts in Biomedical Systems and Modeling, since they provide simple interfaces for manipulating models of some key biomedical systems from the course. These teaching tools are not only being used by the students in my sections, but now are also being used by all of the instructors and students in Biomedical Systems and Modeling.

Future Direction:

Future areas of interest are centered around multi-scale biomolecular engineering, and fall into three main areas: 1) data-driven models of biology and disease for personalized medicine; 2) modeling of protein structure and dynamics across atomistic and systems scales; 3) design of peptides/proteins for therapeutic and diagnostic applications. In area 1, I hope to build upon my experience with models for HIV co-receptor selection, and branch out into areas such as predicting antibody-epitope affinities and proteolytic cleavage of peptides. In area 2, early projects will focus on using data-driven methods to build multi-scale models of large protein complexes, such as though found in the complement immune system, and there will also be an emphasis on using biophysical models of protein interactions to predict parameters of system-scale models. Proposed work in area 3, will be both experimental and computational in nature, focusing on the design on multi-functional peptides and initial characterization using basic assays and structural methods. The proposed multi-scale design approach will account for microenvironmental factors, such as pH, and will incorporate functional characteristics such as binding, solubility, and conformational switching, to name a few. Application areas of interest will include computational immunology and proteases in the context of cancer, which will build off of my prior experience. A major component of the proposed work will center around the development of computational tools, and therefore experimental collaborations will also be actively sought in other related areas.

Selected Publications (34 total co-authored publications/reviews):

CA Kieslich, F Boukouvala, CA Floudas (2018) Optimization of black-box problems using Smolyak grids and polynomial approximations. *J. Glob. Opt.*, In press.

CA Kieslich, P Tamamis, YA Guzman, M Onel, CA Floudas (2015) Highly accurate structure-based prediction of HIV-1 coreceptor usage suggests intermolecular interactions driving tropism. *PLOS ONE,* 11(2), e0148974.

CA Kieslich, J Smadbeck, GA Khoury, CA Floudas (2016) conSSert: Consensus SVM model for accurate prediction of ordered secondary structure, *J. Chem. Inf. Mod.,* 56 (3), 455–46.

CA Kieslich, D Shin, A López de Victoria, G González-Rivera, D Morikis (2013) A predictive model for HIV-1 co-receptor selectivity. *AIDS Res. Hum. Retroviruses,* 29:1386-1394.

CA Kieslich and D Morikis (2012) The two sides of complement C3d: Evolution of electrostatics in a link between innate and adaptive immunity. *PLoS Comp. Bio.,* 8(12): e1002840.

K Chae, **CA Kieslich**, D Morikis, SC Kim, EM Lord (2009) A gain-of-function mutation of Arabidopsis lipid transfer protein 5 disturbs pollen tube tip growth and fertilization. *Plant Cell*, 21: 3902-3914.

Reviews:

GA Khoury, J Smadbeck, **CA Kieslich,** CA Floudas (2014) Protein folding and de novo protein design for biotechnological applications. *Trends in Biotech.*, 32 (2), 99-109.

CA Kieslich*, P Tamamis*, RD Gorham Jr, A López de Victoria, NU Sausman, G Archontis, D Morikis (2012) Exploring protein-protein and protein-ligand interactions in the immune system using molecular dynamics and continuum electrostatics. *Curr. Phys. Chem.,* 20(2):324-343.

RD Gorham Jr, **CA Kieslich**, D Morikis (2011) Electrostatic clustering and free energy calculations provide a foundation for protein design and optimization. *Ann. Biomed Eng,* 39(4): 1252-63.

542538 Design and Development of Point-of-Care Microsystems for Diagnosis of Neurodegenerative Diseases

Abstract:

Research Interests:

Design and Development of Point-of-care Microsystems for Diagnosis of Neurodegenerative Diseases

1) Diagnosis of neurodegenerative diseases by analyzing biomarkers from tear fluids

2) Design and development of microfluidic-based high-throughput mi-RNA analysis system for point-of-care (POC) analytics and diagnosis

3) Eye-on-a-chip

Ocular disease influences human vision and life quality. Worldwide, there are 285 million people who have a visual impairment, and the number of the people increases approximately 7 million every year. Glaucoma and age-related macular degeneration (AMD) are progressive optic neuropathies and are leading causes of blindness in the industrialized countries. Alzheimer's diseases (AD) is a representative neurodegenerative disease causing dementia and eventual death, and 20% of people by the age of 85 are suffering from AD. Aging is known as one of primary factors of those neurodegenerative diseases, and interestingly, those diseases (glaucoma, AMD, and AD) share similar pathological pathway involved with amyloid beta (Aβ) accumulation in the brain and in the retina. Indeed, retinal disorders have been observed in patients who have other neurodegenerative diseases (i.e. Parkinson disease and AD) since the eye is a part of central nervous system. Although diverse pharmaceuticals have been developed, people are still suffering from those diseases because it is difficult to diagnose those diseases at early stage due to the unrecognizable symptoms. The diagnosis of AD at present can be made by physicians when the patients already exhibit early cognitive losses, the irreversible state causing dementia. Even if they have the chance to diagnose the diseases, it requires expensive diagnostic instrumentations such as a fundus camera, OCT (optical coherence tomography), and advanced imaging system. Thus, to overcome those limitations, development of an early diagnostic system which can detect the neurodegenerative diseases with high speed, sensitivity, and cost-effective manner is of paramount importance. In addition, development of point-of-care (POC) system for neurodegenerative disease detection at an early stage is important to demonstrate the self-diagnostic system at home or office. Furthermore, since the pathogenesis of the glaucoma, AMD, and AD is not fully discovered yet, the research investigating pathological pathway of the diseases should be performed simultaneously.

1) Diagnosis of neurodegenerative diseases by analyzing biomarkers from tear fluids

Objective: Early diagnosis of progressive neurodegenerative diseases using tear fluids

As the advance of the genetic and proteomic assay, tear fluid has been grown its interest as a source of biomarker analysis to diagnose the diseases. Tear fluid including various biomarkers such as proteins, lipids, and oligonucleotides, is less complex body fluid than serum or plasma and can be obtained without invasive method. Recently, as it has been known that tear fluid contains a high concentration of microRNA (miRNA), a non-coding and small RNA with approximately 20-22 nucleotides, which has been reported as biomarkers of neurodegenerative diseases. Since conventional diagnosis of neurodegenerative diseases requires expensive instrumentations, thus, miRNA based diagnosis is essential and can provide simple, inexpensive, sensitive, non-invasive method at an early stage of diseases. However, since the roles and pathogenesis of mi-RNAs and the data library among target mi-RNAs and the diseases was not fully investigated, it is paramount of importance to be investigated more quickly and widely.

I have been experienced detecting diverse biomarkers such as DNA/RNA, bacteria, and virus using immune/genetic analytical method. Since my Ph.D. thesis was involved in pathogen detection based on genetic and immune assay, I had obtained knowledge about genetic analysis technique; DNA/RNA amplification (i.e. single/multiplex PCR, isothermal amplification; LAMP, RPA), DNA/RNA purification from human samples. In addition, I have performed immune analysis using diverse antibodies, which were conjugated to graphene oxide, gold/iron oxide nanoparticles, and silica beads using conjugate chemistry, and ELISA reaction. I have knowledge of not only glaucoma and AMD but also ocular anatomy and pharmaceuticals, since I am working on ocular drug delivery in Georgia Tech. Thus, I believe I can successfully investigate miRNAs in the tear fluid based on my comprehensive experience.

2) Design and development of microfluidic-based high-throughput mi-RNA analysis system for point-of-care (POC) analytics and diagnosis

Objective: Design and development of (*Aim #1*) rapid, sensitive, portable and high-throughput mi-RNA diagnostic system based on microfluidics, and (*Aim #2*) POC testing device for self-diagnosis of neurodegenerative diseases at early stage

Diagnosis of neurodegenerative diseases using mi-RNAs in tear would be interesting research. However, tens of down-regulated mi-RNA can be derived from one target mi-RNA from the patients, and moreover, one mi-RNA can be also found in other neurodegenerative diseases. In the end, hundreds of mi-RNA samples should be analyzed to identify and detect target mi-RNA with high sensitivity and reproducibility. Since, the mi-RNA analysis in lab-scale (conventional method) requires much samples and reagents, labor-intensive process, long analysis time with low reproducibility, thus, I would like to exploit 'Lab-on-a-Chip' (LOC) technology to overcome the limitations mentioned above. My objective of this project is that design and development of high-throughput analysis system using on the LOC technology enabling hundreds of reactions simultaneously with high sensitivity and accuracy.

The second objective of this project is demonstrating POC testing device for neurodegenerative diseases for early self-diagnosis. People living in rural area and developing countries are vulnerable to diseases because they have economic and geographical difficulties in getting a medical diagnosis with physicians. In addition, there are some people miss the opportunity to find their disease at an early stage because they are annoyed or afraid of to go to the hospital to be examined. If we could develop a miniaturized portable diagnostic system to solve the above problems, it will contribute to diagnose and treat diseases of people who are not examined with doctors due to economical, geographical, temporal problems or individual's will. Thus, this project aims to develop a portable diagnostic system that can diagnose the diseases more rapidly and accurately based on LOC technology by selecting one to three target mi-RNA that can represent each disease.

Since I fabricated diverse microchips made by glass, silicone, quartz, polycarbonate, and polydimethylsiloxane (PDMS), I have learned various fabrication methods (i.e. photo/soft lithography, by using a CO_2 laser cutting, milling machine cutting). According to the target and detection methods, diagnostic microsystem can be optimized to maximize the detection sensitivity and specificity. In particular, I have mainly demonstrated centrifugal diagnosis microsystems, which do not require external valves or pumps for fluid flow, but fluid transfer can be controlled only by the structure and surface properties of microchannels, and spinning rate of the microchip. By using advantages of the centrifugal microsystems, I am going to develop high-throughput and POC testing diagnostic centrifugal microsystems.

3) Eye-on-a-Chip

Objective: Investigation of pathogenesis of progressive neurodegenerative diseases/ Drug screening for treatment of retinal disorders/ Demonstration of ocular organs on microfluidic devices

Various organs have been mimicked in the microchip, organ-on-a-chip, to investigate various pathological disease mechanisms in the body and to develop effective treatment methods using drugs, but the eyes rarely have been investigated on the microchip. Recently, eye research on the microchip has been started, but more intensive studies are needed and various applications are possible because the research has not been done sufficiently yet. By mimicking the retina on the microchip, "Retina-on-a-chip", time and cost savings can be achieved by replacing some clinical trials in animals and humans, and real-time monitoring can be performed to observe the response and metabolism of retinal cells in a variety of physical, chemical, and biological conditions. Thus, it is effective in identifying pathological mechanisms of onset. In particular, the retina is located inside the eye, so expensive imaging equipment is needed for observation, while a retina-on-a-chip can be observed with a simple microscope. In addition, patient's retinal cells can be cultured on a microchip to allow theragnosis to be diagnosed and treated according to the characteristics of the patient. Moreover, neurodegenerative diseases such as Alzheimer's disease, glaucoma, or age-related macular degeneration are highly related to retinal degeneration, and pathological pathways of the above diseases has not been clearly elucidated. Therefore, by using the retina-on-a-chip, the relationship between the pathogenesis and biomarkers can be identified. In addition, a drug delivery model based on the retina-on-a-chip can be demonstrated, which can be extended to various fields.

The first step is optimization of stem cell differentiation because it is strict to differentiate from stem cell to retinal cells like retinal pigment endothelium. Then, microchip materials capable of stem cell culture is chosen, and design and fabrication of microchips based on the actual structure of the retina is performed. Then, retinal or stem cells are cultured on the microchip by investigating conditions such as temperature, humidity, and pH that the retina cells can grow. The next step is to study the environment of the retina related to various neurodegenerative diseases, then, to demonstrate those conditions that can simulate the disease on the retina-on-a-chip. The factors that can treat the simulated retinal disease are summarized and flowed on the retina-on-a-chip to observe the reaction and denaturation of the cells by real time confirmation. Then, by analyzing the metabolites generated under various conditions, the relationship with the biomarker and diseases can be identified. Furthermore, the optimum conditions of the metabolism and the treatment can be investigated by controlling drugs, dose, and other conditions in the retina-on-a-chip.

Since I have worked on a drug delivery to the posterior eye through a suprachoroidal space (SCS) using a microneedle, thus, I also can mimic the SCS on a microchip, "SCS-on-a-chip", and implement a SCS delivery model on the microchip. Based on my knowledge of microdevice fabrication and ocular drug delivery, SCS can be mimicked not only mechanically but also biologically. By using the SCS-on-a-chip, the pharmacokinetics model of ocular drugs which are injected by a microneedle can be optimized and compared to clinical trials with animals and humans to determine the feasibility of the SCS-on-a-chip.

Research Interests

1. Ocular and dermal drug delivery using microneedles
2. Diagnosis of neurodegenerative diseases by analyzing biomarkers from body fluids
3. Design and development of microfluidic based high-throughput mi-RNA analysis system and point-of-care (POC) system
4. Eye-on-a-chip: Suprachoroidal space-on-a-chip, Retina-on-a-chip

PhD Dissertation

"*Development of a Total Integrated Rotary Genetic Analysis Microsystem for Rapid and Multiplex Pathogen Detection.*"

Under supervision of Dr. Tae Seok Seo, School of Chemical and Biomolecular Engineering, Korea Advanced Institute of Science and Technology (KAIST).

Selected Publications

J.H. Jung, P. Desit and M.R. Prausnitz, **2018.** "Targeted drug delivery in the suprachoroidal space by swollen hydrogel pushing", Invest. Ophthalmol. Vis. Sci. 59, 20469-1079.

J.H. Jung, B. Chiang and Mark R. Prausnitz, **2018.** "Ocular drug delivery targeted by iontophoresis in the suprachoroidal space using a microneedle", J. Control. Release 227, 14-22 (2018).

J.H. Jung, B.H. Park, S.J. Oh, G. Choi and T.S. Seo, **2015**. "*Integrated centrifugal reverse transcriptase loop-mediated isothermal amplification microdevice for influenza A virus detection*". Biosens. Bioelectron., 68, 218-224.

J.H. Jung, B.H. Park, S.J. Oh, G. Choi and T.S. Seo, **2015**. *"Integration of reverse transcriptase loop-mediated isothermal amplification with an immunochromatographic strip on a centrifugal microdevice for influenza A virus identification"*. Lab Chip, 15, 718-725.

J.H. Jung, T.J. Park, S.Y. Lee and T.S. Seo, **2012**. *"Homogeneous Biogenic Paramagnetic Nanoparticle Synthesis Based on a Microfluidic Droplet Generator"*. Angew. Chem. Int. Ed. 51, 5634-5637.

J.H. Jung, D.S. Cheon, F. Liu, K.B. Lee and T.S. Seo, **2010**. *"A Graphene Oxide Based Immuno-Biosensor for Pathogen Detection"*, Angew. Chem. Int. Ed. 49, 5708-5711.

Teaching Interests:

Aside from my research career, I have extensive teaching experience as a TA and RA. During graduate student, I TAed Chemical and Biomolecular Engineering (CBE) Laboratory, Organic Chemistry, and DNA chemistry in School of Chemical and Biomolecular Engineering in KAIST, Korea as a part of government fellowship. All classes were performed in English. I had worked as an RA of undergraduate research program (URP) in KAIST and the research class for undergraduates in CBE during graduate school. Also, I mentored graduate students in my group. In the Georgia Institute of Technology, I have performed ocular drug delivery projects with three undergraduates for 2 year.

Since I have immersed diverse research field I believe those experiences obviously will be helpful to my teaching. My research experiences are focused in the fields including Molecular Biology, Molecular Diagnostics, Bio Engineering. Thus, my teaching interests are also there in the same field. I also have interest in teaching of microfluidics, recent topics in chemical and biomolecular engineering.

In particularly, I would like to develop the class, 'Point-of-care (POC) system design and development'. In my opinion, students should have wide insights in their fields. Thus, I would like to let student develop POC system from design to demonstration in this class. Although pre-requisites, basic classes, will be required to take this class, students can widely comprehend recent technology trends in their fields by literature survey and will understand how to design POC system. Actual development will be limited, but students will learn how to use their knowledge which earned in the class. I also would like to include importance of patents in the system development.

542388 In Vitro Microphysiological Systems for Disease Modeling, Drug Development, and Regenerative Medicine

Abstract:

Research Interests:

My research interests focus on recreating living, functional human tissues for disease modeling, drug development and regenerative medicine. Towards this end, I have been working to develop 3D tissue culture systems, microfluidic bioreactors, and in situ biosensors to achieve precise temporal and spatial control of cell microenvironment that regulates cell behavior. During my doctoral research in Scott Simon's research group at University of California at Davis, I developed a vascular mimetic system emulating the dynamic interactions among vascular endothelial cells, monocytes, and dietary lipids under blood flow to model inflammation and atherosclerosis. I used the model to assay the metabolic perturbations and inflammation of human aortic endothelial cells in response to subjects' triglyceride-rich lipoproteins (TGRL) and identified a regulatory mechanism of endothelial inflammation by TGRL. As a postdoctoral fellow in Michael Shuler's research group at Cornell University, I am combining my expertise in cell biology and flow dynamics with cutting-edge stem cell and organ-on-a-chip technologies to develop in vitro microphysiological systems for drug development. Specifically, I have developed a microfluidic blood brain barrier model that achieves barrier tightness closest to the in vivo levels among current microfluidic models. I am also developing a novel microfluidic platform that facilitates integration of the vasculature and circulating into a multi-organ microphysiological system. As a future faculty member, my research will focus on establishing 3D multicellular neurovascular models for brain drug development, developing automated biofabrication of vascularized human tissue and organs, and creating personalized tissue constructs for regenerative medicine.

Postdoctoral Project: "Advanced Human on a Chip Systems for Drug Discovery" and "An Organotypic Model Recapitulating Colon Cancer Microenvironment and Metastasis"

Under the supervision of Dr. Michael Shuler, School of Chemical and Biomolecular Engineering, School of Biomedical Engineering, and Harold Craighead, School of Applied and Engineering Physics, Cornell University

Ph.D. Dissertation: "Inflammatory Basis of Triglyceride Rich Lipoproteins"

Under the supervision of Scott Simon, Department of Biomedical Engineering, University of California at Davis

Teaching Interests:

I view research and teaching as mutually beneficial. I have been fortunate to participate in classroom teaching and student research project mentoring both in my Ph.D. and postdoc institutes. At Cornell, I lectured on microfabrication for an upper-level undergraduate and graduate-level course "Nanobiotechnology" for two years, and on Physiologically based pharmacokinetic modeling for the "Computer-aided engineering: Applications to biomedical processes" course. In a research lab setting, I have mentored a high school summer intern, many undergraduate students, a Meng student, and a visiting PhD student on their own research projects. I also participated local course and training programs, "The Practice of Teaching in Higher Education" and "Building Mentoring Skills Certificate Program", to reflect on and improve my teaching and mentoring skills.

I had diverse training cross the fields of cell biology, biochemical, biomolecular and biomedical engineering, due to the interdisciplinary nature of my research. I am capable of teaching fundamental classes and graduate level classes in those mentioned fields (e.g. tissue engineering, biomaterials, fabrication of micro/nanodevices, fluid dynamics).

541882 Electrochemical Biotechnology

Abstract:

Research Interests:

Electrochemical Biotechnology

DNA has unique abilities of molecular recognition and self-assembly, which have fostered its widespread incorporation into devices for both science and medicine. In both my graduate and postdoctoral work, I have developed improved chemistries to form DNA monolayers on electrodes. Superior monolayer formation methods led to the direct electrochemical detection of DNMT1, a cancer biomarker, from human tissue samples. Using these chemistries, I have also made devices to pattern non-adherent cells and detect endocrine disrupting compounds. This work has significantly improved the field of DNA-based electrochemistry and broadened the potential applications of electrochemical detection with biomolecules. It has yielded fourteen publications, eleven of which are first authorships, as well as two patents. As I transition to an independent career, I plan to apply the methods developed in this work to improve the study and treatment of infectious disease.

Graduate Research with Prof. Jacqueline Barton at the California Institute of Technology.

Surface modification chemistries for diagnostics. Electrochemical nucleic acid sensors generally involve hybridization events, with DNA serving as a scaffold. I developed monolayers constructed from alkyne-modified DNA that was reacted with azide-terminated thiols on the electrode. The amount of azide in the underlying monolayer was tunable for the desired application. This monolayer assembly method yielded DNA helices with a high solvent accessibility, enabling the significantly more sensitive detection of proteins. The transcriptional activator TATA binding protein (TBP) was detected at 4 nM, its K_D. With this method, I also achieved extremely sensitive and specific monitoring of DNA lesions, mismatches, and hybridization events.

Detection of DNMT1 activity from patient tumor samples as a cancer diagnostic. With the sensitivity and selectivity of my surface-modification chemistries, I developed an electrochemical signal-on assay to measure the activity of the human methyltransferase DNMT1 using a methylation-specific restriction enzyme. Aberrant methylation has been linked to cancer. Prior to my platform, no methods existed to monitor methyltransferase activity in a clinically relevant setting. I successfully measured DNMT1 activity from crude human tumor lysate. In ten sets of tumor and healthy adjacent tissue, a direct correlation was observed between the high DNMT1 activity and tumorous tissue when activity was measured electrochemically. This activity does not correlate with either overexpression or total amount of DNMT1 within the sample.

This work has improved the sensitivity, selectivity and specificity of DNA-modified electrodes, especially for clinical applications. I trained three undergraduates, two of whom are currently pursuing Ph.D.s.

Postdoctoral Research with Prof. Matthew Francis at the University of California, Berkeley.

Reagentless DNA coupling to surfaces. I have developed a reagentless one-pot coupling method to tether biomolecules to surfaces. The reaction, between a catechol and an aniline installed on a biomolecule, proceeds rapidly in the presence of an applied electrical potential (in under four minutes) with the ability to tune the amount of DNA on the surface. I highlighted its application for diagnostics through the electrochemical sensing of the endocrine disruptor bisphenol A. Living, non-adherent cells were also captured on electrode surfaces by DNA hybridization. Both applications were found to be dependent on the surface coverage of DNA. Importantly, I have used this method to immobilize electron transfer-proficient cells and demonstrated efficient current generation on electrodes, which would significantly improve the construction of biofuel cells.

Engineered bacterial biosensor for pollutant detection. Endocrine disrupting compounds (EDCs) are ubiquitous, originating from pesticides, plasticizers, and pharmaceuticals. They have been implicated in diseases such as obesity, diabetes, and cancer. I developed a new detection strategy for EDCs that is both fast and portable. This system is based on an estrogen receptor construct expressed on the surface of *E. coli*, which enables the detection of many detrimental compounds with inherent signal amplification from impedance measurements. I detected sub-ppb levels of estradiol and ppm levels of bisphenol A in complex solutions. This system reports the total estrogenic activity of a sample. As an applied example, estrogenic chemicals released from a plastic baby bottle upon microwave heating were detectable with this technique.

I have trained an undergraduate who received the NSF GRFP fellowship, as well as four graduate students, one of whom also received the NSF GRFP based on our work. In the future, I will establish an independent research lab with a diverse and robust program making use of DNA-based methods to study, detect, and drug pathogenic bacteria.

Teaching Interests:

With interdisciplinary training, I am prepared and excited to teach a range of classes. My teaching program will ideally be centered on upper-level undergraduate and graduate courses, including Thermodynamics, Electrochemical Energy Systems, Device/Diagnostics Design, Biochemical Engineering, Molecular Bioengineering, and Biomaterials.

541901 From Cells to Tissues : Understanding Development, Evolution and Disease Using Single-Cell RNA-Sequencing

Abstract:

Research Interests: Cells in the human body can be subdivided into transcriptomically distinct types, which correspond well to classical distinctions based on morphology, physiology and connectivity. Knowing the full compendium of cell types in a given tissue, and having reliable genetic markers for each type will enable cellular circuits to be studied at unprecedented molecular resolution in health, development, disease and across different species. Together with experimental collaborators, I am applying large-scale single-cell transcriptomic analysis of the vertebrate retina as a model system to address these issues. Building on initial "proof-of-principle" studies (Macosko et al., *Cell*, 2015 ; Shekhar et al., *Cell*, 2016; Pandey, Shekhar et al., *Current Biology*,2016), I am poised to complete a comprehensive neuronal atlas of the mouse retina, which will be among the first of its kind for any brain region. Using state-of-the-art machine learning techniques, we have defined molecular signatures for a total of >120 neuronal types, including 3 photoreceptor, 1 horizontal, 15 bipolar, 45 ganglion, >60 amacrine and ~10 non-neuronal cell types, confirming a large number of them histologically. Using the mouse "retinome" as a foundation, I am to pursue two directions.**First** we are performing large-scale single-cell surveys of two other vertebrate retinas that differ from that of the mouse in key aspects: (a) macaque retina, which contains the fovea, a structure responsible for high acuity vision in primates that is absent in mice, and (b) zebrafish retina, which unlike mice and primates, can regenerate following injury. Using molecular signatures, I aim to computationally compare the taxonomy of retinal neurons across species, and identify "orthologous" neuronal types. Beyond uncovering fundamental biological principles, this also has implications in revealing molecular pathways associated with regeneration, and the unique properties of vision in primates. **Second**, we are transcriptiomically profiling neuronal and non-neuronal cells of the mouse retina following physical injury to the optic nerve. By comparing changes in gene expression among types following injury, we can identify early transcriptional signatures that correlate with, and may underlie, selective resilience. In addition, we are identifying injury responses of interneurons and non-neuronal cells (glia and immune cells), all of which have been shown to influence survival. This project has the potential to uncover strategies to reverse neuronal death in the context of injuries and neurodegenerative diseases like glaucoma, a leading cause of blindness in the developing world.

Teaching Interests: I believe that a sustained teaching practice is a fundamental cornerstone of scientific growth. My doctoral and postdoctoral experience has positioned me to teach a diversity of courses. I describe key highlights of my teaching experience: **First**, as a graduate teaching assistant for the graduate course on thermodynamics and statistical mechanics at MIT Chemical engineering (Instructors : Arup Chakraborty and Bradley Olsen), I was awarded the best TA award in 2011. **Second**, I have been a guest lecturer in the MIT Biological Engineering course "Fundamentals of Biological Networks" targeted towards advanced undergraduates and graduates. **Third,** as a post-doctoral associate, I have designed and delivered short course introduction to single-cell genomics. For e.g. I was invited to deliver a short course on single-cell transcriptomics at the Evomics 2018 conference in Cesky Krumlov, Czech Republic. I believe that I have a strong foundation to teach undergraduate and graduate courses on thermodynamics, statistical mechanics and reaction engineering. I would also like to design graduate electives that brings together physical biology and statistical learning techniques drawing from my own research in single-cell genomics.

541236 Daniel Cook - Understanding and Treating Progressive Diseases at the Levels of Single Cells and Single Patients Though Systems Biology

Abstract:

Research Interests:

I am interested in understanding and predicting treatments for progressive diseases including liver diseases and cancers through high-throughput data analysis, characterization of single cells and cell populations, and genome-scale metabolic modeling.

Briefly, heterogeneity is emerging as a defining feature of biology. This biological heterogeneity can be seen at multiple characteristic scales – from individual patients to single cells – and at multiple regulatory levels – genomic, transcriptomic, proteomic, and metabolomic. Originally perceived as "biological noise", this heterogeneity is now being leveraged to understand fundamental principles of biology and design more complete and effective therapies to treat diseases.

My work uses computational methods to use biological heterogeneity at multiple scales to understand disease progression, understand dynamic regulation of disease processes, and predict more effective disease therapies and personalized therapies. Specifically, I use a combination of "big data" analyses from individual patients, single-cell RNA-seq and proteomic analyses, and personalized genome-scale metabolic modeling to understand dynamic metabolic regulation in and predict therapies for multiple diseases including non-alcoholic fatty liver disease and breast cancer in individual patients and in patient populations.

Teaching Interests:

I am most interested in teaching the following courses: Systems biology and personalized medicine, Controls, Statistics for chemical & life science engineering, Reaction Engineering, Numerical methods and Introduction to programming, and Metabolic Engineering.

I believe that the most important skills an engineering teacher can impart are analytical problem solving, creative thinking, and lifelong learning. Overall, my theory of teaching is that these skills can be developed best by student immersion in engineering thought and repetition of skills. In engineering, as in language studies, immersion and repetition facilitates faster learning, better outcomes, more creative approaches, and an appreciation of how much more one still has to learn. In my view, it is the responsibility of the teacher to create an environment where immersion is possible and repetition of theory and practice is encouraged. It is the students' responsibility to practice repeatedly until "fluency" is reached.

541745 Application of Room Temperature Ionic Liquids in Membrane Based Technology: An Unconventional Green Separation

Abstract:

Research Interests:

Research Interest: In view of the environmental concern associated with the volatile organic compounds, ionic liquids (IL) gained a world-wide acceptance as potential 'green' alternatives to the conventional solvents due to the favourable properties like low vapour pressure, wide liquid range, high degree of chemical, thermal and radiation stability, high degree of solubility for a wide range of organic, inorganic and polymeric materials, large electrochemical window etc. This type of ionic, low lattice energy compounds exhibits unique properties, which otherwise are not observed for conventional chemicals. For example, in hydrometallurgy, conventionally the metal ion gets extracted from aqueous to conventional molecular diluent by predominance of 'solvation mechanism'; while application of IL results the predominance of 'ion exchange mechanism'. In electrochemistry, some uncommon oxidation state of metal ion can be achieved due to large potential window of IL, which otherwise cannot be formed in conventional medium. The most attractive property of IL is the high degree of tunnability. A small change either in cationic or anionic part can bring drastic change in the physicochemical properties of IL and hence its property can be fine-tuned based on its applications, provided the structure-activity relationship is established. In view of such technological importance and unique tunable properties, IL has been considered in present case. On the other hand, membrane technology provides clean energy intensive technology with ease of commercial scale adaptation finds application in aqueous, non-aqueous and biomedical field. The conventional membrane based separations (pressure driven filtration: UF, NF, MF and diffusion based: FO, RO, MD) are mainly based on the size exclusion principle. Fouling is one of the major challenges face by membrane technologies. In view of both these facts, coupling of unique properties of ILs/ PILs with the membrane technology for unconventional separation to be the main aim of future research directions. Broadly, three projects are as follows

1. Pi electron mediated separation and fractionation of aromatics from aliphatic for both aqueous and non-aqueous applications using IL/ PILs. Establishing the structure-activity relationship. Achieving isomeric separation with IL based membranes. The high affinity of ionic liquid towards pi electron cloud density will be utilized in order to separate aromatics from aliphatic compounds, which otherwise is impossible based on size exclusion principle. The influence of inductive and resonance effect on enhancing electron cloud density and subsequent affinity towards ionic liquid will be investigated. The influence of stereo-chemical arrangement can lead to change in planner conformation of particular stereoisomer and hence can have influence on the delocalization of pi electron cloud. This fact will be exploited in order to separate stereoisomers mutually, which is difficult and not possible by conventional membrane based separation.
2. Synthesis of anti-fouling, antimicrobial responsive membrane for mitigating concentration polarization

based on ILs/PILs. To break concentration polarization, micromixing induced by applied oscillating electric field will be exploited by suitably induce electrical responsiveness on membrane surface through grafting of PIL on membrane surface. Oscillating electric field frequency, surface density of responsive elements, length of grafting polymer will be optimized. Exploiting the antimicrobial characteristic of imidazolium moieties, such characteristic will be induced on membrane for task specific applications. Moreover, structure-activity relationship will be established to get optimum desired performance.

3. As an electrolyte in fuel cell applications focusing on Li-S battery to mitigate polysufide shuttle effect. The

Research Experience:

Joined Bhabha Atomic Research Centre, Mumbai, India in 2008 as Scientific Officer and received 2 departmental promotions in 2010 and 2014.

Present: Post-doctoral fellow, Department of Chemical Engineering, University of Arkansas, Fayetteville, Arkansas under Prof. Ranil Wickramasinghe (from 2016 to present)

Ph. D.: Homi Bhabha National Institute, Mumbai, India, 2011-2015

No of Publications journals: 134 (Citation: **1358**, h index 18)

One book chapter on magnetic responsive membrane in Royal Society, coauthored by Xianghong Qian and Dr. Ranil Wickramasinghe

Mentoring: 11 undergraduate, 1 Master and 4 Ph. D students

Teaching Interests:

Each student has an extraordinary talent in some direction. I believe the responsibility of a teacher/mentor is to identify that and makes him/her realize the same. I also believe that not only the bookish knowledge, the mentor should help students for their intellectual growth, enhancement in analytical aptitude, cultivating the leadership quality, making them socially responsible and helpful to others. I would be most interested in teaching classes such as: Separation Science (Solvent extraction, Chromatography, Membrane based separation, Spectroscopy, Statistical analysis of data), Inorganic chemistry (f- block element chemistry and spectroscopy and bonding), Polymer chemistry and Reaction kinetics and thermodynamics

I am open to teach any courses within the chemistry/chemical engineering curriculum. While it will be important to successfully start my research program, I would like to eventually develop a course on green chemistry for the separation of environmental pollutants and toxic elements.

Experience: I was guest lecturer on behalf of Dr. Ranil Wickramasinghe for the Chemical Engineering course: **Advance Reactor Engineering** in University of Arkansas for graduate students

Selected Publications

1. Jebur, **Arijit Sengupta**, Y-H Chiao, M. Kamaz, X. Qian, R. Wickramasinghe, J Membr Sci, 2018, 556, 1-11 (corresponding author)
2. M. Carter, **Arijit Sengupta**, X. Qian, M. Ulbricht, S. R. Wickramasinghe, J Membr Sci, 2018, 554, 109-116
3. **Arijit Sengupta**, A. Bhattacharyya, Sk. M. Ali, W. Verboom, P.K. Mohapatra, J Phys Chem, 2017, 121 (12) 2640-2649
4. **Arijit Sengupta**, Singha Deb, K. Dasgupta, V. C. Adya and M. Ali Sheikh, New J Chem, 2017, 41 (11), 4531-4545 (corresponding author)
5. K. Mohapatra, **Arijit Sengupta**, M. Iqbal, J. Huskens, W. Verboom, *Inorg. Chem.*, 2003, 52 (5), 2533–2541.
6. K. Mohapatra, **Arijit Sengupta**, M. Iqbal, J. Huskens, W. Verboom, *Chem A Euro J*, 2013, 19(9), 3230–3238.

543155 Assessing Role of Signaling Network Properties at the Immune System and Cancer Nexus

Abstract:

Research Interests:

The overall goal of my research career will be setting up *Systems Biology Lab* which focuses on answering the fundamental question of how network architecture of signaling pathways influence cellular behavior. I am interested in signals that activate master transcription factors that respond to cues arising from immune stimulation and inflammatory processes. During my postdoc, I have been looking at the influence of genetic and non-genetic sources of variability and how they affect heterogeneity of response of cancer cells to immune signals. For this I am investigating the causal link between single cell signaling dynamics, transcription factor reporter activity and the nascent transcriptome.

Selected Skills learned during PhD and PostDoc: Molecular Biology WetLab, Time Lapse Fluorescence Microscopy for Single Cell Imaging, Nascent RNA Sequencing using PROSeq, NextGen Bioinformatics Sequencing Analysis, Mathematical Modeling of Biological Systems, Machine Learning

Teaching Interests: I love to work on topics that require mathematical modeling of biological phenomena. Therefore, I would like to teach topics like Systems Biology, Signaling Pathways, Network Analysis and Mathematical Modeling. These reflect the core research areas of my expertise gained during my PhD and Postdoc years. In addition, concepts from Chemical Engineering like Chemical Kinetics, Reaction Engineering and Engineering Mathematics are used extensively for my research and therefore, I am very comfortable with teaching these courses. Most importantly, I am interested in using new software based learning tools to make learning of complex concepts easy and to stimulate critical thinking skills in students.

541933 Leveraging Big Data and Engineering Fundamentals Towards Biological Discovery

Abstract:

Research Interests:

Recent advances in -omics technologies allow us to rapidly collect large amounts of diverse information about biological systems. I am interested in building mechanistic models of biological processes through integration of the -omics data with physical and chemical dynamics that govern biological systems.

My future research will pursue two distinct directions: (I) Gaining mechanistic insights into cell-to-cell variability in mammalian cellular signaling networks and (II) Identifying the forces that shape bacterial genome evolution.

Aim I: Mechanistic insights into cell population heterogeneity. Mammalian signaling networks exhibit extensive cell-to-cell heterogeneity. This heterogeneity has important functional consequences, for example, in emergence of drug resistant subpopulations in tumors. The heterogeneity originates from the cell-to-cell variability in network parameters, such as protein abundances and rates of biochemical reactions. Thus, its mechanistic understanding requires knowledge of the distribution over network parameters consistent with single cell data. This problem is of particular importance now as single cell data is being rapidly generated for diverse cell types, organisms, and biological contexts. I have developed a maximum entropy based computational framework to infer parameter distributions from single cell data. I want to apply this framework to study phenotypic heterogeneity downstream of growth factor signaling in mammalian cells.

Aim II: Quantifying evolutionary forces in bacteria. Bacteria play a significant role in human health as well as biogeochemical cycles. Notably, multiple evolutionary forces such as mutations, recombination (swapping of DNA in the population), and ecological niche dynamics simultaneously shape genetic diversity in bacteria. Notably, we are experiencing an exponential growth in the number of sequenced bacterial genomes. I have developed computational tools to quantify the key evolutionary forces from genomics data. My goal is to better understand how external conditions, such as host diet, and selection dictate the innovation and spread of novel phenotypes in natural bacterial populations. This work will have broad applications in understanding, for example, evolutionary dynamics of host-associated microbial communities, determinants of the spread of antibiotic resistance, and stability of photosynthetic phenotypes in oceanic bacterial populations.

Teaching Interests:

I believe that an essential dimension of an ideal academician is the ability to convey complex ideas to an inquisitive audience. I see teaching and mentoring not as extraneous to my academic career but as one of its vital components.

I can teaching mathematically oriented core chemical engineering subjects such as thermodynamics, transport phenomena, chemical kinetics, and process control. Additionally, my postdoctoral training allows me to teach many non-traditional subjects, for example, computational and systems biology, modeling of biomolecules, and networks in biology. I will ensure that the course material has an extensive online component so that these courses will be easily transformed into MOOCs for students who are not able to physically attend the course.

547303 Modelling the Infectious Microenvironment to Understand Immune Cell Function

Abstract:

Research Interests:

My research has always focused on investigating immune cell behavior using a combination of biological and engineering techniques. Specifically, I use engineered platforms to study the migration and antimicrobial functions of innate immune cells and how these functions are altered by their microenvironment. During my doctoral training with Daniel A. Hammer, PhD at the University of Pennsylvania, I studied the migration and force generation of macrophages and published the first traction force maps generated by primary human macrophages which showed that in contrast to neutrophils, macrophage concentrate their force at the leading edge of the cell. Furthermore, I investigated the differences in the migration and force generation of differentially polarized M1/M2 macrophages and found that M2 macrophages are more motile and produce higher forces than M1 macrophages. In my postdoctoral work, I am studying neutrophil migration in the context of infection with Dr. Anna Huttenlocher, MD at the University of Wisconsin-Madison. As part of this research, I developed a model for studying neutrophil migration through an endothelial-lined blood vessel to a source of the bacteria *Pseudomonas aeruginosa* (Hind et al., 2018).[2] Using this model, I showed that the endothelium significantly increased neutrophil migration. Surprisingly, I found neutrophils, which are thought to be short-lived cells, migrated for up to 24 hours in this model. Additionally, I found that the endothelium, in the presence of *P. aeruginosa,* secreted the inflammatory mediators GM-CSF and IL-6, which contribute to neutrophil survival and migration respectively. Taken together, these findings, recently published in Blood (Hind et al, 2018), show a direct role for multicellular interactions in driving neutrophil function. The combination of my graduate and postdoctoral research highlights the importance of the tissue microenvironment in driving immune cell function and informed my proposed research plan.

Future Research Plan:

The innate immune response to infection is a complicated process that involves a coordinated effort by many cell populations; however, the role of these multicellular interactions in driving immune cell function is unclear. Neutrophils are one of the first cell types to arrive at a site of infection and are critical in limiting pathogen spread but their recruitment must be tightly controlled. Defects in neutrophil recruitment can lead to recurrent, unresolved infections and excessive neutrophil activity can lead to chronic inflammation and tissue damage. A better understanding of how the microenvironment alters leukocyte recruitment and function could provide us with new targets for controlling neutrophil recruitment and improve treatment options for patients. **The overarching goal of my research is to understand how the tissue microenvironment alters innate immune cell function.** *My proposed research plan focuses on immune cell function in the infectious microenvironment; however, the techniques and devices developed throughout the plan could be broadly applied to other contexts in which it is unclear how multicellular interactions alter neutrophil function, including cancer or cardiovascular disease.* I will combine my training in engineering and immunology to design microfluidic devices, inspired by *in vivo* biology, that model the infectious microenvironment *in vitro* and use these models to study how leukocyte function is altered by the environment. Specifically, I will investigate I) how multicellular interactions in the infectious environment modulate neutrophil function and downstream signaling, II) how the physical properties of the infectious environment alter these interactions and their effect on neutrophil function, and III) how myeloid-derived suppressor cells migrate and integrate into the infectious microenvironment.

Teaching Interests:

Teaching students at both the undergraduate and graduate level is an important responsibility. I have learned from exceptional mentors throughout my career that good teaching practices come from deliberate planning and a passion for encouraging student learning. As a scientist, I rely on data-supported conclusions to guide my research. Similarly, as an educator, I will employ the teaching practices shown to be most effective by current education research literature. My overarching goal as an educator is to promote critical thinking skills in my students by providing them the most effective learning environment possible. In addition to teaching in the classroom, I am committed to being a enthusiastic mentor to undergraduates, graduate students, and postdoctoral fellows in the laboratory. **The opportunity to help students develop into independent scientists through research mentoring is a significant reason for my interest in academia.** Overall my goal as a mentor is to help students develop into independent scientists by encouraging them to develop problem-solving and critical-thinking skills

Education:

Ph.D. Biological Engineering 2015, University of Pennsylvania, Philadelphia, PA

B.S. Chemical Engineering 2009, University of Wisconsin-Madison, Madison, WI

Selected Publications:

Hind LE, Ingram PN, Beebe DJ, Huttenlocher A. "Interaction with an endothelial lumen increases neutrophil lifetime and motility in response to P. aeruginosa," Blood, in press.

Hind LE, Lurier EB, Dembo M, Spiller KL, Hammer DA. "Effect of M1-M2 Polarization on the Motility and Traction Stresses of Primary Human Macrophages," Cellular and molecular bioengineering. 2016 September; 9(3):455-465.

Hind LE, Vincent WJ, Huttenlocher A. "Leading from the Back: The Role of the Uropod in Neutrophil Polarization and Migration," Developmental Cell. 2016 Jul 25;38(2):161-9.

Hind LE, Dembo M, Hammer DA, "Macrophage motility is driven by frontal-towing with a force magnitude dependent on substrate stiffness," Integrative Biology, 2015 Apr; 7(4):447-453.

Hind LE, MacKay JL, Cox D, Hammer, DA, "Two-dimensional motility of a macrophage cell line on microcontact-printed fibronectin," Cytoskeleton, 2014 Sep; 71(9):542-554.

Awards/Honors:

Hematology T32 Postdoctoral Fellowship, NIH, 2015-2017

National Science Foundation Graduate Research Fellowship, 2010 – 2013

542429 Computational Insights into Zeolite-Catalyzed Biomass Conversion to Olefins

Abstract:

Research Interests:

Benefited from the shale gas revolution in the United States, the world is supplied with cheap and abundant shale-gas-derived ethane, which drives the shift from heavier to light feedstocks in the North American olefins market. As a consequence, less co-products are generated from ethylene crackers, such as C3/C4 olefins (propylene and butadiene) and aromatics benzene, toluene and xylene (BTX). According to a recent report by ICIS Chemical Business, the US butadiene price has increased dramatically by 60-70% since last July. This opens exciting new opportunities for renewable and on-purpose production routes of these chemicals to make up the supply scarcity and shift to a more sustainable chemical industry in the future. In this regards, my postdoctoral research focused on the mechanistic understanding of catalytic biomass conversion to value-added chemicals, especially dienes such as butadiene and isoprene, both being important monomers in the synthetic rubber industry. Porous zeolite materials are the focus of my studies using a combined density functional theory (DFT) calculations, experiments and microkinetic modeling approach. Unraveling the reaction mechanism in such complex catalytic systems is the primary goal of the study. Additionally, possible reactivity- and selectivity-controlling factors such as zeolite topology, acid type (Brønsted vs. Lewis) and strength were evaluated as part of an effort to identify the key reactivity descriptors and to develop structure-reactivity relationships that could guide the choice of zeolytic materials for these reactions. Future plans will utilize the knowledge obtained from the diene chemistry to find other applications in biobased olefins and aromatics production, to discover new biobased routes with higher transformation efficiency, and to design zeolite materials with superior performance in biobased bulk chemicals production.

Teaching Interests:

Thermodynamics, Chemical Kinetics, Chemical Reaction Engineering, Catalysis.

Engineering catalytic interfaces for the sustainable production of fuels and chemicals

Melis S. Duyar

Associate Staff Scientist, SUNCAT Center for Interface Science and Catalysis
Stanford University, Stanford, CA
SLAC National Accelerator Laboratory, Menlo Park, CA

Research interests:
As the growing world population demands increasing food and fuel resources, the scientific community faces a grand challenge to both develop technologies that can sustain this population, and address the environmental consequences of resource consumption at increasing scales. There is a critical need for better catalysts, adsorbents and processes for the production of liquid fuels, storage of energy and the production of fertilizer through sustainable pathways. My research background and interests focus on the areas of sustainable fuel and chemical production. To address changing needs of the world population, both thermochemical as well as hybrid approaches amenable to electrification of chemicals production sectors are needed, to enable technologies that can deliver high volume of products as well as those that provide necessities to remote locations through distributed production.

Teaching interests:
Learning at the undergraduate and graduate level requires not only the acquisition and recollection of knowledge, but also reflection on and assessment of one's prior knowledge to complete missing pieces. This is ultimately what will bring the engineer, scientist, and any lifelong learner success long after completion of the required course material. My teaching philosophy places emphasis on encouraging self-directed (lifelong) learning by providing an atmosphere of inclusion and creating engaging activities that promote learning and teaching amongst peers both inside and outside of the classroom. At Stanford University I employed this approach to design the undergraduate senior level design course entitled Chemical Engineering Plant Design and have been teaching it since 2017. I have a background in teaching and creating course material for the undergraduate senior laboratory course series, Environmental Engineering Laboratory I-II at Columbia University. I also have experience as a teaching assistant for the Environmental Control & Pollution Reduction and the graduate level Industrial Catalysis courses at Columbia University. I am interested in teaching courses at both the undergraduate and graduate level in chemical engineering and developing courses focusing on environmental catalysis and sustainability.

Research Experience:
Associate Staff Scientist (January 2017-to date)
SLAC National Accelerator Laboratory, Menlo Park, CA
SUNCAT Center for Interface Science and Catalysis, Jaramillo Laboratory

- Responsible for leading experimental thermal heterogeneous catalysis research.
- Current research program is focused on the design of multifunctional catalysts and tandem catalysis schemes for the sustainable production of fuels and chemicals from carbon feedstocks. Catalysts investigated to date include bimetallic alloys, transition metal phosphide/carbide/sulfides, metal organic frameworks and zeolites.
- Co-wrote 4 proposals for funding to date.

Postdoctoral Fellow (August 2015 - January 2017)
Stanford University, Stanford, CA
Chemical Engineering - SUNCAT Center for Interface Science and Catalysis
Supervisor: Prof. Thomas F. Jaramillo

- Conducted experimental research on the theory guided discovery of catalysts for the thermochemical conversion of CO and CO_2 to methanol and higher alcohols
- Synthesized transition metal phosphide, sulfide and mixed anion catalysts
- Characterized catalytic activity, structural properties and adsorbate-catalyst interactions using ex-situ, in-situ and operando techniques (e.g. XRD, XAS, XPS, DRIFTS, TEM).
- Contributed to writing 1 grant which was awarded funding by the US Department of Energy.

Doctoral Researcher (May 2013 - May 2015)
Columbia University, New York, NY
Earth and Environmental Engineering - Catalysis for a Sustainable Environment Laboratory
Supervisor: Prof. Robert J. Farrauto

- Developed catalysts for CO_2 hydrogenation and dual function materials for integrated CO_2 capture and utilization schemes.
- Obtained a provisional patent for a dual function material capable of capturing CO_2 and releasing it as synthetic natural gas (CH_4) upon exposure to hydrogen, with no added heat input.
- Wrote and contributed to 2 successful proposals for funding through industry.

Graduate Student Researcher (January 2012 - May2013)
Columbia University, New York, NY
Earth and Environmental Engineering - Combustion and Catalysis Laboratory
Supervisors: Prof. Marco J. Castaldi, Prof. Tuncel Yegulalp, Prof. Robert J. Farrauto

- Conducted research on CO_2 capture at the catalytic interface for the water gas shift (WGS) reaction.
- Developed monolith-supported catalysts for the production of PEM fuel cell grade hydrogen with zero carbon emissions by removing CO_2 in-situ, as it is produced from WGS.

Undergraduate Research Assistant (September 2010 - January 2012)
Koç University, Istanbul, Turkey
Chemical and Biological Engineering - Computational Systems Biology Laboratory
Supervisor: Prof. Ozlem Keskin

- Conducted research on protein-protein interactions and computational drug design
- Wrote programs in MATLAB and Python to analyze files from the protein databank to identify hot spots for protein-protein interactions.

542452 Catalytic CO_2 Conversion to Clean Fuels and Chemicals: Integration of Traditional Metallic Catalysts and Metal-Organic Frameworks

Abstract:

Research Interests:

Satisfying the demand for efficient energy generation and sustainable chemical production via carbon resources while mitigating adverse environmental effects is a defining challenge of the 21st century. Catalysis offers avenues to satisfy these demands by creating efficient and sustainable ways to produce chemicals using non-traditional carbon resources. My research primarily aims at developing efficient catalysts and adsorbents for clean energy applications. Both my doctoral and postdoctoral studies at Penn State focused on catalytic CO_2 conversion to cleans fuels and chemicals using bimetallic catalysts. My current postdoctoral research at Georgia Tech focuses on developing novel metal-organic framework (MOF) and carbon-based materials for efficient natural gas purification. Going forward, I envision incorporating MOFs materials with traditional metallic catalysts for sustainable value-added chemical synthesis from CO_2 hydrogenation.

As an important branch of Carbon Dioxide Capture, Utilization, and Storage (CCUS), catalytic CO_2 conversion for synthesizing chemical feedstocks and transportation fuels has recently attracted great attention worldwide. Major advances in energy-efficient catalytic CO_2 conversion using renewable energy could reduce both the greenhouse-gas emissions and dependence on non-renewable resources such as petroleum. The synthesis of oxygenates and hydrocarbons from CO_2 hydrogenation is one of the promising approaches for environmentally-friendly syntheses of sustainable chemical feedstocks and fuels. In my Ph.D. work and postdoctoral research at Penn State with Prof. Chunshan Song, I developed a selective, active, and stable Pd-Cu bimetallic catalyst for CO_2 hydrogenation to methanol, and systematically studied the composition-structure-catalytic performance relationship by combining activity tests, kinetic studies, and advanced characterization techniques, including X-ray and infrared spectroscopy. My research also advanced the fundamental understanding of alloy effects, key reaction intermediates, as well as plausible reaction pathways by incorporating both experimental results and theoretical calculations. Currently, I am expanding my research portfolio by working on ZIF-derived catalysts to address the challenge and opportunities involved with CO_2 hydrogenation to value-added hydrocarbons. Through my postdoctoral research at Georgia Tech with Prof. Krista Walton, I have designed carbon- and MOF-based adsorbents with tunable porosities and chemical functional groups for efficient natural gas purification. My most recent work introduced carbide-derived carbon (CDC) materials with tailorable porosity and metallic nanoparticles showing excellent synergetic effect for natural gas purification.

My future work aims to lead the development of novel catalysts by uniquely integrating traditional metallic sites with MOFs for catalytic CO_2 conversion to value-added fuels and chemicals, especially for higher-value alcohol syntheses, where the thermodynamic restraint is limited. These newly designed catalysts represent a class of novel catalysts with excellent tailorability comprising desired active centers, large accessible surface areas, and tunable porosity, as well as adjustable MOF/ZIF-metal site interaction. Design of these catalysts with desired features will be guided by fundamental understandings of adsorption property and structure-property-catalytic characteristic relationship, as well as metal-support interaction. The catalyst preparation method will include both "top-down" encapsulation and "bottom-up" impregnation strategies. With extensive experience in ZIF and MOF materials and strong expertise in catalyst preparation, testing, and characterization, I am uniquely suited to undertake these challenging research efforts. The research in novel MOF-derived catalysts will help to debottleneck current technological issues associated with catalytic CO_2 conversion, and enable the sustainable synthesis of value-added products, such as hydrocarbons and higher alcohols.

Teaching Interests:

I view teaching as an essential and enjoyable part of being a faculty member, so I am excited about the opportunity to lead both undergraduate- and graduate-level classes. As a post-doc at Penn State, I supervised graduate students in the CO_2-conversion project and contributed 3 publications already, and another 2 manuscripts in preparation. Furthermore, I have also served as both lecturer and teaching assistant for a graduate-level course since 2013. Drawing from my previous experiences working with students, I am excited to share my teaching and researching experiences with undergraduate students to help cultivate their interests for science, and contribute to the development of their educational paths. Courses I would particularly enjoy teaching include Chemical Reaction Engineering, Chemical Engineering Fundamentals, and Characterization of Catalytic Materials at both undergraduate and graduate levels. In addition, I also plan to develop a course introducing the fundamentals of catalysis, including reactor design, mass/heat transfer limitations, catalyst preparation and evaluation, and kinetic studies, as well as describing how state-of-the-art characterization techniques can be applied to gain insights into the structure-activity performance relationship and mechanistic study in heterogeneous catalysis.

Selected publications

1. Nie, **X. Jiang**, H. Wang, W. Luo, M. J. Janik, Y. Chen, X. Guo, C. Song, Mechanistic Understanding of Alloy Effect and Water Promotion for Pd-Cu Bimetallic Catalysts in CO_2 Hydrogenation to Methanol, *ACS Catalysis*, 8 (*2018*), 4873-4892. (considered as equal contribution).
2. **Jiang**, X. Wang, X. Nie, N. Koizumi, X. Guo, C. Song, CO_2 Hydrogenation to Methanol on Pd-Cu Bimetallic Catalysts: CO_2/H_2 Ratio Dependence and Surface Species, *Catalysis Today*, (*2018*), https://doi.org/10.1016/j.cattod.2018.02.055.
3. Li, X. Nie, **X. Jiang**, A. Zhang, F. Ding, M. Liu, Z. Liu, X. Guo, C. Song, ZrO_2 Support Imparts Superior Activity and Stability of Co Catalysts for CO_2 Methanation, *Applied Catalysis B: Environmental*. 220 (*2017*), 397-408.
4. Li, A. Zhang, **X. Jiang**, C. Chen, Z. Liu, C. Song, X. Guo, Low Temperature CO_2 Methanation: ZIF-67-Derived Co-Based Porous Carbon Catalysts with Controlled Crystal Morphology and Size, ACS Sustainable Chemistry & Engineering, 5 (*2017*), 7824-7831.
5. **Jiang**, N. Koizumi, X. Guo, C. Song, Bimetallic Pd-Cu Catalysts for Selective CO_2 Hydrogenation to Methanol, *Applied Catalysis B: Environment*. 170 (*2015*), 173-185.
6. Koizumi, **X. Jiang**, J. Kugai, C. Song, Effects of Mesoporous Silica Supports and Alkaline Promoters on Activity of Pd Catalysts in CO_2 Hydrogenation for Methanol Synthesis, *Catalysis Today*, 194 (*2012*), 16-24.

542540 Synergizing Model Surfaces and Real Catalysts for Efficient Electrochemical Energy Conversion

Abstract:

Research Interests:

To meet a growing demand for energy, an economically viable transition to sustainable production of fuels should be made by the nation. Reduction of oil and gas consumption and a widespread implementation of renewable hydrogen technologies require significant investments into the R&D of (photo)electrocatalysts. Studies of the fundamental processes that take place on catalyzing surfaces during selective electrochemical reactions provide crucial guidelines for the design and technological implementation of efficient and inexpensive catalysts.

During my postdoc at Stanford, I focused on fundamental studies of materials for electrochemical water splitting. Using atomically precise growth of hetero-structures, I designed artificial surfaces for model studies of the oxygen evolution reaction and was able to establish a correlation between the activity and stability of the materials. In particular, I demonstrated that a single monolayer of ruthenium can facilitate water splitting when located in the subsurface region of an inert oxide material. Furthermore, I was able to precisely track the emergence of catalytic activity as a function of the Ru monolayer location within the inert oxide. Besides, these findings quantitatively answer the key question of surface stabilization: at what thickness an inert coverage can protect an inherently unstable oxide catalyst during water splitting. Broadly speaking, my work has elucidated the role of subsurface chemical species on the catalytic activity of materials.

During my PhD at Drexel University, I developed a low-temperature and energy-efficient approach to the preparation of epitaxial complex oxides using atomic layer deposition (ALD). Although ALD is extensively used in industry for the deposition of binary oxides, the preparation of functional oxides with a complex chemical composition was not developed. I designed the multi-component ALD process and demonstrated how the strain state of such films develop during crystallization. Besides ALD, I worked on chemical modification of polar perovskite oxides for the induced light absorption in the visible spectrum for photovoltaic applications.

Looking into the future, I plan to develop new approaches to fundamental studies of catalysts and catalytic reactions, with the aim to use model surfaces for the guided design of real catalysts. As the electrochemical phenomena are associated with mass/charge transfer and can involve materials transformations, their *in situ*studies prove most rewarding in providing the fundamental insights. My research interests are best represented by well-controlled synthesis of materials and advanced *in situ* characterization, including a use of diverse synchrotron radiation facilities. I will present a comprehensive overview of my future research at the poster session.

Teaching Interests:

In addition to leading recitations for the course â€œFundamentals of Materialsâ€ for several years at Drexel, I mentored visiting scholars, MS and PhD students at Drexel and Stanford. I am interested in teaching major courses related to electrochemistry (electrocatalysis and energy storage), materials chemistry/science and advanced characterization of materials processes.

Viewing research and learning as intimately connected processes, I will be providing both independence and guidance to my team members. Encouraging creativity and enthusiasm, I will lead my team in the realm of research and technologies.

542142 Mixed-Metal-Oxide Redox-Catalyst for Shale-Oil and Gas Conversion

Abstract:

Luke M Neal

Research Assistant Professor (NC Statue University)

Postdoctoral Research: Chemical Looping Redox Catalyst

Chemical Looping-Oxidative Dehydrogenation (CL-ODH) of Light Paraffins and Naphtha: I co-invented multiple redox-catalysts for the oxidative dehydrogenation of light hydrocarbons and naphtha. This chemical-looping approach permits conversion of low-value/difficult-to-transport shale-gas and oil fractions into valuable commodity chemicals and liquid fuels. The CL-ODH process works by using the lattice oxygen of a redox "catalyst" to selectively combust hydrogen produced by thermal cracking. Regeneration of the redox-catalyst oxygen produces enough heat to drive the reaction without carbon-intensive, high-temperature furnaces. My research has demonstrated that this approach can give high yields and selectivity without significant dilution typically needed in co-feed ODH. I also developed kinetics for the design of a pilot unit. I am currently supporting follow-up research on adapting the material to work at low temperatures, where thermal cracking is kinetically and thermodynamically limited. If successful this will enable cheap, modular reactors that can convert stranded natural gas liquids for easy transport.

Syngas Production: Chemical looping reforming (CLR) replaces the gaseous oxygen needed for partial oxidation reforming with the lattice oxygen of a redox catalyst. CLR can improve the efficiency of Gas-to-liquid (GTL) processes vs. conventional reforming and permit small scale implementation. I formulated and characterized a novel core-shell redox catalyst for CLR of methane to syngas. My research shows that there are four distinct reaction regions for this type of catalyst, which have subsequently been observed in several related systems. My research also shows that the selectivity of a redox catalyst with high internal oxygen mobility is determined by the types of oxygen species available at the redox catalyst surface.

Thesis Research:

PhD Dissertation: "Nano-Particle Supported Catalyst" Advised by Dr. Helena Hagelin-Weaver in the Department of Chemical Engineering at the University of Florida

The oxidative coupling of 4-methyl-pyridines to 4,4′-Dimethyl-2,2′-bipyridine is a green chemistry process and a useful probe reaction for palladium catalysts. Pd on metal oxide supports had previously been reported as inactive for this reaction. I synthesized and characterized multiple palladium catalysts for this selective oxidation. The results showed that "inert" nano-particle metal oxide supports can induce similar behavior in platinum group catalysts typically ascribed to oxides with strong metal support interactions with correspondingly high activity.

Research and Funding Successes: I have authored or co-authored 20 refereed articles on a variety of catalytic reactions. I have supervised multiple research projects, mentoring over 12 graduate and undergraduate researchers, and have participated in the writing of 8 proposals, including as a named Co-PI. I co-invented a novel chemical looping ODH technology which secured a $3.8 million overall plus up from ARPA-E. My work has led to four patent applications, 2 of which have been licensed.

Future Direction: Catalysis is the classic means of "process intensification" in reaction engineering, but new insights into catalyst performance by mechanistic studies of real, practical catalysts is needed to keep pace with growing demands for cleaner energy and chemicals worldwide. My research on both supported platinum-group metal and mixed-metal-oxide redox catalysts will contribute to improved mechanistic understanding of greener, lower-emission catalytic processes. My future research will focus upon bridging the gap between real-world catalyst performance and fundamental studies on model systems. The goal of this emphasis is to guide rational development of practical, cheaper, more active, and greener catalysts. I will also actively pursue multi-dispensary collaboration. For example, system modeling is valuable for evaluating the interaction between process economics and catalyst performance parameters. Model catalyst development and characterization in collaboration with computer simulation for determining catalyst mechanisms is also important.

Research Interests: Building on my experience in practical reaction experimentation and mechanistic characterization of commercially relevant catalysts, I plan to conduct research in three areas: I) Oxygen storage materials for emissions reduction catalyst; II) Mechanisms of oxygenate formation in hydrocarbon systems; III) Unburned hydrocarbon oxidation over true earth-abundant materials.

Rational design of oxygen storage materials for enhanced NOx reduction and advance combustion: Several emissions catalysts rely on "oxygen storage material" (OSM) supports. These materials allow the catalyst to oxidize hydrocarbons and CO in the presence of low pressure oxygen reducing agents for the suppression and or reduction of NO_x creation studies. Rational design of OSM's has focused heavily on oxygen capacity at operating temperatures. Research on direct NO_x reduction catalysts have shown that other properties are impotent, but these studies have tended to limit studies to unrealistically low N_2 and O_2 concentrations, unrealistically high concentrations of reducing gas. My work will look at a fuller range of parameters, including oxygen-vacancy formation energies, internal oxygen transport rates and surface site pH. The principles developed will allow the rational design of improved emissions-abatement catalysts.

Mechanistic study of oxygenate formation: The creation of alcohols and other oxygenates from alkanes is simultaneously a desired energy conversion process and a nuisance. Unwanted oxygenates can entrain water, poison catalysts, and add complexity to separations. However, oxygenate formation in processes such as gas to liquids and steam cracking is not well understood. I plan to use n-butane dehydrogenation as a model system. My research will probe a variety of catalysts with differing selectivities for oxygenate to determine underlying mechanisms. My studies will include in-situ/operando measurements, as well as surface characterizations. This project will yield mechanistic insights that guide suppression of oxygenates, while potentially developing improved catalysts for on purpose oxygenate production and other processes.

True earth-abundant materials for deep oxidation of UHCs and CO: Novel, inexpensive mixed metal oxides will be used for deep oxidation of unburned hydrocarbons (UHCs) and carbon monoxide emitted in flue gases. Some redox catalysts, including Mg_6MnO_8 and $SrMnO_3$, have significant deep oxidation activity. These mixed oxides can exhibit better thermal stability and poisoning resistance than traditional platinum-group metal (PGM) catalysts, while potentially working well in very low partial pressures of oxygen. They can also be made more cheaply than typically proposed rare-earth oxides. Understanding the flexible properties of these earth-abundant metal oxides can guide improvement of coal and biomass energy conversion efficiency, while lowering catalyst costs.

Teaching Interests: I welcome the opportunity to teach graduate and undergraduate courses. I am qualified to teach any core undergraduate chemical-engineering courses and find teaching important to improving my own understanding fundamental aspects of my field. My background in applied, heterogeneous catalysis and redox-materials has given me skills and experiences particularly well suited to teaching undergraduate courses relating to mass balances, thermodynamics, kinetics, and design. I am also interested in teaching graduate-level courses in kinetics and surface science as well as special topic courses in catalysis and energy conversion. My approach to teaching has been molded by my experience as a TA, guest lecturer, as well as my supervision and mentoring of graduate and undergraduates in laboratory research. I have TA'd undergraduate students in thermodynamics and unit operations, and I have trained, mentored, and supervised over a dozen graduate and undergraduate students in their research, including setting research plans for three master students and a Ph.D. student. These mentoring interactions have included numerous international female and minority students. I am committed to pursuing further diversity through teaching and providing research opportunities.

Selected Publications

Seif Yusuf, **Luke M. Neal**, Vasudev Haribal, Madison Baldwin, H Henry Lamb, Fanxing Li. "Manganese silicate based redox catalysts for greener ethylene production via chemical looping–oxidative dehydrogenation of ethane." Applied Catalysis B: Environmental, Applied Catalysis B: Environmental, *In Press* (2018);

Yusuf+, **Luke M. Neal+**, F. Li. "Effect of Promoters on Manganese Containing Mixed Metal Oxides for Oxidative Dehydrogenation of Ethane via a Cyclic Redox Scheme." ACS Catalysis, **7**: 5163–5173 (2017); (+Co-first authors);

Luke M. Neal+, S. Yusuf+, F. Li. "Oxidative Dehydrogenation of Ethane: A Chemical Looping Approach." Energy Technology, **4**: 1200 (2016); (+Co-first authors);

Luke M. Neal, A. Shafiefarhood, F Li. "Effect of Core and Shell Compositions on MeOx@$La_ySr_{1-y}FeO_3$ Core–Shell Redox Catalysts for Chemical Looping Reforming of Methane." Applied Energy **157**, 391–398 (2015);

Luke M. Neal, A. Shafiefarhood, F. Li. "Dynamic Methane Partial Oxidation Using an Fe_2O_3@$La_{0.8}Sr_{0.2}FeO_{3-\delta}$ Core–Shell Redox Catalyst in the Absence of Gaseous Oxygen." ACS Catalysis, **4**: 3560–3569 (2014);

Luke M. Neal, M. Everett, G. Hoflund, H. Hagelin-Weaver. "Characterization of Palladium Oxide Catalysts Supported on Nanoparticle Metal Oxides for the Oxidative Coupling of 4-Methlypyridine." Journal of Molecular Catalysis A: Chemical, **335**: 210-22 (2011);

Luke M. Neal, S. D. Jones, M. L. Everett, G. B. Hoflund, H. E. Hagelin-Weaver. "Characterization of Alumina-Supported Palladium Oxide Catalysts in the Oxidative Coupling of 4-Methylpyridine." Journal of Molecular Catalysis A: Chemical, **325**: 25-35 (2010).

Patent Applications:

Fanxing Li, **Luke M. Neal**, Junshe Zhang. "Redox Catalysts for the Oxidative Cracking of Hydrocarbons, Methods of Making, and Methods of Using." **PCT/US17/51,157** (2017).

Li, **Luke M. Neal**. "Ethylene Yield in Oxidative Dehydrogenation of Ethane and Ethane Containing Mixtures." US **15/422,743** (2017);

J.A. Sofranko, F. Li, F. Li, **Luke M. Neal**. "Oxygen Transfer Agents for the Oxidative Dehydrogenation of Hydrocarbons and Systems and Processes Using the Same." US **62/054,424** (2015).

542167 The Synergistic Effect of Copper and Niobium Species on a Novel Cu/Nb-Ti Mixed Oxide Catalyst for the Selective Catalytic Reduction of NO_x with NH_3

Abstract:

Research Interests: A series of titanium-niobium oxide as the catalyst support was first synthesized by co-precipitation and then Cu/Nb-Ti ternary oxide catalyst was prepared through wetness impregnation. The optimal proportion was explored.And 0.8%Cu/Ti_2NbO_x exhibited the best catalytic activity with NO_x conversion and N_2 selectivity maintaining upon 90% and 97% respectively in the temperature range of 250°C-425°C under a gas hourly space velocity (GHSV) of 177,000h^{-1}. 0.8%Cu/Ti_2NbO_x showed well activity while traditional commercial VWTi catalyst deactivated after 2%K_2O doping. Water vapor or SO_2 would not pose any negative effect on the SCR activity while the catalytic performance showed a slightly reversible inhibition with water vapor and SO_2 existing in the simulated gas stream simultaneously. Also, 0.8%Cu/Ti_2NbO_x had an excellent hydrothermal stability after the process of hydrothermal aging with 10%H_2O, 5%O_2 and N_2 as balance for 7h at 800°C.The activity characterization indicated 0.8%Cu/Ti_2NbO_x was a promising candidate for the NH_3-SCR catalyst in the future practical application.

Teaching Interests: An array of analytical techniques (HRTEM, XRD, LRS, XPS, EPR, in-situ DRIFTS etc.) were employed to elucidate the correlations among the "performance – structure – property" and the reaction mechanism over 0.8%Cu/Ti_2NbO_x catalyst. The introduction of niobium oxide increased the surface area of TiO_2 and decreased the crystallite size, promoting the high dispersion of copper species. Meanwhile, the total acidity and the amount of surface active oxygen over the catalyst were increased by the addition of niobium oxide. Copper species mainly existing in the state of isolated Cu^{2+} and non-isolated Cu^+ enhanced the redox capability of the catalyst and the highest ratio of isolated Cu^{2+}/copper species resulted in an excellent catalytic performance of 0.8%Cu/Ti_2NbO_x. The promotional synergistic effect of copper and niobium in 0.8%Cu/Ti_2NbO_x catalyst gave rise to a better NH_3-SCR performance. Furthermore, the reaction pathway over 0.8%Cu/Ti_2NbO_x catalyst followed both Eley-Rideal mechanism and Langmuir-Hinshelwood mechanism at 225°C. Both coordinated NH_3 and ionic NH^{4+} were active intermediates and monodentate nitrates, bidentate nitrates and bridging nitrates were involved in the reaction over 0.8%Cu/Ti_2NbO_x catalyst at the reaction temperature of 225°C.

https://doi.org/10.1016/j.apcatb.2017.08.021

542181 Frabrication of Fe-ZSM-5@CeO_2 Catalysts with a Core-Shell Structure and the Enhanced Performances for the Selective Catalytic Reduction of NO with NH_3

Abstract:

Research Interests: We have rationally designed and originally developed a high-performance deNOx catalyst based on core-shell structured Fe-ZSM-5@CeO_2, which are synthesized via a solvothermal method. The as-prepared catalysts have been characterized systematically to elucidate their morphological structure and surface properties. As compared with conventional Fe-ZSM-5/CeO_2, Fe-ZSM-5@CeO_2 possess a much better catalytic activity, higher N_2 selectivity, more extensive operating-temperature window, higher stability, and SO_2 tolerance. The feature of hollow and porous structures provides a larger surface area and more active sites to adsorb and activate reaction gases, resulting in the high performance. Moreover, the uniform distribution and strong interaction of Fe-ZSM-5 and CeO_2 species not only enhance the catalytic cycle but also inhibit the deactivation of Fe-ZSM-5, resulting in high catalytic cycle stability.

Teaching Interests: The excellent deNOx performance of Fe-ZSM-5@CeO_2 can be attributed to the core-shell structures, the uniform distribution of active sites, as well as the strong interaction of Fe-ZSM-5and CeO_2 species. The excellent catalytic performance suggests that Fe-ZSM-5@CeO_2 is promising candidates for low-temperature deNOx catalysts. More importantly, the present study indicates that the core-shell architectures and well-dispersed active components can effectively enhance the performance of catalysts.

542190 Combining Theory and Experiment at the Electrode/Electrolyte Interface to Improve Electrochemical Energy Conversion and Storage

Abstract:

2nd Year Post-doctoral Fellow

Post-doctoral project: "*Combining experiment and atomistic scale computational modeling to study the electrode-electrolyte interface*" under the supervision of Dr. Marc Koper, Leiden Institute of Chemistry, Leiden University

Ph.D. Dissertation: "*Understanding the effects of electrolyte pH and spectator ions on electrocatalysis*" under the joint supervision of Dr. Michael Janik and Dr. Michael Hickner, Department of Chemical Engineering, Pennsylvania State University

Awarded proposals: Leading Fellows (Marie Curie COFUND) Post-doctoral Fellowship, EU, 2018; Teaching Fellowship, The Pennsylvania State University, 2015; EPA P3 Awards, US EPA, 2009 ("Farm Waste to Energy: A Sustainable Solution for Small-Scale Farms")

Research Interests:

I am interested in combining detailed electrochemical experiments and atomistic scale computational modeling to study the electrochemical interface. With these methods I will improve our fundamental understanding of electrochemistry and electrocatalysis, as well as develop high performance and low cost energy storage and conversion processes and devices, including batteries and fuel cells.

I began academic research as an undergraduate under Dr. Stefan Grimberg at Clarkson University, where I used both experiment and continuum scale modeling (in MATLAB/Simulink) to probe the effects of staging on the energy efficiency and cost of anaerobic digestion. I further pursued computational modeling under Dr. Michael Janik at the Pennsylvania State University during my Ph.D. in Chemical Engineering, where I learned how to perform density functional theory (DFT) calculations for studying electrocatalysis. I gained an expertise in finding computationally efficient methods to examine the effects of solvation, electrolyte ions, and pH on surface chemistry relevant to fuel cells and electrolysis. I was co-advised by Dr. Michael Hickner, under whom I gained additional skills in performing my own electrochemistry experiments, using cyclic voltammetry to measure reaction kinetics. We made an explicit effort to directly compare our computational results to experiment.

I am continuing to combine experiment and density functional theory modeling during my current post-doctoral work with Dr. Marc Koper at Leiden University in the Netherlands. I perform DFT-based kinetic modeling, as well as detailed electrochemistry experiments, particularly with well-defined single-crystal electrodes. We have shown that not only does the catalyst surface exhibit a significant effect on (electro)catalysis, but the solvent, ions, and pH near the electrode surface also play a significant role in dictating reaction rates and mechanisms. The unique combination of atomistic scale modeling and experiments with well-defined electrodes gives us a mechanistic understanding of the behavior of the electrode/electrolyte interface that would be difficult to obtain with either technique alone.

Over the past four years, I have published 9 peer-reviewed articles (7 first authored) and have recently submitted an additional 5 manuscripts.

Future Research Directions:

1) Benchmarking near-surface solvation models and determining the surface structure/composition of real catalysts in the electrochemical environment. In our prior work, we have used computationally efficient methods to approximate the effects of solvation on the thermodynamics of adsorption of many species, including hydrogen, hydroxide, oxygen, and the halide anions to complex surfaces, and have shown these thermodynamics match experiment. While many techniques have been evaluated for approximating the effects of solvation of the adsorption of more complex species, these adsorption thermodynamics are only rarely compared directly to experiment. I will use density functional theory and experimental cyclic voltammetry measurements with single crystal electrodes to examine the adsorption of many neutral and ionic species to evaluate and benchmark methods of approximating the effects of solvation. These methods include explicit, implicit, and combined explicit/implicit solvation methods. By systematically evaluating the role of the surface structure and the properties of the adsorbate (charge distribution, dipole moment, polarizability) we will determine which methods are best suited for predicting thermodynamics and kinetics in the electrochemical environment, even for adsorbates whose adsorption thermodynamics we cannot directly measure (such as for short-lived reaction intermediates). This work will also allow us to examine the stability of a wide array of different surface facets/steps/defects in the presence of adsorbed electrolyte or reactive species, which have a strong effect on catalyst stability. With these surface structure stabilities, we can determine the preferred size, shape, and surface composition of real, complex metal catalysts in the electrochemical environment.

2) Improved catalysts and processes for the electrochemical synthesis of organic compounds. While significant study has been devoted to organic electrochemistry, and some electrochemical synthesis methods are utilized industrially to produce complex, high-value organic compounds, most of these electrochemical processes have been identified by trial-and-error, and catalysts for these reactions remain under-optimized (in terms of activity, selectivity, and cost). Significant scientific research has been devoted to understanding the electroreduction of carbon dioxide to valuable fuels and chemicals, however, given the complexity and high energy requirements, it is unclear if or when such a process will see commercial use. I will use density functional theory and experiment to better understand structure-activity-selectivity relationships for the electrochemical reduction/oxidation of biomass derived organic compounds (which lie energetically closer to their desired products than carbon dioxide) to high-value chemicals.

3) Bulk, surface, and solution phase properties of electrode materials for both ion-intercalation and redox flow batteries. A wide variety of both inorganic and organic compounds have been investigated for use as energy storage components (anodes/cathodes) in lithium-ion (or other alkali-/alkali earth- metal ion) and redox flow batteries. The solubility of the active components plays an important role in each battery; in an ion-intercalation battery high solubility is undesirable as it leads to poor cyclability, where in redox flow batteries, high solubility is desirable as it leads to high energy and power density. I will use both density functional theory as well as experimental half-cell and whole-cell battery cycling and cyclic voltammetry to examine how the bulk, surface, and solution phase properties of active electrode materials affect their solubility and performance. I am particularly interested in organic active materials, many of which have high capacities and low cost. Altering the molecular structure will allow us to change both the solution phase properties (how the organic compound interacts with the solvent) and the solid-state properties (the crystal structure and binding energy). This will lead to the development of improved active materials for both ion-intercalation and redox flow batteries, suitable for low-cost, grid-scale energy storage.

Teaching Interests:

My previous teaching and research experiences have well prepared me to teach much of the core Chemical Engineering coursework, including thermodynamics, reaction engineering/kinetics, heat transfer, unit operations, and laboratory exercises. While an undergraduate at Clarkson University, I tutored junior and senior students in process modeling and control. While a graduate student at The Pennsylvania State University, I served as both a teaching assistant and instructor (as a teaching fellow) for CHE480W ("Chemical Engineering Laboratory"). As a teaching fellow I was responsible for holding office hours; evaluating, grading, and giving detailed feedback on written technical reports and oral presentations; and for helping students to better understand all of the core chemical engineering principles necessary to carry out the experiments in CHE480W, covering topics ranging from fluid mechanics, mass transport, heat transfer, and unit operations including distillation and liquid-liquid extraction. I have additionally given seminars about the use of atomistic scale (quantum mechanics based) modeling for electrochemistry, surface science, and catalysis. I would like to create a course for undergraduate or graduate students on the use of these ab-initio modelling techniques in chemical engineering. I am passionate about teaching students and am excited to contribute to their future success as chemical engineers in both industry and academia.

542250 Optimizing Electrocatalysts for Energy Storage and CO_2 Conversion

Abstract:

Research Interests:

My research aims to develop electrocatalysts for critical reactions in the energy sector. Analyzing global energy trends can identify catalytic reactions that are central to emerging problems, and mass and energy balances reveal the most promising paths to address these issues. Catalysts can be designed with a combination of density functional theory (DFT) calculations and model surface experimentation. And performance of industrially relevant catalyst morphologies can be evaluated with in-situ vibrational spectroscopy and synchrotron techniques.

Rapid increases in solar energy capacity and sustained use of fossil fuels require electrochemical energy storage and CO_2 conversion technologies. New water electrolysis installation needed to meet energy storage and CO_2-free hydrogen demands could incur up to a 50-percent increase in global iridium (Ir) consumption – an unsustainable spike in one of the world's scarcest metals. During my Ph.D. studies under Prof. Jingguang Chen at Columbia University, I addressed this bottleneck by developing a core-shell Ir-metal nitride particle that reduces Ir loading by half and is near the center of the acidic oxygen evolution reaction (OER) volcano curve, according to DFT calculations. Importantly, the nitride cores are protected from the highly oxidizing environment by the IrO_2 shell, which was observed with in-situ X-ray absorption spectroscopy, and post-reaction electron microscopy. The Ir-metal nitride core-shell framework coupled with the effective DFT descriptor should enable optimization of Ir usage and surface binding energies in order to minimize Ir loading.

Global CO_2 avoidance targets are on the scale of hundreds of megatons per year, which may require fuels and chemicals across many sectors to be produced from recycled CO_2. There must, in turn, be a range of catalysts to convert CO_2 to CO, alkanes, olefins, and alcohols. Metal carbides were identified as robust and versatile electrocatalyst supports, capable of tuning activity for different reactions by matching carbide support with overlayer metal. High-surface area carbides, however, suffer from carbonaceous overlayers that limit synthesis and application. Metal nitrides have similar structure to carbides but are more easily synthesized with a "clean" surface. I am currently studying fundamental electrochemical energy storage and conversion reactions on nitride and metal modified nitride catalysts during a one-year DOE Office of Science Graduate Student Research Program in the lab of Dr. Radoslav Adzic at Brookhaven National Lab. I am using in-situ infrared spectroscopy to observe surface intermediates for electrochemical alcohol oxidation and CO_2 reduction on metal-modified nitrides. The technique helps reveal reaction pathways and can be correlated with DFT calculations to identify critical intermediates to accelerate catalyst development. I am also using in-situ X-ray diffraction and X-ray absorption spectroscopy to observe nitride stability under reaction conditions. This will provide valuable insight for future studies on selecting viable electrocatalyst candidates among this intriguing class of materials.

Moving forward, the methodology of evaluating global trends, performing mass and energy balances, and developing catalysts with DFT calculations, model surface studies, and in-situ techniques can be applied to address a wide range of issues in the energy field. As one example, the growing importance of CO_2-free hydrogen may benefit from substantial cost reduction of membrane-less electrolyzers. This technology presents a new slate of catalyst optimization issues from morphology to ion transport and cell design. These challenges can be met with an approach using fundamental electrochemistry and in-situ techniques, like the cases described above.

Teaching Interests:

I have strong interest in teaching across the breadth of chemical engineering core areas of study and believe that student success is built upon a mastery of mass and energy balances. I believe that developing problem solving skills is vital to an engineer's training, and my teaching philosophy is that instruction should actively guide and foster these skills in a classroom, as opposed to simply lecturing. I have experience in this instruction style from my time as teaching assistant in an undergraduate kinetics course that was operated as a flipped classroom. I am also interested in developing students' written and oral communication skills, which are too often under-stressed in engineering curricula. One of my undergraduate professors once told me that technical brilliance means little without the ability to effectively communicate to broad audiences, and I would emphasize this in any course I teach. I have enjoyed two teaching assistant opportunities as a graduate student and have been an invited lecturer on topics of water electrolysis catalysts for an electrochemical materials course and X-ray photoelectron spectroscopy for a surface science course. For specialized courses, I am most interested in teaching surface reactions and kinetics, scientific writing and communication, and chemical engineering applications of electrochemistry.

542252 Nature of Active O_2-Derived Species in Selective Oxidation Catalysis

Abstract:

Research Interests:

Catalytic selective oxidation is an essential industrial process to produce key chemical intermediates, such as alkenes, alcohols, epoxides, aldehydes, ketones, and organic acids. Rates and selectivities of the oxidation catalysis depend on the type and nature of O_2-derived species, specifically their nucleophilicity and electrophilicity, which determine their H-abstraction and O-insertion abilities and selectivities. In contrast to nucleophilic surface O-atoms in bulk oxides that tend to abstract H-atoms from organic molecules, electrophilic O_2-derived species, such superoxo, peroxo, or hydroperoxo, catalyze O-insertion reactions as in the case of monooxygenase enzymes that convert methane to methanol[2] and titanosilicates that form epoxides from alkenes.[3] Such electrophilic O_2-derived species formed via O_2 activation steps also influence product selectivities in alkane oxidative dehydrogenations (ODH) by converting the primary ODH product, alkene, into more reactive epoxides that ultimately convert to undesired CO/CO_2. Detailed knowledge of catalytic properties that determine the preference of various O_2 activation routes that form different types of active oxygen species and their role in oxidation catalysis will provide a platform in designing more reactive and selective catalysts.

Recent advances in computational chemistry, atomic-level synthetic methods, and characterization tools have brought the Renaissance in catalysis, allowing us to understand and control catalytic reactions on a molecular level. This synergy, however, can be achieved only when these tools are combined rigorously and effectively to solve problems. One of my recent research projects with Prof. Iglesia revealed the involvements of two types of O_2 activation routes, inner and outer sphere pathways, during re-oxidation of Mo-based polyoxometalate (POM) clusters, by combining kinetic and scavenging experiments with density functional theory (DFT) calculations.[4] Kinetic responses of the scavenging reactions of bound peroxo (OO*) species during alkanol ODH reactions provided information about mechanistic details of re-oxidation steps, which cannot be probed using steady-state measurements because of the fast nature of re-oxidation compared to the reduction part of the ODH cycles. In junction with these experimental inquires, DFT calculations on the energetics of intermediates and transition states evaluated the plausible reaction pathways that are otherwise indistinguishable from experiments. These calculations, in turn, revealed that OO* species react with gaseous reductants (alkanols or alkenes) to re-form lattice O-atoms (O*) for next ODH turnovers, instead of with another reduced center via conventionally proposed routes, in order to circumvent kinetic hurdles required for OO* hopping steps to bring OO* next to another reduced center.

A rational design of effective catalytic systems requires detailed understanding of catalytic properties that determine the rates and selectivities. Such assessments, however, are not trivial and need to involve a rigorous use of both experiments and theory. My current project with Prof. Iglesia uses DFT calculations to assess inner and outer sphere O_2 activation barriers for POM clusters with different metal atoms and charge-balancing cations. In doing so, we show that the activation barriers mediating the two routs depend differently on the stability of the reduced centers, leading to their respective rates that strongly depend on the strengths of the M-O bonds, which determine the contribution of inner sphere routes, and of O-H bonds, which determine the contribution of outer sphere routes. The experimental confirmation of such findings is currently underway by synthesizing and analyzing the POM clusters with various metal atoms (Mo, Mo-V, and W).

Controlled synthetic methods with atomic precisions are critical to implement design strategies predicted from DFT methods and to provide catalysts with homogeneity. A part of my graduate research projects with Prof. Stair and Prof. Snurr involved atomic layer deposition (ALD) techniques, which can be applied to synthesize highly dispersed metal oxides (mono oxide or mixed oxides with various combinations) and metal nanoparticles on solid supports. The projects for my own research group will combine these research tools to study the type and nature of active oxygen species involved in selective oxidation catalysis and to identify the catalytic properties that determine the activities and selectivities of the involved oxygen species. The research will include, but not limited to, the effects of mixed M-O bonds in ALD synthesized mixed oxides for O_2 activation pathways, the confinement effects in selective oxidation catalysis, and the role of active O_2-derived species in selective oxidation of light alkanes and alkenes.

Postdoctoral Projects: *"Dioxygen activation routes in Mars-van Krevelen redox cycles catalyzed by metal oxides"* and *"Mechanistic details of formic acid decomposition routes on metals and metal oxides"* under the supervision of Enrique Iglesia, Chemical and Biomolecular Engineering, University of California, Berkeley

PhD Dissertation: *"Gas-phase alkene oxidation by hydrogen peroxide: the nature of active oxygen species in heterogeneous catalysis"* under the supervision of Randall Snurr and Peter Stair, Chemical and Biomolecular Engineering, Northwestern University

Teaching Interests:

Throughout my studies in graduate school and postdoctoral research, I have actively mentored several undergraduate and graduate students, which I found myself truly enjoying and led me to pursue an academic career. I have taught in classrooms and guest lectured undergraduate and graduate level main courses (statistical mechanics, Matlab programming, and computational chemistry) at Northwestern university and UC Berkeley. I am confident in my ability to teach any undergraduate and graduate level course in chemical engineering, with preference for kinetics and reactor design, thermodynamics, and transport, based on my experiences and expertise in heterogeneous catalysis. I am also interested in designing and teaching graduate level courses related to applications in catalysis and in computational chemistry.

References

(1) Serrano-Plana, J.; Company, A.; Costas, M. *O–O Bond Activation in Cu- and Fe-Based Coordination Complexes: Breaking It Makes the Difference*, 1st ed.; Elsevier Inc., 2017; Vol. 70.

(2) Kopp, D. A.; Lippard, S. J. Soluble Methane Monooxygenase: Activation of Dioxygen and Methane. *Curr. Opin. Chem. Biol.* **2002**, *6* (5), 568–576.

(3) Kwon, S.; Schweitzer, N. M.; Park, S.; Stair, P. C.; Snurr, R. Q. A Kinetic Study of Vapor-Phase Cyclohexene Epoxidation by H2O2 over Mesoporous TS-1. *J. Catal.* **2015**, *326*, 107–115.

(4) Kwon, S.; Deshlahra, P.; Iglesia, E. Dioxygen Activation Routes in Mars-van Krevelen Redox Cycles Catalyzed by Metal Oxides. *J. Catal.* **2018**, *364*, 228–247.

542254 Fundamental Understanding of Non-Traditional Feedstock Conversion Processes

Abstract:

Research Interests:

The rapid development of novel processes based on utilization of non-traditional feedstocks for the production of fuels and chemicals is crucial for a sustainable and independent economy. The scientific advancements strongly depend on the systematic and fundamental understanding of these processes. In my future research, I will investigate processes for the direct conversion of shale gas and hydrocarbon polymers to chemicals and fuels by combining data science and experiments. Previously, I performed a systematic investigation on the effect of biomass composition on product composition using a large set of well-defined feedstocks with an engineered lignin composition. Statistical data analysis techniques enabled not only finding correlations within large datasets but also faster processing and evaluation of the results. Currently, I am investigating high-temperature catalytic processes where the interplay between gas phase and surface chemistry is poorly understood. Last but not least, I plan to use my experience in microkinetic modeling and reactor design to obtain detailed insights into the reaction chemistry and provide guidelines for process optimization.

Teaching Interests:

The future of our society lies with young people and well-educated chemical engineers will play a major role among others. As a faculty member, I will give the utmost importance to prepare my students with the right skills that will allow them to succeed in a dynamic, fast and challenging environment. Based on my chemical engineering background, I strongly believe that I can be involved in all chemical engineering courses at an undergraduate level. I will design my courses in a way that the students will actively take part and improve their critical thinking skills during the class via interactive exercises and team projects. I would also like to share my knowledge and research experience via specialized courses such as "Conversion Processes for Non-Traditional Feedstocks" and "Multivariate Data Analysis for Chemical Engineers". The first course aims at setting the scene for challenges and opportunities for the valorization of non-traditional feedstocks for the production of fuels and chemicals whereas the latter will provide an introduction on how to use data science in chemical engineering to advance faster and become more efficient.

542255 Decoding the Complexity of Chemical Reactions on Single Atom Catalysts and Beyond

Abstract:

Research Interests:

The rational selection of catalyst and reaction conditions suitable for a specific application is a challenging task because of the chemical and structural complexity of real catalytic systems. A heterogeneous catalyst is more than a collection of equivalent active sites, traditionally denoted as " " in chemical reaction engineering, while it is rarely the case that a single reaction mechanism operates over different catalytic materials and reaction conditions. For my faculty research, I propose a holistic approach that avoids such oversimplifications and accounts for complex catalyst structures and reaction networks. Computational chemistry and operando spectroscopy can readily provide mechanistic and structural details to explain the catalyst performance, while microkinetic simulations can be used to investigate the effect of reaction conditions on catalyst structure, activity and selectivity in a systematic way. To this end, my past research activities have allowed the development of a first-principles microkinetic model describing the influence of reaction conditions and zeolite framework on reactivity and selectivity for bio-alcohol dehydration. Currently, I am investigating the performance and stability of well dispersed metal catalysts for emission control technologies.

Teaching Interests:

One of the many roles of a professor is to act as a catalyst for learning. Over the years, I have been actively seeking ways to improve this catalytic performance and set up the stage for the development of future competent researchers and engineers. Since engaging the students during lecturing is of paramount importance for their learning experience, I am an avid supporter of game-based classroom response systems (e.g. kahoot) that are perfectly suited to make learning more fun and interactive. Given my scientific background, I always welcome the opportunity to introduce students to the fascinating world of computational chemistry, and to prepare them to conduct innovative research in multidisciplinary areas as well as apply their knowledge and skills in industrially relevant projects. To this end, I would like to share my knowledge and experience in the field of heterogeneous catalysis via specialized courses such as "Multiscale Modeling of Catalytic Reactions" and "Dynamics and Spectroscopy in Catalysis".

542295 Catalysis Informatics: Accelerating Search and Discovery of New Catalysts

Abstract:

Research Interests:

I am interested in the development of tools for catalysis informatics. More specifically, this includes methods for improving the rate at which we search through possible catalytic systems of interest using database mining, machine learning, and high-throughput screening techniques.

During my PhD, I studied computational catalysis at Carnegie Mellon University. I worked with Professor John Kitchin to develop a new kind of atomic potential constructed from neural networks. These potentials have become increasingly popular in the last decade with the advent of fingerprinting schemes, proposed by Behler and Parrinello, which allow them to be trained to a wide variety of atomic input structures. The methods are now capable of reproducing potential energy surfaces (PES) of a multitude of chemical systems with levels of accuracy unrivaled by their physical potential counterparts[1,2]. They are also significantly faster to evaluate than a full quantum chemistry calculation resulting in the ability to study much larger unit cells and more complex chemistry.

For the last year, I have been working as a post-doc at Stanford working with Thomas Bligaard and the catalytic methods development team at SUNCAT. Here I am developing the tools needed to realize high-throughput techniques for the field of computational catalysis. To accomplish this goal, there is a need to enumerate the space of all possible catalytic surfaces of interest. In cheminformatics this is performed through using graph theory tools which allows molecules to be considered as component vertices (the atoms) connected by bonds (edges). With this basic precept, a great number of molecular structures of interest can be represented computationally; a critical first step to any computational high-throughput implementation.

For catalysis, there is a bulk with and surface to be considered along with the adsorbed molecule and producing a unique descriptor is not trivial. In the field of materials informatics, systems are represented in 3D periodically bound boxes, so graphical methodologies are not commonly used. My recent work has focused on beginning from bulk structures and generating graphs from these periodic systems. From a bulk structure, any slab of interest in heterogeneous catalysis can be generated. Finally, unique adsorption sites can be identified through symmetries and used to add molecules to the surface in a systematic way, resulting in a method for producing a unique connectivity matrix for slabs with adsorbates; the basis for a complete catalysis informatics approach. These methodologies are available through the open source Python package, CatKit, for which I am the lead developer.

As available computational resources continue to increase, the rate of data generation in the field of catalysis is quickly beginning to outstrip the ability of those in the field to interpret it all. To improve the ability of all those in the field of catalysis to continue making rapid scientific progress, there is a need to develop better databases which are capable of storing and searching the information; similar to the way AFLOW, Materials Project, OQMD, and others have done for the field of material science. For catalysts, this is significantly more challenging due to the even larger number of possible surfaces and adsorption configurations of interest. Even with a robust database structure capable of storing a breadth of relevant information, tools capable of intelligent and sophisticated queries to the database are also needed. Machine-learning not only benefits from such large repositories of information, it can also essential in making more intelligent selections from a large number of interesting structures to find new and interesting results. These tools will prove to be fundamental for high-throughput research.

Teaching Interests:

As an instructor, I teach: thermodynamics, kinetics, molecular simulation (and underlying quantum theory), reaction engineering, functional coding languages, and general chemical engineering. Chemical engineers fulfill a diverse range of demand in industry positions today. As modern tools become increasingly computationally intensive, there is an ever growing need to shift the emphasis of education to computational tools and techniques. As such, I strongly support the integration of coding into the classroom. I am also interested in perusing educational research to do with code based instruction which I describe in greater detail below.

I have completed instruction on educational techniques for engineering in an evidence-based teaching course specific to STEM fields. I am a strong proponent of the flipped classroom as a pedagogical teaching model. This model requires students to study lecture material before the class and implement it on assignments given during class. Among other benefits, such a system allows the instructor to provide immediate feedback to students which is shown to improve student retention rates and longevity of student knowledge. Further details of my pedagogical teaching strategies and achievements can be found on my website.

I would like to study the potential benefits of teaching code-based chemical engineering tools in an immersive environment. The concept is similar to that of language immersion, in which various subject material is taught through the medium of a second language. Language immersion is known to dramatically improve student comprehension of the second language in technical communication with no discernible negative impact on primary language skills. Teaching students to solve chemical engineering problems can similarly be framed in the medium of a coding language, such as Python or MatLab.

Relevant Work

1. Jacob R. Boes, Mitchell C. Groenenboom, John A. Keith, and John R. Kitchin. Neural network and ReaxFF comparison for Au properties. International Journal of Quantum Chemistry, 116(13):979-987, 2016. DOI

2. Jacob R. Boes and John R. Kitchin. Modeling Segregation on AuPd(111) Surfaces with Density Functional Theory and Monte Carlo Simulations. The Journal of Physical Chemistry C, 121(6):3479-3487, 2017. DOI

Curriculum Vitae

ORCID

Website

Phone: (607) 342-1846

Email: jacobboes@gmail.com | jrboes@stanford

542317 Molecular Modeling and Machine Learning for Catalysis and Separations

Abstract:

Molecular Modeling and Machine Learning for Catalysis and Separations

Research Interests:

As earth's population grows and the standard of living in developing nations improves, demand for energy, water, and agricultural products will increase. These will interact in complex ways as production increases to meet demand; the NSF has even prioritized these challenges in their Innovation at the Nexus of Food, Energy, and Water Systems (INFEWS) initiatives. Human and animal wastes have been viewed as environmental and sanitary liabilities, but they could be viable renewable resources instead. Technologies that economically convert biomass in wastewater to useful fuels and chemicals could conserve energy and clean water, making fresh water available for human or animal consumption or irrigation. In addition, as new biomass pyrolysis, dehydration, and hydrodeoxygenation processes approach commercialization, new water treatment processes will need to be developed to clean up process water, which often contain compounds toxic to microbes and resistant to traditional wastewater treatment.

Chemical transformation and solute-substrate interactions are at the heart of any such future technologies, creating a significant opportunity for applying catalytic science to water treatment. Moreover, in this age of computer-aided catalyst design, progress toward optimal catalyst and reactor designs can be accelerated by utilizing modeling tools to identify and alleviate bottlenecks in reaction networks and in processes, as well as to aid in the search for high-performance materials. As vast libraries of materials are created in lab and *in silico*, extracting actionable information from these big data can be difficult. Techniques from machine learning and data science can be harnessed to identify critical trends and descriptors, as well as build predictive models for discovery.

Developing these technologies will require an interdisciplinary and collaborative approach, bringing together the insights and methodologies of chemical, environmental, and civil engineers. During my PhD, I developed expertise in reaction engineering, catalysis, and computational chemistry for processing biomass into fuels and chemicals, and in my current post doc, I am acquiring new skills for modeling separation processes and designing new materials *in silico*, as well as applying new machine learning techniques. Leveraging these techniques in collaboration with experimentalists will accelerate progress toward a more sustainable future.

Research Experience

During my Ph.D. with Prof. Dion Vlachos, I applied molecular dynamics and electronic structure calculations to investigate fundamental interactions between solvents, solutes, and catalysts in biomass catalysis. In my first project, I discovered how co-solvent interactions protect biomass-derived furans from side reactions in fructose dehydration [1] [2]. I developed structure-activity relationships for homogeneous Lewis acids to probe mechanistic details in glucose isomerization [3] [4]. I also exhaustively screened site geometries for Sn-Beta, identifying the most stable active geometry and proposing a source for its Brønsted acidity [5]. In my final project, which I continued during my post doc, I developed the reaction mechanism for fructose etherification on the hierarchical Sn-SPP zeolite [7].

In my current post doc with Prof. Ilja Siepmann, I am developing and applying Monte Carlo and molecular dynamics simulations in five projects: 1) predicting solution-phase adsorption into hierarchical zeolites under reaction conditions [8], 2) investigating adsorption of organics at the air-water surface [6], 3) calculating the viscosities of novel lubricants, 4) predicting thermophysical properties of hydrocarbons, and 5) developing molecular modeling and machine learning techniques to predict adsorption isotherms.

Teaching Interests:

Excellence in teaching has the potential to multiply progress made toward a more sustainable future through the engineers who will come after us.

I have served as a teaching assistant in both undergraduate and graduate thermodynamics courses, and I have been a guest lecturer in kinetics and reaction engineering at 3 institutions. In addition, I organized a computational chemistry workshop for the graduate reaction kinetics course, to guide students through exercises using the same software I use in my research.

I am particularly excited about introducing machine learning and data science into undergraduate and graduate Chemical Engineering education. By broaching these topics in core courses and going deeper in electives, students will become acquainted with cutting-edge data analysis tools and equipped for careers in data science and its intersection with traditional chemical engineering careers.

Courses of expertise: thermodynamics and statistical mechanics, kinetics and reaction engineering, numerical methods and linear algebra, computational chemistry, and engineering statistics.

First-Year Postdoctoral Fellow

Postdoctoral Project: *Monte Carlo Simulations for Novel Materials and Process Design*

Advisor: Prof. J. Ilja Siepmann, Department of Chemistry and Department of Chemical Engineering and Materials Science, University of Minnesota

PhD Dissertation: *Leveraging First-Principles Simulations for Deeper Insights in Biomass Processing*

Advisors: Prof. Dionisios Vlachos and Dr. Stavros Caratzoulas, Department of Chemical and Biomolecular Engineering, University of Delaware

AIChE Conference Talk:

[8] **Josephson, T. R.**, Joseph, K.E., Dauenhauer, P. J., and Siepmann, J. I. Solution-Phase Adsorption of Furan and Carboxylic Acid in Hierarchical Zeolites: Insights from Molecular Simulation. Molecular Simulation of Adsorption II, Wednesday, October 31, 2018, 3:30 PM.

Selected Publications

[7] **Josephson, T. R.**, DeJaco, R., Pahari, S., Ren, L., Guo, Q., Tsapatsis, M., Siepmann, J.I., Vlachos, D., Caratzoulas, S. "Cooperative Catalysis by Surface Lewis Acid/Silanol for Selective Fructose Etherification on Sn-SPP Zeolite." **2018**. *In Review.*

[6] Minkara, M. S., **Josephson, T. R.**, Venteicher, C. L., Chen, J. L., Stein, D. J., Peters, C. J., Siepmann, J. I., **2018**. "Monte Carlo simulations probing the liquid/vapour interface of water/hexane mixtures: adsorption thermodynamics, hydrophobic effect, and structural analysis." Molecular Physics. doi:

https://doi.org/10.1080/00268976.2018.147123

[5] **Josephson, T. R.**, Jenness, G. R., Caratzoulas, S., Vlachos, D. G., **2017**. Distribution of open sites in the Sn-Beta zeolite." Microporous and Mesoporous Materials. doi: 10.1016/j.micromeso.2017.02.065

[4] **Josephson, T. R.**, Brand, S. K., Caratzoulas, S., Vlachos, D. G., **2016**. 1,2-H versus 1,2-C-shift on Sn-silsesquioxane." ACS Catalysis, 7. doi: 10.1021/acscatal.6b03128

[3] Brand, S. K., **Josephson, T. R.**, Labinger, J. A., Caratzoulas, S., Vlachos, D. G., Davis, M. E., **2016**. Methyl-ligated tin silsesquioxane catalyzed reactions of glucose." Journal of Catalysis. doi: 10.1016/j.jcat.2016.06.013

[2] **Josephson, T. R.**, Tsilomelekis, G., Bagia, C., Nikolakis, V., Vlachos, D. G., and Caratzoulas, S., **2014**. Solvent-induced frequency shifts of 5-hydroxymethylfurfural deduced via infrared spectroscopy and ab initio calculations." Journal of Physical Chemistry A. doi: 10.1021/jp508340p

[1] Tsilomelekis, G., **Josephson, T. R.**, Nikolakis, V., and Caratzoulas, S., **2013**. Origin of 5-hydroxymethyl furfural stability in water/dimethyl sulfoxide mixtures." ChemSusChem. **Cover article**. doi: 10.1002/cssc.201300786

542358 Novel Catalytic Materials for Efficient Chemistry - Elucidation of Fundamental Structure-Activity Relationships

Abstract:

Research Interests:

Our modern world is dependent on a range of chemical transformations that are carried out using heterogeneous catalysts. Monometallic catalysts are limited in their performance while supported bimetallic catalysts offer many advantages in activity, selectivity, and/or stability for a variety of chemistries. In order to further improve performance and develop new bimetallic catalysts for challenging transformations, fundamental structure-activity relationships must be elucidated. By understanding the mechanisms by which reactivity enhancement occurs, we gain insight which informs rational catalyst design of novel materials.

My Ph.D. research under Professor James A. Dumesic has focused on tuning catalytic performance through a precision synthesis approach for bimetallic materials. By using controlled surface reactions to synthesize bimetallic catalysts, I have elucidated structure activity relationships for Pd-based bimetallic catalysts for hydrogenation and amination reactions. Conventional catalyst synthesis techniques, such as impregnation, often result in catalysts with monometallic species within the bimetallic system. I showed that this challenge can be overcome and uniform, bimetallic active sites can be formed by selectively depositing a promoting metal onto a parent catalyst. Using catalysts synthesized by this approach, I demonstrated that Au-Pd active sites are more active than either Pd or Au sites for the gas phase amination of 1-hexanol. Using FTIR, STEM-EDS, and XAS techniques, I concluded that the resulting Pd structure varies with catalyst composition, and that highly active, isolated Pd species can be formed at low Pd loadings.

Additionally, I adapted and applied controlled surface reactions for the synthesis of AgPd catalysts on carbon, SiO_2, and TiO_2 supports and demonstrated that the selective deposition of the promoting metal is dependent on the support. Furthermore, the support influences the bimetallic active site structure and resulting selectivity for the hydrodechlorination of 1,2-dichloroethane. Using spectroscopic techniques, I demonstrated that formation of uniform bimetallic structures was most effective for AgPd/TiO_2, resulting in Pd that is well isolated and diluted in the Ag. Over AgPd/TiO_2, a high selectivity to the desired ethylene is achieved, while a high selectivity to undesired ethane is achieved over AgPd/SiO_2, which was determined to have both monometallic and bimetallic structures. Thus, I showed that by tuning Pd surface structures, the selectivity can be modulated between ethane and ethylene.

My work is currently being expanded to investigate selective hydrogenation reactions using AgPd and AuPd catalysts and develop structure-activity relationships for this chemistry. By combining a highly controlled approach to synthesizing well defined bimetallic active sites and advanced characterization techniques with fundamental reaction kinetics studies, we can develop structure-activity relationships which inform further catalyst improvements. My future research program will use these fundamental approaches to understand the active site and develop novel catalytic materials for a variety of chemical transformations.

Teaching Interests:

During my time as a PhD student, I have developed skills in and explored my passion for teaching in several ways. I served as a teaching assistant for the undergraduate Transport Lab and was awarded a student-nominated TA Award for the course. I also served as the teaching assistant for both undergraduate Reactor Design and Kinetics and graduate Kinetics and Catalysis, for which I lead course discussions and developed homework and exam questions with the goal of fostering a deeper understanding of course material for students.

I further explored teaching and mentoring through the Delta Program in Research, Teaching, and Learning. I studied constructivist pedagogical approaches and active learning techniques, and refined my skills in effective assessments, something I believe to be key in ensuring student learning. I also conducted a teaching-as-research project to identify misconceptions that undergraduate students hold on the topic of catalysis, as misconceptions can be challenging to overcome and replace with correct understanding. Through this project, I gained experience in collecting data to answer questions about student learning and am motivated to continue this practice in my future classrooms to promote deep understanding in my students. This work culminated in a Delta Certificate in Research, Teaching, and Learning.

Additionally, as a WISCIENCE Mentoring Fellow, I cultivated best practices for mentoring in a research setting and have mentored 4 undergraduate students and 2 high school students in the lab. Work from two of these students contributed to publications. I also facilitated Entering Research, a course to help undergraduate students gain communication skills necessary to be successful in research. These experiences have prepared me to be an effective teacher in the classroom and active mentor to students in the lab by giving me tools to continually evaluate and improve my teaching.

From these experiences, my teaching philosophy is focused on three aspects: student engagement, inclusion, and a student-centered classroom. I believe that incorporating active learning approaches is key to ensuring effective learning. The next generation of chemical engineers will face a wide range of challenges in changing feedstocks, the need for improved efficiency, and more, and I aim to guide student development of problem solving skills through hands-on laboratory experiences and project work. I will also encourage students to think critically and take leadership over their own learning through metacognition. I am deeply passionate about teaching and plan to continue to apply best practices from literature on teaching and learning to my own classroom. I am interested in teaching material balances and unit operations courses, as these serve as the foundation for a comprehensive chemical engineering education. I am also interested in teaching kinetics and catalysis courses at both the undergraduate and graduate level and would be interested in developing elective courses on microkinetic modeling or material science for chemical engineers. By incorporating topics from current research into my classroom, I hope to inspire students to build upon their chemical engineering education and develop the skills needed to be the next generation of problem solvers.

542359 Controlled Catalytic Capability through Tailored Nanoporous Materials: For Selective and Sustainable Chemical Processes

Abstract:

Research Interests:

My Ph.D. work with Prof. Wei Fan in the department of Chemical Engineering at University of Massachusetts, Amherst, focused on the synthesis of nanoporous catalytic materials such as zeolites, metal oxides and porous carbon for biomass upgrading processes. In particular, hierarchical Lewis acid zeolites were developed by a seeded growth within mesoporous carbon, and showed >3 times higher catalytic activities for sugar isomerization reactions, compared to conventional Lewis acid zeolites. In addition, the Lewis acid zeolite supported Pt nanoparticles could effectively catalyze glycerol oxidation to form lactic acid with 72% yield through multi-step pathways, whereas no lactic acid was produced over typical metal oxide-supported Pt catalysts. Furthermore, phosphorous-containing zeolite and silica materials achieved 97% yield of p-xylene via a renewable route from biomass. The key to this success was to introduce weak Brønsted acid sites into the catalytic materials, which stems from phosphoric acid on the solid. In contrast, conventional heterogeneous catalysts with Lewis or Brønsted acid sites gave p-xylene at a maximum yield of ~70%, due to catalyst deactivation.

My current postdoctoral work with Prof. Bingjun Xu in the department of Chemical & Biomolecular Engineering at University of Delaware is to develop multicomponent catalysts which mediate a tandem reaction in one-pot. I successfully synthesized a family of zeolite-encapsulated metal catalysts via a novel synthetic approach. The encapsulated catalyst exhibited outstanding catalytic performances, compared to supported metal catalysts, in tandem Aldol condensation and hydrogenation of furfural with acetone. I am also exploring new catalytic system of tandem catalysis mediated by zeolite-encapsulated metal catalysts.

Before starting Ph.D., I had worked for several years in Research and Development (R&D) department of Polymer Industry. During this period, my work aimed at studying and designing polymer properties on a commercial scale via various polymer characterization and processing techniques. This industry experience broadened my perspective towards how polymeric materials should be developed for their use in the real world.

The expertise and experience in the synthesis and characterization of inorganic and organic materials and the heterogeneous catalysis prepared me for my independent research career. I aim to establish a research program on the rational design of advanced nanoporous materials (such as zeolites, metal-organic frameworks (MOFs), porous polymers, etc.) for catalytic applications. My research will be motivated by grand challenges in energy and environmental issues, in particular, shale gas conversion to chemicals, sustainable production of fuels and chemicals, emission control catalysis, CO_2 conversion, and catalysis for waste water treatment.

Selected publications: (22 Total, 632 Citations, h-index 16)

1. **H. J. Cho**, L. Ren, V. Vattipalli, Y.-H. Yeo, N. Gould, B. Xu, R. J. Gorte, R. Lobo, P. J. Dauenhauer, M. Tsapatsis and W. Fan, "Renewable p-xylene from 2,5-dimethylfuran and ethylene using phosphorus-containing zeolite catalysts", *ChemCatChem*, 9, 2017, 398-402

- One of the top 14 most-accessed articles on *ChemCatChem*, October 2016
- Released in the Newspapers *Business Standard* (http://business-standard.com), *Higher Education Tribune* (http://www.highereducationtribune.com) and *Biomass Magazine* (http://biomassmagazine.com)
- Patent (No. PCT/US2017/054558), entitled "Phosphorus-containing solid catalysts and reactions catalyzed thereby, including synthesis of p-xylene"

2. K. Rieger*, **H. J. Cho***, H. Yeung, W. Fan and J. Schiffman, "Antimicrobial activity of silver ions released from zeolites immobilized on cellulose nanofiber mats", *ACS Appl. Mater. & Interfaces*, 8 (5), 2016, 3032-3040 **(* Equally contributed)**
3. C.-C. Chang, **H. J. Cho**, J. Yu, R. J. Gorte, J. Gulbinski, P. Dauenhauer and W. Fan, "Lewis acid zeolites for tandem Diels-Alder cycloaddition and dehydration of biomass-derived dimethylfuran and ethylene to renewable p-xylene", *Green Chem.*, 18, 2016, 1368-1376
4. P. Dornath, **H. J. Cho**, A. Paulsen, P. Dauenhauer and W. Fan, "Efficient mechano-catalytic depolymerization of crystalline cellulose by formation of branched glucan chains", *Green Chem.*, 17 (2), 2015, 769-775

- **Featured on Inside Front Cover**

5. **H. J. Cho**, C.-C. Chang and W. Fan, "Base free, one-pot synthesis of lactic acid from glycerol using a bifunctional Pt/Sn-MFI catalyst", *Green Chem.*, 16 (7), 2014, 3428-3433

- **Featured on Front Cover**
- One of the top 15 most-read articles on *Green Chemistry* from May to July of 2014

6. **H. J. Cho**, P. Dornath and W. Fan, "Synthesis of hierarchical Sn-MFI as Lewis acid catalysts for isomerization of cellulosic sugars", *ACS Catal.*, 4 (6), 2014, 2029-2037
7. **H. J. Cho**, H. S. Jang, S. H. Lim, E. A. Cho, T.-H. Lim, I.-H. Oh, H.-J Kim and J. H. Jang, "Development of a novel decal transfer process for fabrication of high-performance and reliable membrane electrode assemblies for PEMFCs", *Int.J.Hydrogen Energy*, 36, 2011, 12465-12473

Teaching Interests:

I enjoy conveying my knowledge to others as well as exploring new technologies with them. I am comfortable in teaching many of the core chemical engineering courses (Chemical kinetics, Thermodynamics, Mass and Energy balances, and Heat & Mass transfer). I would also be interested in teaching more specialized courses in i) heterogeneous catalysis, ii) nanoporous materials chemistry and engineering, and ii) characterization of nanostructured materials as elective courses at a graduate level.

542380 Advanced Materials for Efficient Energy Conversion Based on Spectroscopic and Mechanistic Study

Abstract:

Research Interests:

Developing catalytic materials and device to enable efficient interconversion of chemical and electrical energies, *e.g.*, fuel cells and electrolyzers, is key to sustainable future. Guiding principles in rational design of electrocatalysts remain elusive. Despite the explosive growth of new materials, few has demonstrated satisfactory performance under realistic conditions in an MEA configuration. Developing advanced materials with enhanced catalytic performance, as well as manipulation of the interfaces between catalysts and membrane, is key to enable efficient energy conversion under realistic conditions in an MEA configuration.

In my PhD research, with Prof. Erkang Wang at Changchun Institute of Applied Chemistry, I developed a selective, active, and stable catalyst for oxygen reduction reaction (ORR), whose sluggish kinetics hinders the commercialization of fuel cells. By decreasing the size of silver nanoparticles to < 2 nm (2 to 5 atoms), the utilization efficiency of catalyst atoms was improved significantly, leading to superior performance as compared to the commercial Pt/C. During my postdoctoral research with Prof. Younan Xia at Georgia Institute of Technology, I developed a type of cost-effective Pt nanocages for ORR with enhanced catalytic activity and durability. Owing to the unique architectures of the synthesized nanoparticles, such as hollow interiors, porous walls, and well-defined facets, the Pt utilization efficiency and specific activity of Pt nanocages were significantly enhanced. During my postdoctoral research with Prof. Bingjun Xu and Prof. Yushan Yan at the University of Delaware, I successfully developed catalysts for electrochemical nitrogen reduction reaction (ENRR) by evaluating them under realistic conditions in an MEA configuration. Further, I have also developed expertise in operando surface enhanced infrared absorption spectroscopy (SEIRAS) to quantify the surface pH changes within the electrochemical double layer using CO_2 reduction reaction as a model system.

In my future research, I plan to develop advanced materials for efficient energy conversion including fuel cell and electrolysis by taking advantage of my unique background in material science, electrolysis, and spectroscopy. Through spectroscopic and fundamental mechanistic studies, the reaction pathway on catalysts can be monitored in MEA configuration, which provides insights for the rational design of nanostructured materials with well-defined structure/composition and manipulation of MEA interfaces to achieve high performance. For example, ORR as the cathode reaction has attracted intensive research interests because of its rate-determining role in fuel cells. Numerous nanomaterials have been demonstrated as catalysts for ORR with enhanced activity and durability in rotating disk electrode (RDE) setup. However, few of them show good performance in fuel cells due to the differences between RDE and MEA configurations. Through operando spectroscopic monitoring of ORR in both RDE and MEA configurations, the interfaces between catalysts and ionomer/membrane can be manipulated, which enables the discovery of advanced materials with enhanced catalytic performance in MEA. With extensive experience in the synthesis of nanomaterials and strong expertise in MEA and operando SEIRAS, I am uniquely suited to undertake these challenging research efforts.

Teaching Interests:

I have extensive experience working with undergraduate and graduate students during my stay at Georgia Institute of Technology and the University of Delaware, and enjoyed teaching them material syntheses, electrochemical measurements, and SEIRAS. I am interested in teaching most of the core Chemical Engineering courses, including thermodynamics and kinetics. I also plan to develop elective courses close to my research interest such as, spectroscopy, catalysis, material sciences, nanoscale fabrication, and electrochemistry.

Selected Publications:

1. Nash, J.*; **Yang, X.***; Anibal, J.; Wang, J.; Yan, Y; Xu, B. Electrochemical Nitrogen Reduction Reaction on Nobel Metal Catalysts in Proton and Hydroxide Exchange Membrane Electrolyzers. Journal of the Electrochemical Society **2017**, 164, F1712–F1716.
2. Dunwell, M.; **Yang, X.**; Setzler, B.; Anibal, J.; Yan, Y.; Xu, B. Examination of Near-Electrode Concentration Gradients and Kinetic Impacts on the Electrochemical Reduction of CO2 using Surface-Enhanced Infrared Spectroscopy. ACS Catalysis **2018**, 8, 3999–4008.
3. **Yang, X.***; Gilroy, K. D.*; Vara, M.; Zhao, M.; Zhou, S.; Xia, Y. Gold icosahedral nanocages: facile synthesis, optical properties, and fragmentation under ultrasonication. Chemical Physics Letters **2017**, 683, 613–618.
4. **Yang, X.**; Roling, L. T.; Vara, M.; Elnabawy, A. O.; Zhao, M.; Hood, Z. D.; Bao, S.; Xia, Y. Synthesis and characterization of Pt–Ag alloy nanocages with enhanced activity and durability towards oxygen reduction. Nano Letters **2016**, 16, 6644–6649.
5. Wang, X.; Figueroa-Cosme, L.; **Yang, X.**; Luo, M.; Liu, J.; Xie, Z.; Xia, Y. Pt-based icosahedral nanocages: Using a combination of {111} facets, twin defects, and ultrathin walls to greatly enhance their activity toward oxygen reduction. Nano Letters **2016**, 16, 1467–1471.
6. **Yang, X.**; Gan, L.; Zhu, C.; Lou, B.; Han, L.; Wang, J.; Wang, E. A dramatic platform for oxygen reduction reaction based on silver nanoclusters. Chemical Communications **2014**, 50, 234–236.
7. **Yang, X.***; Gan, L.*; Han, L.; Wang, E.; Wang, J. High-yield synthesis of silver nanoclusters protected by DNA monomers and DFT prediction of their photoluminescence properties. Angewandte Chemie International Edition **2013**, 52, 2022–2026.

542404 Combining Heterogenous Catalysis and Surface Science. Green Processes and Energy Applications

Abstract:

Research Interests:

My goal as a researcher is to combine heterogenous catalysis and surface science with process intensification to advance fundamental knowledge towards the production of chemicals and fuels, both crucial for our society. I believe that combining experimental results with in situ spectroscopic measurements, and rigorous reaction kinetics analysis will provide fundamental information about catalyst performance and reaction pathways to develop the new chemical processes of the future.

My PhD research focused on the production of biodiesel. The goal was to replace homogenous catalysts (NaOH) by more sustainable heterogeneous catalysts. During this period, I was able to learn and become familiarized with the basic and fundamental tools utilized in the field of catalysis (catalyst preparation and characterization). This work provided me with the background necessary for my next stage as a Postdoctoral researcher at UW-Madison under the supervision of Prof. Dumesic.

During my stay in Prof. Dumesic's group I combined heterogeneous catalysis and process intensification to develop new processes to produce renewable chemicals and fuels. Specially successful was the project to convert gamma-valerolactone (GVL), a biomass derived building block, into butene and high molecular weight olefins to be used as biofuel, which was published in Science and the beginning of a successful journey working with GVL. In the next years, I discovered that GVL can not only be used as platform molecule, but it also can be used as a solvent to improve many chemical reactions. In collaboration with other PhD students, within and outside the group, we demonstrated that GVL can be used as a solvent to produce key platform molecules, such as fermentable sugars, furfural, HMF and levulinic acid with our research being published in Science, Energy and Environmental Science, Angewandte Chemie International Edition and Green Chemistry.

The success of the work led to the publication of 5 patents that were licensed by the start-up company, Glucan Biorenewables LLC, which hired me to commercialize the technology. In the past 5 years, I have divided my research efforts between my work at UW-Madison with Prof. Dumesic and my activities at Glucan Biorenewables LLC as Director of R&D. Working for a company gave me a new perspective for my research. Not only we need to understand the fundamentals of the processes and provide innovative solutions, but these solutions need to make sense economically.

Teaching Interests:

My goal when teaching and mentoring is to create an innovative thinking environment where the students have confidence to develop and to apply their rational and critical thinking skills in problem solving tasks. I believe that to learn, the student needs to be involved in the class and have an active participation.

Nowadays, students are capable to access knowledge by themselves (i.e. internet tutorials, books, youtube videos...) for that reason my classes will focus on demonstrating and teaching how to apply the knowledge acquired to solve chemical engineering problems using a combination of fundamental and practical skill. As mentor of undergraduate and graduate students, lab supervisor and guest lecturer, I understand that a new successful teacher listen and learn from his students. I plan to implement class discussions about the problems solved, so the students can articulate their doubts and suggestions and provide feedback about the teaching methods they find more efficient, so I can improve my skills.

Courses:

My background is in the areas of catalysis, kinetics and process design, and I would enjoy teaching courses in those areas, such as, reaction kinetics and reactor design, introduction to chemical process, and mass and energy balances. At the graduate level, I would like to teach advanced courses focused on heterogeneous catalysis and analytical techniques applied to green processes. Two courses that I want to develop are "Production of chemicals from alternative sources" and "Renewable energies". The first course will focus on chemical processes to produce renewable chemicals.

Funding:

Obtaining funding from Government agencies has been part of my successful activities over the last few years with several proposal funded. This responsibility will also be critical as a Faculty. NSF and DOE are the main funding agencies I will target to fund my research. By my previous interactions with these agencies I have realized that collaboration with other professors and other Universities will be critical to obtain funding for pioneering projects. I would also like to engage companies in research projects. Funding with companies will not only provide additional money for the research, but also a valuable interaction between the students and the companies so they are better prepared for their future work.

Successful proposals as Principal Investigator.

1. **NSF SBIR Phase I:** Green Solvent-Enabled Synthesis of Biobased Furans (ref. 1315356). National Science Foundation (USA). July 2013-December 2013. **$150,000. Principal Investigator**
2. NSF Phase IB: Green Solvent-Enabled Synthesis of Biobased Furans (ref. 1406606) National Science Foundation (USA). January 2014-June 2014**. $29,999. Principal Investigator**
3. **NSF SBIR Phase I**: SBIR Phase I: Catalytic Conversion of Lignocellulosic Biomass into Glucose using Green Solvents (ref. 1519869). National Science Foundation (USA). July 2015-December 2015. **$150,000. Principal Investigator**
4. NSF Phase IB: SBIR Phase IB: Catalytic Conversion of Lignocellulosic Biomass into Glucose using Green Solvents (ref. 1602713). National Science Foundation (USA). January 2016-July 2015. **$30,000. Principal Investigator**
5. SBIR Advance. Matching Grant. Center for Technology Commercialization. UW-Extension. October 2015-September 2016. **$75,000. Principal Investigator**
6. **NSF SBIR Phase II**: SBIR Phase II: Catalytic Conversion of Lignocellulosic Biomass into Furfural and Dissolving Pulp using Green Solvents (ref. 1632394). National Science Foundation (USA). September 2016-August 2018. **$750,000. Principal Investigator**
7. SBIR Advance. Matching Grant. Center for Technology Commercialization. UW-Extension. October 2015-September 2016. **$75,000. Principal Investigator**
8. **DOE SBIR/STTR**: High Purity Cellulose for Low-cost Nanocellulose and Biofuels Production (ref. DE-SC0017939). Department of Energy (USA). June 2017-January 2018 **$150,000. Principal Investigator**
9. **BETO-DOE:** Catalytic Processes for the Production of Alpha Omega Diols from Lignocellulosic Biomass Phase I (ref: DE-EE0006878). Principal award to UW-Madison (Prof. G. Huber). February 2015-January 2017. $2,957,576. **Sub-award Principal Investigator** ($125,000)

Selected publications (total 52):

https://scholar.google.com/citations?user=dQouECQAAAAJ&hl=es

1. **M. Alonso** et al., “Increasing the revenue from lignocellulosic biomass: Maximizing feedstock utilization,” **Science Advances**., vol. 3, no. 5, p. e1603301, May 2017.
2. Q. Bond, **D. M. Alonso**, D. Wang, R. M. West, J. A. Dumesic, "Integrated Catalytic Conversion of gamma-Valerolactone to Liquid Alkenes for Transportation Fuels", **Science**, 2010, 327, 1110-1114.
3. S Luterbacher, Jacqueline M Rand, **D. M. Alonso**, J. Han, J T.Youngquist, C.T Maravelias, B.F. Pfleger, J.A Dumesic, “Nonenzymatic Sugar Production from Biomass Using Biomass-Derived γ-Valerolactone,” **Science** 2014, 343, 6168, 277-280
4. **M. Alonso**, J. Q. Bond, J. A. Dumesic, "Catalytic conversion of biomass to biofuels", **Green Chemistry**, 2010, 12, 1493-1513.
5. **M. Alonso**, S.G. Wettstein, J.A. Dumesic Bimetallic catalysts for upgrading od biomass to fuels and chemicals, **Chemical. Society. Reviews**., 2012, 41, 8075-8098
6. **M. Alonso**, S. G. Wettstein, M. A. Mellmer, E. I. Gurbuz and J. A. Dumesic, “Integrated conversion of hemicellulose and cellulose from lignocellulosic biomass” **Energy & Environmental Science**, 2013, 6 (1), 76-80

541887 Bridging Concepts between Electrochemically and Thermally Activated Catalytic Reactions

Abstract:

Research Interests:

Past and Current Research:

At the atomic level, catalytic reactions involve the making and breaking of chemical bonds between reacting molecules and active sites. This has led to the development of a framework for understanding heterogeneous catalysis in terms of active site-adsorbate interactions. However, as our ability to characterize the complex nature of active sites advances, it has become clear that secondary interactions, resulting from the presence of the supports onto which active sites are dispersed and the solvating media in which the reaction occurs, also strongly impact the reaction energetics. Appreciation of the full complexity of active sites is therefore essential for a full description of a catalytic process, and once understood, can provide additional tools for designing catalysts. Throughout my Ph.D. and postdoc, I have sought to understand the impact of these secondary effects on both electrochemical and thermally activated catalytic reactions.

During my Ph.D. research at the University of California, Berkley with Prof. Alexis Bell, I studied the effect of electrolyte ions on the electrochemical reduction of carbon dioxide. The composition of the electrolyte in which electrocatalysts are tested has been observed to influence reactivity; however, detailed understanding of these effects is still lacking. Through a combination of experimental and theoretical studies, we elucidated the effects of electrolyte cation and anion identity on the activity and selectivity of metal catalysts for this reaction. Kinetic studies, supported by density functional theory calculations, demonstrated that solvated cations interact electrostatically with adsorbed species, having a preferential effect on intermediates with high polarity. Thus, cations in solution act as a built-in promoter with a function analogous to alkali metal promoters for gas phase reactions. Anions can also affect CO_2 reduction rates, acting both as a buffer and a proton source, increasing rates of hydrogenation reactions.

Currently, I am working as a postdoc at the University of California, Santa Barbara with Prof. Phillip Christopher, understanding analogous effects of how the coordination of atomically dispersed platinum catalysts to oxide supports influences activity. Through a combination of atomic level and sample averaged *in-situ* characterization, we demonstrated that the local coordination of the metal to the support determines the ability of the surface site to form additional bonds with adsorbates, altering reactivity.

Future research:

Significant environmental issues associated with the use of fossil fuels, along with advances in technologies to generate renewable electricity have made it likely that future energetic inputs to drive chemical transformations will be electrical in addition to thermal. However, the use of electrocatalytic processes for sustainable fuel and chemical production requires scientific breakthroughs in the design of catalysts and reaction systems to drive these reactions.

Understanding the fundamental concepts that govern reactivity in these systems can enable their rational design and facilitate these breakthroughs. In my research, I plan to establish parallels between well studied thermal catalytic reactions and new electrocatalytic processes to provide design rules for catalyst design and highlight what unique opportunities the electrochemical system provides. I plan to explore the diversity of chemical transformations possible by electrochemical processes, beyond the conversion of small molecules such as water and carbon dioxide. For instance, I plan to study electrochemical processing of biomass derived platform molecules to form fuel additives and polymer precursors, and selective oxidations with application to the conversion of natural gas feedstocks.

Teaching Interests:

I have always had a passion for teaching, both in the classroom and laboratory, and teaching is a major driver in my decision to pursue an academic career. My goal in teaching is to develop a studentâ€™s deep understanding of a few core Chemical Engineering concepts and use them as a guide to approach any problem they are faced with. As an undergraduate I worked as a Chevron Philips Scholar-Mentor for Chemical Engineering Fundamentals. In graduate school, I served as a graduate student instructor (GSI) for undergraduate Thermodynamics and graduate Kinetics and Reaction Engineering courses. I was awarded the departmental Outstanding GSI award for the graduate Kinetics Course. Based on these enjoyable and instructive experiences, I independently developed and taught an elective course *Fundamentals of Electrocatalysis*. As a faculty member, I am excited to teach any of the core courses of Chemical Engineering, particularly courses in Kinetics and Reaction Engineering. I am also eager to develop new courses such as Electrochemical Engineering, or an introductory course on current and emerging energy transformations. These courses would aim to expose students to interesting applications of the foundational principles of thermodynamics, kinetics, and transport phenomena. In the laboratory, I have served as a mentor to undergraduate and junior graduate students. Serving as a mentor and aiding in the development of students in research will be an important focus throughout my career as a faculty member.

541935 Designing Catalysts for Conversion of Alternative Carbon Feedstocks to Fuels and Chemicals

Abstract:

Research Interests: While fuels and chemicals have traditionally been produced from petroleum resources, we face growing challenges including diminishing petroleum reserves and environmentally harmful emissions. These challenges necessitate the development of alternative carbon sources to supply the fuels and chemicals our society needs. Heterogeneous catalysts are a critical component of many chemical processes, transforming lower value feedstocks into higher value products. I am interested in the upgrading of alternative carbon sources to fuels and chemicals using catalysis. In order to design new catalytic technologies to valorize diverse feedstocks, we must gain a deeper understanding of the relevant reaction mechanisms and the relationship between catalyst structure and reactivity.

My PhD research has focused on understanding the mechanisms of biomass conversion to chemicals. Lignocellulosic biomass is an abundant, renewable source of carbon which could be used to produce a variety of oxygenated chemicals, forming the basis for a sustainable chemical industry. Target products include direct replacements for petroleum as well as â€œnewâ€ molecules which cannot be easily derived from petroleum. Co-producing valuable products from biomass could also improve the economics of biofuel production.

Specifically, I have investigated the catalytic conversion of the cellulose-derived intermediate levoglucosenone (LGO) to a variety of valuable chemicals. Relevant reactions include hydrogenation of C=C and C=O bonds over metal catalysts and C-O cleavage over acid catalysts. We used a variety of analytical techniques (HPLC, GC, MS, ^{13}C NMR) in combination with catalyst activity measurements in high-pressure batch and continuous flow reactors, to elucidate the reaction network for these transformations. Mechanisms were additionally studied using ^{13}C radio-labeling and molecular simulations in collaboration with computational chemists. We gained a fundamental understanding of the mechanisms of catalytic conversion of LGO over metal and acid catalysts. Through these insights we have developed new strategies to selectively produce target products with controlled stereochemistry which are being investigated as precursors to bio-based polymers.

Future work should focus on the design of well-defined model catalysts using advanced synthesis techniques to selectively deposit metal oxides onto catalyst surfaces. We propose to design catalysts with i) controlled proximity between different active sites and ii) controlled micropore structure. Ultimately, the goal is to use reaction kinetics measurements over model catalysts to better understand the relationship between catalyst structure and reactivity, leading to control over reaction outcomes.

Teaching Interests: I have numerous teaching experiences including teaching assistantships for two undergraduate courses (Thermodynamics; Materials Science for Chemical Engineers) and a graduate course (Kinetics and Catalysis). I also trained and mentored three undergraduate researchers who made significant contributions to our research efforts.

I am highly involved in informal hands-on STEM education with local elementary and high school students to share my passion for science with the broader community. I look forward to teaching any undergraduate chemical engineering course, especially kinetics and reaction engineering, thermodynamics, and an elective course on alternative energy technologies. I would also be able to teach a graduate course in heterogeneous catalysis.

542138 Catalysts for Sustainable Processes: Understanding and Controlling Active Site Environments

Abstract:

Research Interests:

Producing chemicals and power without affecting the environment will be increasingly important throughout my career. As population and prosperity continue to increase, the environmental impact of greenhouse gas emission, resource extraction, and chemical processing will need to be mitigated. Doing so in economically viable ways will require: **(1)** *natural gas conversion without emitting* CO_2, **(2)** *sequestration of carbon dioxide into chemicals and materials,* and **(3)** *novel processes that take advantage of process intensification.* My research program will address each of these challenges.

Overall Goal: Understand and develop catalysts and processes to produce chemicals, power, and materials in environmentally sustainable ways

The focus of my future research will be to develop high performance heterogeneous catalysts that enable processes for the economic production of chemicals, fuels, and power without producing carbon dioxide. For example, producing solid carbon instead of CO_2 allows for sequestration of the carbon as a useful material, but requires new catalysts (liquid catalysts) to prevent carbon deactivation. Halogens and other indirect oxidants (gaseous catalysts) may also be used as alternatives to oxygen â€" which produces CO_2 - in hydrocarbon dehydrogenation reactions. My specific aims are: **(a)** *understand mechanisms in liquid heterogeneous catalytic processes,* **(b)** *develop new catalysts and processes in which* CO_2 *is consumed or never produced* and **(c)** *model, demonstrate, and analyze new catalytic systems to imagine economically feasible paths to sustainably produce chemicals, materials, and power.*

Scientific Goal: Understand the reaction pathways on the interfaces of gas-liquid and gas-solid catalysts and control the reaction environment of active sites to achieve high performance

The first three projects I intend to work on will be: **(i)** Understanding catalytic melts for the conversion of methane to separable carbon and hydrogen. This will evolve to include the CO_2 reforming of methane and other reactions producing solid carbon. **(ii)** Investigating how catalytic environments affect performance in soft-oxidant (halogens, sulfur, etc.) mediated reactions. This will include molten salt environments for CO_2-free alkane conversion and hydrogen halide reactions, as well as the confined environment of zeolites for methyl halide reactions. **(iii)** Application of (i) and (ii) to chemical looping reactions for integrated chemical conversion and product separation. Over time, these will evolve to include additional reactions and catalytic systems that provide a set of sustainable methods to produce commodity chemicals, fuels, and electrical power from hydrocarbons and carbon dioxide.

PhD Research:

During my PhD with Professors Eric McFarland and Horia Metiu at the University of California, Santa Barbara, I worked on molten metal catalysts[1,2] and molten salt catalysts[3,4] for methane pyrolysis and alkane dehydrogenation. The molten metal catalysts were used in bubble column environments to allow for continuous carbon separation in methane pyrolysis. The molten salt catalyst was used to maintain a neutralizing reaction environment proximate to gas-phase halogen reactions to prevent undesirable hydrogen halide addition. The two catalyst types were used for dehydrogenation reactions to create CO_2-free chemical and power production from natural gas. In collaborations with theorists, we were able to create descriptors for catalytic activity for methane pyrolysis (calculated Bader charge), and model a complete microkinetic pathway for alkane halogenation with mixtures of iodine and bromine. We also completed technoeconomic analyses of methane pyrolysis to assess the impact of different research areas, which lead to focusing on the reaction selectivity.

Postdoctoral Research:

In my work at Stanford University, the confined environment of 8 membered ring pores in mordenite are being used to selectively prevent the formation of intermediates leading to undesirable hydrocarbon formation in the carbonylation of methanol to acetic acid. Modifications of mordenite have resulted in improved selectivity through targeted site passivation. The reaction is part of a novel syngas-to-ethanol process.

Other research:

In addition to the projects mentioned above, I have worked on the conversion of CO_2 to fuels in industry[5] and academia[6]. I directed the catalyst development of a CO_2 reforming company for several years. This included approximately $1,000,000 in lab and benchtop scale testing and IP development.

Teaching Interests:

I look forward to the opportunity to mentor students of all backgrounds as they make choices that will benefit them, those around them, and our society as a whole. I intend to teach each of the core chemical engineering courses eventually, and my research background gives me the ability to teach all levels of kinetics, reaction engineering, physical chemistry, and data analysis courses. My TAing background gives me confidence in teaching chemistry and chemical engineering laboratory courses. My work as a process engineer and chemical engineer gives me the knowledge to teach Separations, Process Design, Chemical Processing, and Material and Energy Balances. Additionally, I would like to develop a course in catalysis.

References

1. **D. Chester Upham,** Alexander Khechfe, Vishal Agarwal, Zachary Snodgrass, Michael Gordon, Horia Metiu, and Eric W. McFarland. *Catalytic molten metals for the direct conversion of methane to hydrogen and separable carbon.* **Science,** 2017 358 (6365) 917-921

2. Brett Parkinson, Mojgan Zavareh, **D. Chester Upham**, Benjamin Ballinger, Simon Smart, and Eric W. McFarland. *Hydrogen production using methane: techno-economics of decarbonizing fuels and chemicals.* **International Journal of Hydrogen Energy**, 2018 43 (5) 2540-2555

3. **D. Chester Upham,** Zachary R. Snodgrass, Mojgan Zavareh, Thomas McConnaughy, Michael Gordon, Horia Metiu, and Eric W. McFarland. *Molten salt chemical looping for reactive separation of HBr in a halogen-based natural gas conversion process.* **Chemical Engineering Sciences**, 2017 160 245-253

4. **D. Chester Upham**, Michael Gordon, Horia Metiu, and Eric W. McFarland. *Halogen mediated oxidative dehydrogenation of propane using iodine or molten lithium iodide.* **Catalysis Letters**, 2016 146 744-754

5. **D. Chester Upham**, Howard Lam-Ho Fong. *Combined steam reforming and dry reforming of hydrocarbons to*

produce synthesis gas. **Carbon Sciences, Inc.** Application number 18832.7.1. US 61/678,498 2013

6. **<u>D. Chester Upham</u>**, Alan R. Derk, Sudanshu Sharma, Horia Metiu, and Eric W. McFarland. *CO2 methanation by Ru-doped ceria: the role of the oxidation state of the surface.* **Catalysis Science & Technology**, 2015 5 (3) 1783-1791

527929 Bimetallic Catalysis for Various Shale Gas and Biomass Conversions

Abstract:

Research Interests: Catalysis Science and Reaction Engineering

Teaching Interests: Chemical Reaction Engineering

To enhance sustainability and environment, it is increasingly attractive to develop alternative resources (e.g. shale gas, biomass) beyond petroleum for production of green fuels and value-added chemicals . Catalysis science and reaction engineering aspects, including catalyst design/synthesis, in-depth understanding of catalyst structure-function relations, reaction kinetics with potential transport phenomena and reactor configuration design, play an important role in these applications. In the present work, we report various bimetallic catalysts for conversion of biomass-derived compounds and shale gas components to alternative fuels and valuable chemicals. It demonstrates that, over these bimetallic catalysts, selectivity to target products may be tuned, and catalyst stability is typically improved. Based on experimental observations and theoretical analysis, structures of these bimetallic catalysts and their relations with catalytic performance are proposed. It is concluded that the primary metal and promoter typically show a cooperative effect, leading to tuning of selectivity and improvement of stability.

531443 Enzymatic Reaction Induced Protocell Motility

Abstract:

Research Interests:

I am a Research Associate in the Department of Chemical and Biomolecular Engineering at the University of Pennsylvania, working under the supervision of Dr. Daniel A. Hammer and Dr. Daeyeon Lee. My research career centers around experimental polymer physics and related applications. Specifically, my current research focuses on the fabrication of microfluidic double emulsion-templated polymeric cellular structures and their biomedical and biological applications. My previous research at Yale University was focused on understanding basic physical principles underlying equilibrium structures and dynamics of diblock copolymer and polymer nanocomposites using synchrotron source small angle x-ray scattering (SAXS) and x-ray photon correlation spectroscopy (XPCS). XPCS uses coherent x-ray beams to study structural dynamics of soft matter on time and length scales in microseconds and nanometers, respectively.

I anticipate my future research direction to conjoin fundamental soft matter physics and biomedical applications. I aim to design and fabricate self-motile microreactors powered by enzymatic or catalytic reactions. To achieve this goal, novel polymer vesicles made from protein-polymer hybrid materials will be developed. When amphiphiles with different chemical constituents are confined within a bilayer membrane, lateral heterogeneities are observed under proper thermodynamic or chemical conditions. In this direction, protein-rich microdomains which can host catalytic or enzymatic reactions will be functionalized. Furthermore, I plan to build artificial cellular network which is biologically functional, including the ability to transport small molecules and communicate inter- and intra-cellularly. Such artificial cellular systems may potentially be applicable for future in vivo studies.

Teaching Interests:

I consider teaching the heaviest responsibility of professorship because the academic preparation of the next generation is the ultimate goal of higher education. Thus, I ardently seek a chance to share my academic accomplishments with young and talented colleagues. In the following, I describe my teaching philosophy, previous teaching experiences within and beyond the academic institutions, and future teaching plans.

Even though the University of Colorado at Boulder and Texas A & M University did not require serving as a Teaching Assistant (TA), I personally sought opportunities on my own to acquire applicable lecturing experience. In the spring of 2003 and 2007, I served as TA and a lab instructor for "Heat Transfer" (University of Colorado at Boulder, MCEN 3022) and â€œDynamic System and Controlâ€ (Texas A & M University, MEEN 364), respectively. Going well beyond traditional TA duties, these positions involved delivering lectures regarding the introduction of each experiment as well as jointly preparing and grading assignments and exams. Additionally, I was also a TA for graduate-level "Solids Mechanics" and "Partial Differential Equations" courses under the supervision of Professors Thomas L. Geers, retired, and Subhendu Datta, retired, respectively. My responsibilities included preparing and grading assignments, holding an hour-long recitation every week and office hours. These TA experiences offered me numerous invaluable insights into teaching from diverse perspectives, especially through interacting with both senior faculty members and students.

I have been actively involved in science and engineering education and outreach both within and beyond the university. I led the annual science and engineering summer program at the University of Pennsylvania in 2014 and 2015. This program offers an opportunity for local high school and undergraduate students to join the graduate-level research. I supervised their experiments and discussed issues regarding the experiments. I also organized the NanoDay demonstration team in 2015; local high school students visited the 'Singh Center for Nanotechnology' and experienced cutting edge researches in science and engineering. I was a member of the preparation committee and my responsibilities included organizing all volunteers and making sure the event ran smoothly. While I sought my PhD degree, I also led a science class for the Texas A & M University Catholic Community, an outreach class geared towards elementary school students and their parents.

Based on my previous teaching and research experiences, I prefer to teach undergraduate and graduate level Thermodynamics, Transport Science, Materials Science including Polymer Physics and Interfacial Science. Additionally, I would be excited to teach data acquisition and control courses since the courses correspond to my particular research background.

I would further be willing to coordinate new courses regarding up-to-date 'Biological Soft Matter' and 'Photon Correlation Spectroscopy.' Biological Soft Matter course will be designed for senior and graduate students. In this course, I will use a physics based approach to study the structure and assembly of a variety materials including polymers, colloidal systems, amphiphiles, and liquid crystal. Rheological behaviors including temperature-dependent phase transition and self assemblies, kinetics, and interfacial interaction will be major topics.

Photon Correlation Spectroscopy course will cover the static and dynamic techniques including small angle x-ray scattering, multi-speckle dynamic light scattering, and x-ray photon correlation spectroscopy. These techniques can capture the static structure and the dynamic behavior of soft matter. The observed microscopic structural behaviors are governed by complex phenomenological laws that couple multiple energy domains. Thus, understanding these structural behaviors would consequently reveal macroscopic material properties and foster developments of new applications.

533193 Environmental Applications of Al/Zr Pillared Clay As Efficient Heterogeneous Catalyst for Catalytic Wet Oxidation of Phenol

Abstract:

Research Interests: Environmental Catalysis

Natural bentonite clay was used in the synthesis of single and mixed oxide Al/Zr pillared clay catalysts by means of ultrasonic treatment during the aging and intercalation steps. The raw bentonite and pillared clay catalysts were characterized using high resolution scanning electron microscopy with energy dispersive system (HRSEM-EDS), powder X-ray diffraction (PXRD), N_2 adsorption/desorption, Fourier transform infrared spectroscopy (FTIR), thermogravimetric analysis (TGA), cation exchange capacity (CEC), zeta potential and all the catalysts evaluated in the catalytic wet air oxidation (CWAO) of phenol. Calcination of the synthesized Al/Zr-PILC at 500 °C for 120 minutes yielded optimum physical properties with BET surface area of 230 m^2/g, pore volume of 0.115 cm^3/g and basal spacing of 1.92 nm and with good thermal stability. Complete oxidation of phenol in aqueous solution using the Al/Zr-PILC catalyst under mild reaction conditions of 100 °C and 10 bars occurred after 120 minutes reaction time. The process obeyed the first-order kinetics by assuming a constant concentration of oxygen. Catalyst reusability studies showed that the Al/Zr-PILC catalyst demonstrated excellent chemical and structural stability with insignificant leaching of Zr^{4+} after six consecutive catalytic runs. The Al/Zr-PILC catalyst therefore demonstrated excellent catalytic properties in the CWAO of phenol in aqueous solution.

Teaching Interests: Material Science and Environmental Catalysis

537349 Electrochemical Strategies for Sustainable Energy Technologies

Abstract:

Electrochemical Strategies for Sustainable Energy Technologies

Joshua M. McEnaney, Postdoctoral Scholar, Stanford University

Research Interests:

Among the biggest challenges facing humanity is the transition to sustainable energy and chemical production processes. The development of clean technologies for energy production, energy storage, and conversion of undesirable chemicals into useful products can mitigate societal dependence on fossil fuels and sustainably provide for societal needs. Electrochemical technologies (i.e. advanced electrolyzers, fuel cells, batteries) are particularly well suited to provide these clean processes as renewable electricity costs are rapidly declining and have already become the cheapest sources of new electricity in many parts of the world. With the ultimate goal of transforming the global energy landscape, the development of materials and methods to directly use renewable electricity is imperative.

It is my career goal to lead and manage a research group at an academic institution that works on research to enable sustainable energy technologies. I am most interested in developing research in electrochemical areas with respect to catalysis, energy transformations, and chemical production processes. Reactions of interest include nitrogen reduction, nitrate reduction, carbon dioxide/monoxide reduction, water electrolysis, oxygen reduction, hydrogen oxidation, water desalination, and hydrogen peroxide production. In the short term, I intend to focus my research group on two major thrusts *1) Electrochemical Nitrogen Reduction* including both dinitrogen and nitrate reduction to form ammonia fertilizer and other N-based products, and *2) Electrochemical Carbon Oxide Reduction* including both CO_2 and CO reduction to fuel and chemical products as a form of artificial photosynthesis. An enveloping theme of the group will be to leverage fundamental electrocatalysis, electrochemistry, materials science, and surface interface science to develop highly active and selective electrochemical systems for these and other reactions. Specific research focus areas we will use to achieve this include targeted electrocatalyst development and characterization, electrolyte engineering, application of unique and advantageous experimental conditions (i.e. pressure, temperature, gas flow, product separations, multi-step processes), and the development and integration of device designs. Over time, we will expand to new electrochemical reactions and related clean energy technologies as the group grows. The overarching goals of the group will be to innovate, understand, and develop new materials solutions and clean energy technologies addressing pertinent energy and environmental challenges facing the world as well as to train the next generation of scientists and engineers to be equipped as creative experts in this important and growing science and engineering field. We will actively foster academic, industrial, and government based collaborations to create an interdisciplinary and highly informed learning environment which will accelerate our opportunities for success at the frontier of sustainable energy.

Research Experience:

Stanford University - Postdoctoral Research Assistant; SUNCAT; Chemical Engineering

Advisor: Thomas Jaramillo — *2015-Present*

- Developed electrochemical and sustainable ammonia production from N_2 and NO_3^-
- Developed methods and materials for selective CO_2 reduction, oxygen reduction, and other sustainable energy relevant reactions spanning materials synthesis and characterization, electrocatalysis, thermocatalysis, and photoelectrochemistry
- Designed and purchased infrastructure ($250K) for new research projects

The Pennsylvania State University - Graduate Research Assistant; CCI Solar Fuels; Chemistry

Advisor: Raymond Schaak — *2011-2015*

- Developed new catalysts for renewable energy applications such as the H_2 evolution reaction and CO_2 reduction reaction using earth-abundant materials
- Introduced and developed novel nanoparticle synthesis methods

SUNY College of Environmental Science and Forestry - UGrad Research Assistant; Chemistry

Advisor: Neal Abrams — *2008-2011*

- Studied solid state synthesis and characterization of photocatalytic materials
- Developed electrochemical educational labs for photovoltaic materials syntheses
- Helped start up new research lab

Undergraduate Research Assistant; Research Foundation of SUNY; Chemistry

Co-Advisors: Neal Abrams, Kelley Donaghy, and Christopher Nomura — *2010*

- Developed new techniques for photocatalytic, biomimetic materials synthesis and characterization
- Developed inorganic chemistry teaching demonstrations
- Studied genetic engineering of E. coli for biodegradable polymer production

Teaching Interests:

Throughout my academic studies, I have been enthusiastically involved as a teaching assistant, a class tutor, a workshop leader, a personal tutor, a research mentor, a conference presenter, a guest lecturer, and the president of a chemical outreach club. These experiences have given me significant teaching and mentoring opportunities across a wide array of audience types and sizes. I have learned that it is important to be well organized and strategic in these teaching scenarios to make sure that the lesson content is received and understood. I believe that classroom and laboratory learning can be strongly enhanced by direct mentoring and engagement, both by myself and by trained student mentors and TAs to reinforce important concepts of the course. It is my goal as a teacher to develop an organized and engaging learning environment that prepares students for the workforce. To become a top caliber teacher, I plan to continuously develop and improve my teaching methods to better engage and educate my students and student researchers.

Select Publications:

¨ S.Andersen, V. Čolić, S. Yang, J. A. Schwalbe, A. C. Nielander, **J. M. McEnaney**, K. Enemark-Rasmussen, J. Baker, A. R. Singh, B. A. Rohr, M. J. Statt, S. J. Blair, S. Mezzavilla, J. Kibsgaard, P. Vesborg, M. Cargnello, S. Bent, T. F. Jaramillo, I. Stephens, J. K. Nørskov, I. Chorkendorff, Assessing the Current State of Catalyst Development for the Electrochemical Reduction of N_2 to NH_3, *Nature*, submitted manuscript **2018**

¨ **J. M. McEnaney**, A. R. Singh, J. A. Schwalbe, J. Kibsgaard, J. C. Lin, M. Cargnello, T. F. Jaramillo, J. K. Nørskov, Ammonia Synthesis from N_2 and H_2O Using a Lithium Cycling Electrification Strategy at Atmospheric Pressure, *Energy Environ. Sci.*, **2017**

DOI: 10.1039/C7EE01126A

¨ P. Chakthranont, T. R. Hellstern, **J. M. McEnaney**, T. F. Jaramillo, Design and Fabrication of a Precious Metal-Free Tandem Core–Shell p^+n Si/W-Doped $BiVO_4$ Photoanode for Unassisted Water Splitting, *Adv. Energy Mater.*, **2017**

DOI: 10.1002/aenm.201701515

¨ L. V. P. Mendes*, J. L. Snider*, S. D. Fleischman, J. Kibsgaard, **J M. McEnaney**, D. A. G. Aranda, T. F. Jaramillo, Polyol Synthesis of Cobalt–Copper Alloy Catalysts for Higher Alcohol Synthesis from Syngas, *Catal. Lett.*, **2017**

DOI 10.1007/s10562-017-2130-5

¨ **J. M. McEnaney**, T. L. Soucy, J. M. Hodges, J. F. Callejas, J. S. Mondschein, R. E. Schaak, Colloidally-Synthesized Cobalt Molybdenum Nanoparticles as Active and Stable Electrocatalysts for the Hydrogen Evolution Reaction Under Alkaline Conditions. *J. Mater. Chem. A*, **2016**

DOI: 10.1039/C5TA07055D

¨ J. F. Callejas, C. G. Read, E. J. Popczun, **J. M. McEnaney**, R. E. Schaak, Nanostructured Co_2P Electrocatalyst for the Hydrogen Evolution Reaction and Direct Comparison with Morphologically Equivalent CoP. *Chem. Mater.*, **2015**

DOI: 10.1021/acs.chemmater.5b01284

¨ E. J. Popczun, C. W. Roske, C. G. Read, J. C. Crompton, **J. M. McEnaney**, N. S. Lewis, R. E. Schaak, Highly Branched Cobalt Phosphide Nanostructures for Hydrogen-Evolution Electrocatalysis. *J. Mater. Chem. A*, **2015**

DOI: 10.1039/C4TA06642A

¨ **J. M. McEnaney**, R. E. Schaak, Solution Synthesis of Metal Silicide Nanoparticles. *Inorg. Chem.*, **2015**

DOI: 10.1021/ic502394u

¨ J. F. Callejas*, **J. M. McEnaney***, C. G. Read*, J. C. Crompton, A. J. Biacchi, E. J. Popczun, T. R. Gordon, N. S. Lewis, R. E. Schaak, Electrocatalytic and Photocatalytic Hydrogen Production from Acidic and Neutral-pH Aqueous Solutions Using Iron Phosphide Nanoparticles. *ACS Nano*, **2014**

DOI: 10.1021/nn5048553

* Note: These authors contributed equally.

¨ **J. M. McEnaney**, J. C. Crompton, J. F. Callejas, E. J. Popczun, C. G. Read, N. S. Lewis, R. E. Schaak, Electrocatalytic Hydrogen Evolution Using Amorphous Tungsten Phosphide Nanoparticles. *Chem. Commun.*, **2014**

DOI: 10.1039/C4CC04709E

¨ **J. M. McEnaney**, J. C. Crompton, J. F. Callejas, E. J. Popczun, A. J. Biacchi, N. S. Lewis, R. E. Schaak, Amorphous Molybdenum Phosphide Nanoparticles for Electrocatalytic Hydrogen Evolution. *Chem. Mater.*, **2014**

DOI: 10.1021/cm502035s

539492 High-Throughput Catalysts Screening of Layered Double Hydroxides for Oxygen Evolution and Reduction Reactions

Abstract:

Research Interests:

1. Computational materials science for catalysts screening
2. Machine learning algorithm for descriptor screening and designing
3. Collaboration with experimental groups to provide theoretical evaluations for electrochemical, electro-optical and other functional applications.
4. DFT functionals and tools development for scientific purposes

Teaching Interests:

- **Fundamentals of Materials Science and Engineering**
- **Computational Materials Science**
- **Heterogeneous Catalysis**
- **Sustainable Energy Technology**
- **Electrochemistry**

HETEROGENEOUS CATALYSTS DEVELOPMENT FOR BENZENE SATURATION IN DIESEL

Shyamal Roy*, Dilip Kumar Mondal, Sayantan Ghosh, Swadhin Chatterjee, Soumyodeep Chaudhuri, Priya Ranjan, Pritam Kumar

Department of Chemical Engineering, Jadavpur University, Kolkata-70032, West Bengal, India

Abstract: A faster growth in the demand for diesel fuels relative to gasoline, together with increasing quality requirements and availability of poorer grade raw materials, has created a need for improved dearomatization processes. An improved knowledge of chemical reactions, kinetic models, and rigorous reactor models is essential to new dearomatization process development and optimization. The major drawback of conventional hydrotreating catalysts includes severe operating conditions such as high temperature, high pressure, low space velocity, and need for excessive hydrogen to feed mole ratios to achieve acceptable aromatic reduction. Therefore, noble-metal based catalysts are preferred for aromatic saturation since they can operate at lower thermochemical severities, thus avoiding the thermodynamic constraints. This work investigated the hydrogenation of benzene using several meso-structured silica supported (HMS, SBA-15, MCM-41) platinum and palladium catalysts in an up-flow catalytic fixed bed reactor. Supports and catalysts were characterized by XRD, SEM, BET surface area analyzer and CO chemisorption techniques and the highest conversion of benzene to cyclohexane was observed 100% for SBA-15 supported Pt-Pd catalyst, 95% for HMS supported Pt-Pd catalyst and 85% for MCM-41 supported Pt-Pd catalyst.

Keywords: Pt-Pd catalyst, HMS, SBA-15, MCM-41 supports, benzene

1. Introduction

More attention focused to catalytic hydrogenation in the oil refining industry is due in part to legislation regarding the maximum contents of sulfur, aromatic compounds, and alkenes in traffic

Email: shyamal.roy@jadavpuruniversity.in

fuels.[1, 2] Aromatics in diesel increase the particle emissions in gases [3] and they have the further effect of lowering the fuel quality. [2] Alkenes in gasoline are undesired in larger amounts because of higher NOx and CO formation and the higher reactivity of evaporative emissions in comparison to their saturated counterparts.[4] Tightening legislation in regards to vehicle exhaust emissions has forced the industry to improve motors and catalytic converters and also the quality of fuels. Lead free and low-sulfur fuels have been introduced and CO, SO*x*, and NO*x* emissions have been significantly reduced. The major concern remaining is particle emissions.[5] Particles are reported[6] to correspond to aromatic compounds, and the reduction of these is the next target. The faster growth in the demand for diesel fuel relative to gasoline, together with today's higher quality requirements and lower grade raw materials, creates a need for improved dearomatization processes.[7] Better knowledge of reactions and rigorous reactor and kinetic models is essential to new process development and optimization. Gas-phase dearomatization has been widely studied, both on metal sulfides (Co/Ni and Mo/W) and on group VIII metals (Pt, Pd, Ni, and Ru).[8] Publications on liquid phase reaction kinetics, in turn, are few,[7, 9] even though industrial dearomatization takes place in the liquid phase in three-phase reactors.

The drawbacks of the conventional hydrotreating catalysts include severe operating conditions such as high temperature, high pressure, low space velocity, and hydrogen to feed mole ratios to achieve acceptable aromatic reduction. The hydrodearomatization reaction is reversible and at normal hydrotreating conditions, complete conversion is not possible because of equilibrium limitations. Furthermore, conventional hydrotreating catalysts comprising sulfided mixed oxides (Ni-Mo, Ni-W, Co-Mo) can only accomplish moderate levels of aromatic saturation under typical hydrotreating conditions in a single-stage operation.[10, 11] Efficient approaches towards performance improvement of the catalysts used in such processes include: changing the active phase (carbide, nitride, phosphide, etc.), using a different catalyst preparation method (co-precipitation, chemical vapour deposition, etc.) or applying new types of catalyst supports (CNTs, mesoporous silicas such as Santa Barbara Amorphous, SBA-15). These ordered mesoporous materials are considered ideal hosts for nanoparticles due to their large surface area and well defined pore structure. Most importantly, their pore sizes can be tuned within the nanometer range and the pore surface can be functionalised with various metal oxides and organic groups. Among these materials, silica HMS, SBA-15 and MCM-41 have been extensively studied due to their 2D structure with interconnected pores.[12] HMS, SBA-15 and MCM-41 are a mesoporous silica molecular sieves with a high surface area, tunable uniform hexagonal channels ranging from 5 to 30 nm and thick framework walls (3–6 nm), which is characteristic for it being thermally and hydro-thermally robust.[13, 14] Such attractive textural properties of HMS, SBA-15 and MCM-41 makes it a potential alternative to the commonly used γ-Al_2O_3 support due to the fact that their pore size can be manipulated via controlling the synthesis parameters, while maintaining a relatively high surface area.

Noble-metal based catalysts are preferred for aromatic saturation since they can work at lower temperatures, thus avoiding the thermodynamic constraints encountered with the sulfided oxides. However, noble-metal hydrogenation catalysts are sensitive and get poisoned by small amounts of sulfur and nitrogen containing hetero organic compounds normally present in the feed.[15] It is reported that high intrinsic activity and sulfur tolerance may be enhanced by deposition of noble metal on acidic supports. Recently, Song [16] proposed a design concept for unique acidic zeolites as supports for noble metals that utilizes shape-selective exclusion, hydrogen spillover, and two types of sulfur resistance. Such zeolite supports can be used to prepare bimodal distributions of noble metal particles. Pt-Pd bimetallic supported on large pore *β*-zeolites have been claimed [17] to display dearomatization/cetane improvement. Lee and Rhee [18] recently reported a simultaneous decrease in the metal dispersion and hydrogenation activity of this catalyst, induced by platinum sulfide formation followed by coke deposition. The large pore zeolites preferably allow fast diffusion and reaction of bulky aromatic compounds. It is known that noble metals deposited on acidic supports show higher turnover frequencies (TOF) compared to those supported on nonacidic supports. This has been related either to the polarization of the metal particles by nearby cations or to a metal-support interaction, which might induce a partial electron transfer between the metal and the oxide ions [19] or between the metal and the zeolite protons.[20, 21] In the latter case, the formation of the more reactive metal-proton adducts $[M_n\text{-}H_z]^{z+}$ has been proposed.[20, 21] The positive effect of the acidity on the noble metal sulfur resistance was explained by the changes in the metal electron deficiency induced by metal/support interactions.[22-24] This effect leads to the formation of electron-deficient metal sites, which, in turn, lowers the strength of the sulfur-metal bond.[21, 22] However, high acidity of zeolite is also known to favor excessive cracking and coke formation. In order to avoid it without suppressing the sulfur tolerance of catalysts, less acidic supports such as silica-alumina or non acidic supports has been recently investigated. For example, Reinhoudt et al.[25] suggested the superior performance of silica-alumina supported Pt-Pd catalyst for second-stage deep hydrodesulfurization process. In contrast, Fujikawa et al.[26] reported that bimetallic Pt-Pd/silica-alumina does not improve sulfur tolerance as compared to the monometallic Pt or Pd catalysts although it significantly enhances the intrinsic activity for hydrodearomatization. Miller et al.[27] investigated the potential for developing nonacidic sulfur tolerant catalyst. Sidhpuria et al.[28] reported rhodium supported on Hβ zeolite catalyst for toluene and benzene hydrogenation in the literature. Recently, Boahene et al.[29] invesitigated hydrotrating of coker light gas oil using Ni/HPMo catalysts.

In the present study noble metals Pt, Pd supported on HMS, SBA-15 and MCM-41 and Ti-SBA-15 have been explored as a hydrodearomatization catalyst. Benzene was used as a model compound to simulate the aromatics in diesel fuels, because the hydrogenation of benzene is known to be more difficult than that of toluene and naphthalene.[29] Hydrogenation activity of benzene over Pt-Pd

/HMS, Pt-Pd/MCM-41 , and Pt-Pd/ SBA-15 and Pt-Pd / Ti-SBA-15 catalysts was studied in a fixed bed reactor at varied hydrogen to toluene mole ratio, temperature, pressure and toluene feed WHSV. A modification of SBA-15 silica was made by frame modification with Ti^{+4} ions. Ti-containing SBA-15 sample was prepared using Ti-isopropaxide as Ti precursor, tri-block copolymer, poly (ethylene glycol)-poly (propylene glycol)-poly (ethylene glycol) (PEG-PPG-PEG) (Aldrich), as the surfactant and tetraehoxysilane (TEOS) as the silica source. Thus, the aim of this work was to study the catalyst activity using Pt-Pd /HMS, Pt-Pd/MCM-41, Pt-Pd/ SBA-15 and Pt-Pd/ Ti-SBA-15 catalysts and compare the catalyst activities.

2. Experimental

2.1. Synthesis of SBA-15 and Ti-SBA-15

HMS and MCM-41 (Merck) were purchased from market. Preparation of SBA-15 and Ti-SBA-15 supports used the following chemicals: tetraethylorthosilicate (TEOS, 98%, Aldrich), tri-block copolymer, poly (ethylene glycol)-poly (propylene glycol)-poly (ethylene glycol) (PEG-PPG-PEG) (Aldrich), isopropanol (Aldrich, >99.5%), titanium isopropaxide (98+%, Acros), tetraamine platinum nitrate (99.9%, Aldrich) and tetraamine palladium (II) nitrate (10 wt% solution in water, Aldrich). Titanium was incorporated into the mesoporous SBA-15 material by direct hydrothermal synthesis routes.

The preparation of mesoporous silica SBA-15 was carried out according to the method of Zhao et al.[13] In the synthesis procedure, 16 g amphiphilic tri-block copolymer, poly (ethylene glycol)-poly (propylene glycol)-poly (ethylene glycol) was first dissolved in 120 ml of water and 480 ml of 2(N) HCl (37 wt %, Merck) solution at 40° C for 2 h under continuous stirring. After addition of 36.4 ml of TEOS the reaction mixture was stirred first at 40°C for 24 h and then at 80°C for 48 h in a Teflon-lined steel autoclave. The resulting solid was filtered, washed with deionized water, dried at

100°C for overnight and calcined in air at 550°C for 6 h at a heating rate of 2°C/min to decompose the tri-block copolymer.

Ti-SBA-15 material with Si/Ti molar ratio 20 was synthesized hydrothermally.[30] TEOS and titanium isopropaxide were been used as silicon and titanium sources, respectively. The tri-block copolymer (PEG-PPG-PEG) was used as the structure-directing agent, whereas concentrated 2(N)HCl was used to adjust the pH. The synthesis was carried out similarly as for pure SBA-15, but with the Ti source added after the introduction of TEOS. The titanium isopropaxide was pre-dissolved in a solution of 2-propanol (99.7 %, Merck). The resulting samples were washed with deionized water and dried overnight at 100°C. Once dry, the resulted white powder was calcined at 550°C for 6 h.

2.2. Synthesis of Pt-Pd /HMS, Pt-Pd/MCM-41 and Pt-Pd/SBA-15 and Pt-Pd / Ti-SBA-15 catalysts

Pt (0.15 wt %) and Pd (0.15 wt %) metal precursors impregnated on HMS, MCM-41, SBA-15 and Ti-SBA-15 supports by excess solution impregnation (ESI) method. In a typical procedure, the calculated quantity of tetraamine platinum nitrate and tetraamine palladium (II) nitrate were dissolved in 100 ml of double distilled water, which was stirred for 30 min at pH ca.9.5, followed by the slow addition of support to the solution. The solution was kept for 4 h at room temperature under stirring conditions. The solvent was evaporated on a rotary vacuum evaporator and then the catalysts were calcined at 673 K for 4 h in an oxygen flow of 2 ml/min.

2.3. Catalyst activity measurements

Hydrogenation of benzene was performed in a continuous-up flow stainless steel catalytic fixed bed reactor (Metal: 316 SS, Diameter: 1.5 cm, Length: 50 cm) with the ability to operate up to 150 bar pressure and 923 K temperature, equipped with high pressure piston pump (Knauer, Germany). The schematic diagram is shown in Figure 1. The catalyst extrudates (Diameter: 1.5 mm, Length: 3 mm) were kept at the centre of the reactor throughout the kinetic study.

The bulk density of the catalyst was 0.785 g / cc and exact length of the catalytic bed was 15 cm when the amount of catalyst is 10 g. The remaining space of the catalyst bed at the top and the bottom of reactor were filled with glass beads (2-3 mm diameter) to avoid the entrance and exit effects. Catalyst was dried at 403 K for 48 h and reduced in situ by hydrogen at 553 K for 6 h.

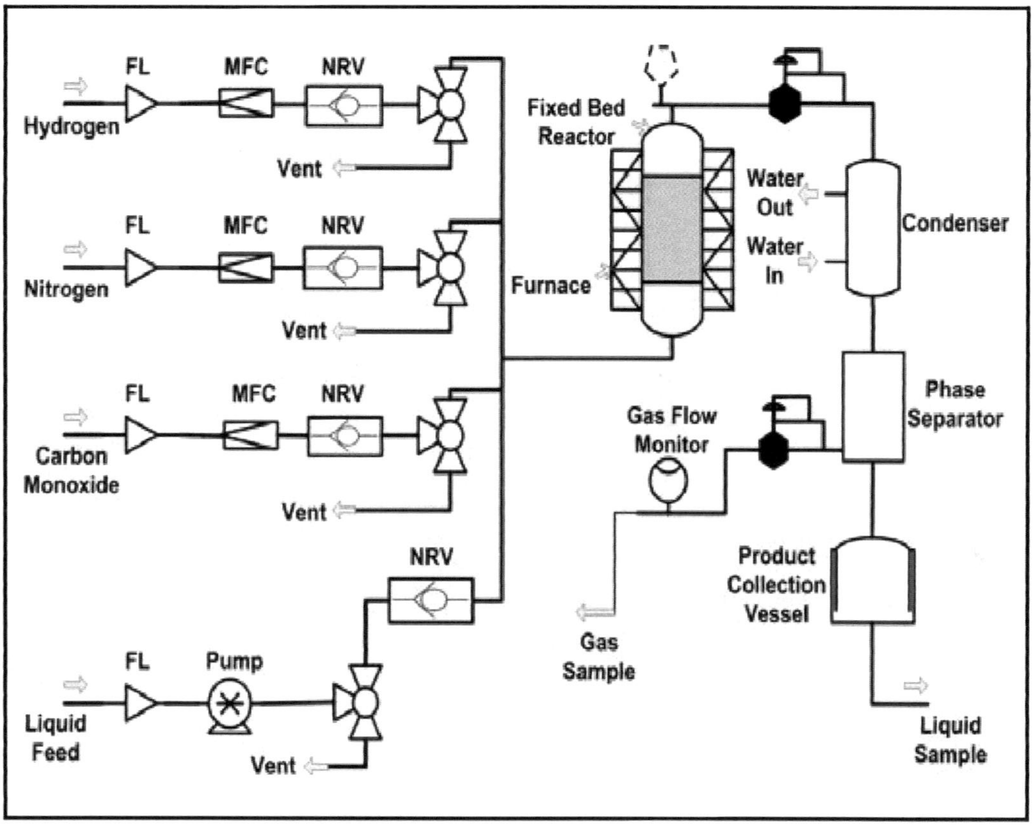

Figure 1. Schematic diagram of the catalytic fixed bed reactor

The reactor was brought to the desired temperature before the reactant was introduced. Benzene hydrogenation was carried out at different temperature (423-498 K), hydrogen pressure (5-35 bar), hydrogen to benzene mole ratio (2-12) and feed WHSV (1-6 h^{-1}). Liquid sample was withdrawn from the sample-collecting vessel and analyzed by gas chromatograph (Agilent Technologies 6890 N, USA) equipped with a flame ionization detector (FID) and DB-Petro, 122-10A6 column (100 m length, 0.25 diameter and coated with 0.50 μm thick stationary phase of 5 % diphenyl and 95 %

dimethyl siloxane) at appropriate times. The initial GC column temperature was increased from 313 to 473 K at the rate of 2 K / min. Helium gas (3.5 mL / min) was used as a carrier gas.

The temperature of injection port and FID were kept constant at 473 K during product analysis. The retention times for different compounds were determined by injecting pure compounds under identical GC conditions. Activity is described in terms hydrogenation of benzene using a specific reaction rate according to Equation (1):

$$r_i = \frac{X_i F}{W_{cat}} \quad (1)$$

where r_i is the specific rate (mol / g. h), X_i the conversion of reactant i (i =benzene), F is the molar flow rate of the reactant i (mol / h), and W is the catalyst weight (g).

Benzene conversion at the end of the 45 min of operation was taken as a steady state conversion. The reactor exhibited no temperature or pressure gradients across the catalyst bed and there are no diffusional limitations for the transport of reactant and product molecules, in this condition.[31]

Hydrogenation of benzene was observed to occur readily within studied temperature range and the selectivity for lcyclohexane was found to be between 25 and 100%. No other products were also observed due to isomerization and cracking reactions. Experimental runs were repeated 5 times and the results compared with those of the previous runs in order to ascertain the reproducibility of the data obtained and the error was found to be ± 2%.

2.4. Characterization

2.4.1. Measurement of N_2 adsorption-desorption isotherms

The BET surface area, pore volume and pore size distribution of the samples were measured with an Autosorb-1C instrument using low temperature N_2 adsorption- desorption isotherms. Before measuring, the sample was degassed in vacuum at 200°C. The surface area was computed from these isotherms using the multipoint Brunauer-Emmett-Teller (BET) method based on the adsorption data in the partial pressure P/P_0 range from 0.01 to 0.2. The value of 0.1620 nm^2 was taken for the cross-section of the physically adsorbed N_2 molecule. The mesopore volume was determined from the N_2 adsorbed at a $P/P_0 = 0.4$. The total pore volume was calculated from the amount of nitrogen adsorbed at $P/P_0 = 0.95$, assuming that adsorption on the external surface was negligible compared with adsorption in pores. The pore diameter and pore volume were determined using the Barret-Joyner-Halenda (BJH) method. In all cases, correlation coefficients above 0.999 were obtained.

2.4.2. CO Chemisorptions

Volumetric CO chemisorptions isotherms at 298 K were recorded separately in order to estimate metal dispersion using an Autosorb-1C instrument.

Table 1. Chemical Compositions,[a] Textural Properties[b] of SBA-15 and Ti-SBA-15 Supported Pt–Pd Catalysts, Average Metal Particle Size,[c] Metal Dispersion,[d] Acidity[e] of Reduced Pt–Pd/SBA-15 and Pt–Pd/Ti-SBA-15 Catalysts.

Catalyst/ support	BET (m^2/g)	Total pore vol (cm^3/g)	Pore dia (nm)	Metal dispersion (%)	Metal particle size (nm)	Acidity using NH_3 (μmol/g)	Pt (wt %)	Pd (wt %)	TiO_2 (wt %)
SBA-15	747	1.488	5–8.6			155.75			
Ti-SBA-15	630	0.88	5–8.7			325.55	-	-	1.85
Pt–Pd/SBA-15_{cal}	560	0.845	5–8.3	88	1.3	340.58	0.145	0.14	
Pt-Pd/Ti-SBA-15_{cal}	590	0.856	5.5–8.65	96	1.4	355.75	0.147	0.145	1.83
Pt–Pd/SBA-15_{spent}	440	0.742	5–7.9		1.7				
Pt–Pd/Ti-SBA-15_{spent}	575	0.822	5–8.58		1.5				

[a]Chemical compositions of the catalysts (Pt, Pd) were determined by inductively coupled plasma atomic emission spectroscopy (ICP-AES)
[b]BET specific area, pore volume, and average pore diameter as measured by N_2 adsorption–desorption isotherms at 77 K.
[c]Average metal particle size of reduced at 573 K samples as measured by CO chemisorptions.
[d]Metal dispersion using CO chemisorptions.
[e]Acidity of reduced samples and bare carriers as measured by TPD of adsorbed NH_3.

For CO chemisorptions catalyst was pretreated and reduced by H_2 as for the catalytic experiments and then out gassed at 9 $\times 10^{-6}$ Torr. After cooling the sample to 303 K, CO was admitted at pressures below 20 Torr. Since in the region of low pressures, the CO isotherm can be described by the Langmuir equation, only one isotherm was recorded. The extrapolation of its linear part to zero pressure gives the amount of CO adsorbed in monolayer. Metal particle size was estimated assuming CO: Pt (Pd) = 1:1 surface stoichiometry. In the bimetallic catalysts, platinum was assumed to behave like palladium and metal dispersion was calculated from the adsorption branch of isotherm by employing the strong/difference extrapolation formalism showed maximum number of active catalyst sites about 96 % for Pt (0.15 wt %)-Pd (0.15 wt %) / Ti-SBA-15 catalyst (Table 1).

2.4.3. X-ray Diffraction Analysis

The broad-angle X-ray diffraction (XRD) patterns of the samples were measured using a Bruker D8 Advance Powder diffractometer with a Gemonochromator producing a monochromatic Cu Kα radiation. The scanning was made from 5 to 80° with a scan rate of 2°/min. In all cases, the generator was operated at 40 kV and 30 mA. To avoid the problem of illuminated areas, all the samples were measured using the same sample holder. The wide-angle X-ray diffraction patterns of the SBA-15, Pt-Pd/SBA-15, and Pt-Pd/Ti-SBA-15 catalyst samples exhibited a broad line between the 2θ values of 20-30°, which was attributed to the siliceous amorphous SiO_2 phase.[32] No reflections belonging to TiO_2 were observed, indicating the homogeneous dispersion of TiO_2 in the framework of mesoporous silica material.[32]

2.4.4. Energy Dispersive X-ray Analysis

Quantitative compositional analysis was carried out with an energy-dispersive X-ray analysis (EDX) system attached to the electron microscope, which was operated at 25 kV. Determination of the chemical composition was based on the average analytical data of individual particles and the result of EDX analyses of HMS, MCM-41, SBA-15, Ti- SBA-15 support and supported Pt- Pt-Pd catalysts in the form of energy peaks at different applied voltages showed the presence of silica (Si), platinum (Pt), palladium (Pd), titanium (Ti), and oxygen in the catalyst and no nitrogen was found. This indicates that during calcinations all the nitrate decomposed completely to their respective state.

3. Result and Discussions

3.1. Effect of hydrogen to benzene mole ratio on conversion

The effect of the hydrogen to benzene mole ratio in the range of 2-12 was studied at 498 K (Figure 2). Benzene conversion increased up to a hydrogen/benzene mole ratio of 10 and then remained constant.

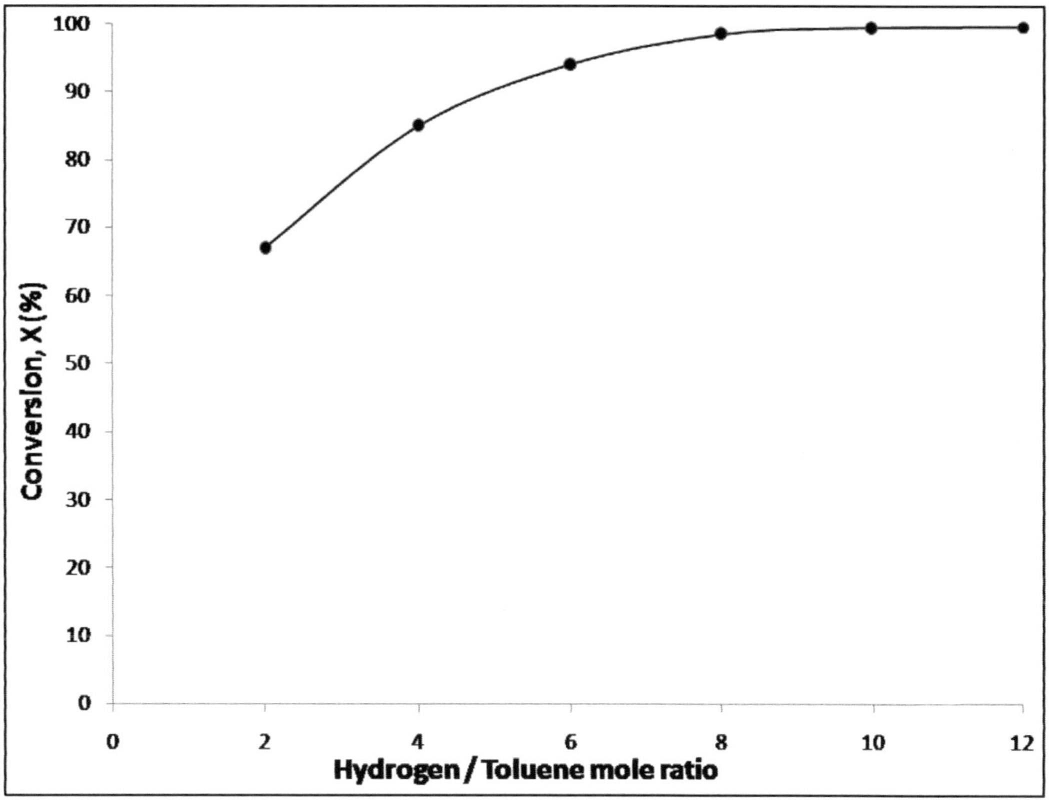

Figure 2. Effect of hydrogen to benzene mole ratio on conversion at T:498 K, P:15 bar, WHSV: $2h^{-1}$, Catalyst: Pt (0.15 wt %)–Pd (0.15 wt %)/Ti-SB-15.

Benzene conversion is increased due to the increase in the concentration of hydrogen on the support surface in the vicinity of the metal particles. Hence, the metal-support interface region could contribute to the overall activity of the benzene hydrodearomatisation if the concentration of adsorption sites for toluene is high enough and the hydrogen transport rate is sufficiently rapid.[28]

3.2. Effect of feed weight hourly space velocity on benzene conversion.

Figure 3 depicts the effect of feed WHSV on the conversion of toluene at 498 K temperature, 15 bar pressure and hydrogen to benzene mole ratio 10. Toluene conversion was found to decrease with an increase in feed WHSV of toluene that can be explained by two different ways.

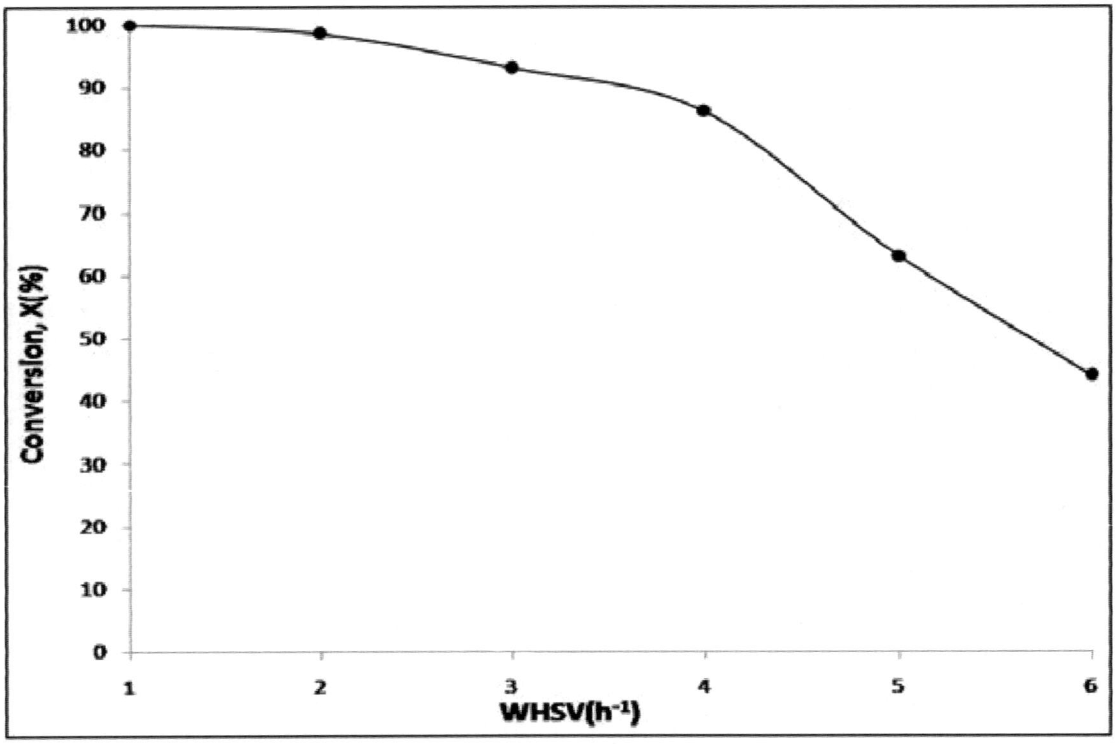

Figure 3.Benzene conversion vs WHSV plot in a fixed bed reactor at T:498 K, P:15 bar.

With increase in liquid feed rate, wetted fraction of the catalyst as well as the gas-liquid and liquid-solid mass transfer coefficients also increase.[33] At lower liquid feed rate, catalyst particles are partially wetted, and under these conditions, conversion increases due to direct transfer of the gas-phase reactant to the catalyst surface, already wetted internally due to the capillary forces. Therefore, with an increase in liquid feed rate an increase in the wetted fraction is expected to retard the conversion, while an increase in the external mass transfer coefficients will enhance the conversion. Another possible reason for observed decrease in conversion is due to the shorter residence time of reactants under high feed WHSV. However, the former explanation is applicable to the large catalyst bed but in this study low amount of catalyst around 10 g (smaller catalyst bed) was used, hence the latter explanation is more appropriate for this study.[28]

3.3. Effect of hydrogen pressure on benzene conversion

Figure 4 shows the plot of benzene conversion at 498 K as a function of pressure using hydrogen as a carrier gas. From the plot, it can be observed that benzene conversion increased with pressure for up to 15 bar beyond which it remains fairly constant up to 35 bar. The initial increase in benzene conversion between 5 and 15 bar can be attributed to the fact that within this pressure range increasing the pressure enhanced the adsorption of benzene on the active sites, leading to increased benzene conversion. Further increase in pressure beyond 15 bar only resulted in a slight increase in conversion probably because the adsorption of benzene had already reached saturation with the occupation of most of the active sites. Hydrogen is adsorbed along with toluene and takes part in the reaction. It should be noted that, because of its smaller size, the diffusion of hydrogen is faster than that of benzene. Thus, there are competing reactions where hydrogen occupies some of the active sites and dissociates into atomic hydrogen which participates in the reactions. This competing adsorption, diffusion, and activation of hydrogen molecule on the active sites may lower benzene

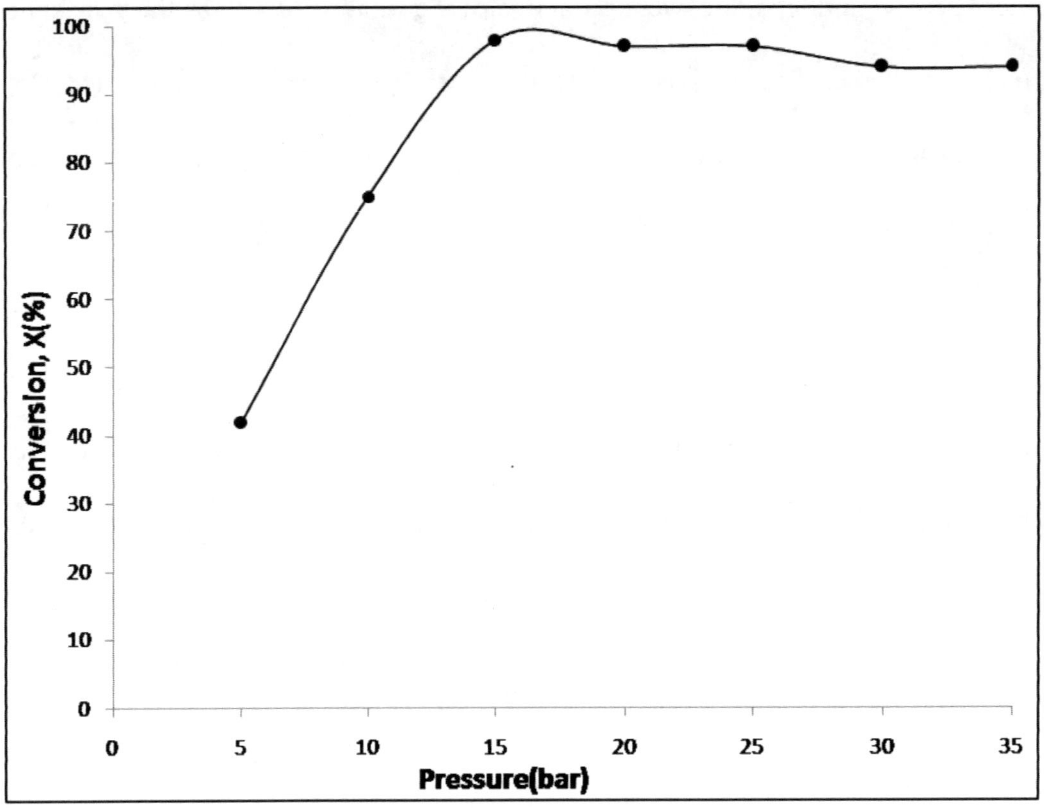

Figure 4. Benzene conversion vs pressure plot in a fixed bed reactor at WHSV: 2 h^{-1}, T: 498 K

conversion. This phenomenon can also be explained on the basis of Bronsted and Lewis acidity. The presence of hydrogen may lead to a decrease in Bronsted acidity and thus decrease in the concentration of intermediates benzylic carbocations, leading to lower benzene conversion.[28, 34]

3.4. Effect of temperature on benzene conversion

Low mechanical strength and collapsing of the structure at elevated temperature in the presence of water are the main drawbacks of mesoporous materials limiting their use as catalysts. In order to investigate the stability of the synthesized materials under real reaction conditions, the Pt (0.15 wt %)-Pd (0.15 wt %) / SBA-15 and Pt (0.15 wt %)-Pd (0.15 wt %) / Ti-SBA-15 were tested in benzene hydrogenation (Figure 5). Only cyclohexane was found as the reaction product. The benzene conversion of 100 % was achieved over the Pt (0.15 wt %)-Pd (0.15 wt %) / Ti-SBA-15 catalyst at 498 K, pressure 15 bar and WHSV = 2 h^{-1}. The catalytic activity was stable during the long-term operation (5 h).

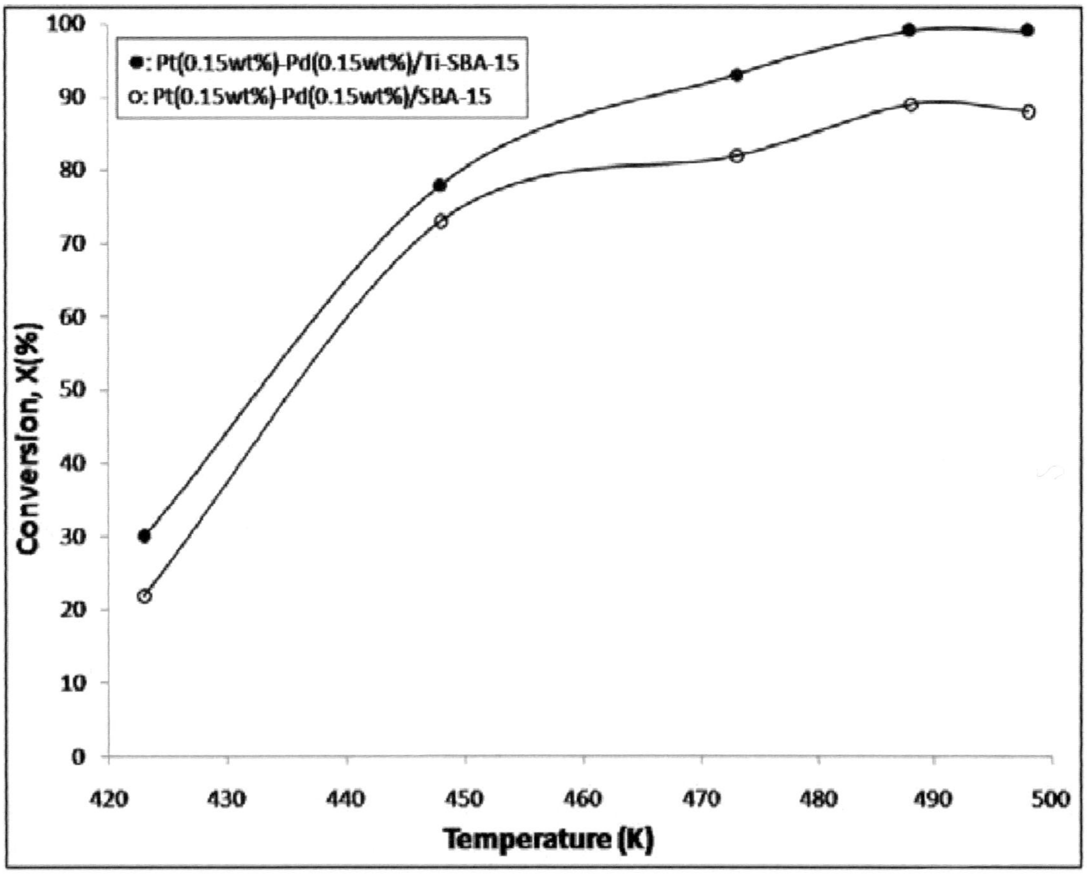

Figure 5. Benzene conversion vs temperature plot in a fixed bed reactor at WHSV: 2 h^{-1}, P:15 bar

The measurements performed after the reaction revealed that the specific surface area of the spent Pt (0.15 wt %)-Pd (0.15 wt %) / Ti-SBA-15 catalyst dropped from 590 m^2/g to 575 m^2/g (Table 1), and the average metal particle size only slightly increased from 1.4 to 1.5 nm (Table 1) but in case of spent Pt (0.15 wt %)-Pd (0.15 wt %) / SBA-15 catalyst the specific surface area dropped significantly from 560 m^2/g to 440 m^2/g and average metal particle size increased from 1.3 to 1.7 nm (Table 1). The Ti-SBA-15 support and Pt, Pd nanoparticles confined within the mesopores under the mentioned reaction conditions were seen to be quite resistant against structure collapsing and metal sintering. The activation energy obtained in the present study is compared with the activation energy reported in the literature for hydrogenation of toluene. Navarro et al.[35] reported that activation energy was found to be 129.2 kJ/mol for hydrogenation of benzene over Pt-Pd / Silica catalyst at 523 K. Coughlar and Keane [36] reported that the activation energy was found to be 78.2 kJ/mol for hydrogenation of benzene over nickel loaded Y-zeolite. It can be noticed that value of activation energies using Pt (0.15 wt %)-Pd (0.15 wt %) / Ti-SBA-15 catalyst for hydrogenation of benzene in the present study are roughly one fourth to half of that reported in literature. This is due to the fact that in the present study hydrogenation was carried out using a catalyst which facilitates a chemical reaction proceeds by a different pathway with a lower energy barrier.

Activity is described in terms hydrogenation of benzene using a specific reaction rate at 498 K, 15 bar pressure and 5 h on-stream according to Eq. (1) are given in Table 1. It is observed from Table 1 that reaction rate for benzene hydrogenation is higher using Pt (0.15 wt %)-Pd (0.15 wt %) / Ti-SBA-15 catalyst than that of Pt (0.15 wt %)-Pd (0.15 wt %) / SBA-15 catalyst. It can be further noticed from Table 1 that addition of Ti on SBA-15 support increases the metal dispersion and acidity. Some authors have reported that Al, Ti, or Zr-containing SBA-15 provided a better dispersion of metal species as compared to the pure SBA-15 and alumina-supported catalysts.[37-39]

4. Conclusions

Benzene conversion was found to be higher when Pt-Pd / Ti-SBA-15 catalyst was used for hydrogenation reaction. In all cases, benzene conversion measured at 498 K was found to be increased with pressure over 5-15 bar, after which it remained practically constant up to 35 bar.

The type of mesostructured material has an important effect on the dispersion of both Pt and Pd species. The highest dispersion is observed for the catalysts prepared on the Ti-SBA-15 material, and their sample had the higher catalytic performance in the hydrogenation of benzene under conditions employed in this work. The higher catalytic activity in this sample was related to the dispersion of both Pt and Pd species and the acidity in this sample. The catalyst Pt (0.15 wt %)-Pd

(0.15 wt %) / Ti-SBA-15 demonstrated stable activity and no metal agglomeration during benzene hydrogenation at 498 K for 5 h. Activation energy was found to be 40.25 kJ/mol using Pt (0.15 wt %)-Pd (0.15 wt %) / Ti-SBA-15 catalyst indicates that the catalyst has higher activity. The effect of feed WHSV, hydrogen pressure, hydrogen to benzene mole ratio and temperature on hydrogenation of benzene into cyclohexane was studied. The high surface area (630–747 m^2/g) and pore diameters in the range of mesopores make them suitable carriers for better and uniform distribution of metal species when the Pt and Pd is used as precursor.

Acknowledgment

The work reported in this article was financially supported by a research grant of Jadavpur University, India. We also acknowledge Prof. A. K. Biswas (Chemical Engineering Department, Indian Institute of Technology Kharagpur, India) for useful discussion.

References

[1] L. Allen, A. Suchanek, *Hart's Fuel Technol. Manage.* **1998**, *8*, 14-16.

[2] M. Hunter, A. Gentry, S. Lee, L. Groeneveld, C. Olivier, D. Pappal, *The NPRA 2000 Annual Meeting, San Antonio*, TX, Mar 26-28, **2000**, Paper AM-00-18.

[3] D. Karonis, E. Lois, S. Stournas, F. Zannikos, *Energy Fuels* **1998**, *12*, 230-238.

[4] S.X. Wang, M. Shao, Y.H. Zhang, *J. Environ. Sci.* **2001**, *13*, 265-271.

[5] M. Walsh, *Platinum Met. Rev.* **2000**, *44*, 22-30.

[6] D. Karonis, E. Lois, S. Stournas, F. Zannikos, *Energy Fuels* **1998**, *12*, 230-238.

[7] B. Cooper, B. Donnis, *Appl. Catal., A* **1996**, *137*, 203-223.

[8] S. Toppinen, T. K. Rantakyl, T. Salmi, J. Aittamaa, *Ind. Eng. Chem. Res.***1996**, *35,* 1824-1833.

[9] U. K. Singh, M. Vannice, *AIChE J.* **1999**, *45*, 1059-1071.

[10] A. Stanislaus, B. H. Cooper, *Catal. Rev. Sci. Eng.* **1994**, *36*, 75-123.

[11] B. H.Cooper, A. Stanislaus, P. N. Hannerup, *Diesel ACS National Meeting, San Francisco*, April **1992**.

[12] R. Ryoo, C. H. Ko, M. Kruk, V. Antochshuk, M. Jaroniec, *J. Phys. Chem. B* **2000**, *104*, 11465-11471.

[13] D. Zhao, J. Feng, Q. Huo, N. Melosh, B.F. Chmelka, G.D. Stucky, *J. Am. Chem. Soc.***1998**, *120*, 6024-6036.

[14] Y. Cao, J. C. Hu, P. Yang, W. L. Dai, K. N. Fan, *Chem. Commun.* **2003**, *7*, 908-909.

[15] J. Barbier, E. Lamy-Pitara, P. Marecot, J. P. Boitiaux, J. Cosyns, F. Verna, *Adv. Catal.* **1990**, 37, 279-318.

[16] C. Song, X. Ma, *Appl. Catal. B: Environ.* **2003**, *41*, 207-238.

[17] S. G. Kukes, F. T. Clark, D. Hopkins, *WIPO Patent*, WO 94/19429, **1994.**

[18] J. K. Lee, H. K. Rhee, *J. Catal.***1998**, *177*, 208-216.

[19] A. D. Mallmann, D. J. Barthomeuf, *J. Chim. Phys. Phys. -Chim. Biol.* **1990**, *87*, 535-538.

[20] S. T. Homeyer, W. M. H. Schatler, *J. Catal.***1989**, *117*, 91-101.

[21] W. M. H. Schatler, A.Y. Stakheev, *Catal. Today* **1992**, *12*, 283-295.

[22] M.V. Landau, V. Y. Kruglikov, N.V. Goncharova, O. D. Konoval'chikov, G. D. Chukin, B. V. Smirnov, V. I. Malevich, *Kinet. Catal.* **1976**, *17*, 1281.

[23] P. Marecot, J. R. Mahoungou, J. Barbier, *Appl. Catal., A* **1993**, *101*, 143-149.

[24] J. T.Miller, D. C. Koningsberger, *J. Catal.***1996**, *162*, 209-219.

[25] H. R. Reinhoudt, R. Troost, A.D. Langenveld, J. A. R. Veen, S. T. Sie, J. A. Moulijn, *Stud. Surf. Sci. Catal.***1999**, *127*, 251-258.

[26] T. Fujikawa, K. Idei, T. Ebihara, H. Mizuguchi, K. Usui, *Appl. Catal., A* **2000**, *192*, 253-261.

[27] J. T. Miller, B. L. Mojet, D. E. Ramaker, D. C. Koningsberger, *Catal. Today* **2000**, *62*, 101-114.

[28] B. K. Sidhpuria, A. P. Parikh, P. Bahadur, V. R. Jasra, *Ind. Eng. Chem. Res.* **2008**, *47*, 4034-4042.

[29] P.E. Boahene, K. Soni, A.K. Dalai, J. Adjaye, *Appl. Catal., B: Environ.* **2011**, *101*, 294-305.

[30] V. Parvulescu, C. Anastasescu, C. Constantin, B.L. Su, *Catal. Today* **2003**, *78*, 477-485.

[31] B. Mitra, J.P. Chakraborty, D. Kunzru, *AIChE J.* 2011, *57*, 3480-3495.

[32] T.A. Zepeda, B. Pawelec, J.L.G. Fierro, T.J. Halachev, *J. Catal.* **2006**, *242*, 254-269.

[33] R.V. Chaudhari, R. Jaganathan, S.P.Mathew, C. Julcour, H. Delmas, *AIChE J.* **2002**, *48*, 110-125.

[34] N. S. Gnep, M. Guisnet, *Appl. Catal.* **1981**, *1*, 329-340.

[35] R.M. Navarro, B. Pawelee, J.M. Trejo, R.M. Mariscale, J.L.G. Fierro, *J. Catal.* **2000**, 189, 184-194.

[36] B. Coughlan, A.M. Keane, *Catal. Lett.* **1990**, *5*, 101-112.

[37] D.P.Sawant, A. Vinu, N.E. Jacob, F. Lefebvre, S.B. Halligudi, *J. Catal.* **2005**, *235*, 341-352.

[38] O.Y. Gutierrez, G.A. Fuentes, C. Salcedo, T. Klimova, *Catal. Today* **2006**, *116*, 485-497.

[39] M. J. Tiernan, O.E. Finlayson, *Appl. Catal., B: Environ.***1998**, *19*, 23-35.

540428 The OH^--Controlled Synthesis of Pt-Ni Nanocatalysts with Different Atomic Distributions for Alkaline Hydrogen Evolution Reaction

Abstract:

Research Interests: My primary research interest is in the area of nanometallic materials and their electrocatalytic properties. Specifically, I work on the system of alkaline hydrogen evolution reaction (HER) and try to prepare platinum-based catalysts by adding some transition metals which could enhance the reaction activity and reduce the cost of the catalysts. I want to explore the relationship between the shape of nanocrystals and their electrocatalytic activities. Moreover, I am interested in the factors which can really influence the alkaline HER activity and intend to use density functional theory method to explore this question.

Teaching Interests: I am interested in teaching knowledge about green chemistry which can solve the global energy crisis and environmental concerns, such as principle and technology of chemical energy saving, synthesis of nanomaterials, technology of fuel cell, electrochemistry and so on.

Abstract: Hydrogen has been regarded as a promising alternative fuel for the future. Nowadays, intense research on pursuing effective platinum-based catalysts for alkaline hydrogen evolution reaction (HER) has been carried out. However, effects of the atomic distribution on the catalytic performance have rarely been investigated. Herein, we demonstrate that a complete alloy-to-hetero phase transformation can be realized under ambient conditions by introducing NaOH in the process of synthesis. Impressively, the Pt-Ni heterostructure with particular atomic distribution exhibits a superior alkaline HER activity, in comparison with the Pt-Ni alloy and commercial Pt-C catalysts. Density functional theory (DFT) calculations reveal the effect of the introduced NaOH to the nanocrystal growth process and explain how the heterostructure with special atomic distribution has been formed. The calculations also demonstrate that the greatly enhanced performance of Pt-Ni heterostructure originates from the superior adsorption ability of H and OH which could reduce the reaction energy barrier. Therefore, we suggest that the alkaline HER performance of Pt-Ni catalysts could be affected by the H and OH binding energies which could be controlled by tuning the atomic distribution.

540460 Catalytically Active and Hazardous Gas Adsorbent Polymer Fibers Functionalized By Atomic Layer Deposition and Metal-Organic Framework Thin Films

Abstract:

Research Interests: Metal Organic Framework Thin Film Synthesis and Practical Application

Noxious compounds, including toxic industrial chemicals (TICs) (e.g., ammonia, chlorine, hydrogen sulfide, nitrogen dioxide) and chemical warfare agents (CWAs) (e.g., sarin, soman, mustard gas) can pose a substantial threat to human health and the environment. Increased international threats and broader industrial usage is drawing research attention to practical and effective means for compound decontamination and isolation. Current technologies for ambient and environmental purification technology mainly relies on filtration and adsorption via impregnated activated carbon-based porous materials and metal oxides. However, such active components possess limited shelf-life, target specificity and overall effectiveness.

Metal-organic frameworks (MOFs) are a new class of porous materials featuring highly crystalline and exceptional high surface area[1] Some MOFs are known to be stable, versatile and rapidly functional for removing and catalytically decomposing noxious compounds. MOF materials are also well regarded for applications in gas storage and separation, drug delivery, light harvesting, electronics and others. In addition, the facile tunability in functional groups and metal clusters of MOFs allows them to be tailored and designed adsorb and abate specific target chemicals.[2] Most current studies of MOFs focus on synthesized power materials. For most applications, however, powders lead to practical limitations in application development.

My interest is in formation, characterization and evaluation of MOF thin film materials formed on functional and active surfaces for applications in environmental protection, personal safety and security, energy and other advanced fields. To overcome current critical challenges, I started working on a MOF-related project funded through DoD, primarily focused on integrating various MOFs onto polymeric non-woven textiles (MOF-functionalized textiles) by capitalizing on atomic layer deposition (ALD) technique, and investigating their catalytic and adsorbent performance against a range of CWAs, their simulants, and TICs.

The first research achievement I made is entitled **Catalytic MOF-Cloth Formed via Directed Supramolecular Assembly of UiO-66-NH_2 Crystals on Atomic Layer Deposition-Coated Textiles for Rapid Degradation of Chemical Warfare Agent Simulants**.[1] In this work, I came up with a novel approach to assembling as-synthesized Zr-based UiO-66-NH_2 crystals onto ALD-treated nonwoven polypropylene (PP) fibrous scaffolds, followed with demonstrating a promising catalytic performance of the products against a CWA simulant, dimethyl 4-nitrophenyl phosphate (DMNP) with a half-life of less than 5 minutes. The facile assembly method using surfactant assembly agents facilitates a swift MOF integration under mild assembly environment (a few hours at 25 ^{o}C), otherwise it generally takes more than 20 hours at higher than 80 °C.

The second research project I completed is **<u>UiO-66-NH$_2$ Metal-Organic Framework (MOF) Nucleation on TiO$_2$, ZnO, and Al$_2$O$_3$ Atomic Layer Deposition-Treated Polymer Fibers: Role of Metal Oxide on MOF Growth and Catalytic Hydrolysis of Chemical Warfare Agent Simulants</u>**.[3] Herein, I explored a conventional solvothermal integration of UiO-66-NH$_2$ onto different ALD thin films (i.e., Al_2O_3, TiO_2, or ZnO), and unveiled how the inorganic ALD layer with different compositions influences the MOF nucleation mechanism and the quality of the surface-bound MOF. The results are proved by correlating the material properties with catalytic performance of MOF-functionalized textiles towards degrading DMNP. Through this work, we found that TiO_2 surface enables to form a robust and more catalytically active surface-anchored MOF crystals compared to the other inorganic films tested.

In addition to the research progress above, I have been controlling the mass loading of UiO-66-NH$_2$ onto ALD TiO_2 coated PP fibrous mats and investigating catalytic and permeation properties of MOF-functionalized textiles to provide more pragmatic and quantifiable insights into protective equipment field.

Finally, I was not only limited to working on Zr-based MOFs, but also have recently achieved growing a robust and water-stable porphyrin-based Cu-TCPP MOF onto non-woven textiles, consisting of a copper paddle-wheel structure of metal clusters interconnected by porphyrin-based organic linkers. I also controlled the orientation of two-dimensional structure of the MOF onto fiber substrates by utilizing a modified route of a hydroxy double salt (HDS) mediated approach and, for the first time, found that the oriented MOF exhibited a promising removal performance for 2-chloroethyl ethyl sulfide (CEES), a blistering agent simulant, as well as ammonia even under humid environment (80% RH). Besides, I successfully fabricated other M-TCPP [M = Cu, Zn, Co, and Al] functionalized textiles and observed substantially improved CEES absorption capacity ($mol_{(CEES)}/kg_{(adsorbent)}$) in Al-TCPP MOF in comparison with Cu-TCPP.

Future Research Interests

My near future research interest lies in sustainable and clean energy and the environment using multiple nanomaterials like semiconducting, metallic, and porous structures (including MOFs) for CO_2 reduction, gas storage/separation, water desalination, toxic gas removal, water harvesting, and designing advanced membrane devices. In this regard, I believe that my research experience of thinking out feasible ideas, material synthesis, characterizing and analyzing the results, and making a rational conclusion would make me an independent researcher to contribute to sustainable and clean energy-relevant community.

Teaching Interests:

While I have been being served as a research- and teaching assistant under MS and PhD programs in the Chemical Engineering Department, I believe I have developed not only teaching, but also communicating with students to have them effectively learn both in class and out of class. Especially such a research and teaching experience makes me confident of teaching chemical reaction engineering, thermodynamics, and mathematical approach to solving problems in transport phenomena. In addition, I can also teach more advanced or elective courses covering, core chemical engineering topics, for example reaction kinetics, separations, and thermodynamics, couple together in sample industrial systems. My teaching goal is to inspire students to actively learn and apply their knowledge to engineering problems that they are facing and intrigued in, thus fostering the academia as well as the industry in the end.

Reference

1. Lee, D. T.; Zhao, J.; Peterson, G. W.; Parsons, G. N., Catalytic ¡°MOF-Cloth¡± Formed via Directed Supramolecular Assembly of UiO-66-NH2 Crystals on Atomic Layer Deposition-Coated Textiles for Rapid Degradation of Chemical Warfare Agent Simulants. *Chem. Mater.* **2017,** *29* (11), 4894-4903.

2. Zhao, J.; Lee, D. T.; Yaga, R. W.; Hall, M. G.; Barton, H. F.; Woodward, I. R.; Oldham, C. J.; Walls, H. J.; Peterson, G. W.; Parsons, G. N., Ultra-Fast Degradation of Chemical Warfare Agents Using MOF-Nanofiber Kebabs.

Angew. Chem. Int. Ed. **2016,** *55* (42), 13224-13228.

3. Lee, D. T.; Zhao, J.; Oldham, C. J.; Peterson, G. W.; Parsons, G. N., UiO-66-NH2 MOF Nucleation on TiO2, ZnO, and Al2O3 ALD-treated Polymer Fibers: Role of Metal Oxide on MOF Growth and Catalytic Hydrolysis of Chemical Warfare Agent Simulants. *ACS Applied Materials & Interfaces* **2017**.

540838 Nife Layered Double Hydroxide/Hollow Prussian Blue Via Alkaline Etching As an Efficient Electrocatalyst for Oxygen Evolution Reaction

Abstract:

Research Interests: electrochemical catalysis of water splitting

Teaching Interests: electrochemical catalysis

Oxygen evolution reaction (OER) is an imperative electrode reaction in many electrochemical applications, such as metal-air batteries, fuel cells and electrochemical water splitting. Much effort has been made to overcome its sluggish kinetics and decrease the overpotential. Recently, NiFe layered double hydroxide (NiFe-LDH) has been reported frequently as a new benchmarking OER electrocatalyst, even surpassing the activity of noble metal electrocatalysts such as Ir and Ru-based electrocatalysts [1]. In order to make further progress, creating NiFe-LDH with novel 3D morphologies is becoming a topical strategy [2].

Recently, we successfully synthesized a NiFe-LDH in-situ grown on hollow Prussian blue (PB) via alkaline etching as an efficient OER electrocatalyst. In this presentation, we will show a hollow-structured electrocatalyst denoted as LDH/PB-pH11-50 ^{o}C. It has an incompletely etched hollow PB as the inner framework and an in-situ grown NiFe-LDH as the main catalytic surface layer. When applied as OER electrocatalyst, it shows a low overpotential of ~242 mV to drive a current density of 10 mA cm^{-2} in 1 M KOH. This result is obviously superior to that of pure NiFe-LDH (~257 mV) and pure PB (~334 mV) synthesized by the similar method. The highlights of this work are shown below. Firstly, because the NiFe-LDH is synthesized from PB via a weakly alkaline etching, the residual PB may improve the performance of NiFe-LDH in the alkaline electrolyte. Besides, the hollow structure of residual PB is beneficial to the mass transfer and the enlargement of specific surface area, which can further enhance the catalytic activity of the electrocatalyst. Secondly, the main OER active component NiFe-LDH can be synthesized and in-situ imbedded onto the hollow PB framework simultaneously via the alkaline etching. This strategy significantly simplifies the synthesis of the material. In order to optimize the performance of the as-prepared electrocatalyst, the influences of pH and temperature towards structure-electrocatalytic activity relationship will be investigated in detail in the following work.

Reference

[1] Dionigi F and Strasser P. NiFe-Based (Oxy)hydroxide Catalysts for Oxygen Evolution Reaction in Non-Acidic Electrolytes. *Advanced Energy Materials* 1600621, 1-20 (2016).

[2] Xiaowen Y U, Zhang M, Yuan W and Shi G. High-performance three-dimensional Ni-Fe layered double hydroxide/graphene electrode for water oxidation. *Journal of Materials Chemistry A* 3, 6921-6928 (2015).

540841 Synthesis of 5-Hydroxymethylfurfural from Disaccharides Using Niobium-Modified Montmorillonite

Abstract:

Research Interests: the design of heterogeneous catalyst for the production of 5-HMF from carbohydrates

Teaching Interests: biomass conversion

Abstract: The technology to converse carbohydrates into 5-hydroxymethylfurfural (5-HMF) effectively is a very significant and potential way. In this work, niobium-modified montmorillonite (Nb-MMT) was synthesized by a one-pot, simple, ion exchange method and used as an efficient catalyst for the direct transformation of disaccharide into 5-HMF, which showed high catalytic activity and selectivity. A 5-HMF yield of 58.82% from cellobiose in a methyl isobutyl ketone (MIBK)/saturated salt water (SSW) biphasic solvent was obtained at 160 °C after 3 h. With sucrose as substrate under the same condition, 5-HMF yields were 50.45%. Moreover, 5-HMF with satisfactory yields were directly achieved from disaccharides in water. Pyridine-FTIR showed Lewis acid sites and Brønsted acid sites existed on Nb-MMT, which might facilitate the hydrolysis of disaccharide and play essential roles in the subsequent isomerization of glucose monomer and the dehydration of fructose monomer. SSW in the inorganic phase can increase the disaccharide conversion and 5-HMF selectivity, which might be attributed that the chloride ion has salting-out effect to increase the distribution coefficient of 5-HMF in the solvent. A catalytic mechanism for conversion of disaccharide to 5-HMF was proposed. In addition, Nb-MMT was recycled at least for four times without a significant decrease in the catalytic activity. This study provides a promising catalytic route for the heterogeneous Nb-containing catalysts used in the effective production of 5-HMF from carbohydrates.

540862 Synthesis of Organometallic Single-Site Heterogeneous Catalysts for Sustainable Chemistry

Abstract:

Research Interests:

Amid environmental concerns associated with fossil fuels and the ever-increasing trend in worldwide energy consumption, interest in the industrial-scale production of fuels and products from biomass and carbon dioxide (CO_2) is rapidly growing. The primary challenges in current approaches include low reaction rates, low product yields and high costs. All of these challenges could be overcome through the development of active, selective, and robust catalysts. To this end, I am interested in studying the synergistic effect between active supports and homogeneous catalysts in single-site heterogeneous catalysts (SSHC) to achieve a more economically feasible and sustainable conversion of biomass. In contrast to traditional heterogeneous catalysts that are designed using rare transition metals, my research will focus, on: (1) developing SSHC from relatively abundant transition metals (Fe, Cu, Ni) and active supports such as zeolites, tunable acidic & basic clays (montmorillonite & hydrotalcite respectively), and sulfonated SiO_2 , (2) studying catalyst valorization of lignocellulosic sugars, lignin, chitin, and CO_2 to fuels and higher-valued products and (3) investigating novel base-catalyzed dehydration products of glucose. Computer simulations will complement experimental studies to probe mechanisms and rationally design catalysts with optimal activity, selectivity and lifetime.

Teaching Interests:

Given my research and educational background, I am most interested in teaching undergraduate and graduate level kinetics and reaction engineering, as well as undergraduate unit operations, product design, and introductory courses in chemical engineering that introduce students to fundamental concepts. I am interested in developing graduate and undergraduate elective courses in green chemistry and engineering, catalysis, and a course focusing on various commonly used instrumentations such as mass-spectroscopy. I am particularly drawn to the concept of flipped classrooms in which students study the course material outside the classroom while class time is devoted to more in-depth discussion of particular concepts, problem-solving, and engaging hands-on activities to further support course material. I was introduced to this concept during an undergraduate kinetics and reaction engineering course.

Research Experience:

My graduate studies focused on fundamental reaction engineering and heterogeneous catalysis. More specifically, I pursued conversion of cellulosic sugars to value-added products (5-hydroxymethylfurfural, levulinic acid, and GVL) which could be used for fuels, polymers, and specialty chemicals. During this time I learned how to operate and fabricate high pressure batch and flow reactors, design and characterize heterogeneous catalysts, and perform ab-initio quantum chemistry calculations (Gaussian-3 (G3)). For my post-doctoral studies, I joined The George Washington University's chemistry department to expand my research breadth and knowledge in organic synthesis. I learned how to synthesize homogeneous catalysts that could be immobilized onto active supports to create SSHC with a synergistic effect between the support and immobilized organometallic catalyst.

Teaching Experience:

As a scholar and researcher, I believe it is extremely important to impart knowledge and enjoy mentoring, training, and working with researchers of all levels. Throughout my graduate studies, I served as a Teaching Assistant for an undergraduate/graduate course in Material Engineering and Science and for undergraduate courses in Organic Chemistry I and Organic Chemistry II in the Department of Chemical and Biological Engineering and in the Department of Chemistry, respectively. In addition, as a postdoc, I have given guest lectures for a graduate course in Green Chemistry and Engineering. Over the course of my Ph.D. studies I trained a postdoctoral fellow and mentored over 2 dozen undergraduates, 3 master students, and a high school senior. It was most valuable to see the progress students made towards becoming critical thinkers and independent researchers, integrating classroom knowledge with additional literature search and curiosity to form hypotheses, design experiments to test hypotheses, and interpret results.

Selected Publications:

Heltzel J, Finn M.T, Ainembabazi D, and Voutchkova-Kostal A. Transfer Hydrogenation of Carbon Dioxide and Bicarbonate from Glycerol Under Aqueous Conditions. RSC ChemComm, 2018. DOI: 10.1039/C8CC03157F

Finn M.T, Ridenour J, **Heltzel J**, Cahill C, and Voutchkova-Kostal A. Next-Generation Water-Soluble Homogeneous Catalysts for Conversion of Glycerol to Lactic Acid. ACS Organometallics, 2018. DOI: 10.1021/acs.organomet.8b00081

Karwa, S., Gajiwala, V., **Heltzel J.,** Patil S. K.R., and Lund C. R.F. *Reactivity of Levulinic Acid during Aqueous, Acid-Catalyzed HMF Hydration.* Catalysis Today., doi:10.1016/j.cattod.2015.06.020

Heltzel J, and Lund C. R.F. *Glucose Formate Conversion in gamma-Valerolactone. Catalysis Today.. 2016 26, 88-69*

Patil S. K.R., **Heltzel J.,** and Lund C. R.F. *Comparison of Structural Features of Humins Formed Catalytically from Glucose, Fructose and HMF.* Energy & Fuels., 2012. **26** (8), 5281-5293

Braden D. J., Henao C. A., **Heltzel J.**, Maravelias C. C., and Dumesic J. A. *Production of liquid hydrocarbon fuels by catalytic conversion of biomass-derived levulinic acid.* Green Chem., 2011.**13**:1755-1765.

Heltzel J, Patil S.K.R., and Lund C. R.F. *Temperature-Programmed Pyrolysis and Oxidation of Glucose-Derived Humins.* (manuscript under review at Industrial & Engineering Chemistry Research)

Book Chapters:

Heltzel, J., S. R. Patil and C. F. Lund (2016). Humin Formation Pathways. Reaction Pathways and Mechanisms in Thermocatalytic Biomass Conversion II. M. Schlaf and Z. C. Zhang, Springer Singapore: 105-118.

Patents:

Heltzel, J., Finn, M.T., Voutchkova-Kostal, A..Catalytic Transfer Hydrogenation of CO_2 to Potassium Formate using Glycerol as a Hydrogen source. (provisional application number 62/508,509)

Submitted Proposals:

National Research Council (NRC) Postdoctoral Research Fellowship, recommended for funding, unfunded

Arnold O. Beckman Postdoctoral Fellowship, unfunded

Awards and Fellowships:

Innovation & Entrepreneurship Prize, The George Washington University, Spring 2017

Eastman Kodak Scholars Program, SUNY University at Buffalo, Spring 2014

Honorable Mention, SUNY University at Buffalo School of Engineering and Applied Sciences,

Graduate Student Poster Competition, Buffalo, NY, April 23, 2014

Best professional poster, SUNY University at Buffalo Department of Chemical & Biological Engineering Symposium, Buffalo, NY, October 18, 2013

Mark Diamond Research Award: Non-Aqueous Production of Biofuel, Fall 2013

Postdoctoral Project: "*Glycerol Catalytic Transfer Hydrogenation of CO_2 using Ir and Ru Single-Site Heterogeneous Catalysts Immobilized on Hydrotalcites*."

Under supervision of Adelina Voutchkova-Kostal, Chemistry, The George Washington University

PhD Dissertation: "*A Study of Selectivity During Acid-Catalyzed Dehydration of Cellulosic Sugars*."

Under supervision of Carl R.F. Lund, Chemical and Biological Engineering,

540888 Catalysis for Sustainability: Probing the Fundamentals of Chemical Conversion Using Synthetic, Kinetic, and Electrocatalytic Approaches

Abstract:

Research Interests: The petrochemical and energy industries are currently tethered to carbon-based feedstocks for the foreseeable future in the transition towards reaching sustainable, carbon-neutral solutions. Chemical conversion is facilitated and enhanced through the use of both electrocatalytic and thermocatalytic processes, and sources of renewable-derived electrical energy will be critical to the future of these processes. In the medium-term, there is a crucial demand for the development and improvement of technologies for fuel and chemical production from carbon-based feedstocks including natural gas and biomass, and reduction of the global carbon footprint needs to be pursued via CO_2 conversion technologies. My research aims to fundamentally understand and optimize electrochemical and thermocatalytic petrochemical conversion pathways as a means to facilitate the transition from fossil fuel dependence towards a sustainable future founded upon efficient, electrocatalytic conversion methods. My research aims to understand how catalysts function in order to develop novel processes and to improve upon existing methods of chemical conversion for a sustainable future, specifically with regards to the utilization of electrochemical energy, natural gas, and CO_2.

Postdoctoral Projects: *Catalytic partial oxidation of methane to methanol over metal-exchanged zeolite catalysts*, and *Non-Faradaic electrochemical promotion of catalytic alkane oxidation and CO_2 reduction*, Department of Chemical Engineering, Massachusetts Institute of Technology. (Adviser: Associate Professor Yuriy Román-Leshkov).

PhD Dissertation: *Evolution in structure and function of transition metal carbides during the catalytic activation of C-O, C-H, and C-C bonds*, Department of Chemical Engineering and Materials Science, University of Minnesota-Twin Cities. (Adviser: Associate Professor Aditya Bhan).

Research Experience: During my PhD, I focused on the study of transition metal carbides as catalysts for biomass hydrodeoxygenation and alkane oxidative dehydrogenation. Specifically, my research prioritized the use of kinetic characterization and active site titration techniques as tools to quantify and comprehend the dynamic nature of the working catalyst in situ. Using site-specific titrations, I was able to manipulate and measure the number of active acid sites in situ, directly attributing co-feed induced changes in reaction rate to site density alterations [1,4]. I also developed detailed kinetic models of propane oxidative dehydrogenation reactions using CO_2 to transiently model the changing nature of the catalyst surface as a function of extent of oxidation, and this model could directly describe changes in dehydrogenation, hydrogenolysis, and reverse water-gas-shift reactions [2]. During my postdoctoral research, I focused upon learning how to incorporate electrochemical techniques into catalytic systems, specifically non-Faradaic electrochemical promotion of catalysis (EPOC or NEMCA), and I learned catalyst synthesis methods to direct active site speciation. I also gained experience with in situ x-ray absorption spectroscopy as an additional technique to identify the nature and local environment of the active site in the working catalyst. In order to characterize catalysts both in situ and ex situ, I have experience using X-ray diffraction, Fourier-transform infrared spectroscopy (FTIR), UV-Vis spectroscopy, liquid nitrogen physisorption, and chemisorption techniques. I have also worked closely with collaborators on transmission electron microscopy and X-ray photoelectron spectroscopy.

Future Research Directions: As a principal investigator, my philosophy is that a catalyst is best understood when we watch it work; in situ and operando kinetic and spectroscopic characterizations are at the heart of fundamental catalysis science, and the incorporation of electrical energy into catalytic systems offers enormous potential. This philosophy will be applied to petrochemical conversion of carbon-based feedstocks, specifically CH_4 and CO_2, via thermochemical and electrocatalytic pathways using selective catalyst synthesis methods. The Sullivan lab will pursue these goals by three interrelated avenues: (i) the incorporation of in situ electrochemical catalytic modification as a method of tuning reaction energetics and modifying rates and selectivity, (ii) targeted catalyst synthesis for the purpose of optimizing active site density, speciation, and local environment, and (ii) detailed kinetic and mechanistic studies of catalytic reactions in order to probe active site requirements and chemical conversion pathways. This work will incorporate the kinetic and mechanistic investigative tools I gained from my PhD as well as the electrochemical and synthetic techniques I learned during my postdoctoral research.

Teaching Experience: In my time at the University of Minnesota, I was a teaching assistant (TA) for three semesters of classes: (i) one undergraduate class (Mass Transport and Separations Processes, CHEN 3006), (ii) one graduate class (Physical Rate Processes, CHEN 8301), and (iii) one semester as a recitation instructor for two separate sections of ~25 undergraduate students that I taught twice a week (Reaction Kinetics and Reactor Engineering, CHEN 3102). I was awarded the Outstanding TA Award for my work with both CHEN 3102 and CHEN 8301 while working with graduate students and while working as a recitation instructor. Serving as a recitation instructor for CHEN 3102 afforded me the opportunity to work with the lead professor in order to plan recitation lectures, to write homework and exam problems, and to gain feedback from the undergraduate students. I lectured the entire class of ~130 students on multiple occasions when filling in for the lead instructor both by using his notes and by writing my own lecture from scratch, and I also filled in as an instructor for lectures of CHEN 3102 the following year when I was not a recitation instructor for the class. I also had the privilege of training and working with two undergraduate researchers from the University of Minnesota as well as a visiting foreign graduate student who worked in the Bhan lab for one summer.

Teaching Interests: I prefer to teach courses pertinent to the fundamentals of chemical engineering. Specifically, I am interested in teaching classes related to catalysis and reaction engineering, heat and mass transport, fluid mechanics, thermodynamics, mass and energy balances, or mathematics relevant to chemical engineering.

Selected Publications:

1. **M. Sullivan**, J.T. Held, A. Bhan, "Structure and site evolution of molybdenum carbide catalysts upon exposure to oxygen." *Journal of Catalysis*, 326, 82-91 (2015).
2. **M. Sullivan**, A. Bhan, "Effects of oxygen coverage on rates and selectivity of propane-CO_2 reactions on molybdenum carbide." *Journal of Catalysis*, 357, 195-205 (2018).
3. **M. Sullivan**, C.-J. Chen, A. Bhan, "Catalytic deoxygenation on transition metal carbide catalysts." *Catalysis Science & Technology*, 6 (3), 602-616 (2016).
4. **M. Sullivan**, A. Bhan, "Acid site densities and reactivity of oxygen-modified transition metal carbide catalysts." *Journal of Catalysis*, 344, 53-58 (2016).
5. **M. Sullivan**, A. Bhan, "Acetone hydrodeoxygenation over bifunctional metallic-acidic molybdenum carbide catalysts." *ACS Catalysis*, 6 (2), 1145-1152 (2016).
6. T. Dinh*, **M.M. Sullivan***, M. Dinca, Y. Román-Leshkov, "A viewpoint on partial oxidation of methane to methanol using Cu- and Fe-exchanged zeolites." *In revision*, (2018). (*co-first)

541067 Photocatalytic and Electrocatalytic Reduction Process of CO_2 with H_2O to CH_3OH over Bismuth-Promoted Perovskite-Based $BaTiO_3$ Catalyst

Abstract:

Research Interests: Dr. Venkata D. B. C. Dasireddy (m) (Research Associate, Department of Catalysis and Chemical Engineering) received his PhD in Applied Catalysis in 2013 at University of KwaZulu Natal, South Africa. He has been working at NIC for 3 years (a member of NIC since May 2015). He started his career as a teacher in chemistry (2006-2007) and then a research assistant in a pharmaceutical industry (2008) and in Indian Institute of chemical technology (2008). He worked as a tutor for undergraduate students (BSc and BTech) at University of KwaZulu Natal from 2009-2012. He was also a postdoctoral researcher at the University of KwaZulu Natal from 2013-2015. His expertise lies mainly in Green Chemistry and Heterogeneous nano-catalysis. During his PhD years, his research is focused on activating of C-H bonds through heterogeneous catalysis. After PhD, his research is focused in heterogeneous catalysis for green chemistry applications, chemical (process) engineering and reactor design. He published over 41 scientific research papers (28 as the first author) up to now and has an h-index of 8 (221 citations). He published his results with his co-workers in the Applied Catalysis A and Applied Catalysis B as an original research article and wrote two books on his research area. He also served as a Reviewer for various journals like Applied Catalysis A, ChemcatChem, Fuel, Journal of Taiwan Institute of Chemical Engineers, Energy Technology, and Catalysis Letters. He presented his results in Gordon Research Conference and Seminar, a highly prestigious conference series in the USA, as well as in European Catalysis Conference, EUROPACAT and other international conferences in Europe, India, Australia, and USA. Dr. Venkata D. B. C. Dasireddy will be responsible for heterogeneous catalysis materials.

Teaching Interests: I have taught a number of courses since 2006 as a tutor, lab technician, Demonstrator and Lecturer. These courses included (a) Inorganic and Analytical Chemistry, (b) Analytical Chemistry (both quantitative and qualitative analysis with labs), (c) Instrumentation Analysis (with lab work), (d) Physical Chemistry (with lab work), (e) Chemical Literature and (f) Lab preparation and demonstration of Inorganic Chemistry, Physical Chemistry, Analytical Chemistry and Instrumentation Analysis. I would feel very comfortable to teach the following courses at both undergraduate and graduate levels: (a) Analytical Chemistry (lab work), (b) Instrumental Analysis (lab work), (c) Physical Chemistry or Inorganic Chemistry, (d) Chemical Literature. I could also teach Heterogeneous Catalysis which is my main research interest.

541519 Rational Design of Pt-Ni Catalysts for the Oxygen Reduction Reaction By Building Atomic-Scale Structure-Property Relationships

Abstract:

Research Interests: Pt-Ni nanostructured materials are promising catalysts for the oxygen reduction reaction (ORR), but they suffer from Ni dissolution under cell operating conditions. To enable rational design of catalysts with improved stability and activity, I present the use of *ab-initio* calculations, cluster expansions, and Monte Carlo methods to study atomic-scale structure-property relationships for Pt-Ni surfaces and nanoparticles Using cluster expansions of Pt-Ni(111) surfaces, I built a direct bridge between the atomic structure and catalytic properties of Pt-Ni alloy surfaces at continuously varying strains, compositions and chemical environments. I investigated the synergy of strain and ligand effects in determining the catalytic activity, and demonstrated how this synergy can be leveraged to rationally design Pt-Ni surfaces with optimized ORR activity by searching for surfaces with targeted oxygen binding energy. For Pt-Ni nanoparticles, it has recently been shown that it is possible to stabilize octahedral Ptâ€"Ni nanoparticles by alloying them with transition metals (e.g. Mo, Cu, Ga, and Rh). I discuss two examples of alloyed Pt-Ni nanoparticles: Mo-Pt-Ni and Cu-Pt-Ni. Using a kinetic Monte Carlo (KMC) model based on cluster expansions, I demonstrated that surface Mo oxides preferentially located on low-coordination sites help protect Ni in sub-surface layers against acid dissolution. KMC simulations reveal that the enhanced stability of Cu-Pt-Ni is likely due to the reduction of the number of Ni and Cu atoms on the surface during synthesis, reducing the opportunity for Ni and Cu atoms in sub-surface layers to move to the surface and dissolve.

Teaching Interests: During my Ph.D. period at Johns Hopkins University, I worked as a teaching assistant for two courses, General Physics and General Physics Lab for undergraduate with class sizes of 24. As a TA, I accumulated rich experience in teaching by preparing and giving weekly presentations, and holding weekly office hours. For my teaching interests, provided I have B.S and Ph.D. degrees in Physics (major in Condensed Matter Physics), I believe I am qualified for teaching courses related to condensed matter physics, such as Solid State Physics, Electromagnetics, Quantum Mechanics, and Statistical Mechanics, at both undergraduate and graduate levels. I also plan to develop a graduate-level lab-oriented course to introduce commonly used computational methods closely related to my research on nanomaterials for energy storage and conversion.

541620 Enabling Concepts in Catalysis Science

Abstract:

Research Interests:

Catalysis has served as the critical enabling technology in the evolution of modern-day fuels and chemicals and in addressing environmental concerns associated with their use. The complexity of catalytic solids arises from varied architectures which range from isolated sites within crystalline porous materials to metal particles supported on amorphous metal oxides augmented by promoters, from dynamic restructuring of these surfaces under relevant process conditions, and from contributions of solvent phases or secondary environments to observed reactivity. Controlled materials synthesis, structural and chemical characterization of these materials in situ, and rigorous kinetic and isotopic studies are required to probe the structure and function of these materials under working conditions. These efforts in my research group will be augmented by collaboration with computational catalysis experts in order to leverage theoretical tools critical to modern catalysis research. My research objectives are to enable the description of catalytic processes at the level of rate constants and to determine requisite active site properties in an effort to guide next generation materials synthesis and to effect improvements in reactivity, selectivity, and catalyst lifetime.

Graduate Studies (Purdue University; Advisors: Fabio H. Ribeiro and Rajamani Gounder)

PhD Dissertation Title: "*Titration and Characterization of Lewis Acid Sites in Siliceous Oxides for the Catalysis of Oxygenates*"

As chemical transformations of bio-derived oxygenates primarily occur in the liquid-phase, development of methods to promote favorable solvent-solid interactions is paramount to successful catalytic upgrading of renewable feedstocks. In my graduate studies, I gained expertise in materials synthesis, including oxide-supported metal nanoparticles and hydrothermal synthesis and post-synthetic modification of molecular sieves. I developed methods to quantify Lewis acidic active sites in metallosilicate molecular sieves via chemical titrations combined with measurement of infrared spectra. These techniques were instrumental in characterizing metallosilicates of various crystal topologies and in quantifying their active site evolution and degradation during reaction. These efforts allowed for comparison of glucose isomerization reaction rate constants between hydrophobic and hydrophilic catalysts and provided a clear demonstration of the impact of the polarity of the secondary environment around catalytic sites on reactivity in aqueous media, resulting in higher reaction rates over sites confined in hydrophobic pores.

Postdoctoral research (University of Minnesota; Advisor: Aditya Bhan)

Topic: "*Selective oxidation of ethylene over promoted Ag catalysts*"

My postdoctoral research focuses on the kinetics of ethylene epoxidation over a promoted silver catalyst. This catalyst system is employed to produce ~27 million tons of ethylene oxide per year, yet precise descriptions of the surface chemistry required for production of ethylene oxide at high selectivity (>85%) have not been developed. In industrial practice, co-fed alkyl chlorides deposit chlorine adatoms on Ag surfaces resulting in dramatic effects on the kinetics and increases in ethylene oxide selectivity, while co-fed alkanes remove chlorine and prevent catalyst poisoning. I have developed techniques for quantifying the chlorine coverage and determining mechanisms of chlorine deposition onto and removal from Ag surfaces, and correlated these changes in surface coverages to observed changes in ethylene oxide rate and selectivity. Through this experience, I have broadened my skillset to new classes of catalysts and reactions, and refined the mechanistic descriptions of catalytic processes occurring over industrially-employed, complex catalytic materials. My future research group will utilize strengths in materials synthesis, spectroscopic characterization techniques, and kinetic studies to guide rational catalyst design in applications critical to sustainable energy transformations, such as the conversion of multifunctional oxygenates to specialty chemicals and the control of toxic gas emissions.

Teaching Interests:

My interest in teaching, strengthened while designing and teaching a course during my undergraduate studies, motivated me to pursue both a graduate degree and a position in academia. Successful research in heterogeneous catalysis requires a thorough understanding of the core chemical engineering disciplines of transport, thermodynamics, and reaction engineering. The bulk of my teaching experience relates to the latter topic, as I served as a teaching assistant for undergraduate and graduate level reaction engineering courses as a graduate student. In these roles, I designed homework and exam exercises, created lesson plans for discussion sessions, gave guest lectures, and supervised laboratory sessions. I also served as the lead instructor for a section of a senior laboratory course at Purdue. I devoted significant effort to improving my teaching and preparing for a future in education by taking a graduate level engineering education course and completing the requirements for the Graduate Teaching Certificate from the Purdue Teaching Academy. These experiences, along with my mentorship of undergraduates and junior graduate students during my graduate and postdoctoral studies, have prepared me for successful teaching both in the laboratory and the classroom.

541772 Photocatalytic and Electrocatalytic Reduction Process of CO_2Â with H_2O to CH_3OH over Bismuth-Promoted Perovskite-Based $BaTiO_3$Â Catalyst

Abstract:

Research Interests: Dr. Venkata D. B. C. Dasireddy (m) (Research Associate, Department of Catalysis and Chemical Engineering) received his PhD in Applied Catalysis in 2013 at University of KwaZulu Natal, South Africa. He has been working at NIC for 3 years (a member of NIC since May 2015). He started his career as a teacher in chemistry (2006-2007) and then a research assistant in a pharmaceutical industry (2008) and in Indian Institute of chemical technology (2008). He worked as a tutor for undergraduate students (BSc and BTech) at University of KwaZulu Natal from 2009-2012. He was also a postdoctoral researcher at the University of KwaZulu Natal from 2013-2015. His expertise lies mainly in Green Chemistry and Heterogeneous nano-catalysis. During his PhD years, his research is focused on activating of C-H bonds through heterogeneous catalysis. After PhD, his research is focused in heterogeneous catalysis for green chemistry applications, chemical (process) engineering and reactor design. He published over 41 scientific research papers (28 as the first author) up to now and has an h-index of 8 (221 citations). He published his results with his co-workers in the Applied Catalysis A and Applied Catalysis B as an original research article and wrote two books on his research area. He also served as a Reviewer for various journals like Applied Catalysis A, ChemcatChem, Fuel, Journal of Taiwan Institute of Chemical Engineers, Energy Technology, and Catalysis Letters. He presented his results in Gordon Research Conference and Seminar, a highly prestigious conference series in the USA, as well as in European Catalysis Conference, EUROPACAT and other international conferences in Europe, India, Australia, and USA. Dr. Venkata D. B. C. Dasireddy will be responsible for heterogeneous catalysis materials.

Teaching Interests: I have taught a number of courses since 2006 as a tutor, lab technician, Demonstrator and Lecturer. These courses included (a) Inorganic and Analytical Chemistry, (b) Analytical Chemistry (both quantitative and qualitative analysis with labs), (c) Instrumentation Analysis (with lab work), (d) Physical Chemistry (with lab work), (e) Chemical Literature and (f) Lab preparation and demonstration of Inorganic Chemistry, Physical Chemistry, Analytical Chemistry and Instrumentation Analysis. I would feel very comfortable to teach the following courses at both undergraduate and graduate levels: (a) Analytical Chemistry (lab work), (b) Instrumental Analysis (lab work), (c) Physical Chemistry or Inorganic Chemistry, (d) Chemical Literature. I could also teach Heterogeneous Catalysis which is my main research interest.

541810 Single-Molecule Organometallic Catalysis, and Fluorescent Materials Preparation and Application

Abstract:

Research Interests:

The chemical industry relies on catalytic processes, which are multibillion-dollar industry and involve heterogeneous and homogeneous catalysts. Combining the advantages of heterogeneous catalyst and homogeneous catalyst, surface organometallic chemistry attaches molecular catalysts to heterogeneous supports, and endows the e cacy and selectivity of homogeneous molecular catalysts with the reusability of heterogeneous catalysts. Interactions between the surface and the supported molecular catalyst add new functionality, and these supported molecular catalysts have become important tools in catalysis technology, particularly for in-flow reactors and electrochemical transformations. It is essential to comprehend activity species, active sites, dynamic order and static order of organometallic catalysts, so we can understand how molecular catalysts work during catalysis reactions and further rational design molecular catalysts.

My research objective is to investigate surface organometallic catalysts through novel fluorogenic reactions in single-molecule (SM) level. Palladium (Pd) organometallic catalysts are chosen as my target system due to their extensive applications in modern organic synthesis such as Suzuki, Negishi, Heck, Sonogashira, and Stille coupling or cross-coupling reactions, etc. Utilizing these reactions, I thus propose three research directions: (1) Single-molecule Pd catalysis of Heck-fluorogenic reactions, (2) Pd and Cu co-catalysis of Liebeskind-Srogl cross-coupling fluorogenic reactions in single molecule level, and (3) Fluorophore tagged alkene living polymerization in single molecule level using Pd catalysts.

My research will combine organic synthesis, catalysis, fluorescent materials and single-molecule imaging techniques. At the University of Connecticut, I developed a series of fluorescent nanomaterials (including polymer doped fluorophore materials and carbon dots with different emission colors) for sensing nitro-explosives. Following that, I synthesized small molecule fluorophores with green, yellow and red emissions at Cornell University. These molecules were used as monomers for living polymerization in both ensemble and single molecule level using Grubbs' molecular catalysts. Total internal reflection fluorescence (TIRF) microscopy has been extensively applied to investigate single nanoparticle catalysis by Peng Chen et al. and also for single molecular polymerization studies in my research. Now I propose apply these approaches to make significant advancements and to lead new research directions in catalysis field.

Biological structure and dynamics have been extensively explored using single molecule techniques, while abiotic applications are mostly limited to studies of heterogeneous catalysts such as gold and copper nanoparticles, platinum silica core-shell particles, and inorganic nanocrystals. The reactive behavior of individual molecules is rarely observed.

Average properties of billions of molecules are usually measured, however, the catalytic activity of a few percent of molecules in homogeneous catalysis or active sites in heterogeneous catalysis can dominate the outcome of a chemical reaction in ensemble level, and even stoichiometric chemical reaction mechanisms must be determined by indirect methods that potentially miss competing pathways. Study of molecular catalysts by SM fluorescence microscopy promises to improve our understanding of chemical reactions by providing insights into catalyst kinetics and heterogeneity within catalyst populations. Catalytic polymerization is a key process in making synthetic polymers, which are one of the most important materials of our modern world. Living polymerization through molecular catalysts in single-molecule level is helpful to understand the relation between the macroscopic properties and microscopic structure and dynamics, to identify the origin of polymer heterogeneities, and to better polymer properties.

One big challenge in using SM fluorescence imaging to visualize molecular catalysts is the so-called "concentration barrier" for fluorescent species. There is currently no satisfactory way to probe chemical systems with commercial fluorophores, thus design and synthesis of novel fluorophore is highly necessary. Besides the fluorophore, the type of chemical reactions that could be studied by SM fluorescence method is still in infancy or discovery stage. I will design and develop a series of fluorogenic reactions to investigate molecular catalysts through Heck Reactions, Liebeskind-Srogl cross-coupling reactions and living polymerizations in single-molecule level.

My research utilizing new analytical tools that provide a microscopic, molecular-level understanding of molecular catalysis will lead new theories and research dimensions in catalysis. This research approach can ultimately provide a roadmap to develop high efficient catalysts for different applications with great stability and low cost and to obtain (polymeric) materials with well-controlled and improved properties.

Teaching Interests:

One reason why I want to be a faculty and to teach is to share my knowledge and to interact with students. I will teach by connecting the lecture to cutting-edge research and to practice. I will share my learning, industrial and research experiences that closely to chemical engineering field to educate students the attitude, the approaches and the core to learn in each course. I believe that caring for students and engaging them as collaborative authors of classes can improve their performance.

I have had extraordinary teachers throughout my education, especially in U.S. as a graduate student. During these classes, I was always wondering, if I were a teacher, which kind of methods could be taken and applied in my classroom. At the University of Connecticut, I was a teaching assistant in two courses (Transfer Operations and Chemical Engineering Analysis), to grade the students' homework and exams, to explain the problem solution for their homework, and to answer the students' questions. I also tutored several undergraduates for lab research, and even guided one undergraduate to present his scientific research under my supervision in AIChE national conference. I assigned research work to them, encouraged them to read scientific papers, let them express and try their research ideas in the lab, etc. Two undergraduates got co-authored scientific papers with me. I found these experiences enjoyable and gratifying.

For an engineering class, it is especially important and necessary to keep lessons relevant and to link the scientific concept to the outside world. I will design course projects, which will be either closed related to real life examples, my research and industrial experiences or projects. Some course projects are delicately decomposed and prototyped from my ongoing research projects. My research will be integrated into my teaching, because this could provide a vision of relationship between the knowledge imparted in courses and practical problems, update students the latest advance in the field and cultivate their interest in pursuing advanced study. These could help students understand that they work to solve problems in "real world", not spending their time on something useless.

If I teach, various teaching methods could be combined and applied, such as projectors for powerpoint presentations with handwriting on blackboard; images, and videos with classic textbooks. Students have different background and learning styles. Teaching style and speed should be adjusted to the needs of students. A variety of methods could help examine and know the speed and progress of the class, such as quizzes, homework, surveys, in-class discussions, projects, writing, presentations, etc.

I would like to develop three courses, one introductory course – Introduction to Chemical Engineering, one advanced course for senior undergraduates – Separation Processes, and the third one for graduates – Chemical Kinetics.

543805 Catalyst Studies on the Conversion of Biobased Intermediates to Biobased Products

Abstract:

Ph.D. Dissertation:

"Catalyst Studies on the Conversion of Biobased Intermediates to Biobased Products"

Under the supervision of Dr. Dennis J. Miller, Department of Chemical Engineering and Materials Science, Michigan State University

Research Interests:

Catalytic reactions and chemicals separations are of quite high industrial significance among chemical engineering researches and the principle for the production of a wide variety of chemicals in the market. My work at Michigan State University has been focused on optimizing the conversion of bio-based raw materials and selectivities of desired products through catalytic reactions. I have been able to work on several projects in this area such as the production of acrylate esters from 2-acetoxy propanoate esters (APA esters) and *n*-butanol from ethanol. In both projects, I improved the selectivity of desired products to up to 80%. Through this work, I have been able to hone my experimental skills and develop a foundation of conceptual knowledge in the areas of reactions, process and reactor design, catalyst characterization, and kinetic and thermodynamic modeling. Perhaps the most rewarding part of my graduate school experience was learning the "ins and outs" of scientific research through practice. The objectives I met in both projects revealed new questions that I am interested in pursuing as a faculty member.

My research on the production of acrylate esters [1] was conducted in a fixed bed reactor at atmospheric pressure and the temperature of 500-600 ^{O}C. In this reaction, acrylates are formed by the elimination of acetic acid group of the APA ester molecules. While multiple contact materials were tested, best results were obtained using non-porous silica support with low surface acidity/basicity. It is suspected that porous materials can trap the reactants and provide the contact time required for their decomposition to smaller compounds. Furthermore, our studies showed the dramatic effect of using carbon dioxide as the carrier gas instead of nitrogen on acrylate yield. This was due to the occurrence of the Boudouard reaction ($CO_2 + C \leftrightarrow 2CO$), which in the presence of excess CO_2 facilitated the removal of carbon deposits from the surface of the contact material, therefore enhancing the available sites for the reactant to produce the desired product.

In the second research project, I investigated the continuous condensed phase conversion of ethanol to *n*-butanol. The first stage involved reproducing results previously obtained using nickel supported lanthanum-alumina catalyst in a batch reactor [2] with a packed bed reactor. Next, several bimetallic catalysts were characterized and tested to optimize the performance of the reaction. These characterizations and experiments entailed several analytical techniques such as spectroscopy and chromatography, and material surface analyses including microscopies, surface area measurements, and molecules adsorption/ desorption techniques. Our studies showed the beneficial effect of hydrogen uptake of the catalyst on ethanol conversion, suggesting that the first step of the reaction, the dehydrogenation of ethanol to acetaldehyde, is the rate limiting step for the catalysts with low hydrogen adsorption capability. Moreover, addition of copper to the nickel catalyst improved *n*-butanol selectivity while decreasing the C_6+ alcohols selectivity, therefore increasing the *n*-butanol purity in the outlet stream while maintaining the same C_4+ selectivity. Another catalyst variation, cobalt-nickel bimetallic catalyst, showed superior activity compared to the nickel catalyst despite lower selectivity to higher alcohols. Finally, the economic viability of *n*-butanol production process with the same mechanism was analyzed using experimental data, kinetic and thermodynamic modeling, and process simulations. The results indicated that the *n*-butanol selling price of $1.33-$1.58 provides a return on investment of 25% for different cases of ethanol conversion and alcohols selectivities. This price is reasonably competitive with current *n*-butanol market price.

Upon continuing my research, I aim to continue improving the efficiency of *n*-butanol formation in the Guerbet reaction. The Guerbet mechanism consists of three main reactions: ethanol dehydrogenation to acetaldehyde, aldol condensation of acetaldehyde to crotonaldehyde and finally, crotonaldehyde hydrogenation to *n*-butanol. Several studies report a selectivity of >97% for the production of acetaldehyde from ethanol with an ethanol conversion of >37% [3–5]. However, little work has been performed on distinctly improving the last two steps of the Guerbet mechanism. Optimizing these stages will provide a sustainable process to produce bio-derived *n*-butanol with high selectivities. Several discussions have approved the need for an acid-base supported catalyst for the second step, and a metallic catalyst for the third step [6,7]. Our initial screenings on the aldol condensation reaction show that while acetaldehyde is significantly more active compared to ethanol, the selectivity to crotonaldehyde is within the same range as the *n*-butanol selectivity from ethanol in the successive Guerbet reaction using a multi-functional catalyst. This suggest that the crotonaldehyde production could be the limiting step of the Guerbet mechanism regarding the selectivity to desired products. I aim to improve the total selectivity of *n*-butanol from ethanol by optimizing the catalyst required for the aldol condensation reaction, experimental analyses on the *n*-butanol selectivity of the last step of the Guerbet reaction, and detailed kinetic modeling of the reaction tree.

Teaching Interests:

Throughout my academic and professional careers, I have continuously tried to improve my leadership and mentoring skills by learning from multiple resources and collaborating in numerous activities. For instance, I pursued and acquired certification through the Engineering Research Mentor Training program developed by the National Research Mentor Network (NRMN). In this program, I participated in a series of workshops that involved multiple aspects of mentor-mentee relationship such as aligning expectations, maintaining effective communication, and addressing equity and inclusion. I have tried to develop these concepts during the time that I have had as a research mentor for training undergraduate research assistants in our group's laboratory. I also had the opportunity to contribute to the business development of an innovative technology as a venture fellow. During this experience, I was a part of STEM-MBA student team that investigated the market, developed financial analysis and created a business model to assess the viability of bringing the idea into the market. Recently, I have been able to participate in the A2H2 health hackathon team project to develop an innovative idea regarding the transition into independence for individuals with disabilities. All these experiences along with other groupwork activities that I have had in recent years have helped me grow a confident personality in leading projects and coordinating group works.

Aside from my mentoring capabilities, I have developed an evolving passion in teaching throughout my career. I served as a teaching assistant (TA) for the sophomore-level Materials and Energy Balances, junior-level Modeling and Analysis of Transport Phenomena, and junior-level Pascal Programming courses at Michigan State University and Sharif University of Technology. In this role, my responsibilities included (but were not limited to) leading recitations, designing exam questions, grading assignments and exams, and holding office hours. To further improve my teaching capabilities, I took a course on pedagogical concepts and instructional design. The class served to further familiarize me with effective pedagogical practices, including collaborative learning and using technology in the classroom. I am currently in the process of completing the requirements for the certification in college teaching program. Moreover, my research deals directly with principles normally taught in Material and Energy Balances, Chemical Engineering Kinetics and Thermodynamics, Transport Phenomena, Unit Operations, Separation Sciences, and Chemical Engineering Design. Therefore, I am confident in my ability to teach these courses. However, I am also interested in teaching several other areas such as Computational Courses, Data Analysis, Process Control, Fluid Mechanics, and Physical Chemistry.

References

[1] I. Nezam, L. Peereboom, D.J. Miller, Enhanced Acrylate Production from 2-Acetoxypropanoic Acid Esters, Org. Process Res. Dev. 21 (2017). doi:10.1021/acs.oprd.7b00047.

[2] T.L. Jordison, C.T. Lira, D.J. Miller, Condensed-Phase Ethanol Conversion to Higher Alcohols, Ind. Eng. Chem. Res. 54 (2015) 10991–11000. doi:10.1021/acs.iecr.5b02409.

[3] W.-D. Lu, Q.-N. Wang, L. He, W.-C. Li, F. Schüth, A.-H. Lu, Copper Supported on Hybrid C@SiO_2 Hollow Submicron Spheres as Active Ethanol Dehydrogenation Catalyst, ChemNanoMat. 4 (2018) 505–509. doi:10.1002/cnma.201800021.

[4] P. Zhang, Q.N. Wang, X. Yang, D. Wang, W.C. Li, Y. Zheng, M. Chen, A.H. Lu, A Highly Porous Carbon Support Rich in Graphitic-N Stabilizes Copper Nanocatalysts for Efficient Ethanol Dehydrogenation, ChemCatChem. 9 (2017) 505–510. doi:10.1002/cctc.201601373.

[5] M. V. Morales, E. Asedegbega-Nieto, B. Bachiller-Baeza, A. Guerrero-Ruiz, Bioethanol dehydrogenation over copper supported on functionalized graphene materials and a high surface area graphite, Carbon N. Y. 102 (2016) 426–436. doi:10.1016/j.carbon.2016.02.089.

[6] D. Gabriëls, W.Y. Hernández, B. Sels, P. Van Der Voort, A. Verberckmoes, Review of catalytic systems and thermodynamics for the Guerbet condensation reaction and challenges for biomass valorization, Catal. Sci. Technol. 5 (2015) 3876–3902. doi:10.1039/C5CY00359H.

[7] X. Wu, G. Fang, Y. Tong, D. Jiang, Z. Liang, W. Leng, L. Liu, P. Tu, H. Wang, J. Ni, X. Li, Catalytic Upgrading of Ethanol to n-Butanol: Progress in Catalyst Development, ChemSusChem. 11 (2018) 71–85. doi:10.1002/cssc.201701590.

544964 Process Intensification Driven Catalysts Development for CO_2 Utilization and Drop-in Fuels Production from Renewable Feedstock

Abstract:

Research Interests:

Process intensification has been receiving increased attention and importance because of its potential to obtain innovative and more sustainable process alternatives. Novel catalysts development or improvement of the existing catalysts is one of the key elements of application of process intensification at the early stage of novel process development or retrofitting of the existing processes. My research initiates with conducting techno-economic, process safety, process energy and exergy efficiencies, environmental impact analyses of different value-added products production routes. These analyses provide a benchmark for the catalyst performance targets and are used to guide catalyst development efforts. Also, these analyses are conducted periodically to track the progress of the process development. The catalyst development and testing phase includes the use of machine learning approach (e.g., Artificial Neural Network (ANN) model) to extract valuable information regarding catalyst type requirements from the previous literature, catalyst synthesis, catalyst characterization using various techniques, including X-ray Absorption Fine Structure (XAFS) technique, catalytic activity testing, deployment of machine learning techniques to establish the catalyst structure-activity relationship, understanding the reaction mechanism of the selected catalysts (screened through machine learning techniques) using *in-situ* operando Raman or Fourier Transform Infrared Spectroscopy (FTIR) techniques and utilization of six sigma approach to optimize the reaction process parameters and detailed reaction kinetics development.

CO_2 emissions into the atmosphere is a global concern and responsible for global warming and climate change. The CO_2 levels should not exceed 15 gigatons annually in order to limit the temperature increase within 2°C by 2050. Comparing the CO_2 generation vs. utilization scenario, the Carbon Capture and Storage (CCS) is no longer adequate to mitigate the CO_2 generation challenge and the CCS has to replace with CO_2 utilization. Thermodynamic analysis indicates that using only CO_2 as a reactant is more energy demanding than used as a co-reactant with another substance that has higher Gibbs free energy, such as methane (CH_4), ethane (C_2H_6) and hydrogen (H_2). My research interests include development and retrofitting of the existing processes for dry reforming of methane and ethane (using CO_2) to produce syngas, CO_2 hydrogenation to produce methanol and oxidative dehydrogenation (ODH) of ethane to produce ethylene by using CO_2 as a soft oxidant. At Idaho National Laboratory (INL), I am working on oxidative dehydrogenation (ODH) of ethane to produce ethylene by using CO_2 as a soft oxidant. The project involves novel catalysts development, understanding the reaction mechanism using *in-situ* operando Raman technique, optimizing the process, and conducting techno-economic, process safety, process energy and exergy efficiency, and gate-to-gate life cycle analysis. I am leading this project as the Principal Investigator (PI) and this project yielded a provisional patent based on the novel catalyst composition within eight months of starting the project.

The current increase in global energy demand, as well as the negative impact petroleum-based energy sources are having on the environment, has led to a renewed interest in renewable energy resources. My research interests include conversion of second and third generation of biomass feedstock (e.g., waste grade vegetable oils, microalgae, fat, oil, grease (FOG)) into green diesel and biodiesel through hydrodeoxygenation, hydrocracking, hydrothermal liquefaction, transesterification (in presence of expandable gas solvents) routes. During my Ph.D., I have worked on heterogeneous catalyst development for biodiesel production from Green Seed Canola (GSC) oil (waste grade canola oil). The research involved development of an Artificial Neural Network (ANN) model based on the literature review for pattern recognition of different catalysts and their activities in biodiesel production, novel catalysts development, characterization of the catalysts using various techniques including XAFS, testing the catalytic activities, reaction process parameter optimization using Response Surface Methodology (RSM), reaction kinetics development and integration of biodiesel and glycerol to value added products production processes using the same catalyst. I have concluded my Ph.D. research by conducting techno-economic, process safety, energy efficiency and environmental impact analyses of my developed processes. My Ph.D. research led eight journal papers and forteen conference presentations.

Successful proposal:

Chinmoy Baroi (PI), Frederick F. Stewart, Rebecca Fushimi, John Klaehn, Chris Orme, "Production of Ethane to Ethylene using Carbon dioxide as a soft oxidant" research proposal submitted and approved from US Department of Energy (DOE) funded Laboratory Directed Research and Development (LDRD) project office, Idaho National Laboratory (INL) August, 2017

Selected publications:

<u>Patent:</u>

Chinmoy Baroi, Harry W. Rollins, "Catalyst structures, methods of forming catalyst structures and methods of oxidizing hydrocarbon", Patent application no. 62/683,425, 11 June, 2018

<u>Peer-reviewed journal papers:</u>

Chinmoy Baroi, Anne M. Gaffney, Rebecca Fushimi, "Process economics and safety considerations for the oxidative dehydrogenation of ethane using the M1 catalyst," Catalysis Today, 298, 138-144, (2017)

Soe Lwin, Weijian Diao, Chinmoy Baroi, Anne M. Gaffney, Rebecca Fushimi, "Characterization of MoVTeNbOx Catalysts during Oxidation Reactions Using In Situ/Operando Techniques: A Review," Catalysts, 7 (4), 109, (2017)

Chinmoy Baroi, Ajay K. Dalai, "Process sustainability of biodiesel production process from Green Seed Canola oil using homogeneous and heterogeneous acid catalysts", Fuel Processing Technology, 133, 105-119, (2015)

Chinmoy Baroi, Ajay K. Dalai, "Review on Biodiesel Production from Various Feedstocks Using 12-Tungstophosphoric Acid (TPA) as a Solid Acid Catalyst Precursor", Industrial & Engineering Chemistry Research, 53, 18611 – 18624, (2014)

Teaching Interests:

I strongly believe knowledge is far more valuable when shared. One of my passions is teaching. In my view, a great teacher is like a great artist who uses basic tools to transform raw materials into valuable assets of society. My teaching experience in the university (during M.Sc. and Ph.D.) as a teaching assistant has not only helped me to develop my teaching skills, but also have encouraged me to pursue my teaching as a part-time course instructor at the "Saskatchewan Polytechnique (Canada)" consecutively for three years with a very good reputation. After completing my Ph.D., I taught "Chemical Engineering Thermodynamics" as a course instructor for a semester at the University of Saskatchewan, Canada. In addition, I have mentored two Ph.D. students, two master students, four undergraduate students in conducting Ph.D. and master thesis research, novel and green process design, simulation and techno-economic analysis projects using Aspen HYSYS and Aspen Plus. Now that I am equipped with the knowledge of fundamentals of Chemical Engineering as well as research experience at the catalysis, reaction engineering, process simulation in the green process and natural gas processing areas.

I am flexible and open-minded in teaching courses. I would like to teach chemical engineering process design projects, chemical engineering laboratories, catalysis, chemical engineering thermodynamics, heat transfer, mass transfer, fluid mechanics, unit operations, and reaction engineering courses for both undergraduate and graduate students.

*** Formatting Issue - Best Available Paper/Graphic ***

542357 New Frontiers in Process Systems Engineering for Large Multiscale Chemical and Energy Networks

Abstract:

Research Interests: My research lies in the field of process systems engineering (PSE), and the overarching theme of my proposed research plan is to develop computational decision making tools to design and operate chemical, energy, biochemical, and agricultural systems in a sustainable and economic manner. Modern chemical plants contain nonlinear phenomena which occur on a wide range of spacial and temporal scales, and my future research aims to develop new theory and methods which can solve large, nonlinear, and multiscale optimization problems in a computationally e cient manner. Such multiscale problems often have multiple layers of optimization, and by recognizing the parallels between an optimization problem and a dynamic system, which both have inputs, state (dynamic system) or decision (optimization problem) variables, and outputs which are functions of these variables, one may be able to predict *a priori* the relationships between inputs and outputs of the inner optimization problems and reduce the difficulty required to solve the overall problem.

Building understanding of the interactions present in large scale systems through the PSE theory and tools I aim to develop is essential in bridging the so-called "valley of death" in scientific funding of new processes and products, thus enabling the transition of these new ideas from something that works great in a lab to something that can be made at scale sustainably and e ciently to make a difference in people's lives. As such, I will always aim to forge collaborations with my future experimental colleagues to find relevant applications for the theory I develop. PSE tools are widely applicable to, for example, the production of new life-saving drugs, the incorporation of a new type of catalyst or chemical reactor in a chemical plant, or the implementation of a new type of battery into the power grid.

My previous research at the University of Minnesota demonstrates my ability for developing decision making tools for the specific application of combined chemical and renewable energy production. I have developed powerful optimization models which enable decision making for plant design[1], facility placement within a supply chain[2], and system operation[3]. Through this work, I was able to elucidate synergies between the chemical and renewable energy subsystems and take advantage of them for e cient operation and design, as well as better understand the tradeoffs of centralized vs. distributed production within a chemical supply chain. I also developed multiple new methods to improve the computational tractability of these and other optimization problems: first, I developed a community based approach which breaks down a large problem into smaller subproblems by finding groups that tightly interact internally but weakly interact with other groups[4,5]. Next, I developed a method for designing systems with time-varying operation whereby operating decisions are made independently from and used to inform design decisions [6]. This work has been recognized by multiple conference travel grants and best presentation awards.

Teaching Interests: My passion for teaching goes back to my childhood years where I would always check over my brother's homework every night and explain to him when he did things incorrectly. I believe an e ective teacher in chemical engineering needs to give each student two things: the knowledge of a

"toolbox" that they can use to approach any problem that they may face in their career, and the wisdom to know which tool to use in which situation. I believe e ective written and verbal communication is essential in achieving both of these goals and aim to make developing e ective communication skills a key outcome for students in all classes I teach and the research group I lead. In my previous experience, I have TA'ed graduate courses in both linear algebra and transport phenomena. In both courses I have developed a comprehensive text document of lecture notes and problem solutions and been recognized with the department outstanding TA award. I have also taught recitations for the undergrad process control course, where I helped to develop recitation problems, taught two sections two times a week, and received an SRT average rating of 5.57/6. For my future teaching plans, I would feel confident teaching any of the core chemical engineering courses, although I have the most expertise in applied mathematics and control. I would also be interested in developing an "applied math and optimization in chemical engineering" course to be taken by seniors and first year graduate students for departments that do not already have a similar o ering.

References

[1] Allman, A., Daoutidis, P. Optimal design of synergistic distributed renewable fuel and power systems. Renewable Energy (100), 2017, 78-89.

[2] Allman, A., Ti any, D., Kelley, S., Daoutidis, P. A framework for ammonia supply chain optimization incorporating conventional and renewable generation. AIChE Journal (63), 2017, 43904402.

[3] Allman, A., Daoutidis, P. Optimal scheduling for wind-powered ammonia generation: e ects of key design parameters. Chemical Engineering Research and Design (131), 2018, 5-15.

[4] Tang, W., Allman, A., Pourkargar, D.B., Daoutidis, P. Optimal decomposition for distributed optimization in nonlinear model predictive control through community detection. Computers and Chemical Engineering (111), 2018, 43-54

[5] Allman, A., Tang, W., Daoutidis, P. Towards a generic algorithm for identifying high-quality decompositions of optimization problems. In Proc. of the 13th International Symposium on Process Systems Engineering (2018).

[6] Allman, A., Palys, M.J., Daoutidis, P. Scheduling-informed optimal design of systems with time-varying operation: A wind-powered ammonia case study. AIChE Journal, 2018, submitted.

542466 Toward Autonomous Molecular Discovery: Machine Learning and Automation for the Rational Design and Optimization of Novel Compounds

Abstract:

Research Interests:

The identification and synthesis of molecules that exhibit desired functions are an essential part of addressing contemporary problems in science and technology. The typical paradigm of molecular discovery is an iterative cycle of design, synthesis, and testing. The rate at which this process yields successful compounds can be limited by bottlenecks at all three stages and is plagued by inefficiencies in resource allocation, including the need for frequent manual intervention.

My future research will address the unification of chemical knowledge and empirical data to support autonomous molecular discovery.

The overall goal of my work will be to develop novel modeling strategies and computational approaches that in combination with more traditional automation techniques will improve the efficiency of small molecule discovery. Molecular discovery is a problem of inference from incomplete and imperfect information, for which techniques in machine learning and artificial intelligence are well-suited. My research will build upon methodologies in machine learning, laboratory automation, and synthetic chemistry to design, implement, and apply AI-driven experimental platforms for the rational design and optimization of novel molecular agents, e.g., as chemical sensors, antibiotics, high performance polymer precursors, organic semiconductors, organocatalysts, and protein binders.

My graduate research with Klavs F. Jensen and William H. Green has focused on addressing one aspect of this broad challenge—streamlining the synthesis of small molecules in the context of pharmaceutical discovery—from two perspectives: one experimental, the other using techniques in data science and machine learning.

My experimental research interests are driven by the goal of increasing the time, material, and experimental efficiency of data collection. To this end, I have developed several automated microfluidic reactor platforms for studying physical and chemical processes at the micromole scale. This has included the design and construction of an oscillatory droplet platform for the on-demand synthesis and purification of compound libraries including those requiring multi-phase, multi-step, or photochemical transformations [1-2]. The platform is capable of efficient closed loop self-optimization using an optimal design of experiments to bridge the gap between research-scale and production-scale synthesis [3]; my industrial collaborators have demonstrated that conditions optimized at the microliter scale are suitable for scaled-up continuous flow synthesis [4].

As a complement to identifying new opportunities for and applications of automated experimentation, my computational research interests have been driven by the goal of meaningfully generalizing existing data to new problems in synthesis and synthesis design. Specifically, in addition to contributing to the quintessential cheminformatics problem of developing structure-activity relationships [5], I have developed new data-driven methodologies for the design and validation of small molecule synthetic routes [6]. Among those methodologies is an automated workflow to propose retrosynthetic disconnections for novel molecules based on analogy to precedents in reaction corpora, intended to mimic manual approaches to the same task [7]. Published reactions also contain an implicit definition of synthetic complexity that a neural model can be trained to quantify for the evaluation of potential synthetic targets [8]. As a more significant undertaking, I have developed novel strategies to predict the reaction outcomes *in silico*—for virtual screening, impurity prediction, or structural elucidation—leveraging the flexibility in pattern recognition afforded by neural networks [9, 10]. My ongoing work in this area focuses on extracting new insights into chemical reactivity that support and supplement expert chemist intuition.

Teaching Interests:

My primary aims in teaching are to instill in students an inquisitiveness about why physical phenomena behave the way they do and to equip them with the ability to articulate and answer their questions. As an undergraduate at Caltech, I TAed a broad range of core chemical engineering subjects including separation process principles, thermodynamics I and II, a synthetic chemistry laboratory, and heat transport. I strongly support the cross-disciplinary use of programming and computational techniques and, after twice teaching the introductory computer science course at Caltech, TAed the first-year graduate course at MIT on numerical methods as applied to chemical engineering. Most recently, I completed the Kaufman Teaching Certificate Program to help formalize the best practices I have observed in the classroom both as a student and as a teacher.

From my formal education and past teaching experience, I am confident in my ability to teach any core chemical engineering course at the undergraduate or graduate level. I am particularly interested in teaching courses that combine chemical engineering and numerical methods, data science, or machine learning, whether as an expansion of an existing course or as a new course, *Introduction to Statistical Inference for the Chemical Sciences*.

Selected Awards:

- DARPA Riser, 2018
- William C. Rousseau Award in Leadership and Ethics in Chemical Engineering Practice, 2016
- Robert T. Haslam Presidential Graduate Fellowship, MIT, 2014
- NSF Graduate Research Fellowship, 2014
- Frederic W. Hinrichs, Jr. Memorial Award for Student Leadership, Caltech, 2014
- HHMI Undergraduate Teaching Assistant Fellowship, 2013
- Edward Richter Memorial Undergraduate Research Fellowship, 2011

Selected Publications:

[1] Hwang, Y.-J*; Coley, C. W.*, Abolhasani, M., Marzinzik, A.L., Koch, G., Spanka, C., Lehmann, H., Jensen, K.F. "A segmented flow platform for on-demand medicinal chemistry and compound synthesis in oscillating droplets." *Chem. Comm.* **53**(49), 6649–6652 (**2017**).

[2] Coley, C. W., Abolhasani, M., Lin, H. & Jensen, K. F. "Material-efficient microfluidic platform for exploratory studies of visible-light photoredox catalysis." *Angew. Chem. Int. Ed.* **129**(33), 9979–82 (**2017**).

[3] Baumgartner, L.*, Coley, C. W.*, Reizman, B., Gao, K., Jensen, K. F. "Optimum catalyst selection over continuous and discrete process variables with a single droplet microfluidic reaction platform." *React. Chem. Eng.* **3**(3), 301-311 (**2018**).

[4] Hsieh, H.-W., Coley, C. W., Baumgartner, L., Jensen, K. F., Robison, R. "Photoredox iridium-nickel dual catalyzed decarboxylative arylation cross-coupling: from batch to continuous flow via self-optimizing segmented flow reactor." *Org. Process Res. Dev.* **22**(4), 542-550 (**2018**).

[5] Coley, C. W., Barzilay, R., Green, W. H., Jaakkola, T. S., & Jensen, K. F. "Convolutional embedding of attributed molecular graphs for physical property prediction." *J. Chem. Inf. Model.* **57**(8), 1757-1772 (**2017**).

[6] Coley, C. W., Green, W. H., Jensen, K. F. "Machine learning in computer-aided organic synthesis." *Acc. Chem. Res.* **51**(5), 1281-1289 (**2018**).

[7] Coley, C. W., Rogers, L., Green, W. H., Jensen, K. F. "Computer-assisted retrosynthesis based on molecular similarity." *ACS Cent. Sci.* **3**(12), 1237-1245 (**2017**).

[8] Coley, C. W., Rogers, L., Green, W. H., Jensen, K. F. "SCScore: Synthetic complexity learned from a reaction corpus." *J. Chem. Inf. Model.* **58**(2), 252-261 (**2018**).

[9] Coley, C. W., Barzilay, R., Jaakkola, T. S., Green, W. H. & Jensen, K. F. "Prediction of organic reaction outcomes using machine learning." *ACS Cent. Sci.* **3**(5)**,** 434–443 (**2017**).

[10] Jin, W., Coley, C. W., Barzilay, R., & Jaakkola, T. "Predicting organic reaction outcomes with weisfeiler-lehman network." *NIPS* (**2017**).

542494 Understanding and Exploiting the Tunability of Long-Range Electrostatic Interactions in Soft Materials

Abstract:

Research Interests:

My research interests lie in understanding the tunability of long-range interactions for controlling soft matter behaviors by multiscale computational simulations. The electrostatic interactions are sensitive to various tuning parameters including solvent permittivity, electrolyte valence, size and concentration, and interfaces, and provide tunability for the properties of soft materials. Despite long-term usage of electrostatics in controlling the soft matter properties, such as buffer system in biology and the self-assembly of metamaterials, the understanding of electrostatics interactions is still limited. Here we discuss about researches on the effects of surface polarization on the electrostatic interactions between ion-containing emulsions, the effects of electrolyte concentration on the electrostatic interactions between colloids, and the effects of solvent permittivity on the aggregation of macrocycles. The researches shed light on why classical electrostatic theories such as the Debyeâ€"HÃ¼ckel theory and even the non-linear Poisson Boltzmann theory break down in recent electrostatic mediated experiments. The poster will then proceed to the prospective of tunability of general long-range interactions in manipulating soft matter properties.

Teaching Interests:

1. Classical and statistical thermodynamics

2. Kinetics of materials

3. Electrostatics

4. Nanoscale heat transfer

5. Transport phenomena

6. Condensed matter physics

542515 Molecular Modeling of Anti-Microbial Peptides at a Water-Lipid Bilayer Interface

Abstract:

Current Research

In the last 20 years, interest in peptide folding mechanisms at water-membrane interfaces and peptide-membrane interactions has increased in diverse biological fields. A example is the interaction of anti-microbial peptides (AMP) against bacterial membrane. Research studies show future potential of AMPs as an applicable antibiotic drug. Although experimental studies have explored the details of AMP-membrane interactions by suggesting significant roles of the structure of peptide, the kind of amino acid sequences, and the structure of membrane, there is still much uncertainty in the exact mechanism of cell death by AMPs and the relationship between the structure of AMPs and its antibacterial functionality. Molecular modeling provides atomistic details and 3D structure of peptide-membrane interaction that contributes to investigation of this mechanism and also can help to find which biophysical parameters and AMPs characteristics can affect the performance of AMPs. Prior computational studies demonstrate the crucial effect of hydrophobicity, charge and the helical structure of peptides, and important role of aromatic amino acids such as Tryptophan (TRP) and Tyrosine (TYR) during interactions with membranes.

In my research under supervision of Professor Michael Greenfield, as a general aspect, I attempt to understand how changes in amino acid residues have an effect on peptide folding and peptide-membrane interactions. To that end, I am focusing on the structures of novel hybrid AMPs LM7-1 and LM7-2 at water-membrane interface. These peptides were designed previously using combinations of naturally occurring AMPs, and they differ in sequence only at the 15th residue. Prior experimental results indicate different performance of these two AMPs against diverse bacteria. One of the potential explanation for this issue can be the different folded structure of these AMPs at a water-membrane interface. Molecular dynamics (MD) simulation is used as an appropriate tool to investigate the details of AMP and membrane structure exactly in the moment of interaction. The hydrophobicity distribution on the folded structure of peptide, the deviation of peptide from helical structure at the moment of binding with membrane, and the pore formation structure on the membrane are all key points that can be analyzed using the results of MD simulation. However, all of these kinds of studies depend on the accuracy of the simulation.

The accuracy of biomolecular modeling results is based in part on force field (FF) parameters. Several simulation studies have proved that applying different FFs leads to various peptide folded structures and consequently results in different functionalities of peptide. The results from the first steps of these simulations on the basis of Charmm36 FF show that the structures of aromatic rings (TRP and TYR) have deviations from planarity. These aromatic amino acids have an effective role in the peptide-membrane interaction. Hence, I investigated the geometry and inherent nonplanar vibrations of these aromatic rings in more detail. I applied all-atom Normal Mode Analysis (NMA) on the results of the MD simulation to calculate the inherent vibrations of atoms. On the basis of this analysis, some deficiencies in calculated nonplanar mode frequencies from Raman spectroscopy and the results of quantum calculation have been observed.

We found that Charmm36 has not considered improper torsion for atoms in the ring of TRP and TYR. Including improper torsion in simulation brings atoms in the ring of TRP and TYR to a planar structure. However, the nonplanar vibrational modes shifted to higher frequencies compared to Raman Spectra and to results of quantum calculation. Beside the improper torsion, torsion angle that include atoms in the ring have a direct effect on nonplanar vibrations. By including the improper torsion with specific force constants and applying a specific set of amended torsion angles parameters, we could fix the pattern and frequencies of nonplanar modes that can match with Raman spectra and the resutls of quantum calculations. In summary, we modified charmm36 FF to provide proper dynamics and structures for atoms of TRP and TYR.

Investigating the journey of peptide from coil structure in solution to folded structure at a water-membrane interface can help us to find how the one different residue in the sequence of LM7-1 and LM7-2 leads the peptide to a different path of the folding process. MD simulation can explore the conformational space for a peptide folding process. Since large-scale motions in peptide are a consequence of small-amplitude fluctuations, it is also essential to investigate the individual atomic fluctuations over MD simulation. NMA is applied on the configuration results of MD simulations of peptide in solution or at water-membrane interface. This calculation provides the vibrations in a peptide and also provides ta surface energy map that shows which specific vibrational modes lead a peptide to be folded or unfolded. Comparing analysis for both LM7-1 and LM7-2 can help us understand how one difference in the peptide sequence affects their performance against a bacterial membrane. The outcome of the computational studies will contribute to explaining atomistic detail in the process of peptide folding at water-membrane interfaces, and also will provides a proper perspective in understanding biophysical characteristics of peptide-membrane interactions.

Future Research Interests:

I would like to expand my research to study the structure and the interaction of biomolecules at interfaces by applying molecular modeling. Besides peptide or protein folding at a water-membrane interface, which is my current research and I explained it above, I am interested in nanoparticle-membrane interactions, especially nowadays that the implementation of nanoparticles (NPs) in drug delivery is increasing. Investigating the effect of physical parameters such as hydrophobicity and charge distribution provides insight to the design of biocompatible NP structure for pharmaceutical applications. In addition, the interaction of protein with NPs and understanding the formation mechanism of protein corona in atomistic detail is another significant phenomena from the perspective of NPs toxicity.

Teaching Interests:

Beside my research career, I have extensive teaching experience of about 12 years that shows my intensive interest in teaching. I believe teaching helps me to achieve the deeper level of understanding. I started teaching at high school and institutes when I was an undergraduate student. As a PhD student, I continued teaching as a teaching assistant (TA) at University of Rhode Island (URI) for 4 years. During these four years, I have been chosen to be a TA for diverse courses. Even as a Ph.D student, I had the opportunity to be the sole instructor of “Chemical Engineering Thermodynamics 1” course (CHE313-spring 2017) at URI.

My previous experiences in teaching:

1- Instructor of “ Chemical Engineering Thermodynamics 1” at URI – Spring 2017

2- TA for Senior Chemical Engineering Lab – 9 semesters at URI

3- TA for courses: Chemical Kinetics and Reactor Design, Thermodynamics 1 and 2, Engineering Materials, Chemical Process Calculations.

4- Teacher in Iran of Chemistry and Geometry, Analytical Geometry, Linear algebra for 8 years at high school and university entrance exam preparation institutes.

In the future, I am most interested in teaching:

1- Applied mathematics in chemical engineering

2- Molecular modeling

3- Thermodynamics

4- Mass transfer

5- Heat transfer

6- Fluid Mechanic

7- Unit Operations

542520 Computational Design of Functional Materials and Their Interfaces

Abstract:

Research Interests:

One of the greatest challenge that engineers currently face is to design new materials with enhanced properties for targeted applications. To efficiently realize this aim, we must precisely understand and control the structure-property relationship of materials at the atomic level. My goal is to take part in this material design challenge by using computational techniques to not just interpret experimental findings, but also serve as a guide towards important discoveries.

As a postdoctoral researcher, I have gained experience in a wide range of computational surface science and catalysis methods, as well as tackling problems related to soft matter. For example, I built computational models to predict the interfacial orientation of liquid crystals on solid surfaces. Through high-throughput screening applications, these models have been successfully implemented to design liquid crystal-based chemoresponsive systems that can detect various chemical analytes selectively down to the parts per billion level. I have also developed computational frameworks to identify catalytic active sites, calculate diffusivity, and explore catalytic reaction networks in an automated fashion which allows one to easily investigate realistic model systems with high complexity.

In the future, I plan on designing materials to solve problems where the solid-liquid interface plays a key role. Specifically, I will focus on computationally predicting surface modifications to (i) alter solid-liquid interactions, (ii) increase the stability, activity, and selectivity of catalysts, and (iii) improve the stability and cyclability of electrochemical storage systems.

Teaching Interests:

I passionately believe that the inclusion of state-of-the-art research findings in teaching is one of the best ways to foster an engineering mindset, as well as promote and maintain the curiosity and open-mindedness of students to novel ideas. Integrating research with teaching allows students to develop self-motivation and drive to search for new interesting problems and potential solutions on their own. For this reason, I would like to develop undergraduate and graduate level electives involving the use of computer simulations which are close to my own research interests. Additionally, I will teach most chemical engineering courses due to my chemical engineering background.

542612 Multi-Scale Modeling of Biophysical Systems and Soft Matter

Abstract:

Research Interests:

Processes that span multiple length scales and timescales are common in nature, and this multiscale nature often limits our knowledge of many biological systems. Systematic models based on a strong fundamental basis often serve as powerful tools to probe such systems in aspects that lie outside the current experimental limits, and complement our understanding of these systems. My research interests lie in developing systematic modeling frameworks to understand and manipulate the structure and dynamics of biological macromolecules such as DNA, RNA and proteins and synthetic polymers by manipulating their environments, and to use such modeling frameworks for designing commercial applications.

My graduate work at University of Massachusetts Amherst (UMass Amherst) introduced me to the idea of using nanopores as rapid low-cost DNA sequencing devices and the design challenges involved in transforming this idea into a commercial technique. Using extensive coarse-grained simulations of DNA as a polyelectrolyte translocating through biological and synthetic nanopores, I was able to highlight the important role of electrostatic interactions in the kinetics involved. I also developed simple design principles for using nanopores to characterize polyelectrolytes based on their architectures. I implemented the Multi-Particle Collision Dynamics technique on a GPU architecture as a part of this work. I constructed a theoretical model for the translocation process based on the Fokker-Planck formalism and established the parameter window for validity of the model. After advancing to a postdoctoral position at the University of Chicago, I focused on developing systematic coarse-grained models for biologically relevant polymers. I developed a multi-scale coarse-grained modeling framework for studying filament networks in the cell. I discovered the underlying mechanism of hydrolysis of the nucleotide bound to the filamentous protein actin using this framework, and used enhanced sampling simulation methods to aid modeling of actin associating proteins.

In the near future, I plan to combine the expertise I have developed through these research experiences to propose application based nanopore design strategies. I will start by developing an efficient multi-scale modeling and simulation framework to allow for rapid screening of nanopore designs to control the transport of important biological and synthetic polymers. I will then extend this framework to study the design of nanopores as commercial devices to characterize protein folding. My broader goal is to develop a strong research program in the area of biophysics and soft matter using systematic modeling and simulation techniques, with a balance between efforts aimed at developing a fundamental understanding of biological systems and those aimed at developing technologies of commercial importance.

Teaching Interests:

I have had a pleasant experience as a teaching assistant for many courses while I was a graduate student in the Departments of Chemical Engineering at both, UMass Amherst and at Indian Institute of Technology Bombay (IIT Bombay). At IIT Bombay, I conducted laboratory sessions involving heat exchangers for the undergraduate Chemical Engineering Laboratory and was involved in constructing problems aimed at understanding the design principles of heat exchangers for the laboratory sessions and the for final examination. I also conducted experiments and examinations for pneumatic conveying of granular materials for the undergraduate Laboratory. I graded assignments and exams for undergraduate courses at both these institutions and enjoyed the experience of helping individual students to confidently solve assignment problems by providing them just enough information to trigger their thought process. At UMass Amherst, I had the additional opportunity to conduct weekly, hour-long tutorial sessions that complemented the lectures by the course instructor for the undergraduate Fluid Mechanics course, and had a first-hand experience conducting tutorials for the graduate Transport Process course. I also value the opportunity I had as a mentor at both these institutes, to my colleagues and to undergraduate summer interns.

Through my experience as a student of some outstanding teachers, and as a teaching assistant for the courses mentioned above, I have learnt simple principles that make a course easy to learn. I will provide a course outline and make references to it during my lectures. I will provide peer-learning opportunities through the use of group assignments. I will strive to make efficient use of chalk boards or white boards for the delivery of suitable parts of the course material, especially those containing important mathematical derivations. Based on my teaching experience and my research expertise, I propose to teach the following courses.

Core Courses: Transport Processes, Fluid Mechanics, Chemical Reaction Engineering, Chemical Engineering Analysis, Numerical and Computational Methods in Chemical Engineering.

Advanced Courses: Modeling and Simulations, aimed at imparting Chemical Engineers with basic skills required to construct systematic models and perform simulations from molecular scale up to the process scale.

542353 Toward Emergent, Adaptive, and Hierarchical Bio-Inspired Materials

Abstract:

Research Interests:

Precise structural control over materials design has enabled unprecedented technological developments to address some of our most pressing global challenges, which include climate change, sustainability, medicine, and water. Many of these advances can be attributed to the foundational paradigm of structure-property relationships that dictate the fundamental chemistry and physics of materials. However, the prescriptive nature of the thermodynamics (and kinetics) that dictate such materials are also self-limiting. While designer materials are finely tuned for a unique purpose, it is attractive to consider an alternative in which materials are encoded with multiple functions that can be modulated on-the-fly during operation.

To this end, I have always been fascinated by the opportunities of out-of-equilibrium phenomena as a means to both expand and tune the possible materials design space. Nature is an exemplary model of this principle as most biological machines, e.g., macromolecular complexes, often undergo highly-dynamic and regulated processes for construction, modification, and recycling. Intriguingly, these complexes are assembled from many copies of individual macromolecules, which may further transition between diverse sets of hierarchical architectures. Therefore, these systems are an ideal platform to understand how collective molecular mechanisms may enable emergent and adaptive functionality.

I am particularly interested in (1) understanding the principles used by biological building blocks (i.e., proteins, lipids, and nucleic acids) to regulate the structure and function of macromolecular complexes and (2) adapting these principles to create functionally-diverse bio-nanomaterials. Through these efforts, I anticipate that the fundamental structure-kinetic-function relationships that control these inherently dynamic processes can be leveraged toward materials design in applications including energy storage, catalysis, drug targeting and delivery, and water purification. My strategic approach is to utilize the computational methodologies that I have developed, ranging from quantum mechanical calculations to systematic coarse-graining (coupled with enhanced sampling approaches), to probe these biophysics, which are inherently hierarchical and far-from-equilibrium. Therefore, I envision a research program that exploits my extensive background in multiscale computational simulations and unites my interests in nanotechnology, biophysics, and sustainability.

Teaching Interests:

Throughout my training, my mentors have been invaluable for my growth as an independent scientist and engineer. As a faculty member, I eagerly look forward to my responsibility as a mentor for the next generation of young scientists. My core teaching philosophy can be described by the following three principles. My foremost goal is to instill a passion for learning by fostering both the curiosity and the confidence to be inquisitive without fearing failure. I have thoroughly enjoyed preparing interactive activities, which highlight interesting concepts from my research, to engage local students (from elementary to high school levels). My second goal is to create tangible connections between what students learn in academic settings and real-world problems. For this reason, I firmly believe in the value of immersive, project-based learning as a supplement to lectures. Not only do these projects translate engineering skills toward problem solving, but also establish communication and collaborative skills. In fact, one of my undergraduate courses followed this model and eventually inspired me to pursue computational research. Finally, I aim to create an environment for my students to question underlying assumptions and to develop physical intuitions. As a teaching assistant, I have found that one of the best forums for this discourse emerges from thought-experiment challenges, which are first debated within small student clusters with analysis and conclusions presented to the rest of the class.

While I am prepared to teach any core chemistry and chemical engineering course, I am most interested in fluid mechanics, thermodynamics, or numerical methods. I am also looking forward to developing new courses based on my core competencies, such as a broad introduction to the growing field of computational sciences.

541888 Multiscale Simulations of Nonequilibrium Mechanisms in Aqueous Solutions

Abstract:

Research Interests:

My scientific research revolves around aqueous mechanisms on the molecular level (changes in polymer conformations, colloid aggregations, etc. in water). Importantly, such systems exhibit multiscale behavior: For example, a minor mutation in a protein sequence can cause a major error in a cell network. Aiming at clarifying the nuances of such issues, I employ my computational skills in multiscale simulations, which are algorithms for correctly and efficiently connecting microscopic and macroscopic models in soft matter. My most distinguished scientific achievements are the following:
â€¢ While working with Prof. M. Scott Shell, I resolved a paradox between numerical predictions and empirical observations that has existed for decades, convincingly showing that the force between small hydrophobes is weak oscillatory and between large hydrophobes is strong monotonic.
â€¢ While employed by Prof. Kurt Kremer, I formulated a multiscale strategy that gains an efficiency of as much as an order of magnitude, while describing the correct static and dynamic behavior of molecular liquids across their phase diagram.

I am currently moving in the direction of nonequilibrium systems relevant for environmental and pharmaceutical nanotechnologies, particularly focusing on adjusting their microscopic features for optimizing their macroscopic functions. At the onset of my career, I will pursue two specific routes in my research. The first one deals with impurity filtration via graphene nanochannels: My main goal here will be in optimizing a nanochannel in being permeable for water but selective against contaminants. The second one deals with drug delivery via liposome nanocarriers: My main goal here will be in optimizing a nanocarrier for releasing the bulk of a certain medicine in nominal time. Over time, I will incorporate these two projects in a research program that also explores multiscale phenomena in biology: In particular, I am interested in how microscopic mutations in residue sequences influence macroscopic errors in signal networks.

Teaching Interests:

I have a unique background in pedagogy. At university, I volunteered earlier on as a Teaching Aide, and I worked later on as a Teaching Assistant and as a Teaching Associate. While the former position involved the regular duty of a student, the latter position comprised the entire role of an instructor. In fact, it was because of my distinguished performance as a Teaching Assistant that I received a prestigious award that provided me with the opportunity of being a Teaching Associate. Consequently, I have correspondingly completed an official certificate in instruction. In my spare time, I am also employed as a mathematics instructor at a local academy.

I can lead many courses in Chemical Engineering. Considering my mathematical expertise, I am especially suitable for

the fundamental courses (e.g., Thermal Physics, Transport Phenomena, Reaction Kinetics, Quantum Physics, etc.). Via my scientific experience, I also intend to develop two elective courses (i.e., one about molecular simulations and one about aqueous systems).

541952 Computational Design and Characterization of Nanoscale Materials for Applications in Energy, Separations, and Catalysis

Abstract:

Research Interests:

In my research I use molecular modeling and *ab initio* calculations to predict the structural, electronic, and thermodynamic properties of materials. This allows us to accelerate the design of new materials with broad applications in energy, catalysis, and chemical separations. In order to move our society from our current reliance on fossil fuels to sustainable energy sources, we will need to develop advanced materials for energy production and storage, as well as more energy-efficient chemical processing methods.

My graduate research focused on using density functional theory (DFT) to predict vibrational spectra for Si and Ge nanoparticles with impurities and heterojunctions. By understanding how changes to the geometry and electronic structure affect the Raman response, we can interpret experimental spectroscopic data to infer the size, dopant concentration, and interfacial strain in these nanoparticles. My postdoctoral work centers on the design of nanoporous materials for gas storage, separations, and catalysis, with particular emphasis on hydrogen storage and the capture of toxic gases. As a postdoc, I have worked closely with experimental researchers to use computational modeling to offer additional insights into experimental results.

In the future, I plan to build on my background in molecular modeling, materials science, and high-performance computing to discover new materials with applications related to energy, separations, and catalysis. I have identified three areas of research I would like to focus on in my research group:

1. Adsorption-based separations using MOFs, zeolites, and other nanoporous materials
2. Structure and applications of metal nanoparticles supported within porous materials
3. Electronic structure of organic crystals and their application in electronic devices.

Many chemical separations rely on energy-intensive thermal processes like distillation. However, it is difficult and expensive to separate similar molecules, such as close-boiling isomers, using distillation. For example, the C4 aldehyde isomers n-butyraldehyde and isobutyraldehyde are produced in a mixture of n- and iso- isomers that must be separated before they can be used. This separation is currently done by distillation, but sorbent-based separations using nanoporous materials, such as zeolites or metal-organic frameworks (MOFs) offer a better option that is more energy efficient and highly selective. There are a huge number of possible MOFs with diverse textual properties and chemical functionality. This means we can screen thousands of structures and choose a MOF with high selectivity and capacity for separating the specific molecules we are studying. High-throughput screening generates a large amount of data that must be analyzed, so we will use machine learning and data mining techniques to evaluate this data more efficiently and make meaningful predictions about materials properties.

Metal nanoparticles embedded in nanoporous support can be used for many catalysis applications. In many cases, the choice of support influences the catalytic activity of the nanoparticle. For example, there is experimental evidence that some MOFs interact with metal nanoparticles (e.g. Cu in MOF-5) and affect the catalytic process in a way that does not happen in other supports like silica. I plan to use DFT calculations to study this interface between the MOF support and the metal nanoparticle in detail. I am interested in how the electronic structure of the metal nanoparticle is altered by various supports and how this affects the catalytic activity of the particle. I also will use simulations to study the formation of metal nanoparticles in supports to help give insight into experimental synthesis.

I am interested in the use of MOFs and organic crystals in electronic and photonic devices. Some MOFs are conductors or semiconductors and have potential to be used in light-harvesting, chemical sensing, and energy storage applications. However, relatively little research effort is devoted to the electronic structure of these materials. Most device applications will require the MOF to be grown in a thin film on a metal or metal-oxide support. The nature of the interface between these supports and the MOFs is not well understood. I plan to use DFT to study the interface between MOFs and metal or metal oxide supports and how this influences electronic properties such as band structure and electron mobility.

Teaching Interests:

My goal as an instructor in chemical engineering is to give students necessary foundational knowledge and practical skills, while also imparting a sense of the broader connections between various courses they will take throughout their education. Therefore, I am interested in teaching any of the fundamental courses such as thermodynamics, kinetics, transport, and mass and energy balances. Since my background is primarily in molecular modeling and computation, I would also be interested in teaching courses such as numerical methods or MATLAB programming. I am also intrigued by the opportunity to teach more applied courses such as chemical engineering design, which draw fromâ €"and synthesizeâ€"all elements of chemical engineering.

In addition to these courses, I would be interested in designing and teaching an elective course for graduate students taken from my own research, which focuses on computational materials science and molecular modeling. The course would offer an introduction to density functional theory, molecular dynamics, Monte Carlo methods, and high performance computing. This would be a valuable primer for graduate students who plan to pursue computational research and also provide an accessible overview for students in experimental research who want to gain a sense of how modeling is done and what advantages and limitations various techniques offer.

My interest in teaching extends beyond lecturing in a classroom. The primary role of a professor is mentoring students either in a lecture hall or in a laboratory. As a postdoc, I have supervised several undergraduate students and one masters student doing research. Several of these students have graduated and gone on to jobs in industry or graduate school. I find mentoring students in research to be a rewarding part of academia, and I enjoy building relationships with the students through working together on a challenging research problem.

531950 Data-Driven Modeling in Chemical Engineering and Molecular Science

Abstract:

Recent advances in data-driven modeling and deep machine learning have revolutionized the fields of computer vision, natural language processing, and game playing. Inspired by these recent success, I demonstrate the application of several methods from deep generative modeling and reinforcement learning on problems in molecular design and discovery, transport phenomena, and process control.

Research Interests: Data-Driven Modeling and Machine Learning, Molecular Design and Discovery, Artificial Intelligence in Chemical Process Design and Control

Teaching Interests: Thermodynamics, Transport Phenomena, Chemical Reaction Engineering & Kinetics, Data Science in Chemical Engineering

540473 Process Systems Engineering and Artificial Intelligence for Advanced Manufacturing: *Including Applications to Biopharmaceuticals*

Abstract:

Highlights:

- Research interests
  1. Development of high-fidelity models that can reliably predict production and product quality (for the purpose of model-based process design and control)
  2. Detection and prevention of unintended negative consequences of manufacturing using process systems engineering, risk management, and artificial intelligence techniques
  3. Advanced decision-making framework through "human-in-the-loop" computing that integrates both human and machine intelligence
- Background
  1. Doctoral (Columbia University): solving sociotechnical problems with process control, risk management, data science, and game theory
  2. Postdoctoral (University of Delaware): applying system identification, process design, design of experiments, and multiscale modeling to biopharmaceutical processes for the improvement of production and product quality
- Teaching interests
  1. Core: process dynamics and control, process safety, design project, simulation, and mathematics
  2. Electives: computing, project management, data science, and optimization

Research Interests:

Achieving high product quality has become more challenging as the complexity and scale of manufacturing rapidly increase. The U.S. Food and Drug Administration (FDA), for instance, has been shifting the pharmaceutical industry away from the traditional quality by testing (QbT) approach, which involves univariate analyses of product quality by trial-and-error and heavily relies on heuristics. Instead, more systematic and model-based methodologies such as quality by design (QbD) and quality by control (QbC) are gradually being adopted. My research interests revolve around addressing core challenges for QbD and QbC such as efficient processing of manufacturing and R&D data, developing reliable models, and designing optimal processes and control strategies with the help of process systems engineering (PSE) and artificial intelligence (AI).

PSE is a chemical engineering discipline that focuses on developing systematic and computer-assisted approaches to the design, operation, optimization, and control of processes. At its core, PSE is about systems thinking, which considers a system, its interacting constituents, and competing objectives. QbD and QbC could benefit from both the technological advances in PSE—progress in model predictive control

(MPC), multi-objective optimization, global optimization, nonlinear control and optimization, etc.—and its unique systems perspective.

Due to the significant role that computing plays in supporting PSE, PSE also has a long history of applying AI techniques. The field of AI research was born more than sixty years ago and early PSE adoptions date back to the 1980s (e.g., expert systems). Recent advances in computing technologies (power, storage, and infrastructure) have led to rapid developments in certain AI fields such as machine learning and natural language processing (NLP), providing new tools for PSE to adopt in the age of big data. The abundance of data in manufacturing creates both opportunities and challenges: On the one hand, the scale of process data is enormous and new data are created every day. On the other hand, much of the data are unorganized, not machine-readable, prone to human error, or contain proprietary information. These limitations make it difficult to generate useful information from the data algorithmically. The challenges are especially pronounced in certain manufacturing fields such as biopharmaceuticals where large-scale computing has yet to be widely adopted.

I come from a diverse and unusual research background: My doctoral training from Columbia University with Prof. Venkat Venkatasubramanian and Prof. Garud Iyengar focused on solving sociotechnical problems with process control, risk management, data science, and game theory. I have collaborated with leading experts and executives in chemical engineering, computer science, operations research, finance, economics, and public health. Such diverse research experience gives me profound understanding in risk management and decision-making as well as the capacity to conduct high-quality interdisciplinary research. My current research with Prof. Babatunde Ogunnaike and Prof. Kelvin Lee at the University of Delaware involves a close collaboration with a major pharmaceutical company and the application of system identification, process design, design of experiments (DoE), and multiscale modeling to biopharmaceutical processes for the improvement of production and product quality. Our hybrid model (knowledge-based and data-driven) achieves high prediction accuracy; its modular structure can be easily updated with new process inputs such as experimental conditions, media/feed compositions, and sampling/feeding strategies. I have also optimized the computer codes, making them run sixty times faster than the previous version. The deliverable for my first six-month collaboration was a fully functional user interface that can predict production and product quality and also solve process variables to meet certain production and product quality goals thus achieving QbD.

My understanding of the key challenges in advanced manufacturing is as follows. First, one of the most important tasks central to achieving the objectives of QbD and QbC is the development of high-fidelity models that can reliably predict production and product quality. I will take advantage of the computing breakthroughs (e.g., deep learning, GPU computing, and NLP) to analyze (structured and unstructured) process data and extract information regarding process dynamics for the purpose of building data-driven models. Furthermore, I will use global optimization tools such as genetic algorithm, simulated annealing, and particle swarm optimization to design and discover optimal processes and control strategies. This will be a natural extension of my postdoctoral work.

Second, the positive benefits of rapid technological developments sometimes come with negative unintended consequences to society. Unintended consequences, by definition, are difficult to detect. For example, efficient light bulbs could increase overall energy consumption due to the low cost. I will apply PSE techniques (e.g., fault detection and diagnosis), risk management techniques (e.g., failure mode and effects analysis), and AI techniques (e.g., applied ontology and various search algorithms) to algorithmically identify hazards and reduce negative unintended consequences of manufacturing.

Last but not the least, decision-making process in manufacturing can benefit from the "human-in-the-loop" computing—algorithms generate timely recommendations based on models and data while human decision-makers evaluate, accept, modify, or reject such recommendations. This arrangement, if done right, would benefit both human decision-makers and algorithms. On the one hand, algorithms are quick, exhaustive, and objective. On the other hand, human judgment and feedback in return can serve as supervised learning to improve algorithms. In addition to human–machine interactions, according to my doctoral research, the collective intelligence of individual human decision-makers can be potentially enhanced through certain types of information sharing. Developing mechanisms that integrate both human and machine intelligence requires insight beyond engineering. My prior experience with social science gives me an edge on creating advanced decision-making framework through “human-in-the-loop” computing.

Teaching Interests:

Given sufficient preparation, I can teach any chemical engineering course. That said, my first priority is teaching courses related to my research background, such as process dynamics and control, process safety, design project, simulation, and mathematics (differential equation, numerical analysis, linear algebra, etc.). I also plan to introduce new content about relevant and current technological advances to enhance students' understanding and motivate them to learn. Examples of topics include but are not limited to MPC, risk management, and AI.

In addition to teaching the core curriculum, I am also interested in designing technical electives on topics such as computing, project management, data science, and optimization, either as standalone courses, or packaged together as a single course that covers the basics of each topic. The purpose of these electives is to guide students to use tools and techniques outside chemical engineering—code documentation, code optimization, coding style, version control, project planning, project monitoring, SQL, deep learning, linear programming, convex optimization, etc.—to conduct research more efficiently.

541830 Optimization in Three Process System Engineering Problems: Inventory Routing, Product Scheduling and Design of Experiments

Abstract:

Research Interests:

I am interested in using mathematical optimization to solve practical chemical engineering problems. Mathematical Optimization is the selection of the optimal elements from many, possibly infinite number of, alternatives, based on a proposed model for the process to be studied, which will provide timely and intelligent aids in decision-making. Over the past few decades of study, optimization has been proved to improve many process performances, including but not limited to: product quality, process robustness, environment sustainability, capital and operational cost, etc. To use optimization properly for solving chemical engineering problems, one should (i) model the process effectively and accurately enough, and (ii) solve the models in a timely fashion.

Previously, I studied the applications of optimization in three process system engineering problems, including (i) inventory routing, (ii) production scheduling, and (iii) design of experiments for the pharmaceutical industry. In the future, I plan to study how to solve the online scheduling and inventory routing problem through the perspectives of finding the â€œbetterâ€ solution from the pool of symmetric solutions. I also plan to combine the knowledge of optimization and design of experiments together, which will not only provide the optimal solution, but also offer statistical insights.

Teaching Interests:

I am very interested in teaching courses on (i) process synthesis, design and optimization, and (ii) process dynamics and control. First, I have a solid background on process synthesis, mathematical optimization, design of experiments, and artificial intelligence. Therefore, I will love to teach how to analyze and synthesize the chemical processes, as well as how to design and optimize them using the modern optimization or artificial intelligence tools (e.g., integer programming, neural networks, etc.); additionally, design of experiments will also be a very tool for many chemical engineering students to include in such a course. Second, I have studied process dynamics and control, including the advanced model predictive control (MPC), in both undergraduate and graduate years; I have also assisted the teaching of such a course for many semesters (including guest-lecturing). Thus, I would love to teach students how to study the dynamics of process, and how to control the dynamic processes, using the important ideas such as feedback and close-loop; moreover, topics such as MPC, the comparison of knowledge-driven and data-driven models will be interesting to introduce in this course. I am open to teaching other courses too.

Overall, I wish I can lead and guide the students to learn the knowledge, as well as to obtain the tools for self-education, because learning should be a lifelong process, especially considering the rapid growth of technology.

Previous Research Topics and Results:

First, inventory routing problem (IRP) is a supply chain optimization problem that arises in a wide range of manufacturing sectors, including industrial gases and petrochemicals. To improve the efficiency of their supply chains, vendors in a wide range of manufacturing sectors switch to vendor-managed inventory (VMI) practices. VMI refers to an agreement between a vendor and a customer in which the latter allows the vendor to choose the timing and volume of deliveries, while the former agrees to ensure that the customer does not run out of product. A switch to such a practice has a distinct advantage because the vendor can combine deliveries to make more efficient use of the resources and can therefore reduce the distribution cost. However, executing VMI is non-trivial, which requires the solution of IRP, a difficult combinatorial problem. We have conducted studies for IRP in the following three aspects: propose mixed-integer programming models that handle different features and constraints that appear in IRP, especially, driver constraints (1); develop solution methods that can solve IRP efficiently (2); and study how reoptimization should be carried out for IRP (3).

Second, production scheduling has been widely studied, because it is a complex and practical problem that needs to be addressed for many chemical manufacturers. In the face of new information and uncertainties, online scheduling is required, and terminal constraints should be added in the model. However, how to write the terminal constraints is not apparent. The traditional approach would require that the terminal inventory level of each material is required to be greater than a lead-time-based threshold, but this method does not exploit the relationships of inventory levels among materials. Considering this issue, we propose new types of terminal constraints on inventory levels for different network structures (4). The proposed terminal constraints lead to better closed-loop solutions in two aspects, compared to the traditional approach: they prevent stockout and save inventory holding cost. Additionally, we proposed different formulations, and compared their theoretical tightness and computational performances for modeling the sequence-dependent changeovers (5).

Third, Dynamic Response Surface Modeling (DRSM) is a recently proposed data-driven method that models time-resolved outputs, and it can be used for the classical Design of Experiments (DoE) and the generalized Design of Dynamic Experiments (DoDE). Compared to the classical approaches, DoDE and DRSM methodologies allow time-varying inputs, accommodate time-resolved output measurements, and model their interrelationship. Because DRSM quantifies the effects of experimental conditions on the process outputs over time, it can enable more convenient and comprehensive process optimization. We apply the DRSM model in a pharmaceutical problem to obtain the optimal operating condition. The optimization is aimed to maximize the length of an operational window, which is defined based on the species concentrations; within this window, the concentrations of all starting materials, intermediates and by-products should be below their specifications. Because DRSM provides the models of the concentrations under different conditions and at different times, it is very convenient and useful to apply it for this optimization problem. Moreover, we also incorporate two types of robust constraints in the optimization to explicitly handle the uncertainty in the operating condition as well as in the DRSM model .

Reference

1. Dong, Y., Pinto, J.M., Sundaramoorthy, A., and Maravelias, C.T. MIP Model for Inventory Routing in Industrial Gases Supply Chain. *Industrial & Engineering Chemistry Research,* 2014, 53, 17214-17225.
2. Dong, Y., Maravelias, C.T., Pinto, J.M., and Sundaramoorthy, A. Solution Methods for Vehicle-based Inventory Routing Problems. *Computers & Chemical Engineering*, 2017, 101, 259-278.
3. Dong, Y., Maravelias, C.T., and Jerome, N.F. Reoptimization and Policy Analysis for Maritime Inventory Routing under uncertainty. *Optimization & Engineering,* 2018, doi: 10.1007/s11081-018-9383-8.
4. Dong, Y. and Maravelias, C.T. Terminal Constraints for Single-stage Multi-product Scheduling Problems. In *Foundations of Computer Aided Process Operations (FOCAPO)*, Tucson, AZ, January 8-12 2017.
5. Velez, S., Dong, Y., and Maravelias, C.T. Changeover Formulations for Discrete-Time Mixed-integer Programming Scheduling Models. *European Journal of Operational Research.* 2017, doi: 10.1016/j.ejor.2017.01.004.

543704 Designing Chemical Reactivity at the Nanoscale using Molecular Simulation

Abstract:

Research Interests:

There is great need for methods that can simulate both reactivity and condensed phases at the molecular level. Quantum mechanical (QM) methods can model the changing electronic environment as a reaction occurs, but are generally too small— O(100) atoms—to represent the condensed phase. Classical molecular mechanics (MM) simulations can be much larger, but cannot model reactivity. I am interested in employing and developing methods that can address both aspects—including QM/MM simulations, reaction ensemble Monte Carlo, and quantum-accurate reactive classical force fields.

Some specific research problems include developing methods to assess ionization in strong electrolytes (e.g. protic ionic liquid formation); modeling nucleation under reactive conditions (e.g. corrosion of plutonium); adding functionality to solid supports for reactive absorption (e.g. in amine-functionalized metal-organic frameworks); and developing quantum-accurate reactive force fields compatible with classical solvents (e.g. SPC water).

Teaching Experience:

I am energized by teaching and interacting with students at all levels. As a teaching assistant, I volunteered to give lectures while professors were away on travel. My goal was to help students understand the foundational assumptions of each equation we derived. They recognized my hard work by voting me TA of the Year.

As a grad student and in both my postdocs, I've sought out opportunities to lead elementary, middle, and high school students in hands-on STEM demos that connect molecular-level visualization to classroom experiences. For example, I've developed videos of molecular dynamics (MD) to pair with the following demos:
(*) Demo: crush a metal can by pulling a vacuum inside
MD video: a box of air that shows how the molecules move and collide

(*) Demo: pour super-cooled water and watch it freeze instantaneously
MD pictures: nucleation of an ice crystal in super-cooled water

(*) Demo: pour liquid nitrogen on the floor
MD Video: a nano-sized droplet of liquid nitrogen boils after contacting a hot surface

Teaching Interests:

I will teach anything. I would be particularly interested in teaching chemical reaction engineering, thermodynamics, and mass transfer, but would be happy to teach introductory classes, too. I am eager to develop elective courses on molecular simulation at both the undergraduate and graduate levels, and graduate courses in statistical mechanics, reaction rate theory, and mass transfer.

Research Experience:

My research career has been focused on modelling rare events and chemical reactions. As a grad student, I developed transition path sampling methods for modeling dynamics of rare events. My methods were used in a collaboration, published in *Nature*, to study ice nucleation. In my first postdoc, I developed general Monte Carlo methods for modeling reaction equilibria and investigated CO2 absorption in ionic liquids. In my second postdoc, I am modeling the corrosion of plutonium.

Selected Publications:

L Lupi, A Hudait, B Peters, M Grunwald, **RG Mullen**, A Nguyen, V Molinero, "Role of Stacking Disorder in Ice Nucleation" *Nature* 551.7679 218-222 (2017)

http://dx.doi.org/10.1038/nature24279

ZA Levine, MV Rapp, W Wei, **RG Mullen**, C Wu, GH Zerze, J Mittal, JH Waite, JN Israelachvili and J-E Shea, "Surface force measurements and simulations of mussel-derived peptide adhesives on wet organic surfaces" *Proceedings of the National Academy of Sciences* 113.16 4332-4337 (2016)
http://dx.doi.org/10.1073/pnas.1603065113

RG Mullen, EJ Maginn, "Reaction ensemble Monte Carlo simulation of xylene isomerization in bulk phases and under confinement" Journal of Chemical Theory and Computation 13.9 4054-4062 (2017)
http://dx.doi.org/10.1021/acs.jctc.7b00498

JK Shah, E Marin-Rimoldi, **RG Mullen**, ..., EJ Maginn, "Cassandra: An open source Monte Carlo package for molecular simulation." *Journal of Computational Chemistry* 38.19 1727-1739 (2017)
http://dx.doi.org/10.1002/jcc.24807

RG Mullen, J-E Shea and B Peters, "Transmission coefficients, committors, and solvent coordinates in ion-pair dissociation" *Journal of Chemical Theory and Computation* 10.2 659-667 (2014)
http://dx.doi.org/10.1021/ct4009798

543798 Theories and Simulations for Liquid-Liquid Phase Separation in Biology

Abstract:

Research Interests:

In molecular and structural biology, liquid-liquid phase separation underpins the formation of various newly discovered mesoscopic, liquid-like, functional assemblies/droplets in the cell not encapsulated by lipid membranes. These assemblies consist of multiple protein, DNA, and RNA molecules, termed “membraneless organelles” or “biomolecular condensates”. They compartmentalize intracellular environment into condensed and dilute phases, respectively, inside and outside of the droplet. The membraneless assemblies are critical in cellular signalling, homeostasis, gene regulation, and cell growth. Despite the recent flurry of experiments, their underlying biophysical mechanisms are rarely understood. **I am particularly interested in the "sequence-specific" phenomena in biological phase separation**. Biomolecules are heteropolymers with unique, precisely regulated sequences of amino acids or nucleic acids. Their phase separation has brought a new degree of freedom to polymer science: sequence specificity. To build a comprehensive understanding of the liquid-liquid phase separation in biology, **new polymer physics has to be developed.**

Review: **Y.-H. Lin**, J. D. Forman-Kay, and H. S. Chan (2018) Theories for sequence-dependent phase behaviors of biomolecular condensates, *Biochemistry,* 57, 2499

Research Experience:

I have been a joint postdoc of a theoretical group and an experimental group, closely collaborating with experimental biochemists and computational biologists. I am focusing on theoretical polymer physics and have been developing sequence-specific theories and computer simulation tools for the phase separation of charged intrinsically disordered proteins. In addition, in my PhD, I have studied the theoretical biophysics in the sequence-dependent structural effects in protein-RNA interaction. My research has provided a general theoretical framework for investigating phase separation and molecular recognition mechanisms of biomolecules.

Postdoctoral Project: "Theories for sequence-dependent phase behaviors of biomolecular condensates", under supervision of Hue Sun Chan, Department of Biochemistry, University of Toronto, AND Julie D. Forman-Kay, Molecular Medicine, The Hospital for Sick Children

PhD Dissertation: "The interplay between single-stranded binding proteins on RNA secondary structure", under supervision of Ralf Bundschuh, Department of Physics, Ohio State University

Future Directions:

Based on my achievements on researching phase separation in biology, I will investigate this topic by the following four approaches:

1. Advanced sequence-specific polymer theories: Based on the success of my polymer theories for charged biopolymers, I will build theories for other interactions in proteins and RNA, such as hydrophobicity, pi-electron interactions, and hydrogen bonds. More molecular and environmental details, for example protein side chains, pH values, and salt concentration, will be addressed. The kinetic properties of the phase separation droplets will also be studied by dynamic models.

Accomplished works: **[1] Y.-H. Lin**, J. D. Forman-Kay, & H. S. Chan (2016) Sequence- specific polyampholyte phase separation in membraneless organelles, *Phys. Rev. Lett.* 117, 178101 **[2] Y.-H. Lin**, J. Song, J. D. Forman-Kay, & H. S. Chan (2017) Random- phase-approximation theory for sequence-dependent, biologically functional liquid-liquid phase separation of intrinsically disordered proteins, *J. Mol. Liq.* 228, 176 **[3] Y.-H. Lin** & H. S. Chan (2017) Phase separation and single-chain compactness of charged disordered proteins are strongly correlated, *Biophys. J.* 112, 2043 **[4] Y.-H. Lin**, J. P. Brady, J. D. Forman-Kay, & H. S. Chan (2017) Charge pattern matching as a "fuzzy" mode of molecular recognition for the functional phase separations of intrinsically disordered proteins. *New J. Phys.* 19, 115003

2. Assess analytical theories using computer simulation: Computer simulations can complement the conceptual formulations proposed by analytical theories. Based on my experience in coarse-grained simulations for charged polymers, I will conduct simulations with more molecular details to assess and improve the theoretical approach so as to better rationalize phase separation phenomena.

Accomplished works: **[1]** S. Das, A. Eisen, **Y.-H. Lin**, and H. S. Chan (2018) A lattice model of charge-pattern-dependent polyampholyte phase separation, *J. Phys. Chem. B*, 122, 5418 **[2]** S. Das, A. Amin, **Y.-H. Lin**, and H. S. Chan (2018) Coarse-grained residue-based models of disordered protein condensates: utility and limitations of simple charge pattern parameters (in preparation)

3. Collaboration with experimentalists: Experiments are essential for evaluation of theories. I will keep my current successful mode of closely collaborating with experimental groups to receive first-hand experimental results which will help me develop relevant theories; at the same time I will contribute theoretical concepts and predictions to help develop new experimental targets. In addition I will also directly develop and apply my theories to the numerous published experiments of biological phase separation.

Accomplished work: J. P. Brady, P. J. Farber, A. Sekhar, **Y.-H. Lin** et al. (2017) Structural and hydrodynamic properties of an intrinsically disordered region of a germ-cell specific protein upon phase separation, *PNAS* 114, E8194

4. Bioinformatics: High-throughput database search requires simple but concise quantities and parameters. I will apply analytical theories to the various existing DNA, RNA, and protein databases to propose statistical ideas for categorizing and predicting phase separation phenomena.

Teaching Interests:

As a theoretical physicist working on polymer theory in the Department of Biochemistry in a Faculty of Medicine, I have been well experienced in conveying knowledges to people from different scientific and engineering disciplines. My teaching goal will be to prevent students from being constrained by a specific, fixed thinking context, but rather to be able to adapt various perspectives. My students will be able to find multiple strategies for the same problem and pick up the best-fit solution for different requirements and constraints, which, I believe, is one of the most precious skills conferred by engineering training. I have mentored two talented undergraduate research students during my postdoc, and have been a teaching associate in Department of Physics during my PhD study. My teaching strategy will be to urge students to study by themselves: before the class or research project, I will first assign study materials, textbook chapters and/or research papers; students will have to read the materials and write a summary about what they have learned and what they have not fully understood. My lecture will then be based on their summaries and questions, addressing those most complicated topics. I am particularly competent to teach physics-related engineering classes, for example thermodynamics and fluid dynamics. I am also competent to teach biophysics and polymer/biopolymer sciences classes due to my research background. Meanwhile, as I am well-experienced in computational methods for polymer and biomolecule solutions, I can also teach numerical methods and programming techniques.

542810 Materials Discovery for Energy and Environmental Applications Using First-Principles Multiscale Simulations

Abstract:

Persistent growth in global energy demands, depletion of fossil fuel resources and climatic changes are of great environmental concern which necessitates the development of innovative strategies for energy production and utilization. To address these grand challenges, advanced materials with superior performance will be central to future technologies for energy conversion (renewable) and storage. Today, the public and industrial sectors are investing billions of dollars for the discovery of superior materials that can meet our societyâ€™s high energy demand. Multiscale computer simulations enable accelerated materials discovery with offering a detailed atomic-level understanding of the structure and materials properties, guiding the experiments towards identifying promising candidates, thereby decreasing the use of human resources and efforts. Computational methods like Density Functional Theory (DFT), Molecular Dynamics, Kinetic Monte Carlo, and Atomistic Thermodynamics play a central role in these efforts and lead the *in-silico* discovery of next generation materials.

In this poster, I will give an overview of examples which i) unravel competing reaction mechanisms and develop structure-activity relationship in alkane dehydrogenation on metal oxides (using periodic DFT and Molecular Dynamics), ii) elucidate the role of oxygen coverage on Mo_2C for CO_2 reduction iii) demonstrate the cause of structural degradation during cycling and low thermal stability of Ni-rich layered transition metal-based electrode materials.

Research Experience:

My research addresses several challenges in energy conversion and storage following a multiscale approach. Using ab-initio quantum chemical, periodic density functional, Molecular Dynamics and hybrid QM/MM methods, I elucidate the underlying mechanisms (and underpinning physics) for chemistries of practical interest, identify transition states and reaction energies with high accuracy under realistic experimental conditions, provide insights into complex physiochemical processes (phase transformations, atomic ordering, electronic structure and electrochemical properties) as a function of material composition, structure and surface coverage. Specifically, my research has been focused in three major directions: (i) design advanced catalysts and understand the reaction mechanisms on these catalysts (such as metal oxides and metal carbides) towards the production of high-value chemicals through industrially relevant reactions (e.g. CO_2 reduction and alkane dehydrogenation), (ii) understand basic properties of lithiated mixed metal oxides (and phosphates) for electrochemical energy storage (Li/Na ion batteries), towards elucidating the role of electronic structure, spin-orbit coupling, lattice doping, phase transitions and covalency on cell voltages and structure of cathode materials and (iii) demonstrate the catalytic control of trichodiene synthase (terpene synthase), and in particular, electrostatic guidance of enzyme to avoid the formation of byproducts.

Research Interests:

As an assistant professor, my research efforts will focus on the accelerated design of materials for energy and environmental applications. Specifically, I will focus on the following directions:

1. Design lithiated oxide-derived electrocatalysts to improve the activity (and selectivity) through electrochemical tuning (intercalation and de-intercalation) of metal oxides.
2. Elucidate the growth mechanism of metal oxides on metal surfaces (inverse metal-oxide/metal catalysts), and understand the relationships between morphology, size, and composition of oxides supported on metals under realistic experimentalconditions.
3. Develop general structure-activity relationships for different classes of materials (metal oxides, single metal catalysts, metal carbides and perovskites) as a function of metal composition with explicit solvents.
4. Understand the degradation mechanism of Li/Na ion batteries by investigating interface reactions at cathode-electrolyte interfaces

Teaching Interests:

Teaching is about inspiring the students, it is much more than just providing the information. Effective teaching involves telling interesting and inspiring stories along with providing a comprehensive and clear understanding of the concepts. A good teaching approach can effortlessly accelerate the learning process by maximizing student interest and engagement. Critical thinking of students can be ignited by asking pro-active questions and pushing the limits of students' thinking by challenging them with different problems. I have a diverse teaching experience. I have taught the courses of Quantum Chemistry, Physical Chemistry and Computational Chemistry and delivered lectures on the topic of Computational Chemistry, Molecular Dynamics, and Electrochemical Energy Storage. I have gained experience in teaching at both the undergraduate and graduate levels as a teaching assistant, mentor, and instructor. In addition, I have mentored several undergraduate and graduate students, both in the Chemical Engineering (University of Pittsburgh, USA) and Chemistry (Bar-Ilan University, Israel) Departments. The teaching opportunities I have had on my career have enriched my philosophy as a teacher and will guide me towards succeeding in education. Finally, I interact with all the students in the class frequently. Along with regular board-based teaching I use multimedia material (pictures, videos, animations) to enrich my lectures and clearly portray the information I need to pass on to my students. I am willing to teach any core Chemical Engineering course. Based on my solid foundation on fundamentals of Physical Chemistry, I express a special interest in teaching Thermodynamics, Kinetics, Molecular Spectroscopy and Statistical Thermodynamics. I plan to enrich the curriculum with courses focusing on Energy, Electrochemistry and Computational Modeling (of solids, molecules, and enzymes). My strong background and teaching experience in computational and quantum chemistry, as well as my supervising experience in theoretical chemistry projects, will aid me in developing such courses.

542302 Engineering Electrocatalysts for Sustainable Energy Technologies: From Theory to Rational Design through in-Situ Characterization

Abstract:

Engineering Electrocatalysts for Sustainable Energy Technologies: From Theory to Rational Design through In-Situ Characterization

Mohammad Norouzi Banis

Department of Mechanical & Materials Engineering, University of Western Ontario, London, Ontario, Canada

Research Interests:

The ever-increasing demand for energy and the excessive dependence on depleting fossil fuel reserves has resulted in two major challenges; a shortage of energy and serious environmental pollution (including high CO_2 concentration). CO_2 electrochemical reduction (CER) technology has attracted considerable attention in recent years and is believed to be a viable route capable of addressing these challenges by producing chemical fuels while reducing CO_2 emissions. However, the development of CER technologies face several issues, originating from: (i) insufficient fundamental understanding of the underling mechanisms and behaviour of material in their working environment, and (ii) difficulties related to the fabrication of materials at the molecular and nanoscale level.

Combining theoretical studies (such as DFT modeling) and advanced characterization techniques (including in-situ/ex-situ synchrotron-based methods - XAS, XES, CT) with novel material designs can provide the breakthroughs required for the advancement of energy industries including CER technologies. This approach can provide a detailed picture of the electronic, chemical, and atomic structure of the materials in their working environment. Subsequently, the study and correlation between the performances of nanomaterials in CER systems with their various properties can provide a road map for the development of efficient electrocatalysts.

Successful Proposals:

(1) Canadian Light Source Beamtime Proposals: Over 20 Beamtime proposals accepted – Granted ~40% of beamtime in July – December 2018 in three beamlines with more than 1000 hours of beamtime

(2) Advanced Photon Source Beamtime Proposal: PNC beamline - In-situ study of electrochemical energy conversion systems – 96 hours

(3) Co-authored NSERC Discover Grants:

a. Design and development of 3D nanowire-based composite electrodes for Low-Cost and High-Performance PEM

Fuel Cells - over 100,000 CAD

b. Atomically precise cluster Pt-based electrocatalysts with ultra-low loading and extremely high stability for PEM fuel cells – 500,000 CAD

Postdoctoral Projects:

(1) In-situ soft XAS: Understanding the underlying mechanisms of charge/discharge processes in Na-O_2 batteries

(2) Design and development of novel interfaces of cathode/sulfide solid-state electrolyte: Revealing the nature of interfacial resistance in solid-state batteries

(3) Single atom catalysts through ALD: Modeling and experimental study of the behaviour of noble single atom catalysts for fuel cell reactions

PhD Dissertation:

"Controlled synthesis of one dimensional nanostructured materials and their applications as catalyst supports in proton exchange membrane fuel cells"

US Patent. US20120241192 A1: M. Cai; X. Sun; Y. Zhang; **M.N. Banis;** and R. Li (2012) Microfiber supported metal silicide nanowires.

Research Experience:

My research career revolves around multidisciplinary collaboration with various researchers and industries for energy storage and conversion systems. Through these collaborations, I have gathered valuable tools ranging from the design of synthetic processes and advanced characterization techniques to theoretical calculations. My research has been mainly focused on understanding the correlation between the electronic and atomic structure of electrocatalysts and their performance in PEMFCs and next-generation batteries.

I have developed a strong background in the application of vapor deposition methods such as atomic layer deposition (in collaboration with General Motors) and magnetron sputtering (in collaboration with Ballard and Canadian Space Agency) for the controlled synthesis of nanomaterials and thin films. As a materials scientist, I have worked extensively on the application advanced characterization techniques for energy-related materials with a focus on synchrotron-based methods. I have developed close collaborations with various national and international laboratories including Advanced Photon Source and Canadian Light Source, where I developed various setups for the in-situ study of electrocatalysts in their working environment. Furthermore, with intensive training, I have become proficient in Density Functional Theory and XAS fitting and modeling, which have contributed to over 60 authored and co-authored publications.

Teaching Interests:

Over the course of my research career, I have been extensively involved in the teaching and mentoring of graduate and undergraduate students. As a graduate student, I was a teaching assistant for several courses including Kinematics/Dynamics and corrosion. I also was a guest lecturer in the introduction to Nanomaterials course, teaching second and third year undergraduate students in the engineering faculty. I was further involved in the development and revision of the curriculum and provided students with support for the end of course projects.

As a postdoctoral fellow and research scientist, I organized annual and semi-annual workshops on the application of advanced synchrotron-based techniques in material science. I developed course material and taught over 50 graduate students and postdoctoral fellows the (1) Fundamentals of X-ray absorption spectroscopy, (2) Principles of EXAFS modeling and fitting and (3) Introduction to X-ray diffraction experiments and data analysis.

In my role as a faculty member, I am interested in teaching undergraduate and graduate introductory and advanced courses in topics related to catalysis, material science and mechanical engineering which align well with my research experience. I also plan to create courses on characterization techniques for materials scientists and chemical engineers covering techniques used in research and industry.

Future Direction:

As a faculty member, I would like to continue my work in understanding the correlation between the properties of unique materials and their performance in energy-related applications. In particular, I would like to combine theoretical studies with novel in-situ characterization techniques that I have designed to develop low cost, highly active catalysts for CO_2 electrochemical reduction with good control over the product selectivity.

Initially, I will continue my post-doctoral work on the development of single and double atom catalyst for application in CER technologies. I intend to establish a lab for the synthesis and characterization of single atom catalysts through my unique atomic layer deposition-based method and widely reported solution-based processes. Additionally, in collaboration with other researchers, with extensive theoretical modeling and advanced characterization (XAS, XES, XPS) in various research institutes and laboratories, I will develop a general guideline for the application of sub-nanometer catalytic (single atom) CER processes.

Benefiting from my approach towards selection, characterization and implementation of nano/sub nanostructured electrocatalysts, I plan to expand my research towards other electrochemical based energy conversion/storage systems.

My main goal is to build a world-leading multidisciplinary group in the area of advanced characterization and large-scale implementation of nanostructured materials as solutions to the challenges in energy applications. This will provide me with ample opportunities to collaborate with researchers of different backgrounds while contributing to my university and community.

Selected Publication (over 60 authored, co-authored)   h-index: 33, i10-index:44

Ø **M. N. Banis**, H. Yadegari, Q. Sun, T. Regier, T. Boyko, J. Zhou, Y. M. Yiu, R. Li, Y. Hu, T. K. Sham, X. Sun, Revealing the charge/discharge mechanism of Na-O_2 cells by in situ soft X-ray absorption spectrsocopy, ***Energy & Environmental Science***, *2018*

Ø **M. N. Banis**, H. Yadegari, A. Lushington, Q. Sun, R. Y. Li, T. K. Sham and X. Sun, A bifunctional solid state catalyst with enhanced cycling stability for Na and Li-O_2 cells: revealing the role of solid state catalysts. ***Energy & Environmental Science***, 2017, **10**, 286-295.

Ø **M. N. Banis**, Y. Zhang, Q. F. Xiao, M. Cai, R. Y. Li and X. Sun, Tailoring of Single-Crystalline Complex Ta5Si3 Nanostructures: From Networked Nanowires to Nanosheets. *Crystal Growth & Design*, 2014, **14**, 436-441.

Ø **M. N. Banis**, S. Sun, X. Meng, Y. Zhang, Z. Wang, R. Li, M. Cai, T. K. Sham and X. Sun, $TiSi_2O_x$ Coated N-Doped Carbon Nanotubes as Pt Catalyst Support for the Oxygen Reduction Reaction in PEMFCs. *Journal of Physical Chemistry C*, 2013, **117**, 15457-15467.

Ø **M. N. Banis**, X. Meng, Y. Zhang, M. Cai, R. Li and X. Sun, Spatially Sequential Growth of Various WSi_2 Networked Nanostructures and Mechanisms. *Journal of Physical Chemistry C*, 2013, **117**, 19189-19194.

Ø **M. N. Banis,** Y. Zhang, R. Li, X. Sun, X. Jian and D. Nikanpour, Vanadium oxide assisted synthesis of networked silicon oxide nanowires and their growth dependence. *Particuology*, 2011, **9**, 458-464.

Ø **M. N. Banis**, Y. Zhang, H. N. Banis, R. Li, X. Sun, X. Jian and D. Nikanpour, Controlled synthesis and characterization of single crystalline MnO nanowires and Mn–Si oxide heterostructures by vapor phase deposition. *Chemical Physics Letters*, 2011, **501**, 470-474.

Ø N. Cheng, S. Stambula, D. Wang, **M. N. Banis**, J. Liu, A. Riese, B. Xiao, R. Li, T. K. Sham, L. M. Liu, G. A.

Botton and X. Sun, Platinum single-atom and cluster catalysis of the hydrogen evolution reaction. ***Nature Communcations,*** 2016, **7**, 13638.

Ø N. Cheng, **M. Norouzi Banis**, J. Liu, A. Riese, S. Mu, R. Li, T.-K. Sham and X. Sun, Atomic scale enhancement of metal–support interactions between Pt and ZrC for highly stable electrocatalysts. ***Energy & Environental Science***, 2015, **8**, 1450-1455.

Ø J. J. Wang, J. L. Yang, Y. J. Tang, J. Liu, Y. Zhang, G. X. Liang, M. Gauthier, Y. C. K. Chen-Wiegart, **M. N. Banis**, X. F. Li, R. Y. Li, J. Wang, T. K. Sham and X. Sun, Size-dependent surface phase change of lithium iron phosphate during carbon coating. ***Nature Communications,*** 2014, **5**, 8.

542527 Novel Electrokinetic Solutions for Energy and Environmental Problems

Abstract:

Research Interests: Electrokinetics, Transport Phenomena, Computational Science and Engineering, High Performance Computing

Teaching Interests: Transport Phenomena, Thermodynamics, Numerical Analysis, Mathematical Physics

Ion transport phenomena and electrochemical processes in porous media are relevant to many technologies including energy storage in batteries and supercapacitors, water desalination through capacitive deionization, electrokinetic remediation of contaminated soil, and monitoring of fluid flow in underground reservoirs. Although these problems are inherently multi-scale and multi-physics in nature, they are fundamentally described by similar processes. My future research will focus on understanding the underlying mechanisms of ion transport and their implications for optimizing existing technologies and possible novel applications.

During my PhD at UC Santa Barbara, and under the joint supervision of Prof. Frederic Gibou and Prof. Todd Squires, I developed efficient computational techniques for studying ion transport in porous media. My research led to the discovery of the significance of pore microstructure and surface conduction in accelerating the charging kinetics and has inspired new guidelines for designing porous electrodes, e.g. based on CNT or graphene. As a postdoctoral researcher at MIT in Prof. Martin Bazantâ€™ lab, I am currently investigating novel applications of electrokinetic phenomena to two-phase flow problems in porous media. My work in this area has led to the surprising discovery that electro-osmotic flows can control, and under certain conditions suppress, viscous fingering. This finding has exciting implications for environmental problems, e.g. improving the extraction efficiency during secondary oil recovery. I am further investigating the possibility of using electrokinetic-assisted flows as a new and efficient fracturing technique for shale gas harvesting.

As a new faculty, I will apply my combined knowledge of transport phenomena and mechanics as well as computational skills to challenging problems for energy and environmental applications. In particular, my research group will focus on understanding and optimizing the performance of CDI technology for water desalination where recent developments in flow-electrode designs has promising prospects for grid-scale energy storage. I am also interested in further exploring novel applications in the area of geomechanics and two-phase electrokinetic flows for enhanced oil and gas recovery. For instance, while recent experimental observations suggest low-salinity water injection can substantially improve secondary oil recovery, the exact physical cause is still debated. This is inherently a multi-scale and multi-physics problem involving ion transport, fluid flow, and surface chemistry processes that span sub-micron to kilometer in length-scale. I am confident that my expertise in computational and theoretical modeling of ion transport put me in a unique position to tackle these problems and help me to advance this field in collaboration with experimental colleagues.

542537 Electrocatalysis for Sustainable Energy Storage and Conversion

Abstract:

Research Interests:

Electrochemistry offers a versatile toolkit for enabling a sustainable energy economy. Through liquid fuel synthesis, energy storage and conversion devices it provides a pathway to fully engage renewable energy into the grid. From device scale catalyst integration, to precise control and characterisation of catalyst surfaces, my research interests are diverse but converge with the pursuit of uncovering fundamental understanding of catalytic activities. Through careful experimental design coupled to precise control of catalyst surfaces, my emphasis will be to determine intrinsic catalytic activity, selectivity and corrosion mechanisms. This fundamental insight will provide a feedback loop for the design and engineering of new, industrially relevant catalysts. Beyond fuel cell and electrolyser applications, I am very interested in applying these catalytic developments to environmental remediation technologies for water purification and reducing atmospheric pollution.

Teaching Interests:

I believe the future of our society lies in the hands of the next generation of scientists and therefore, as educators, teaching is of the uttermost importance. From lecturing undergraduates to mentoring postgraduate and undergraduate level students and from supervising undergraduate laboratory experiments to leading outreach in the community, I am dedicated to enthusing and inspiring students in chemical engineering. As a substitute lecturer for 3 separate courses (both at Stanford University and at the University of Wyoming) I have taught over 20 lectures. In addition to lecturing, for each course, I heavily participated in designing course content, as well as writing and grading exam questions and homework problems. In all aspects of my teaching and student interactions I strive to convey both the fundamental science, as well as, where appropriate, its impact on technology, highlighting the importance of the science. The success of engaging all students is certainly challenging, yet it is always my priority throughout my teaching. This can be in the context of fundamental scientific research, medicinal chemistry, chemical industry, the energy sector and other topics of pertinence to students.

To date, I have had the pleasure of supervising over 15 undergraduate, graduate students and postgraduates. My role has varied from student to student, but has included designing research projects, laboratory and safety training, and generally assisting data acquisition and analysis. I have been fortunate enough to interact with, inspire and lead students and am excited to transfer this skillset to designing laboratory work for Undergraduate (and Master) level students. Overall, I believe I am an extremely dedicated, patient and engaging educator and am extremely well suited to teaching in a chemical engineering department.

Research Experience

Research Engineer (Staff Scientist)

Stanford University, Chemical Engineering, USA August 2017 – present

Supervisor: Professor T.F. Jaramillo

- **Synthesis of stable and active** non-precious metal oxygen reduction reaction (ORR) catalysts for fuel cell application (e.g. transition metal nitrides, boron nitrides and oxides)
- **Leader and coordinator** of the experimental team (>10 researchers) working with the Toyota Research Institute at Stanford University in ORR catalysis development. This role involves mentoring students and postdocs working on our project within the Jaramillo laboratory via regular one-on-one meetings and discussions. Organise, schedule and lead regular meetings with our industrial sponsors.

Postdoctoral Scholar

Stanford University, Chemical Engineering, USA August 2015 – Jul 2017

Supervisor: Professor T.F. Jaramillo

- **Stabilised photocathodes from corrosion**. Investigated protection schemes to prevent corrosion of silicon photocathodes by an over layer of molybdenum disulfide for >1500 h stable PEC water splitting. To date, this is the most stable photocathode to be reported.
- **Discovered new catalysts**. Designed and synthesised a previously unexplored family of materials (transition metal arsenides) as active catalysts for the hydrogen evolution reaction. Led collaboration with Norskov Group at Stanford to compare DFT calculations with experimental results.
- **Synthesised and characterised** transition metal phosphides and sulfides at >1 g scale for use by our industrial collaborators (Proton Onsite) in membrane electrode assemblies for electrolysis of water.

Postdoctoral Researcher

University of Wyoming, Chemistry Department, USA March 2013 – July 2015

Supervisor: Professor B.A. Parkinson

Developed model systems for fundamental analysis of quantum dot (QD) and dye sensitization at single crystal surfaces

- **QD sensitized single crystals**. Synthesised, characterised and sensitised single crystal wide band gap semiconductors with near infrared absorbing QD materials.
- **Designed attenuated total reflection setup**. Developed an in-house attenuated total reflectance (ATR) absorption setup to probe simultaneously the spectral absorbance and photocurrent response of dye sensitization at a single crystal surface uncovering the correlation between adsorbed species and their ability to contribute to photocurrent signal.
- **Outreach kit manager**. Solar to Hydrogen Activity research Kit (SHArK) II Outreach Project. SHArK facilitates high school and undergraduate students for the discovery of new materials for photoelectrochemical water splitting. Coordinated software and electronic engineers to finalise designs and test our kits for use in the classroom.

Graduate Student

Imperial College London, Materials Department, UK September 2009 – December 2013

Supervisor: Professor D.J. Riley

Fabricated and characterised QD and molybdenum disulfide materials for sensitized solar cells

- **Synthesised and sensitized mesoporous electrodes.** Synthesised, characterised and tested cadmium selenide quantum dot sensitized photoelectrodes. Fundamental studies into (i) the influence of QD purification procedure and (ii) the effect of polysulfide electrolyte on CdSe sensitized solar cells photoelectrochemical performance.

- **Investigated the photoelectrochemical properties** of atomically thin (monolayer and few-layer) molybdenum disulfide materials.

Select Publications [11 published total. 8 first author. 4 in preparation. 1 book chapter]

1. L. A. King, T. R. Hellstern and T. F. Jaramillo, Highly Stable Molybdenum Disulfide Protected Silicon Photocathodes for Photoelectrochemical Water Splitting. *ACS Applied Materials and Interfaces,* 2017, 9, 36792 – 36798. DOI: 10.1021/acsami.7b10749
2. Kubie, L. A. King, M. E. Kern, J. R. Murphy, S. Kattel, Q. Yang, J. T. Stecher, W. D. Rice and B. A. Parkinson. Synthesis and Characterization of Ultrathin Silver Sulfide Nanoplatelets. *ACS Nano*, 2017, 11, 8471-8477. DOI: 10.1021/acsnano.7b04280
3. L. A. King and B. A. Parkinson, *Photosensitization of ZnO Crystals with Iodide-Capped PbSe Quantum Dots, J. Phys. Chem. Lett.,* 2016, 7, 2844-2848, **DOI:**10.1021/acs.jpclett.6b01133
4. L. A. King, Q. Yang, M. L. Grossett, Z Galazka, R. Uecker, and B. A. Parkinson, Photosensitization of Natural and Synthetic SnO2 Single Crystals with Dyes and Quantum Dots, J. Phys. Chem. C, 2016, 120, 15735-15742, DOI: 10.1021/acs.jpcc.5b11071
5. L. A. King, W. Zhao, M. Chhowalla, D. J. Riley, and G. Eda, *Photoelectrochemical properties of chemically exfoliated* MoS_2. *J. Mater. Chem. A*, 2013, 1, 8935–8941. DOI: 10.1039/c3ta11633f

Proposal Writing Experience

- Led, coordinated, organised and wrote 3 proposals at Stanford University from various funding agencies (including Department of Energy, EERE, DoD SBIR as well as from industrial partners) (2016 – 2018).
- Successfully organised, wrote and edited 2 grants that were submitted and awarded funding (National Science Foundation, USA) (2014 – 2015)
- Competed for and successfully awarded various travel and attendance bursaries as a graduate student, including: "International Centre for Materials Research Summer School on Inorganic Materials for Energy Conversion and Storage" – University of California, Santa Barbara, USA (2012) "IDEAL Winter School Photovoltaic Solar Energy", Paris, France (2011). Additional funding awarded from the "Energy Futures Laboratory", Imperial College (2009 – 2011)

542144 Understanding and Controlling Multielectron Transfer Electrochemistry Toward Sustainable Energy Technologies

Abstract:

AIChE Meet the Faculty Poster Session

Title: Controlling Multielectron Transfer Electrochemistry for Sustainable Energy Technologies

Research Interests:

The sustainable generation of the fuels and chemicals that power our daily lives is one of the key challenges facing the chemical sciences today. The large-scale generation of chemical fuels (such as hydrogen, methanol, and ammonia) from renewable electricity sources (e.g. wind, solar) is a tantalizing prospect, but requires the careful coupling of electrical energy and chemical reaction.

Electrical energy can be a powerful tool for catalysis. A change in reaction driving force that would require the application of hundreds of atmospheres of pressure can be achieved with less than a volt of applied potential. However, the ability to control the spontaneity of a reaction does not always come with concomitant control over the selectivity of the electron transfer. Instead, a multitude of products as well as corrosion processes are often observed. My research interests center around understanding and promoting efficient, selective electron transfer, particularly with respect to the multielectron charge transfers often necessary to synthesize chemical fuels. Efficiency and selectivity will be derived from both improved catalyst design and improved electrolyte engineering, with a strong focus on understanding the structure/function relationship at catalyst surfaces. The relationship of catalyst surface structure to its activity and selectivity will be elucidated by careful *in situ* study of catalyst surface structure and electrochemical measurements, and the knowledge we gain will be used to rationally develop next-generation catalysts. These new electrocatalysts will then be integrated into device structures to evaluate them under practical conditions. The knowledge gained from each of these thrusts: catalyst surface chemistry study, catalyst design, and device integration, will inform the others to help us achieve our ultimate goal of sustainably generating fuels.

Teaching Interests:

My training has prepared me to ably serve a Chemical Engineering department in the instruction of undergraduate and graduate students both in the laboratory and in the classroom. My teaching experience includes the mentorship of multiple graduate and undergraduate students in the laboratory. Additionally, I have acted as the head instructor for advanced graduate courses in photoelectrochemistry during my PhD, and have served in teaching assistant roles.

I would be comfortable teaching a range of undergraduate-level chemical engineering and chemistry courses, including kinetics, thermodynamics, and other physical/analytical courses. In the future, I hope to have the opportunity to develop (1) a course directed toward graduate students and senior undergraduates aimed at understanding

electrochemistry and its practical application to electrocatalysis of energy-rich molecules, as well as (2) a course designed around semiconductor electrochemistry, photoelectrochemistry, and applications to emerging renewable energy technologies.

Research Experience:

Postdoctoral Scholar 2016-Present

Department of Chemical Engineering, Stanford University

Advisor: Thomas F. Jaramillo

Project: Synthesis and Characterization of N_2 Reduction Electrocatalysts

- Investigated the dependence of electrocatalytic NH_3 production rate on electrolyte composition and catalyst identity
- Developed and applied analytical techniques and best practices for identifying electrochemically generated NH_3 in non-aqueous electrolytes

Graduate Research Assistant 2010-2016

Division of Chemistry and Chemical Engineering, California Institute of Technology

Advisor: Nathan S. Lewis

Ph.D. Thesis: *Chemical and Photoelectrochemical Behavior of Graphene-Covered Silicon Photoanodes*

- Investigated the chemical and photoelectrochemical stability of Si and GaAs photoelectrodes covered by atomically thin graphene layers in aqueous electrolytes
- Explored the effect of graphene interfacial layers on the electronic properties of silicon/electrolyte junctions related to solar cell device performance
- Developed methods and classification scheme for the characterization of the efficiency of photoelectrochemical and photovoltaic solar-to-fuels devices

Undergraduate/Graduate Research Assistant 2005-2010

Department of Chemistry, University of Virginia

Advisor: W. Dean Harman

M.S. Thesis: *Laying the Groundwork for a Tantalum Dearomatization Agent*

B.S. Thesis: *Synthesis and Application of an Î·^{2} N-acetylpyridinium Complex*

- Synthesized tungsten- and tantalum-based organometallic reagents used to modify and enhance the reactivity of stable aromatic arenes and pyridines
- Investigated the relationship between reactivity enhancement in aromatic ligands and electrochemical properties of the associated organometallic complex

542157 Electrochemical Plasma Reactions and Supersonic Printing: A Route Towards Multi-Component Materials Discovery and Scalable Device Manufacturing

Abstract:

Research Interests:

I am interested in the complex interactions between multi-component materials systems through heat-, light- and charge-transfer reactions. Flames and plasmas present these elements in the form of phonons, photons and reactive charges. These reactions are clean from chemical interferences and can be created under controlled and atmospheric-pressure conditions. They provide the ideal media for studying fundamental chemical reactions. I will incorporate *in situ* and *ex situ* techniques including velocimetry, mobility analyses, spectroscopy and materials characterization to understand the underlying chemical reaction phenomena. The fundamental understanding of these reactions at the electrochemical-interfaces is the key to the development of a sustainable and energy efficient future.

My current research scope has shaped my interests in materials discovery and stoichiometry-controlled multi-component nanoparticle synthesis. In particular, doped complex oxides and their thin films are promising candidates for energy efficient devices. Existing additive manufacturing techniques are unable to deposit high temperature materials on soft and temperature-susceptible substrates such as polymers. Researchers usually rely on substrate-specific complex chemical techniques such as self-assembly for the deposition of complex oxides, carbides and nitrides. Supersonic aerosol impaction can deposit a wide variety nano- to micro- sized powders and is not limited to factors such as viscosities and surfactants that otherwise plague conventional printing techniques. Using my expertise on plasmas, flames, and supersonic aerosol printing, I look forward to opportunities to discover new materials, analyze their electrochemical transport properties and develop devices for an energy efficient future.

Post-doctoral Project: "*High-throughput Discovery of Thermoelectric Materials via Supersonic Aerosol Impaction.*" Supervisor: Prof. Chris Hogan (Department of Mechanical Engineering, University of Minnesota, Minneapolis). Funding agency and program: ARPA-E IDEAS

Ph. D. Dissertation: "*Atmospheric-Pressure in situ Plasma Reduction and Patterning of Metal-ion Containing Polymers.*" Supervisor: Prof. R. Mohan Sankaran (Department of Chemical and Biomolecular Engineering, Case Western Reserve University). Funding agency and program: NSF Scalable Nano-Manufacturing

Research Experience: My academic path has traversed the fields of physics, materials science and chemical engineering. In the process I have developed plasma-assisted direct-write printing, electrochemical ion and gas conversion, and nanoparticle nucleation phenomena in plasmas. The work exposed me to challenges of interfacial phenomena, electrochemical nucleation phenomena, charge fluence and energy transfer through interfaces, and electron-induced reactions. My research showed residence-time and gas-expansion induced control and nucleation of metal nanoparticles in plasmas, with near-100% faradaic efficiency during plasma-electrochemical reactions. I also developed and refined a micro-plasma based direct-write printing technique. Study findings from this research showed a novel phenomenon of plasma-induced migration of nanoparticles through polymers. I was awarded the Coburn and Winters Plasma Science award at the American Vacuum Society meeting (AVS 2016) for this work.

Currently with ARPA-E funding I am engaged in a multi-university effort on scalability and rapid feedback based growth of thermoelectric thin films. A key finding from my work was the effects of the physical parameters such as momentum and energy flux under supersonic conditions on the growth of crystalline thin films. I gained considerable experience in scalability and manufacturing-centric project that relied strongly on inter-university collaborations. I became accustomed to the intricacies required to address stringent reporting and deliverable guidelines for an ARPA-E project. Most importantly, I learned how to negotiate for systematic tangible deliverables within a given time-frame with my funding agency, while keeping my collaborators on board.

Future Direction: Understanding the surface modifiability, reactivity, and electrical and chemical transport through the interface, and subsequent engineering of the same is the key to developing nanoparticle printing as an industry standard. My prior work indicates that strain delocalization and electrodiffusion are two promising surface engineering routes for printing flexible and stretchable structures. However, the underlying physics and chemical processes that can occur in tandem with these phenomena are unknown. As a natural extension of my research, I anticipate that my work will have two primary aims. First, a plasma-assisted materials conversion and printing, and second, supersonic impaction printing of powders and nanomaterials for the additive manufacturing of 2D and 3D structures. In the long term, I envision a printing technique that incorporates *in situ* combinatorial (stoichiometric control of multiple components) nanoparticle synthesis concomitantly with deposition of thin films or patterns. *In situ* feedback mechanisms will control the stoichiometry and properties of the materials. Sequential in-line measurement of physical properties will predict and guide the multi-material synthesis process. This approach will optimize the stoichiometric composition of the material based on a desired physical property. By adopting this methodology, I am excited to delve into research on materials discovery and prediction based on transport properties as well as discovering alternative materials and innovative devices.

Taking cues from my post-doctoral experience, I understand that manufacturing is an integrative process. The translation of scientifically proven research from bench-top to industry can be best implemented when scientific analyses is executed at par with system modelling, design, and integrating industry standards. My research will adopt a proactive approach towards low-cost manufacturing, integration of smart electronics, and *in situ* process-feedback mechanisms. I am confident that this two-pronged approach of scientific discovery in the context of manufacturing scalability is much needed for translating academic research to industrial practice.

In parallel to the research theme of interfacial phenomena, materials discovery, scalable synthesis and additive manufacturing, I foresee myself collaborating closely with theoreticians. Plasma-reactors are scalable and industrially viable, yet most laboratory research either focuses on existing materials to optimize devices or rely on trial-and-error methods to synthesize new materials, as opposed to first-principles approaches. Theoretical simulations, machine learning and predictive modelling are the key to the discovery of materials for next generation energy-efficient applications. Plasma-properties that induce chemical reactions are challenging to predict due to the inherent variability of the electric field intensity and coupling based on reactor geometries. Further, depending on the requisite chemical reaction, the active reactant species can vary from being the electron to ions or radicals and metastables for the same plasma gas. Therefore, I anticipate working closely with collaborators for prediction and validation of my work.

Teaching Interests:

I am comfortable teaching at undergraduate and graduate level courses. As a Ph. D. candidate I served as a teaching assistant for the courses Process control, Process analysis and design and Measurements laboratory. I was responsible for grading and problem solving during these courses except the laboratory course where I was required to supervise, debug and grade experiments. I also mentored 2 graduate, 12 undergraduates, and 2 high school students. In addition, I have published 8 peer-reviewed publications with 7 of them as my coauthors.

Significant Publications:

1. Ghosh, B. Bishop, I. Morrison, R. Akolkar, D. Scherson, and R. M. Sankaran, "Generation of a direct-current, atmospheric-pressure microplasma at the surface of a liquid water microjet for continuous plasma-liquid processing," *J. Vac. Sci. Technol. A* **33**, 021312 (2015).
2. Ghosh, T. Liu, M. Bilici, J. Cole, I-M. Huang, D. Staack, D. Mariotti, and R. M. Sankaran, "Atmospheric-pressure dielectric barrier discharge with capillary injection for gas-phase nanoparticle synthesis," *J. Phys. D* **48**, 314003 (2015).
3. Ghosh, E. Klek, C. A. Zorman, and R. M. Sankaran, "Microplasma-induced in situ formation of patterned, stretchable electrical conductors," *ACS Macro Lett.* **6**, 194 (2017).
4. Ghosh, R. Hawtof, P. Rumbach, D. B. Go, R. Akolkar, and R. M. Sankaran, "Quantitative study of electrochemical reduction of Ag^+ to Ag nanoparticles in aqueous solutions by a plasma cathode," *J. Electrochem. Soc.* **164**(13), D818 (2017).
5. Hawtof, S. Ghosh, Cheyan Xu, R. M. Sankaran, J. Renner, "Catalyst-free, high selectivity electrolytic synthesis of NH_3 from N_2 and H_2O by plasma-produced solvated electrons," *Science Adv.* (2018) (*Accepted*).

542307 Development of Devices and Selective Catalysts for the Solar-Driven Electrochemical Reduction of CO_2 to Fuels

Abstract:

The electrochemical and photoelectrochemical reduction of CO_2 presents an attractive route towards fuel formation from renewable energy. However, the dearth of selective, active and inexpensive catalyst materials remains a key challenge on the way to renewable production of hydrocarbon fuels. Here, we address these challenges through the design of novel devices and selective earth-abundant catalysts for the electrochemical[1â€"3] and solar-driven[1,4â€"6] reduction of CO_2. In addition, we provide insight into the origin of selectivity on electrocatalysts based on copper, paving the way towards more rational design of electrochemical fuel forming devices.

Copper-based catalysts derived from the reduction of CuO are known to reduce CO_2 but simultaneously exhibit poor selectivity. In tackling this issue, we pioneer the application of Atomic Layer Deposition (ALD) to direct the catalytic selectivity of copper materials. ALD coating of CuO nanowires with thin layers of SnO_2 leads to a catalyst which produces CO with high selectivity even at low overpotentials. This compares favorably to conventional gold catalysts, known to selectively convert CO_2 to CO. Unmodified CuO nanowires, in stark contrast, lead to a wide product distribution and yield considerable parasitic hydrogen. To gain insight into the origin of these striking changes, we perform kinetic studies, high resolution transmission electron microscopy and gas chemisorption measurements. These data serve to explain the observed selectivity changes on the basis of decreased binding strengths for both H and CO intermediates on the Sn-containing catalyst surface.

We subsequently combine our ALD-derived catalyst with an Earth-abundant oxide anode to effect overall splitting of CO_2 to CO and O_2. Driven by a 3-junction solar cell, this device achieves long-term solar splitting of CO_2 into CO and O_2 at an efficiency of 13.4 %, presently the state of the art.[1]

Going beyond the production of CO, we investigate the further reduction of CO to hydrocarbons. Insights into the factors controlling the branching between different hydrocarbon species are scarce, precluding any rational design of selective catalysts. To remedy these shortcomings, we devise a platform allowing for the systematic study of the relevant reaction parameters. Exploiting non-aqueous electrolytes, we isolate activation-controlled kinetic data in electrochemical driving force, proton availability, solvent dielectric and reagent concentration. Our observations provide unique insight into the mechanism of hydrocarbon production on copper electrodes. We demonstrate that the competition between H and CO for surface sites plays an important role in governing the product selectivity between methane, ethylene and hydrogen. These results establish a new paradigm in controlling hydrocarbon product selectivity in electrochemical fuel formation.[2]

Research Interests:

Future research directions will be presented, focusing on the design and careful study of electrochemical systems and surface processes that mediate the activation and selective catalytic turnover of weakly-reactive, industrially-relevant substrates. This research aims to employ electrochemical methods to understand the interaction of n-alkanes with catalyst surfaces, leading to their stepwise dehydrogenation to alkenes, and ultimately their total oxidation to CO_2. Scaling challenges related to the need for ion transport in electrochemical systems will be investigated both trough theory and experiment, leading to the development of more scalable electrochemical reaction systems, enabled by the advent of additive manufacturing technologies. This research holds the promise to provide substantial efficiency gains for industrial processes and will provide a cornerstone for the coupling of intermittent and renewably sourced electrical energy, decreasing the carbon footprint of industrial chemical processes. It will furthermore lead to novel insights into the factors governing bonding and reactivity of hydrocarbons on catalyst surfaces.

Teaching Interests:

Chemical engineering is diverse. Incorporating knowledge from a large variety of fields of study, it spans the scale from the reactions of molecules on a surface to the operations of entire an petrochemical production plant. My training as a chemical engineer, in combination with my Ph.D. and postdoctoral research in inorganic chemistry give me a broad overview of the pieces which constitute chemical systems and processes. Drawing from this background and my diverse interests, my students will learn to cross the bridge between fundamental understanding and real-world applications. I firmly believe that a high degree of integrated understanding provides the groundwork to truly innovative engineering solutions.

My main teaching interests lie in the field of electrochemistry, electrochemical engineering, catalysis, kinetics and photochemistry, as such, I look forward to developing programs in electrochemistry, electrocatalysis and solar energy conversion.

542411 Electrochemical Ion Insertion: Mechanisms and Applications in Energy Storage and Computing

Abstract:

Research Interests:

Electrochemical ion insertion is a powerful method to tune the properties of materials. In such devices, an electronic current and voltage is used to move an ion in and out of the interstitial or layered sites of a crystal host [1]. By injecting electrons into the host using an external circuit, a positively charged ion like Li^+or H^+follows the electron from the ion reservoir into the host and vice versa, restoring electroneutrality. This changes the chemical stoichiometry of the host material, and allows us to conduct solid-state chemistry at room temperature by changing the current and voltage. This "materials switch" continuously and reversibly tune the stoichiometry of the host by electrically controlling the injected charge.

My research aims to use advanced characterization and microfabrication to further elucidate mechanisms of ion insertion as well as expand the range of ion insertion devices, with a focus on devices for energy applications. Two of my core interests today are batteries for energy storage and low-energy computing using artificial neural network. In the batteries space, I aim to understand the mechanisms of ion insertion through advanced characterization and precise quantification. In my doctoral work, I used synchrotron X-ray microscopy to observe the insertion and migration of lithium within individual particles *in situ* [2]. I correlated such nanoscale, time-resolved measurements with phase-field modeling accounting for both reaction and diffusion within solids [2-3]. I intend to further utilize advanced characterization, model microfabricated systems, and precise geometries to elucidate reaction, diffusion, and phase transformation processes in lithium insertion materials used for batteries. In particular, I want to treat individual battery particles as nanodevices in order to understand the intercalation mechanisms.

I also aim to create new, low-energy electronic devices that operate on the basis of electrochemical ion insertion. Here, electrochemical lithium insertion dynamically modifies the doping concentration, enabling such devices to be used as analog non-volatile memory artificial synapses. Inspired by the human brain, this post-Moore computing architecture can reduce the energy consumption of artificial neural networks by 2-3 orders of magnitude compared with an optimized silicon-based system, and is compatible with many of the deep learning algorithms used today. Future work here entails identifying and optimizing materials that improve the switching speed through engineering bulk diffusion and surface reaction, improving the endurance and non-volatility, and designing architectures for learning and training. My current work as a Truman Fellow at Sandia National Laboratories investigates the potential of TiO_2 for use as an artificial synapse.

Select publications:

[1] Li, Y. and Chueh, W. C. Electrochemical and chemical insertion for energy transformation and switching. *Annual Reviews of Materials Research* 48, 137-165 (2018).

[2] Lim. J*, Li. Y.* *et al.* Origin and hysteresis of lithium compositoinal spatiodynamics within battery primary particles. *Science*.353,566–571 (2016). Highlighted in DOE, BES Basic Research Needs 2017.

[3] Li, Y. *et al.*Current-induced transition from particle-by-particle to concurrent intercalation in phase-separating battery electrodes. *Nature Materials*.13,1149–1156 (2014).

*Equal contribution author

Teaching Interests:

I have broad teaching interests within chemical engineering and materials science, including transport, reaction, thermodynamics, electrochemistry, renewable energy, crystal structures, and electronic properties. I will flavor my classes with an increased emphasis on applications towards energy technology, when appropriate. For example, in a new thermodynamics class that I designed with my PhD advisor, we structured thermodynamic concepts like entropy around important technologies like carbon capture and sequestration. Students engagements dramatically increased when they saw the connection between the core concepts and real-world applications. I will also aim to structure my classes around projects as much as possible, in the vein of my undergraduate, projected-based engineering curriculum at Olin College of Engineering. As the principal instructor for a batteries class at Stanford, students were tasked with proposing a new energy storage business, with representatives from local industry helping judge the novelty and feasibility of such new technologies. Students devised ideas including freeze-casting electrodes, batteries for space, and new battery charging algorithms. I was awarded the schoolwide Walter Gores Award at Stanford in 2016 for my teaching.

541135 A Fundamental Understanding of CO_2 Electrolysis Using Synchronous X-Ray Studies

Abstract:

Research Interests:

The limited efficiency with which renewable fuels and feedstocks are synthesized from electrical sources at present is determined by the sluggish electroreduction of CO2 (CO2RR) and water oxidation reaction (WOR). Herein, we reported a strategy to create high active sites for CO2RR in tin catalysts by incorporating copper and non-metal sulfur in an intimately integrated fashion (1, 2). We also developed a room-temperature synthesis to produce gelled FeCoW and NiCoFeP oxy-hydroxide materials with high WOR activity (3, 4). In situ X-ray absorption and computational studies, taken together, reveal that the enhanced energetics for CO2RR and WOR result from the ready promotion of favorable local coordination environments and electronic structures.

1. Zheng X L, De Luna P, Sargent E H, et al. Sulfur-Modulated Tin Sites Enable Highly Selective Electrochemical Reduction of CO2 to Formate, Joule, 2017, 1, 794-805.
2. Zheng X, Tang J, Ji Y, Wang J, Liu B, Steinrück H, Lim, K, Li Y, Toney M, Chan K, Cui Y, Theory-guided discovery of electrocatalysts with high formate selectivity at low overpotentials. Nature Catalysis, in revision.
3. Zhang B, Zheng X, Voznyy O, Sargent E H, et al. Homogeneously dispersed multimetal oxygen-evolving catalysts. Science, 2016, 352(6283): 333-337.
4. Zheng X L, Zhang B, Sargent E H, et al. Theory-driven design of high-valence metal sites for water oxidation confirmed using in situ soft X-ray absorption. Nature Chemistry, 2018, 10, 149-154.

Teaching Interests:

My goal of teaching will be to create a classroom where students find eager, enthusiasm, confidence and patience to actively participate in the challenging yet rewarding coursework. I would be very happy to develop the following courses. 1) X-ray Techniques for Chemistry and Materials Science. The unique properties of synchrotron radiation, including continuous spectrum, high flux and brightness, and high coherence, provide unprecedented experimental tools to study materials science, bioscience, geoscience, physical and chemical sciences, among many others. The goal of the course is to get students familiar with the advanced X-ray techniques for their research in chemical engineering, materials science, chemistry, physics, etc. 2) Science & Policy of Materials Sustainability. The course, designed as a combined graduate and undergraduate class, provides important concepts for the critical relationships between materials, environment, energy and sustainability.

542401 Nanostructured 2D Carbides and Nitrides for Electrochemical Energy Storage and Conversion

Abstract:

I am in my 2^nd^year of postdoctoral training in the Chemistry and Nanoscience Center at the National Renewable Energy Laboratory (NREL). I received my Ph.D. in Chemical Engineering in August 2016 from the University of Michigan under the direction of Professor Levi Thompson. My doctoral thesis involved understanding the charge storage mechanisms of high surface-area carbides and nitrides for electrochemical supercapacitors and batteries. Prior to joining NREL, I led a team of engineers at a startup company in Michigan to advance the commercialization of electrochemical supercapacitors based on metal nitrides. My current research at NREL with Nathan Neale and John Turner focusses on the structure-function relationship of the electrocatalytic and photocatalytic behavior of semiconductor and metal 2D carbides and nitrides MXenes for applications in energy storage and conversion. I am passionate about research and teaching. My career goal is to develop a world-class and innovative research laboratory that explores novel materials and chemistries for next-generation energy storage and conversion systems.

Research Interests:

Energy storage systems

- Batteries
- Supercapacitors

Energy conversion

- Fuel cells
- Photovoltaics
- Photoelectrochemical systems

Teaching Interests:

Traditional courses of interest

- Thermodynamics
- Reaction engineering
- Heat and mass transfer
- Fluid dynamics

Non-traditional courses of interest

- Electrochemical engineering
- Renewable energy and energy storage systems
- Nanostructured materials synthesis and characterization

Professional Links

- The Djire Foundation: https://www.djirefoundation.org
- LinkedIn:https://www.linkedin.com/in/abdoulaye-djire-phd-a596b37a/
- Google Scholar: https://scholar.google.com/citations?user=z4LvoUkAAAAJ&hl=en
- NREL:https://www.nrel.gov/research/abdoulaye-jire.html
- Research Gate: https://www.researchgate.net/profile/Abdoulaye_Djire

542414 Flame-Made Nanoparticles: Morphology, Optical Properties and Climate Impact

Abstract:

Combustion-generated nanoparticles during material-synthesis, energy generation (e.g. for transportation and power generation) as well as open fires form fractal-like structures. Yet, their optical properties are calculated typically by Rayleigh or Mie theories for spheres impeding accurate estimation of their environmental impact, characterization by optical diagnostics and selective sensing by fire detectors. For example, the direct radiative forcing of soot given by global climate models (90 % uncertainty) is 0.44 - 1.32 W/m^2. As soot constitutes 19 - 58 % of the total anthropogenic forcing, there is a clear need for an accurate estimation of its contribution. Similarly, fire detectors are not selective enough to soot light scattering and thus produce false alarms costing up to 1 billion £/y in United Kingdom alone. This motivates research for understanding the evolution of nanoparticle morphology and optical properties during flame synthesis and processing.

Discrete element modeling (DEM) for soot dynamics by agglomeration and surface growth[1] describes a realistic pathway for the evolution from nascent spheres to mature soot agglomerates. So, right after inception by reactive dimerization,[2] soot nuclei coagulate into small clusters of primary particles in point contact (i.e. agglomerates). However, surface growth by acetylene reactions forms covalent bonds between soot primary particles, compacting them into hard aggregates.[1] The size distribution and morphology of these nascent soot aggregates found excellent agreement with data from different combustion groups, explaining several experimental observations made in the last 15 years. This DEM is also extended to account for coagulation of nascent single and aggregated primary particles into mature soot agglomerates after the complete conversion of gaseous hydrocarbons.[3] This allowed the distinction of nascent and mature soot morphology with simple and easy-to-use scaling laws, finding good agreement with both premixed and diffusion flame data.

The DEM is then coupled with discrete dipole approximation (DDA) to simulate the evolution from nascent to mature soot light absorption by agglomeration and surface growth.[4] These robust DEM-DDA simulations found excellent agreement with laser induced incandescence[4] and light scattering data.[5] In contrast, using Rayleigh Debye Gans (RDG) theory for soot agglomerates consisting of monodisperse and non-aggregated primary particles underestimates soot light scattering up to 60 %.[5] Coupling RDG theory with a newly-derived scaling law accounting for primary particle aggregation and polydispersity[3] results in good agreement with the soot scattering cross sections estimated by more accurate but laborious models. Thus, the revised RDG theory can be used to estimate climate forcing and sharpen optical diagnostics for soot characterization. For example, accounting for the detailed soot morphology yields an average direct radiative forcing of 0.63 ± 0.05 W/m^2 in a cloud-free sky.[6] This is in between the predictions of current global climate models for cloudy skies (0.44 - 1.32 W/m^2) using Mie theory for spheres. Such results indicate that the soot contribution on global warming might be overpredicted substantially by these models that need to account for soot structure and polydispersity.

The nanoparticle structure is also essential for the performance of inorganic materials in different applications, including biosensors.[7] The morphology of flame-made nanoparticle agglomerates changes drastically during atmospheric aging, storage or industrial processing by granulation, spray drying or fluidization in the presence of humidity. So, the impact of humidity on the structure of inorganic nanoparticle agglomerates is quantified experimentally for the first time.[8] This research showed that ramified agglomerates of flame-made silica primary particles and aggregates restructure to smaller and compact structures by water condensation-evaporation, similarly to what had been observed for decades only for soot. Most importantly, it was found that agglomerate compaction by condensation-evaporation depends on the size of their constituent aggregates that is determined in the early stages of flame synthesis.[8]

In sum, the development of detailed models for nanoparticle dynamics and their thorough validation with experiments provides insight into the morphology and optical properties of soot and other nanomaterials during flame synthesis, atmospheric aging and industrial processing. This facilitates the mitigation of particulate emissions, control of their environmental impact and detection by robust fire sensors, as well as the design and optimization of nanomaterial performance in different applications.

References

[1] Kelesidis, G.A., Goudeli, E., and Pratsinis, S.E. "Flame synthesis of functional nanostructured materials and devices: Surface growth and aggregation" (2017) *Proc Combust Inst* **36**, 29-50.

[2] Kholghy, M.R., Kelesidis, G.A., and Pratsinis, S.E. "Reactive polycyclic aromatic hydrocarbon dimerization drives soot nucleation" (2018) *Phys Chem Chem Phys* **20**, 10926-10938.

[3] Kelesidis, G.A., Goudeli, E., and Pratsinis, S.E. "Morphology and mobility diameter of carbonaceous aerosols during agglomeration and surface growth" (2017) *Carbon* **121**, 527-535.

[4] Kelesidis, G.A., and Pratsinis, S.E. "Soot light absorption during agglomeration and surface growth" (2018) *Proc Combust Inst* **37**, under review.

[5] Kelesidis, G.A., Kholghy, M.R., Zurcher, J., Robertz, J., Allemann, M., Duric, A., and Pratsinis, S.E. "Light scattering from nanoparticle agglomerates" (2018) *Powder Technol*, submitted.

[6] Kelesidis, G.A., Kholghy, M.R., and Pratsinis, S.E. "Soot morphology and composition dominate radiative forcing" (2018) *Environ Sci Technol*, in preparation.

[7] Pratsinis, A., Kelesidis, G.A., Zuercher, S., Krumeich, F., Bolisetty, S., Mezzenga, R., Leroux, J.C., and Sotiriou, G.A. "Enzyme-Mimetic Antioxidant Luminescent Nanoparticles for Highly Sensitive Hydrogen Peroxide Biosensing" (2017) *ACS Nano* **11**, 12210-12218.

[8] Kelesidis, G.A., Furrer, F.M., Wegner, K., and Pratsinis, S.E. "Impact of humidity on silica nanoparticle agglomerate morphology and size distribution" (2018) *Langmuir*, accepted.

Research Interests:

- **Accurate estimation of nanoparticle climate impact using detailed models for their morphology and optical properties.**
- **Design of robust fire detectors with high selectivity to smoke and limited false alarms.**
- **Control of flame-made nanoparticle structure, light scattering and absorption for optimization of their performance in different applications, including biomedicine.**

Teaching Interests:

- **Engineering courses, including "Heat and Mass Transfer", "Fluid Mechanics", "Thermodynamics" and "Numerical Analysis".**
- **MSc level classes, such as "Aerosol Science and Technology" and "Computational Chemistry and Physics".**

542438 Multi-Scale Modeling of the Structure and Dynamics of Bio-Inspired Light-Harvesting Technologies

Abstract:

Research Interests:

Natural photosynthetic systems have engineered the ability to efficiently absorb and transport energy over long distances to the reaction center, where charge separation occurs. These energy transport properties are efficient due to the precise placement of optically-active chromophores, both weakly- and strongly-coupled, scaffolded in place by protein sub-units. When designing synthetic mimics of natural photosynthetic systems, an alternative to protein scaffolding is DNA nanotechnology, which has been used to scaffold chromophores by covalent and non-covalent interactions. Common closely-packed structural motifs seen in natural photosynthetic systems are strongly interacting J-aggregates, which extend the exciton delocalization length and increase the energy transport rate, and H-aggregated dimers, which increase the radiative lifetime of the exciton and can lead to charge separation. I have studied both FRET-based energy transport in DNA-intercalated thiazole orange dyes, and exciton delocalization in J-aggregated pseudoisocyanine fibers which self-assemble. These photosynthetic and DNA-templated systems require a multi-scale approach to modeling, so that dynamic motion of the complex can be captured (all-atom molecular dynamics), and so that the electronic structure of the excited states and excitons can be probed.

I am interested in studying the structural and electronic basis for the emergent photophysical behavior found in natural photosynthesis, and in applying these design principles to synthetic constructs. Long-range, or incoherent energy transfer between chromophores is well-studied, and can be adequately described by a Coulombic interaction between donor and acceptor transition charge densities, also known as FRET. Short-range energy transfer, on the other hand, is described by both Coulombic interactions as well as the electronic wavefunction overlap between the donor and acceptor chromophores. Short-range interactions lead to electronic properties such as exciton delocalization and charge separation. These short-range interactions are difficult to characterize, but are necessary for an adequate understanding of energy transfer properties in natural and synthetic constructs. In my future work, I will study the electronic structure properties of closely-packed chromophore systems, both in natural and synthetic complexes, to better understand how to build an energy capture device that can compete with the efficiencies seen in nature. Whether the properties of natural and synthetic light-harvesting complexes will converge or diverge will be of intense interest.

Postdoctoral Project: "Structure and function of DNA-templated chromophoric materials." Under the supervision of Dr. Mark Bathe at Massachusetts Institute of Technology, in the Biological Engineering Department.

Ph.D. Dissertation: "A computational study of excitation energy transfer in photosynthetic light-harvesting complexes." Under the supervision of Dr. Cynthia Lo at Washington University in St. Louis, in the Energy, Environmental, and Chemical Engineering Department.

Teaching Interests:

I have experience in teaching and mentoring over my entire scientific career, starting as a TA for *Organic Chemistry* laboratory while an undergrad at Bucknell University and several Chemical Engineering courses as a graduate student at Washington University, to teaching the undergraduate course entitled *Modeling and Computing in Chemical Engineering* after my graduate tenure as the instructor of record at Washington University. My teaching interests are interdisciplinary, as is my research, at the boundary between chemistry, biology, physics, and engineering, and I would feel comfortable teaching core Chemical Engineering courses as well as topics in multi-scale modeling and quantum chemistry.

Selected Publications:

WP Bricker & CS Lo. Excitation energy transfer in the peridinin-chlorophyll *a*-protein complex modeled using configuration interaction. *Journal of Physical Chemistry B* **118**, 9141-9154 (2014).

WP Bricker & CS Lo. Efficient pathways of excitation energy transfer from delocalized S_2 excitons in the peridinin-chlorophyll *a*-protein complex. *Journal of Physical Chemistry B* **119**, 5755-5764 (2015).

J Chmeliov, WP Bricker, C Lo, E Jouin, L Valkunas, AV Ruban & CDP Duffy. An 'all pigment' model of excitation quenching in LHCII. *Physical Chemistry Chemical Physics* **17**, 15857-15867 (2015).

WP Bricker, PM Shenai, A Ghosh, Z Liu, MGM Enrinquez, PH Lambrev, H-S Tan, CS Lo, S Tretiak, S Fernandez-Alberti & Y Zhao. Non-radiative relaxation of photoexcited chlorophylls: Theoretical and experimental study. *Scientific Reports* **5**, 13625 (2015).

K Pan, WP Bricker, S Ratanalert & M Bathe. Structure and conformational dynamics of scaffolded DNA origami nanoparticles. *Nucleic Acids Research* **45**, 6284-6298 (2017).

PD Cunningham, WP Bricker, SA Díaz, IL Medintz, M Bathe & JS Melinger. Optical determination of the electronic coupling and intercalation geometry of Thiazole Orange homodimer in DNA. *Journal of Chemical Physics* **147**, 055101 (2017).

WP Bricker, JL Banal, MB Stone & M Bathe. Molecular model of J-aggregated pseudoisocyanine fibers. *Journal of Chemical Physics* **149**, (2018). *In Press.*

542444 Techno-Economic Analysis and Optimization for Energy Storage Systems

Abstract:

Research Interests:

The overall theme of my research is to develop advanced computational models and decision-support tools for solving problems in energy storage and manufacturing systems. My immediate focus is on reducing the cost of lithium-ion batteries for automotive applications. In order to accomplish this, I plan to apply a combinatorial approach (Fig. 1) of the following research problems. (1) Techno-economic analysis of automotive lithium ion battery manufacturing processes, (2) Integrated supply chain optimization and operations analysis, and (3) Improving operational efficiency and design. My PhD and postdoctoral training in modeling, optimization, supply chain analysis, and developing decision-support tools will be critical for the aforementioned research plan. Furthermore, my postdoctoral work at Argonne National Laboratory has enriched my knowledge of lithium ion battery design, characterization, manufacturing, and supply chain. This experience and relevant expertise puts me in a good position to undertake the proposed research.

In a preliminary work, the major cost contributing components â€" ranging from materials costs to the cost of unit operations, have been studied to identify the cost and energy drivers and to explore options that enable the reduction of battery manufacturing cost. As the cost of materials is a major contributor to the overall battery cost, a primary focus of this work has been to identify production routes to reduce material costs. On one hand, detailed techno-economic models are developed to estimate the cost and energy required to produce cathode materials such as nickel-cobalt-manganese oxide (NCM) and lithium manganese oxide (LMO) and electrolyte materials such as lithium hexafluorophosphate ($LiPF_6$). On the other hand, process models are developed for electrode coating step to study solvent drying and recovery in further detail and identify factors to reduce the cost of operations and improve the quality of product. In this poster, a collage of results will be presented highlighting the key learnings and outcomes from the aforementioned studies. The results will show the impact of various process, operating, and economic parameters on the overall battery cost. Also, the pathways to minimize the battery manufacturing cost will be outlined. The poster will conclude with a discussion on the lessons learned, and the research needs and trends.

In future, I wish to continue my work in the area of techno-economic analysis of battery material production. Furthermore, I intend to design and optimize integrated production and recycling facilities for manufacturing lithium ion battery materials. In addition, I intend to develop models for supply chain optimization including decisions such as capacity planning, material sourcing, inventory management, international material transfers, and material recycling.

Teaching Interests:

I am comfortable to teach any of the core chemical engineering courses such as Process Design & Modeling,

Mathematical Methods in Chemical Engineering, Transport Phenomena, Mass & Heat Transfer, Separation Processes, and Reaction Engineering. In addition, I am interested in developing elective courses stemmed from my research experience in the area of energy systems modeling and optimization. Here, I plan on introducing students to techno-economic and environmental impact analysis techniques and their application to energy systems such as power systems design, evaluating low carbon energy supply alternatives, and challenges in waste disposals and emission control.

542447 Atomistic Modeling of Energy Storage Materials

Abstract:

Overview:

A societal shift towards greater adoption of renewable energy is underway. To accelerate this transition, new approaches for energy storage are needed to address the intermittent nature of these resources. My work employs computational methods to investigate two categories of energy storage: electrochemical and thermal. With regard to electrochemical storage, the lithium-ion battery has revolutionized portable electronics. Nevertheless, batteries that possess energy densities much larger than state-of-the-art technologies are highly desirable for emerging applications such as grid-based storage and electric vehicles. Batteries employing metallic anodes have gained considerable attention, as they can improve safety and theoretically boost energy densities by as much as ten times. However, battery performance is highly susceptible to the interfacial properties at the surface of the metallic anode. Using atomistic techniques such as density functional theory and molecular dynamics, my work probes the chemical origin of these limitations.[1] Thermal energy storage materials, on the other hand, store energy in the form of heat. These systems are able to capture 'lost' energy that is typically rejected from a device or process. A central question is: 'which storage materials are best?' My research addresses this question by computing the theoretical energy capacities of hundreds of hydration reactions.[2] The materials database generated by these computations is being examined with machine learning algorithms to more rapidly relate the structure and composition of the materials to their performance. Through collaborations with experimentalists, these relationships can be used to pinpoint specific thermal energy storage materials for synthesis and subsequent characterization.

Research Interests:

I am broadly interested in modeling processes/phenomena related to energy storage materials, and collaborating with experimental colleagues. Some specific examples from my research background and interests include:

- Analyzing electrolyte decomposition reaction pathways at anode surfaces
- Probing structural properties of the solid electrolyte interphase layer in lithium metal batteries
- Analyzing the performance of core-shell nanoparticles as electrocatalysts
- Predicting energy storage capacities of current and new materials
- Applying machine learning to the discovery of new materials

Teaching Interests:

I am interested in teaching a variety of topics including introductory, core, and elective courses. However, based on my interests and expertise I would be best suited to teach: General Engineering, Engineering Thermodynamics, Chemical Reaction Engineering, Scientific Computing, and Battery Fundamentals.

References:

1. **J.S. Lowe** and D.J. Siegel, "Reaction Pathways for Solvent Decomposition on Magnesium Anodes," *The Journal of Physical Chemistry C* 2018, 122, 10714-10724.
2. S. Kiyabu, **J.S. Lowe**, A. Ahmed, and D.J. Siegel, "Computational Screening of Hydration Reactions for Thermal Energy Storage: New Materials and Design Rules," *Chemistry of Materials* 2018, 30, 2006-2017.

542460 Fit Batteries to the Grid or Grid to the Batteries?

Abstract:

The CapEx of batteries accounts for 60% of the entire microgrid components, which is the most significant portion as a single component,[1] but current approaches to integrating batteries into the entire microgrid components do not make batteries economically feasible. Typically, batteries are treated as a black box that does not account for their internal states in current microgrid simulations.[2-5] This might lead to under-utilization and over-stacking of batteries. In contrast, detailed physics-based battery models, accounting for internal states, can save a significant amount of energy and cost, utilizing batteries with maximized life and usability.[6,7] It is important to identify how efficient physics-based models of batteries can be included and addressed in grid control strategies.[8]

In this talk, simple examples for Battery-PV microgrids and the direct simulation of the same including physics-based battery models will be presented.[6] It is straightforward to implement other batteries in this efficient equation-based microgrid framework. For example, one might visualize a situation in which lithium-ion batteries with different chemistries being used simultaneously, sometimes even flow batteries in combination with lithium-ion batteries. The model of the battery as a pack can be considered as well (e.g., a set of cells in series and parallel with slight inherent variation in SOC and SOH). The results of the proposed approach are compared with the conventional approaches and improvements in performance and speed are reported.

References

1. N. Ivanova, Energy storage update, London, UK, 2015. Available: http://analysis.energystorageupdate.com/lithium-ion-costs-fall-50-within-five-years
2. D. A. Beck, J. M. Carothers, V. Subramanian, and J. Pfaendtner, *AIChE Journal,* **62** (5), 1402-1416 (2016).
3. X. Li, *IET Renewable Power Generation,* **6** (5), 340-347 (2012).
4. X. Li, D. Hui, and X. Lai, *IEEE Transactions on Sustainable Energy,* **4** (2), 464-473 (2013)
5. P. Naderi, *Journal of Solar Energy Engineering,* **135** (2), 024506 (2013).
6. S. B. Lee, C. Pathak, V. Ramadesigan, W. Gao, and V. R. Subramanian, *Journal of The Electrochemical Society,* **164** (11), E3026-E3034 (2017).
7. M. Pathak, D. Sonawane, S. Santhanagopalan, R. D. Braatz, and V. R. Subramanian, *ECS Transactions,* **75** (23), 51-75 (2017).
8. V. R. Subramanian, (A. Staller, ed.), ECS redcat blog, The Electrochemical Society, USA, 2017. Available: https://www.electrochem.org/redcat-blog/free-batteries-towards-bottom-renewable-grid-control/

Research Interests:

My research interests are centered around battery simulation, model-based experimental validation, and battery and renewable application integration (e.g., renewable grids and electric vehicles). These challenging problems are multi-disciplinary and require collaboration with scholars in other engineering or industrial fields. These types of collaborations tend to be strongly encouraged by funding agencies. My research background in the renewable energy field of modeling (Ph.D.) combined with my previous experimental experience (M.S.) will help to facilitate active collaboration. If I can continue my research as an assistant professor, I plan to submit several proposals to funding agencies in government and industry, including NSF, DOE, ECS Amazon Catalyst, and Samsung global research outreach. My research capabilities include:

- Developing physics-based battery models for the present and the next-generation batteries (e.g., lithium-ion, lithium metal, and flow batteries)
- Validating physics-based models with experimental data and analyzing chemical/physical phenomena with mathematical models
- Combining data science tools with model-based approaches
- Identifying efficient ways to integrate batteries with renewable energy applications for cost-effective systems (e.g., renewable grids and electric vehicles).

Teaching Interests:

I have degrees in Chemical Engineering (Bachelor and Ph.D.) and Chemistry (M.S.). I am confident in my abilities to teach the core Chemical Engineering curriculum from the perspectives of engineers as well as scientists.

âˆ™ *Core chemical engineering curriculum:* mathematical methods in engineering, chemical reaction engineering, process control, transport phenomena, and thermodynamics at the undergraduate and graduate levels.

Also, my research background makes me well placed to develop and teach technical electives, which are invaluable from the perspective of current employment in industry and performing academic research (e.g., data science and battery/ grid integration).

âˆ™ *Technical electives:* Electrochemical engineering, microgrid architectures and controls, renewable energy storage/conversion systems, and data science for chemical engineering and material science.

542530 Advanced Materials and Nanotechnologies for Efficient, Solution Processable Energy Devices

Abstract:

Research Interests:

In the past century, the extent to which discoveries of new molecules and materials have impacted technology, particularly in the energy sector, has been dramatic. These materials have quietly transformed the way we live, finding their way into chemical production, energy production, transportation, and information technologies. However, as the world energy consumption continues to grow, we are still mainly replying on limit resources such as petroleum for energy generation. How to meet the ever-increasing energy consumption in a sustainable way is extremely important and challenging for human society's long term development.

My main interests lie in exploring and developing next generation materials (from 3D to 0D crystal structures, and from bulk materials to nano-structures) with an emphasis on their potential applications especially in energy harvesting (Photo-energy and thermal energy to electricity), electronic and optoelectronic devices, as well as energy storage using cost and energy efficient solution processes. I aim to 1) design and synthesize multifunctional materials with low dimensional crystal structures; 2) investigate the fundamental properties of these materials; 3) develop high performance energy harvesting devices as wellas other electronic and optoelectronic devices and 4) incorporate flexibility, stretchability, and transparency into high performance thin film devices for ubiquitous electronics.

During my postdoctoral research, I have successfully explored several material systems for photovoltaic and optoelectronic devices. One of such material systems is environmental friendly organic-inorganic hybrid perovskite materials for photo-energy harvesting. The sun is continuously supplying us with photo-energy that is freely available by nature, but harvesting the photo-energy efficiently and cost-effectively is challenging at present due to the high processing cost, complex processing steps and material recycling. I have designed a universal processing protocol for synthesizing Sn-based halide perovskites that have a great potential to achieve over 30% efficiency for PV theoretically. Through molecular precursor design, I successfully achieved over 5 times improvement on the solar cell performances. I have also explored these materials' potential application in X-ray detectors and thermoelectric devices. Furthermore, I have designed and built an integrated in-situ characterization system in order to unveil the fundamental material formation mechanisms and their optical properties. This approach can provide much faster characterizations that will benefit future in-line process monitoring and enable us to explore the material properties in non-equilibrium states.

I have also worked on various materials, including organic and inorganic, for PV and optoelectronic devices during my Ph.D research. I was able to achieve high performance devices using different semiconductor materials which required a well designed device structure, proper energy band alignment, precise process control and a deep understanding of the material properties. I was the first to demonstrate a high

performance fully solution processed solar cell with over 10% efficiency using $CuInSe_2$ as light absorber. In this work, an unique process design combining metal nanowires and semiconducting nanoparticles was demonstrated to replace traditional ITO as the solution processable transparent electrode. Moreover, the local nanoscale welding effect in solution processed metallic nanowires were investigated, which resulted in the invention of an energy efficient fabrication process that overcomes the challenges for their applications into thermal sensitive devices. In another project, I redesigned the synthesis route of Cu_2ZnSnS_4 nanoparticle precursors as an earth abundant PV absorber and demonstrated a significant improvement (over 40%) of device performance. To overcome the thermodynamic limit of PV efficiency, I also designed and fabricated the tandem PV device structure in collaboration with IBM scientists and successfully achieved over 10 % efficiency for the first time with the combination of organic and amorphous Si solar cells. Last but not least, I investigated the formation process of organic-inorganic hybrid perovskite materials and achieved precise control of the film formation process. Power conversion efficiencies of over 19% was demonstrated within a year effort.

My unique background has prepared me to work in this multidisciplinary research area of material science, chemistry and device physics to develop advanced materials and nanotechnologies and bring in new perspectives and better solutions for energy technologies.

Teaching Interests:

I am interested in teaching most of the major courses for Chemical and Materials Engineering including thermodynamics and kinetics, spectroscopy, microscopy, material design and synthesis, thin film deposition, semiconductor physics, photovoltaic and optoelectronic devices. I have experience as TA and guest lecturers for both graduate and undergraduate level courses including semiconductor devices and fabrication, introduction to solar cells, material physic at UCLA. I am also actively mentoring several UC Berkeley undergraduate students on their research courses in LBNL.

Selected Publications:

1. **T.-B. Song**, T. Yokoyama,S.Aramaki, M. Kanatzidis "Performance Enhancement of Lead-free Tin-based Perovskite Solar Cells with Reducing Atmosphere-assisted Dispersible Additive" ACS Energy Letter 2, 897-903. (2017)
2. **T.-B. Song**, T. Yokoyama, C. Stoumpos, J. Logsdon, D. H. Cao, M. R. Wasielewski, S. Aramaki, M. Kanatzidis "Importance of Reducing Vapor Atmosphere in the Fabrication of Sn-based Perovskite Solar Cells" Journal of the American Chemical Society 139 (2), 836-842. (2017)
3. **T.-B. Song**, Y.S. Rim, F. Liu, B. Bob, S. Ye, Y. Yang "Robust Silver Nanowire Network for Transparent Electrode" ACS Applied Materials & Interface 7 (44), 24601-24607. (2015)
4. **T.-B. Song**, Q. Chen, H. Zhou, S. Luo, Y. Yang, J. You, Y. Yang "Unraveling Film Transformations and Device Performance of Planar Perovskite Solar Cells" Nano Energy 12, 494-500. (2015)
5. **T.-B. Song**, Y. Chen, C.-H. Chung, Y. Yang, B. Bob, H.-S. Duan, G. Li, K.-N. Tu, Y.Huang, Y. Yang "Nanoscale Joule Heating and Electromigration Enhanced Ripening of Silver Nanowire Contacts" ACS Nano 8 (3), 2804-2811. (2014)
6. Zhou*, **T.-B. Song***, W.-C. Hsu, S. Luo, S. Ye, H.-S. Duan, C.-J. Hsu, W. Yang, Y. Yang "Rational Defect Passivation of $Cu_2ZnSn(S,Se)_4$ Photovoltaics with Solution-Processed Cu_2ZnSnS_4:Na Nanocrystals" (**These authors contributed equally*) Journal of the American Chemical Society 135 (43), 15998-16001. (2013)
7. Zhou, Q. Chen, G. Li, S. Luo, **T.-B. Song**, H.-S. Duan, Z. Hong, J. You, Y. Liu, Y. Yang "Interface Engineering of Highly Efficient Perovskite Solar Cells" Science 345 (6196), 542-546. (2014)
8. C H. Chung, **T. B. Song**, B. Bob, R. Zhu, H. S. Duan, Y. Yang "Silver Nanowire Composite Window Layers for Fully Solution Deposited Thin Film Photovoltaic Devices" Advanced Materials 24 (40), 5499-5504. (2012)

542541 Design and Development of Materials and Electrolytes for Energy: From Fundamental Mechanisms to Applications

Abstract:

Research Interests:

The development of new solutions to satisfy the growing demand for energy in areas ranging from the portable electronics to grid energy storage is becoming increasingly important. Current research efforts are directed toward improvement in the device performance (i.e. charging times, energy, capacity, cycling stability), cost and safety. Traditional batteries require prolonged charging and therefore have a limited high rate energy harvesting capacity. Supercapacitors on the other hand can provide fast charging times of below one minute but require development of the new materials in order to achieve energy densities comparable with batteries at much higher rates.

My research interests focus on the design of new supercapacitor materials and electrolytes guided by understanding of the charge storage processes and mechanisms to push the performance of energy storage devices while ensuring their safety and low cost.

During my postdoc in Prof. Zhenan Bao group at Stanford University, I worked in two different research directions. First, I explored electrochemical behavior of the new conducting metal-organic frameworks and demonstrated exceptionally high areal and volumetric capacitance of these materials. This work showed for the first time that metal-organic frameworks (MOFs) can outperform traditional supercapacitor materials such as porous carbons, establishing a new research direction. Currently, I am studying fundamental principles underlying the charge storage mechanism in MOFs. The second direction that I developed is related to superconcentrated water-based electrolytes for green and safe energy storage. In particular, I demonstrated that potassium acetate-based highly concentrated electrolytes can provide the same benefits of the extended voltage window as imide-based electrolytes and, once combined with lithium acetate, demonstrate compatibility with traditional Li-ion battery electrode materials while being low-cost and environmentally benign.

The focus of my Ph.D. work in Prof. Yury Gogotsi group was on (1) the investigation of a new family of two dimensional materials, MXenes (2D transition metal carbides), for high-rate energy storage applications and (2) understanding the charge storage mechanism in them. My studies revealed that a variety of cations of various charges and sizes such as sodium, magnesium, potassium and even aluminum can readily intercalate from aqueous solutions into the MXenes and participate in charge storage. Moreover, I demonstrated that spontaneous chemical and electrochemically-assisted intercalation of cations in between the MXene layers leads to extremely high volumetric capacitances. Volumetric capacitance is proportional to the amount of energy stored for a unit of volume of electrode and its high values are particularly crucial for portable electronics applications, where device volume is a main constraint. Also, combining *in situ* characterization techniques, I provided strong evidence that the mechanism of the electrochemical charge storage in MXenes is pseudocapacitive, i.e. due to changes in the oxidation state of transition metal during charge/discharge. Using the knowledge of the mechanism of charge storage, I demonstrated that the rate performance of MXenes can be pushed to extreme limits with charging times of just a fraction of a second.

Teaching Interests:

At Stanford University, I served as an instructor for a laboratory class as a part of the Chemical Engineering Laboratory course and mentored undergraduate and MS students. During my PhD, at Drexel University I served as a recitation and laboratory instructor in a number of undergraduate courses, mentored several students, led a Senior Design project and participated in the STEM outreach activities. I am interested in teaching various courses in Chemical Engineering at both the graduate and undergraduate levels, and in particular to teach/develop course in the field of electrochemistry.

Selected publications:

1. **R. Lukatskaya**, B. Dunn, Y. Gogotsi, "Multidimensional materials and device architectures for future hybrid energy storage", **Nature Communications**, 2016, 7, 12647
2. Feng*, T. Lei*, **M.R. Lukatskaya***, J. Park, Z. Huang, M. Lee, L. Shaw, S. Chen, A.A. Yakovenko, A. Kulkarni, J. Xiao, K. Fredrickson, J.B. Tok, X. Zou, Y. Cui, Z. Bao, "Robust and Conductive Two-Dimensional Metal–Organic Frameworks with Exceptionally High Volumetric and Areal Capacitance", **Nature Energy**, 2018, 3, 30–36
3. **R. Lukatskaya***, S. Kota*, Z. Lin*, M.-Q. Zhao, N. Shpigel, M.D. Levi, J. Halim, P.-L. Taberna, M.W. Barsoum, P. Simon, Y. Gogotsi, "Ultra-high-rate pseudocapacitive energy storage in two-dimensional transition metal carbides", **Nature Energy**, 2017, 6, 17105
4. **R. Lukatskaya***, S.M. Bak*, X. Yu, X.Q. Yang, M.W. Barsoum, Y. Gogotsi, "Probing the mechanism of high capacitance in two-dimensional titanium carbide using in-situ X-Ray absorption spectroscopy", **Advanced Energy Materials,** 2015, 5 (15), 1500589
5. Ghidiu*, **M. R. Lukatskaya***, M.Q. Zhao, Y. Gogotsi, M. W. Barsoum, "Conductive two-dimensional titanium carbide clay with high volumetric capacitance", **Nature**, 2014, 516, 78-81
6. **R. Lukatskaya**, O. Mashtalir, C.E. Ren, Y. Dall'Agnese, P. Rozier, P. L. Taberna, M. Naguib, P. Simon, M.W. Barsoum, Y. Gogotsi, "Cation Intercalation and High Volumetric Capacitance of Two-dimensional Titanium Carbide", **Science**, 2013, 341 (6153), 1502-1505.

542587 Exploring the Solid-Electrolyte Interface and Interphase By Surface-Plasmon Resonance Spectroscopy

Abstract:

Education

- Ph.D., Chemical Engineering, Florida State University (Supervisor: Dr. Daniel Hallinan), 2017
- M.S., Chemical Engineering, Florida State University, 2016
- M.S., Biomedical Engineering, Southeast University, 2012
- M.S., Advanced Materials, Ulm University (Germany), 2012
- B.S., Biomedical Engineering, Southeast University (China), 2009

Postdoc Training (Oak Ridge National Laboratory, supervisor: Dr. Jagjit Nanda)

- Probing the Nanoscale Heterogeneity of SEI on Silicon Thin-film Anode Using Tip Enhanced Raman Spectroscopy (TERS)
- Developing Solid Polymer Electrolyte for Non-aqueous Redox Flow Batteries

Research Interests:

The growing need for energy storage systems requires batteries with high power and energy density as well as longer life and improved safety. This calls for innovations in high energy density electrode materials as well as in design of robust and efficient charge transfer interfaces without electrolyte degradation or chemical side reactions. Charge transport across the solid electrode-liquid electrolyte interface (SLI) is believed to be one of the rate limiting steps. In lithium-ion batteries, the thermodynamic instability of the electrolyte at the SLI leads to formation of a passivation layer, commonly referred to as the solid-electrolyte interphase (SEI). Although the composition, morphology, and structure of SEI of lithium-ion batteries have been extensively studied, probing its evolution *in situ* at nanoscale is still challenging to a large extent. Surface-plasmon resonance (SPR) spectroscopy is promising to solve this challenge. SPR is the resonant oscillation of conduction electrons at the interface between negative and positive permittivity molecules stimulated by incident light. It includes surface-enhanced Raman spectroscopy (SERS) and tip-enhanced Raman spectroscopy (TERS). The prerequisite of employing SPR to probe the surface molecules is to have a highly ordered plasmonic nanostructured SPR substrate, which is extremely challenging.

In my Ph.D. research, I developed two interfacial self-assembly techniques to fabricate gold nanoparticles (Au NPs) into large area (cm^2) monolayers, which ensured the formation of long-range ordered nanogap arrays. The local electromagnetic field is extremely intense within the metallic nanogap (<10 nm) due to the coupling effect among adjacent nanoparticles, which allows for probing the molecules at immediate vicinity of SLI (distance from the solid surface < 20 nm). The Au NP monolayers exhibit high SERS sensitivity even down to single-molecule level (enhancement factor $> 10^7$). This allows for probing a broad range of trace electrolyte components, such as lithium hexafluorophosphate ($LiPF_6$), fluoroethylene carbonate (FEC), ethylene carbonate (EC) and diethyl carbonate (DEC), etc. The investigation of a commercial lithium-ion battery electrolyte ($LiPF_6$ in EC+DEC binary solvents) using SERS allows for the determination of the solvent coordination numbers, which ranges from 2 to 5, in sharp contrast to those calculated from bulk liquid electrolytes by standard confocal Raman and infra-red (IR) spectroscopy from 3 to 6.

SERS is promising to probe molecular fingerprints in nanoscale through sample plane. However, it lacks the nanoscale resolution within sample plane. This makes it extremely hard for getting spatial heterogeneity of the SEI, which can be only on the order of 10 nm. Tip Enhanced Raman Spectroscopy (TERS) combines SERS with atomic force microscopy (AFM), capable of providing the chemical vibrational information and topography of the sample in the nanometer spatial resolution simultaneously. My postdoc training is focused on using TERS to understand the SEI evolution on amorphous silicon (a-Si) thin film electrode. TERS analysis on cycled a-Si indicates that the nanometer scale SEI "islands" are unevenly distributed. Even for the same SEI "island", the composition is different from point to point with inter-point distance smaller than 10 nm. The local chemical information studied by TERS is intrinsically different than that collected from the standard confocal Raman and IR spectroscopy.

My near future research involves in continually developing *in operando* SPR techniques based on SERS and TERS to understand the interfacial phenomenon *in situ* in energy storage units. The model systems of interest include the high voltage cathode/electrolyte interface, electrode/solid-state electrolyte interface, and electrode/polymer electrolyte interface. Such surface sensitive probing techniques can also be readily applied in water desalination, catalysis, corrosion, energy storage and ion distribution/transport through biological membranes, etc.

Teaching Interests:

Served as a teaching assistant for more than 4 years in chemical engineering department at FSU, I got rich hands-on experience in teaching and greatly enjoyed sharing my knowledge with the undergraduates. The core courses include 'Mathematics in Chemical Engineering', 'Mass and Energy Balance', 'Process and Analysis Design', and 'Unit Operations Lab', etc. I truly enjoyed mentoring students during recitation and office hour, helping with preparing the teaching materials and grading their homework. I see teaching more than an integral part of an academic career. I much look forward to having the opportunity to teach chemical engineering core course for undergraduate or graduate students.

Selected Publications (27 in total, with 10 first author publications and 2 filed patents)

1. Yang, G., Ivanov, I., Ruther, R., Sacci, Robert., Subjakova, V., Hallinan, D. T. & Nanda, J. (2018). Electrolyte Solvation Structure at Solid/liquid Interface Probed by Nanogap Surface-enhanced Raman Spectroscopy. ACS Nano (under review).
2. Ruther, Rose, Guang Yang, Frank M. Delnick, Zhijiang Tang, Michelle Lehmann, Tomonori Saito, Yujie Meng, Thomas A. Zawodzinski, and Jagjit Nanda (2018). Mechanically Robust, Sodium-Ion Conducting Membranes for Nonaqueous Redox Flow Batteries. *ACS Energy Letters*.
3. Yang, G., Kim, K., Wang, W., Chen, B., Mattoussi, H., & Hallinan Jr, D. T. (2018). Scaling Laws for Polymer Chains Grafted onto Nanoparticles. *Macromolecular Chemistry and Physics*, *219*(8), 1700417.
4. Yang, G., Nanda, J., Wang, B., Chen, G., & Hallinan Jr, D. T. (2017). Self-assembly of large gold nanoparticles for surface-enhanced Raman spectroscopy. *ACS Applied Materials & Interfaces*, *9*(15), 13457-13470.
5. Yang, G., Hu, L., Keiper, T. D., Xiong, P., & Hallinan Jr, D. T. (2016). Gold nanoparticle monolayers with tunable optical and electrical properties.*Langmuir*, *32*(16), 4022-4033.
6. Yang, G., & Hallinan Jr, D. T. (2016). Self-assembly of large-scale crack-free gold nanoparticle films using a 'drain-to-deposit' strategy.*Nanotechnology*, *27*(22), 225604.
7. Yang, G., & Hallinan, D. T. (2016). Gold Nanoparticle Monolayers from Sequential Interfacial Ligand Exchange and Migration in a Three-Phase System. *Scientific Reports*, *6*, 35339.
8. Yang, G., Chang, W. S., & Hallinan Jr, D. T. (2015). A convenient phase transfer protocol to functionalize gold nanoparticles with short alkylamine ligands. *Journal of colloid and interface science*, *460*, 164-172.

542278 Hydrothermal Technologies forÂ Valorizing Biomass and Producing Valued-Added Chemicals

Abstract:

5th Year PhD Candidate

Research Interests:

I am interested in exploring hydrothermal technologies for synthesizing industrially attractive chemicals, degrading environmentally harmful pollutants, and valorizing biomass or organic wastes into fuels, fertilizers, and novel bio-derived materials. Hydrothermal technologies use water at elevated temperatures and pressure and offer environmentally benign routes for chemical processing. The properties of hydrothermal media deviate from those of ambient water and can be â€˜tunedâ€™ to create unique reaction environments. For instance, pressurized water heated to temperatures above its boiling point, yet kept below its critical temperature, forms a hot, compressed-liquid phase that effectively solvates organic molecules and promotes acid- and base-catalyzed reactions. Furthermore, at temperatures and pressures exceeding its critical point, water forms a supercritical fluid that supports free-radical chemical reactions. The tunable properties of hydrothermal media can provide reaction environments capable of supporting diverse applications ranging from organic synthesis, converting biomass into bio-crude oils, or gasifying organic wastes into fuel gases. My research will explore the applications of hydrothermal technologies for providing solutions for environmental remediation, sustainable fuel production, and synthesis of chemical from renewable resources.

Research Experiences:

My PhD research has been combined approach of experimental work and kinetic modeling for investigating the conversion of proteinaceous biomass in hydrothermal media. Initial projects investigated the use of sub- and supercritical water for converting model proteins into bio-crude oils and nutrients. Kinetic models were developed for correlating the experimental data and identified optimal operating conditions for producing bio-crude oils and recovering nutrient enriched aqueous-phase products. Similarly, I developed kinetic models for investigating the influence of the biochemical composition of microalgae on resultant yields of bio-crude oil product yields. Additionally, I performed fundamental experimental studies for investigating the influence of pH on the hydrothermal reaction pathways of peptides. The experimental results were evaluated with thermodynamic and chemical kinetic models to elucidate the effects of the dissociation states of peptides on their reaction rates and selectivity. This complementary approach of thermodynamic and chemical kinetic modeling provided a framework for understanding the influence of the properties of hydrothermal environments on chemical reactions. This approach could be used for tailoring the properties of hydrothermal media for unique green chemistry applications. Further, I received training in statistical and chemical kinetic modeling of batch reactor kinetics during an industrial summer internship.

Teaching Interests:

My teaching interests include core chemical engineering courses such as reaction engineering, heat transfer, and mass and energy balances. I have further interests in developing curricula for elective courses in chemical kinetic modeling and introduction to computing environments (e.g., Mathematica, Matlab) tailored for chemical engineering undergraduates.

Teaching Experiences:

During my academic tenure, I have enjoyed participating in several teaching experiences. As an undergraduate chemical engineering student at the Illinois Institute of Technology, I was employed as a tutor at the Academic Resource Center and provided tutoring in thermodynamics, transport phenomena, reaction engineering, and process modeling and my work garnered a tutoring award. The summer after graduating from IIT, I was hired by After School Matters as a STEM Instructor and developed and instructed a sustainable energy management course for 23 high school students with backgrounds under-represented in STEM related fields. As a PhD student at Penn State, I have served as a teaching assistant for CHE 430: Reaction Engineering. As a TA, I created interactive Mathematica notebooks for enhancing lectures, tutored students on modeling chemical reactions and sizing reactors with computing software (e.g., Matlab, Mathematica) and gave lectures on differential and integral reactors and multiple reactions in continuous reactors. In addition, I have been a Teaching Fellow under the mentorship of Professor Ali Borhan for CHE 350: Process Heat Transfer. As a Teaching Fellow I developed midterm and final exam reviews, provided class lectures on solving example problems, and closely mentored students working on honors projects in which students had to apply numerical methods for simulating 2-dimensional conductive heat transfer on solid surfaces.

542297 All-Solid-State Batteries for Next Generation Electrochemical Energy Storage

Abstract:

Research Interests:

The goal of my research is to realize safe, high-performance, and cost-effective all-solid-state batteries for powering electric vehicles and renewable energy storage through rational design of materials and interface. I aim to build a research group that uses an integrated theoretical and experimental approach to define and address the key challenges of all-solid-state batteries towards a robust energy storage technology. Specifically, my research interests include, but are not limited to: (i) the underlying mechanisms for high interfacial resistances between electrodes and electrolyte and lithium dendrite formation; (ii) the design principles of electrode and electrolyte materials as well as their interfaces; (iii) advanced techniques for large-scale fabrication and device integration.

During my Ph.D. at University of Maryland, I initiated all-solid-state battery project in Prof. Chunsheng Wang's lab. As one of the main findings of my Ph.D. work, I demonstrate the electrochemical stability windows of solid electrolytes are overestimated from the conventional measurement, and the decompositions of solid electrolytes occur and cannot be ignored when developing all-solid-state batteries. My efforts have led to several papers, patents and proposals funded by Department of Energy, Department of Defense and National Science Foundation. I have also had the opportunity to work with a number of world-famous experts in battery field from universities, national labs, and industry. I will leverage my experience and established collaborations in my future research.

Successful Proposals:

(1) National Science Foundation (2018-2021, $340K, Portable)

(2) Army Research Office (2014-2018, $450K)

Selected Awards:

(1) Battery Division Student Research Award, Electrochemical Society, 2018

(2) Graduate Student Gold Award, Materials Research Society, 2017

Teaching Interests:

I am interested in teaching *Thermodynamics*, *Kinetics*, *Transport Phenomena*, and *Process Design* at both the undergraduate and graduate levels. Based on my education and research background, I would also like to develop a new course named "*Energy Materials Chemistry*", which will cover the material and chemical fundamentals underlying the disciplines of sustainability and energy.

Teaching Experience:

I was serving as a teaching assistant (TA) on two undergraduate courses: *Chemical & Biomolecular Engineering Thermodynamics* (CHBE 301) and *Process Engineering Economics and Design* (CHBE 446). Part of my duties is to lead 1 hour discussion every week. The most important thing I learned from teaching is the gap between knowing and teaching out. Through the National Science Foundation Research Experiences for Undergraduates (REU) program, I have mentored three undergraduate students to do battery research in our lab. One of my mentees defended his research in the Engineering Honors Program at University of Maryland. In addition, I have also been assisting my advisor to mentor our *Chem-E Car* team for the regional and national competitions since 2014.

I was awarded "*TA of the Year*" from our department for my teaching efforts. I am also a recipient of the University of Maryland *All S.T.A.R. Fellowship* that honors graduate students who are both outstanding scholars and outstanding teaching assistants. My teaching experience also enabled me to join the prestigious *Future Faculty Program*, a three to five semester long program that aims to improve the students' professional skills (research, teaching/mentoring, service, networking, ethics etc.) to obtain and succeed in a university faculty position.

Selected Publications:

1. **F. Han**, A. S. Westover, J. Yue, X. Fan, F. Wang, M. Chi, D. N. Leonard, N. J. Dudney, H. Wang, C. Wang, Origin of Lithium Dendrite Formation within Solid Electrolytes. *Nature Energy,* Under Review.
2. **F. Han**[†], J. Yue[†], C. Chen, N. Zhao, X. Fan, Z. Ma, T. Gao, F. Wang, X. Guo, C. Wang, Interphase Engineering Enabled All-Ceramic Lithium Battery. *Joule*, 2 (2018) 497.
3. **F. Han**[†], J. Yue[†], X. Zhu, C. Wang, Suppressing Li Dendrite Formation in Li_2S-P_2S_5 Solid Electrolyte by LiI Incorporation. *Advanced Energy Materials*, 8 (2018) 1703644.
4. **F. Han**, J. Yue, X. Fan, T. Gao, C. Luo, Z. Ma, L. Suo, C. Wang, High-Performance All-Solid-State Lithium–Sulfur Battery Enabled by a Mixed-Conductive Li_2S Nanocomposite. *Nano Letters*, 16 (2016) 4521.
5. **F. Han**[†], Y. Zhu[†], X. He, Y. Mo, C. Wang, Electrochemical Stability of $Li_{10}GeP_2S_{12}$ and $Li_7La_3Zr_2O_{12}$ Solid Electrolytes, *Advanced Energy Materials*, 6 (2016) 1501590.
6. **F. Han**, T. Gao, Y. Zhu, K. J. Gaskell, C. Wang, A Battery Made from a Single Material, *Advanced Materials*, 27 (2015) 3473.
7. **F. Han**, Y. Bai, R. Liu, B. Yao, Y. Qi, N. Lun, J. Zhang, Template-Free Synthesis of Interconnected Hollow Carbon Nanospheres for High-Performance Anode Material in Lithium-Ion Batteries, *Advanced Energy Materials*, 1 (2011) 798.
8. **F. Han**, B. Yao, Y. Bai, Preparation of Carbon Nano-Onions and Their Application as Anode Materials for Rechargeable Lithium-Ion Batteries, *Journal of Physical Chemistry C*, 115 (2011) 8923.
9. J. Yue[†], **F. Han**[†], X. Fan, X. Zhu, Z. Ma, J. Yang, C. Wang, High-Performance All-Inorganic Solid-State Sodium-Sulfur Battery, *ACS Nano*, 11 (2017) 4885.

542299 Organic Molecular Electrocatalysts for Energy-Water Applications

Abstract:

Xi Yin, *Ph.D.*

Los Alamos National Laboratory, Materials Physics & Applications,
Los Alamos, New Mexico 87545

Phone: (505) 667-3238; Fax: (505) 665-4292; E-mail: xiyin@lanl.gov; xiyinuiuc@gmail.com
Personal website: https://sites.google.com/view/xiyin

Expertise

(i) Materials chemistry, coordination chemistry, and inorganic chemistry; (ii) molecular simulation and computational modeling; (iii) physical chemistry, electrochemistry, and electrocatalysis; (iv) organic molecular electrocatalysis; (v) energy conversion and storage, polymer electrolyte fuel cells.

Education

- Ph.D., Chemical Engineering (advisor: Prof. Hong Yang), University of Illinois at Urbana-Champaign
- M.Sc., Materials Science, University of Rochester
- M.Sc. and B.Sc., Materials Science and Engineering, Tsinghua University

Employment

- Postdoctoral Research Associate (mentor: Dr. Piotr Zelenay), Los Alamos National Laboratory, Los Alamos, NM (2015-present)

Proposal Writing Experience

- "Organic molecular electrocatalysts for hydrogen evolution reaction" to LANL Laboratory Directed Research and Development, Exploratory Research (LDRD-ER), Co-PI, 2018 (funded, **$ 990K** for three years)
- "Metal-free molecular catalysts for hydrogen evolution reaction" to LANL Laboratory Directed Research and Development, Exploratory Research (LDRD-ER), Co-PI, 2017
- "A new class of metalorganic structure containing infinite one-dimensional metal chain" to National Science Foundation, draft, 2013

Awards & Honors

- MRS Graduate Student Award Silver Medalist, Materials Research Society (2015)
- SCS Science Image Challenge Finalist, UIUC School of Chemical Science (2014)

Journal Reviewer

- *Nano Energy, Nano Letters, Carbon, Fuel Cells, ACS Applied Materials & Interfaces, Journal of Materials Chemistry A, Journal of Electroanalytical Chemistry, Nanotechnology, The Journal of Physical Chemistry, Journal of Nanoparticle Research, etc.*

Summary of Previous Works

Throughout my research, I have sought to combine experimental studies with computational modeling and theory to accelerate discoveries that will have extensive impact on society. My dissertation work at the University of Illinois at Urbana-Champaign focused on a fundamental understanding of the nucleation and growth of metal nanocrystals for catalysis. Different from most observation-based studies, my work emphasized the chemistry of metal-ligand complexation during pre-nucleation and designed ligand environments to tune nucleation and growth *via* thermodynamic and kinetic leverages. By employing density functional theory (DFT) calculations and reaction kinetic models, I developed new theories on nonclassical nanocrystal growth. These works led to the world's first atomically thin palladium nanosheets and a quantitative understanding of the roles of surfactant ligands in nanocrystal synthesis for electrocatalysis, gas sensing, and biological applications.

My postdoctoral work at Los Alamos National Laboratory has focused on the development of platinum group metal-free (PGM-free) catalysts for fuel cell cathodes, driven by the need to replace scarce and expensive PGM materials with compounds based on earth-abundant elements, such as C, N, Fe, and other transition metals. Through molecular-level design of catalyst templates, kinetic study of catalyst degradation, and optimization of the electrode structure, my work has covered a broad spectrum of fuel cell research to boost the performance of state-of-the-art PGM-free catalysts.

While expanding my research portfolio from precious metal-based materials to carbon-based electrocatalysts, I noticed that most electrocatalysts are built on metal-centered active sites that have been electronically modified *via* metallic, ionic, or covalent bonds. The activity improvement of conventional catalysts may intrinsically be limited by the possible electronic structures of these metal sites. To overcome this limitation, I conceived of an idea to use organic molecules as metal-free electrocatalysts, which are attractive due to their highly tailorable geometry, tunable electronic structures, and chemical robustness. With these factors in mind, I designed and modeled a metal-free N-containing molecule as a bifunctional catalyst for H_2O_2 production and the hydrogen evolution reaction. *This discovery is the first demonstration of high catalytic activity from an inexpensive, small organic molecule in a practical polymer electrolyte membrane (PEM) device for crucial electrochemical reactions.* Building upon these findings, I plan to establish a world-leading program that aims to develop an entirely new class of electrocatalysts based on rationally designed organic molecules for clean energy and environmental applications, which goes beyond conventional electrocatalysts that rely on precious metal or transition metal elements.

Research Interests:

My future research aims to unravel the structure-activity relationships in novel *Organic Molecular Electrocatalysts* (**OMECs**) and apply that fundamental knowledge to the molecular-level design of next-generation electrocatalysts for renewable energy and environmental applications. Different from conventional metal-containing electrocatalysts, OMECs are solely based on the earth abundant non-metal elements, *e.g.*, C, N, H, etc. The covalent bonding in organic structures ensures the robustness of OMECs in harsh electrochemical environment. The highly-tailorable structures of OMECs will allow the tuning of electrochemical functions in a large organic chemical space. Although millions of organic compounds have been designed and synthesized, no systematic works have been done to fully explore the electrocatalytic activities of the metal-free organic molecules.

My proposed research will encompass three main thrust areas: (1) distributed electrochemical production of hydrogen peroxide (H_2O_2) for water treatment; (2) electrochemical hydrogen production and energy storage; and (3) carbon dioxide (CO_2) reduction into high-value liquid fuel. The first thrust aims to address the increasing need for distributed on-site production of H_2O_2 as a water disinfection agent for use in remote areas and disaster mitigation during the disruption of regular infrastructure services. The second and third trusts will provide solutions for the integration of hydrogen facilities with smart grids, as well as the production of clean fuels.

Teaching Interests:

My research background is in chemical engineering and materials science, with expertise in electrochemistry, materials chemistry and computational modeling. My research experience allows me to confidently teach the following courses:

- Thermodynamics of Chemical Processes
- Electrochemical Engineering and Fuel Cell
- Electrochemistry
- Computational Design of Energy Materials
- Fundamentals of Materials Science and Engineering
- Reaction Kinetics
- Synthetic Nanomaterials
- Materials Chemistry

Publications

1. **X. Yin**, L. Lin, H. T. Chung, U. Martinez, A. Baker, S. Maurya, P. Zelenay, "Organic molecular electrocatalyst for hydrogen evolution reaction in acidic polymer electrolyte", *Science*, 2018, **under review**.
2. **X. Yin**, P. Zelenay, "Kinetic models for the degradation mechanism of PGM-free ORR catalysts", *ECS Transactions*, 2018, 85, **accepted**.
3. **(Front Cover) X. Yin**, M. Shi, K. S. Kwok, D. L. Gray, J. A. Bertke, H. Yang, "Dish-like higher-ordered nanostructures of palladium through metal ion-ligand complexation", *Nano Research*, 2018, 11, 3442–3452
4. **X. Yin**, M. Shi, J. Wu, D. L. Gray, J. A. Bertke, H. Yang, "Quantitative analysis of different formation modes of platinum nanocrystals controlled by ligand chemistry", *Nano Letters*, 2017, 17 (10), 6146–6150
5. **X. Yin**, L. Lin, H. T. Chung, S. K. Babu, U. Martinez, G. M. Purdy, P. Zelenay, "Effects of MEA fabrication and ionomer composition on fuel cell performance of PGM-free ORR catalyst", *ECS Transactions*, 2017, 77 (11), 1273–1281
6. J. Kim, P.-C. Shih, K.-C. Tsao, Y.-T. Pan, **X. Yin**, C. J. Sun, H. Yang, "High-performance pyrochlore-type yttrium ruthenate electrocatalyst for oxygen evolution reaction in acidic media", *Journal of the American Chemical Society*, 2017, 139 (34), 12076–12083
7. Y.-T. Pan, J. Wu, **X. Yin**, H. Yang, "*In situ* ETEM study of composition redistribution in Pt-Ni octahedral catalysts for electrochemical reduction of oxygen", *AIChE Journal*, 2016, 62 (2), 399–407
8. **(Front Cover) X. Yin**, J. Wu, P. P. Li, J., H. Yang, "Self-heating approach to the fast production of uniform metal nanotructures", *ChemNanoMat*, 2016, 2 (1), 37–41
9. **X. Yin**, X. H. Liu, Y.-T. Pan, K. A. Walsh, H. Yang, "Hanoi tower-like multilayered ultrathin palladium nanosheets", *Nano Letters*, 2014, 14 (12), 7188–7194
10. **X. Yin**, S. A. Warren, Y.-T. Pan, K.-C. Tsao, D. L. Gray, J. Bertke, H. Yang, "A new motif for infinite metal atom wires", *Angewandte Chemie International Edition*, 2014, 53 (51), 14087–14091
11. Y.-T. Pan*, **X. Yin***, K. S. Kwok, H. Yang, "Higher-ordered nanostructures of flexible palladium two-dimensional nanosheets for fast hydrogen sensing", *Nano Letters*, 2014, 14 (10), 5953–5959 (***Equal contributors.**)
12. J. Kim, **X. Yin**, K.-C. Tsao, S. Fang, H. Yang, "$Ca_2Mn_2O_5$ as oxygen-deficient perovskite electrocatalyst for oxygen evolution reaction", *Journal of the American Chemical Society*, 2014, 136 (42), 14646–14649
13. J. Wu*, M. Shi*, **X. Yin**, H. Yang, "Enhanced oxidation reduction stability of (111) surface dominant $Pt_3Ni@Pt_3Pd$ core-shell nanoparticle catalysts", *ChemSusChem*, 2013, 6 (10), 1888–1892 (*Equal

contributors.)

14. J. Wu, P. Li, Y.-T. Pan, S. Warren, **X. Yin**, H. Yang, "Surface lattice-engineered bimetallic nanoparticles and their catalytic properties", *Chemical Society Reviews*, 2012, 41, 8066–8074
15. **X. Yin**, K. X. Chen, H. P. Zhou, X. S. Ning, "Combustion synthesis of Ti_3SiC_2/TiC composites from elemental powders under high gravity conditions", *Journal of the American Ceramic Society*, 2010, 93 (8), 2182–2187.
16. **X. Yin**, K. X. Chen, H. P. Zhou, X. S. Ning, "Combustion synthesis of Ti_3SiC_2 from Ti/Si/C powders under high gravity conditions", *Rare Metal Materials and Engineering*, 2008, 37 (9), 1606–1609.

Conference Presentations

1. **(Invited) X. Yin**, U. Martinez, S. Komini Babu, H. T. Chung, G. M. Purdy, P. Zelenay, "Kinetic insight into the degradation mechanism of PGM-free ORR catalysts", *233rd ECS Meeting*, Seattle, WA, 2018
2. **X. Yin**, L. Lin, H. T. Chung, U. Martinez, A. Baker, S. Maurya, P. Zelenay, "Organic molecular catalyst for hydrogen evolution reaction", *232nd ECS Meeting*, National Harbor, MD, 2017
3. **X. Yin**, L. Lin, U Martinez, H. T. Chung, P. Zelenay, "Organic molecular catalyst for electrochemical production of hydrogen peroxide", *232nd ECS Meeting*, National Harbor, MD, 2017
4. **X. Yin**, L. Lin, H. T. Chung, S. Komini Babu, U Martinez, G. M. Purdy, P. Zelenay, "Effects of porosity and ionomer composition on fuel cell performance of PGM-free ORR catalysts", *231st ECS Meeting*, New Orleans, LA, 2017
5. **X. Yin**, H. T. Chung, U. Martinez, L. Lin, P. Zelenay, "Magnetic purification of PGM-free catalysts", *PRiME 2016/230th ECS Meeting*, Honolulu, HI, 2016
6. **X. Yin**, H. T. Chung, U. Martinez, J. H. Dumont, G. M. Purdy, P. Zelenay, "Non-PGM ORR catalysts prepared from polyaniline-type polymers with strong affinity to iron", *229th ECS Meeting*, San Diego, CA, 2016
7. **X. Yin**, J. Wu, Y.-T. Pan, H. Yang, "Quantitative analysis of kinetically controlled growth of metal nanoparticles based on ligand-metal ion interactions", *2015 MRS Spring Meeting*, San Francisco, CA, 2015
8. **X. Yin**, X. Liu, Y.-T. Pan, K. Walsh, H. Yang, "Hanoi tower-like multilayered 2D Pd nanosheets and the molecular level understanding of the formation mechanisms", *2015 MRS Spring Meeting*, San Francisco, CA, 2015
9. **X. Yin**, J. Wu, P. Li, H. Yang, "Formation of nanoparticles in self-regulated reaction systems", *246th ACS National Meeting*, Indianapolis, IN, 2013
10. **X. Yin**, J. Wu, W. Zhou, H. Yang, "Development of facet-controlled Pt and Pt alloy nanocatalysts for oxygen reduction reaction", *246th ACS National Meeting*, Indianapolis, IN, 2013

542305 Experimental Investigation on Different Baffles of Shell-and-Tube Heat Exchanger

Abstract:

Research Interests: different baffles of shell-and-tube heat exchanger

Teaching Interests: flow and heat transfer

Experimental investigation on different baffles of shell-and-tube heat exchanger

T Cheng, J Chen, M Zeng

Key Laboratory of Thermo-Fluid Science & Engineering, MOE, School of Energy and Power Engineering, Xi¡¯an Jiaotong University, Xi¡¯an 710049, Shaanxi, China

Shell-and-tube heat exchanger is a common heat exchange equipment in industry. Among them, the segmental baffle heat exchanger is the basic shell-and-tube heat exchanger, which has simple structure, convenient processing and manufacturing features. With the development of the times and the progress of science and technology, people have higher requirements for energy efficiency, and have developed various types of heat exchangers to compensate for the insufficient on heat transfer performance and pressure drop characteristics of the segmental baffle heat exchanger. However, the study of new heat exchangers is mostly limited to numerical simulation methods, and there is no systematic study and comparison of the performance of different heat exchangers.

In this paper, the experimental system is verified by the traditional segmental baffle heat exchanger. An experimental study is carried out on four heat exchangers with different baffles, in which the inner diameter of the shell side and tubes are the same and the adjacent baffle distance is equal.

The results indicate that the heat transfer coefficient of the rod baffle, the orifice plate, the small hole heat exchanger and the segmental baffle heat exchanger are all increased with increasing flow rate at the shell side while the flow rate at the tube side remains constant; and the pressure drop of the four kinds of heat exchangers increased rapidly with the increase of the shell side flow rate. In general, the pressure drop of the other three heat exchangers is lower than that of the segmental baffle heat exchanger. In terms of the shell side heat transfer coefficient, the orifice plate is the best, the segmental baffle is the next, the small hole is the third, and the rod baffle is the worst. However, the effect of the rod baffle heat exchanger is the strongest, when taking the shell side heat transfer coefficient per unit pressure drop into consideration.

Key Words: Shell-and-tube heat exchanger; Experimental investigation; Different baffles

542413 Electrolyte Design and Fundamental Studies of Battery Systems for Better Energy Storage Media

Abstract:

Research Interests:

My research group will be at the interface of chemical engineering and chemistry. Using experimental and computational tools, we will design and synthesize new electrolytes, study ionic transport processes, and probe the kinetics of electrochemical reactions in multiple battery systems.

Motivation

Increased use of portable electronics, and the need to move from fossil fuels to intermittent renewable energy technologies have generated intense interest in energy storage media such as batteries. Lithium-ion (Li-ion) batteries are the most energy-dense batteries commercially available. They have dominated the portable electronic market, helped spur the growth (and expected boom) in electric vehicles, and are envisioned as an integral part of the future grid that is dependent on solar and wind. However, current Li-ion batteries have reached their capacity limitations, are expensive, and are dangerous.

Batteries can be broken down into roughly three components: cathode, electrolyte, and anode. Current Li-ion battery cathodes use transition metal oxides with limited capacities (e.g., $LiCoO_2$ with 145 mAh/g) and whose cost scales with the cobalt content. Secondly, electrolytes currently use carbonate-based solvents that are volatile and flammable; responsible for high fire risks of these batteries. In addition to cost and safety concerns, the choice of anode (graphite) and the heavy intercalation cathodes limit the energy densities that can be obtained. Therefore, researchers have focused on shifting from a graphite anode (372 mAh/g) to lithium metal (3860 mAh/g); designing electrolytes that are safer and with a wider electrochemical stability window; and cathodes that do not depend on intercalation chemistry but conversion reactions (oxygen and sulfur reduction).

Research interests

1. Design new small molecule and polymer electrolytes

All rechargeable battery systems (lithium-ion, lithium-metal, lithium sulfur etc.) need better electrolytes. Higher energy densities can be obtained in these battery chemistries if the electrolytes were stable across a wider electrochemical range, had higher ionic conductivities, were more selective (preferential solubility of lithium salt compared to the electrochemically active specie such as sulfur), and safer. My research interest involves the design, synthesis, and characterization of new electrolytes (small molecules and polymers). Furthermore, density functional theory (DFT) and molecular dynamics (MD) tools will be used to obtain structure-property relationships, better understand transport properties in these systems, and study the functionalities responsible for ionic conduction.

2. Develop new lithium and multivalent-ion battery chemistries

My lab will study lithium-based battery chemistries such as lithium metal and lithium-sulfur, and perform fundamental electrochemical studies on multivalent ion chemistries such as magnesium to better understand magnesium deposition and stripping in non-Grignard-based electrolytes.

3. Design *in situ* tools to better study batteries during operation in real world settings

Current battery management systems depend on simple measurements of temperature and pressure to develop their models to predict the state of health and lifetime of battery cells. We will design sensors that can be inserted into batteries to better monitor the state and health of batteries during real world use.

Doctoral research (PI: Paula Hammond, MIT)

My PhD work focused on a new class of batteries – lithium-air – with theoretical energy densities at least three times that of Li-ion (*1–6*). Conventional lithium-air batteries use lithium metal as the anode, an ether based electrolyte, and oxygen as the active specie in the cathode. Despite the high energy density promise, Li-air batteries have poor battery capacity retention, and pose a greater safety hazard compared to Li-ion because of the presence of oxygen. Hence, my work explored nonflammable polymer electrolytes that can be used for lithium-air batteries. I studied polymer stability in the presence of highly reactive reduced oxygen species (such as Li_2O_2), and developed selection rules for designing stable polymers for lithium-air use (*3*). Using these rules, we fabricated gel polymer electrolytes that controlled the oxygen reduction process, and yielded a previously unreported 1 mol e /mol O_2 process; showing that careful selection of the gel polymer electrolyte can finely tune the battery chemistry (*4*).

Postdoctoral research (PI: Zhenan Bao, Stanford)

During my PhD, I obtained skills in electrochemistry, electrolyte design, spectroscopic characterization, battery fabrication among others. However, my postdoctoral work has delved more into the synthesis and characterization of new electrolyte compounds. Furthermore, my work incorporates MD/DFT-based tools to study the molecular properties, lithium-ion transport, and kinetics of sulfur reduction in these new electrolytes.

Teaching Interests:

Teaching interests

I love to teach and mentor, and have actively sought these opportunities. I am interested in teaching chemical engineering reaction kinetics, thermodynamics, fundamentals of polymer science, and elective graduate courses (e.g., electrochemistry). During my graduate career, I was an instructor for an MIT undergraduate course "Polymer Science Laboratory" where I taught polymer synthesis, characterization, theory and applications. I have also tutored math and chemistry classes at a local high school, and have mentored multiple undergraduates pursuing research under my guidance. On behalf of MIT, I attended multiple conferences and events to recruit minority engineers to pursue graduate studies.

Selected publications

1. **Chibueze V. Amanchukwu**, Hao-Hsun Chang, and Paula T. Hammond. "Influence of Ammonium Salts on Discharge and Charge of $Li–O_2$" *J. Phys. Chem. C.* 2017, 121, 17671-17681.
2. Michal Tulodziecki, Graham M. Leverick, **Chibueze V. Amanchukwu**, Yu Katayama, David G. Kwabi, Fanny Barde, Paula T. Hammond, and Yang Shao-Horn. "The Role of Iodide on the Formation of Lithium Hydroxide in Lithium-Oxygen Batteries." *Energy Environ. Sci.* 2017, 10, 1828-1842.
3. **Chibueze V. Amanchukwu**, Jonathon R. Harding, Yang Shao-Horn, and Paula T. Hammond. "Understanding the chemical stability of polymers for lithium-air batteries." *Mater.* 2015, 27, 550-561
4. **Chibueze V. Amanchukwu**, Hao-Hsun Chang, Magali Gauthier, Shuting Feng, Thomas P. Batcho, and Paula T. Hammond. "One electron process in a gel polymer electrolyte $Li-O_2$" *Chem. Mater.* 2016,

28, 7167-7177.

5. **Chibueze V. Amanchukwu**, Magali Gauthier, Thomas P. Batcho, Chanez Symister, Yang Shao-Horn, Julio D'Arcy, and Paula T. Hammond. "Evaluation and stability of PEDOT electroactive polymer electrodes for Li-O_2" *J. Phys. Chem. Lett.* 2016, 7, 3770-3775
6. Jonathon R. Harding, **Chibueze V. Amanchukwu**, Paula T. Hammond, and Yang Shao-Horn. "Instability of Polyethylene Oxide upon oxidation in Lithium-Air Batteries." *J. Phys. Chem. C.* 2015, 119, 6947-6955.

541863 Systems Approaches to Design Sustainable Food-Water-Energy-Waste Nexus Processes and Systems

Abstract:

Research Interests: sustainability, food-water-energy-waste nexus, process systems engineering, mathematical modeling, optimization, ecology, life cycle assessment, smart and sustainable manufacturing

Sustainable Design of Food-Water-Energy-Waste Nexus Systems

Water, energy, and food are all fundamentally connected in a concept termed the Food-Water-Energy-Waste Nexus (FWEWN), and it is difficult to think of any human activity that does not depend on these resources. However, these systems are currently unsustainable. Over 650 million people do not have safe access to water, and expected increases in severe weather wrought by climate change will increase stresses on freshwater sources. Nearly 800 million people are undernourished, and mounting evidence suggests there is not enough land to support a projected 10 billion people by 2050 with a typical Western/American diet, especially if we also choose to produce biofuels from edible crops like corn or soy. The production, distribution, and consumption systems of water, energy, or food cannot be studied independently of the corresponding systems of the other resources. Thus, chemical engineers, and process systems engineers especially, are well-positioned to handle the challenge of designing sustainable processes and systems in the FWEWN.

Interconnected systems of food, water, and energy are often multi-scale both in space and time and involve multiple stakeholders with multiple objectives. Each of these properties defines significant research challenges for design of sustainable FWEWN systems. I have developed frameworks that systematically address these challenges for large biofuel and biochemical product and process networks, global biofuel production, and sustainable production networks. I presented the first ever framework integrating global, computable general equilibrium (CGE) models from economics to model global environmental impacts of production, like land use change. With our collaborators in mechanical engineering, I designed a framework to leverage emerging smart manufacturing technologies to not only increase economic efficiency, but also to limit energy consumption and emissions of a multi-stakeholder manufacturing network. In the future, I plan to integrate the developed multi-objective, multi-stakeholder approach into design of sustainable FWEWN systems and processes.

Ecosystem Services Valuation in Sustainable Design

Many systems including those in the FWEWN are intimately connected with ecological systems and processes (especially food production). However, design of chemical processes and supply chains based on interactions with natural ecosystems and their services is poorly understood and rarely considered. To address this challenge, I presented the first work that integrates different types of ecosystem service valuation methodologies into sustainable supply chain design, maximizing Green GDP (nominal GDP plus natural capital) as an objective. Three different valuation methods from ecological economics - the aggregated benefit transfer method, the hedonic method, and the avoided cost method

- were integrated into a spatially-explicit supply chain decision-making model. Ecosystem services from forests, grasslands, rivers, wetlands, etc. were all considered. More tangible results surface when the values of ecosystem services lost or gained are calculated and presented for various supply chain decisions. Economically-driven decisions are familiar to both public and private actors, so such approaches can leverage existing ecological valuation methods to guide policy and supply chain design decisions that resonate directly with all stakeholders. I will use this foundational work to galvanize my future research to consider ecosystem services in multi-scale, multi-stakeholder decision-making models. The method also integrates GIS data analysis with mathematical programming, a research approach which I will also focus on moving forward.

Quantitatively Considering Social Sustainability

Sustainability is made of three pillars: economic, environmental, and social. We process systems engineers have devised sophisticated and effective means to design systems based on economic and environmental objectives. However, rigorous, creative approaches to model social sustainability in process and supply chain design are less developed. I pose that social sustainability can be modeled systematically with emerging methodology in social network analysis (SNA). New frameworks using SNA could help determine where facilities, plants, or processes should be located to maximize social good (perhaps through focusing on job creation in highly socially-connected areas), identify key stakeholders along the value chain, and evaluate how current facilities and supply chains are operating in terms of social sustainability. This new area of research will leverage my background on network optimization and large-scale data analysis.

Teaching Interests:

I am qualified to teach all core chemical engineering courses, but I am especially interested in teaching the introductory chemical engineering course, thermodynamics, mathematical optimization (especially geared towards chemical process or supply chain design), and courses on environmental/sustainability topics. At the graduate level, I will develop courses based on sustainable design of chemical processes and supply chains, as well as a course that defines, quantifies, and integrates problem solving in the Food-Water-Energy-Waste Nexus from a chemical engineering perspective.

Presentations at the 2018 AIChE Annual Meeting

Sustainable Engineering Forum (SEF) Plenary, David L. Lawrence Convention Center - 315, Wednesday, October 31, 9:30 AM: *Addressing global environmental impacts including land use change in life cycle optimization: Studies on biofuels*

Session: Design, Analysis, and Optimization of Sustainable Energy Systems and Supply Chains I, Tuesday, October 20, 12:52 PM: *Methods for Quantitative Consideration of Ecosystem Services in Supply Chain Design and Optimization*

Session: Sustainable Energy Generation and Utilization in System Design, Wednesday, October 31, 12:30 PM: *Addressing Uncertainty in Large-Scale Bioconversion Product and Process Networks with Two-Stage Adaptive Robust Optimization*

Selected Peer-Reviewed Journal Publications

Garcia, D., Lovett, B., You, F., Applying Ecosystem Service Valuation Methods in the Food-Water-Energy-Waste Nexus, *In Preparation*.

Garcia, D., Mozaffar, M., Ren, H., Correa, J., Ehmann, K., Cao, J., You, F. Sustainable Manufacturing with Cyberphysical Manufacturing Networks: Overview and Modeling Framework, *Submitted.*

Garcia, D. & You, F. (2018) Addressing global environmental impacts including land use change in life cycle optimization: Studies on biofuels. *Journal of Cleaner Production*, 182, 313-330.

Garcia, D. & You, F. (2017). Systems Engineering opportunities for agricultural and organic waste management in the food-water-energy nexus, *Current Opinion in Chemical Engineering*, 18, 23-32.

Gong, J., **Garcia, D.,** You, F. (2016). Unraveling Optimal Biomass Processing Routes from Bioconversion Product and Process Networks under Uncertainty: An Adaptive Robust Optimization Approach, *ACS Sustainable Chemistry & Engineering,* 4, 3160-3173. **Cover article, June 2016 issue**

Garcia, D. & You, F. (2016) The Water-Energy-Food Nexus and Process Systems Engineering: A New Focus, *Computers & Chemical Engineering,* 4, 3160-3173.

Garcia, D. & You, F. (2015) Network-based Life Cycle Optimization of the Net Atmospheric Carbon Ratio (NACR) of Fuels and Chemicals Production from Biomass, *ACS Sustainable Chemistry & Engineering,* 3, 1732-1744.

Garcia, D. & You, F. (2015) Supply Chain Design and Optimization: Challenges and Opportunities, *Computers & Chemical Engineering,* 81, 153-170.

Garcia, D. & You, F. (2014) Multiobjective optimization of product and process networks: General modeling framework, efficient global optimization algorithm, and case studies on bioconversion, *AIChE Journal*, 61, 530-554.

541928 Sustainable Fuel and Chemical Synthesis Via Catalytic Valorization of Abundant and Renewable Resources

Abstract:

Research Interests:

Sustainability is a fundamental challenge of our society and will remain one, since we rarely limit our consumption until the impacts of diminishing resources become both clear and present. A sustainable, independent economy operates more effectively when taking advantage of abundant feedstocks including both grown and waste resources such as sugars, lignocellulose, plant/algal oils, industrial off-gases, and commercial wastes. My interest is in understanding the fundamental catalysis behind the conversion of such feedstocks to fuels and chemicals to rationally develop new materials and processes, practically influenced by current and projected demands and constraints imposed by society. My research has thus far targeted production of heavier fuels such as jet and diesel, demand for which is projected to increase while that for lighter fuels such as gasoline is projected to decrease. Commercial technologies for lignocellulose-derived jet and diesel fuels thus far have not been developed, creating a large need for research into this area.

I synthesized a low-cost base metal catalyst for the conversion of sugars to mono-oxygenates which can be used to produce renewable diesel and jet fuels. Previously only precious metal catalysts were used for this reaction. The cobalt catalyst used in this study is over 50 times cheaper than previously used precious metal catalysts and was used to achieve higher mono-oxygenate yields than previously reported. I also found evidence of a possible oligomer-mediated pathway for production of these species.

I have additionally developed a novel technology for the conversion of ethanol to distillate-range fuel molecules. I developed an extensive kinetic model to elucidate the reaction chemistry that occurs in a key stage of this process using experimental data I collected for ethanol condensation over hydroxyapatite. Alcohol structure (i.e. branching and chain length) and undesired olefin production are related to both fundamental catalytic parameters and natural consequences of condensation (e.g. increasing water pressure, decreasing alcohol pressure). This model shows the role dehydration of C_{4+} alcohols plays in limiting alcohol yields at increasing conversions. We have used this model to aid in the development of more selective catalysts and to better understand how this step affects the overall ethanol-to-distillate technology.

Teaching Interests:

As a professor, the greatest impact I can have is in educating and inspiring future generations of scientists and engineers. My passion is fueled by the reward of seeing understanding develop in a studentâ€™s eyes, knowing that I was the difference. I am a classically trained Chemical Engineer and am excited to share my passion for chemical engineering with future generations. I am most interested in teaching the following classes: *Reaction Kinetics and Reactor Design*, *Thermodynamics I*, *Thermodynamics II*, and *Chemical Engineering Fundamentals*. In the classroom I have served as a teaching assistant twice for *CBE 250 Process Synthesis* and once for *CBE 311 Thermodynamics of Mixtures*, receiving two department awards for my teaching. I used my discussion sections to reflect upon practices proven to be effective in literature to find myself as a teacher. Here I have been guided by three main philosophies: (1) teaching as a relatable, collaborative facilitator, (2) providing a learning environment that addresses different learning styles and backgrounds while emphasizing collaborative efforts and real-world applications, and (3) rationally designing curriculum based on backward design with clearly-established learning outcomes, feedback-directed adaptation, and holistic examination design. As an instructor I will apply these philosophies in the classroom and continuously reflect upon and refine my teaching to inspire my students to become the next generation of innovators, armed with mindsets and life-long skills that will serve them far beyond the classroom. Additionally I have been a research mentor to four undergraduate students and a high school student. Each of these students learned differently and possessed a unique perspective on research, which has given me an ability to adapt my mentoring style to best benefit my students.

541989 Kinetics and Reliability of Thermo-Electro-Chemical Processes for Energy Conversion and Chemical Production

Abstract:

Xiao-Yu Wu, PhD'17, Massachusetts Institute of Technology

Postdoctoral Associate, Reacting Gas Dynamics Laboratory/Center for Energy and Propulsion Research, MIT

http://xywu.scripts.mit.edu/xywu/

Research Interests:

My research interests focus on the kinetics and reliability in the thermo-electro-chemical processes under extremely-harsh operating conditions, e.g., high temperature/pressure (depending on the materials), supercritical conditions, and corrosive and radioactive environment for combustion, advanced energy conversion and chemical production. My lab will aim at investigating the reacting flows under these harsh conditions in energy conversion systems, drawing on the *operando* experimental methods. We will also explore numerical methods to study the relationship between the performance and economics on a system-scaled level. We will then apply the understanding about kinetics to avoid performance degradation and improve reliability. My personal aim is to closely collaborate with scientists and industrial partners to develop the scientific findings in my lab into engineering applications in industry.

My experience in high-temperature heterogeneous reactions, electrochemical reactions and enhanced heat transfer has motivated the following proposed research topics. Two topics on membrane-supported H_2O/CO_2 to value-added chemicals and supercritical hydrocarbon pyrolysis regenerative cooling are developed from my previous research topics. And two other topics, (i.e., ultra-lean methane utilization and sensing in harsh environments) are under development to be my future directions.

Teaching Interests:

I am planning to teach courses in the fields of thermodynamics, transport theory, separation, chemical kinetics.

Selected Publications:

1. X.Y. Wu, A.F. Ghoniem, "CO_2 reduction and methane partial oxidation on surface catalyzed $La_{0.9}Ca_{0.1}FeO_{3-\delta}$ oxygen transport membranes", *the Proceedings of the Combustion Institute*, vol. 37, 2018, *in press*
2. X.Y. Wu, A.F. Ghoniem, "Hydrogen-assisted Carbon Dioxide Thermochemical Reduction on $La_{0.9}Ca_{0.1}FeO_{3-\delta}$ Membranes: A Kinetics Study", *ChemSusChem*, 11 2018 483 - 493 (featured at MIT News, and picked up by other media)
3. Y. Luo, X.Y. Wu, Y. Shi, A.F. Ghoniem, N. Cai, "Exergy Analysis of an Integrated Solid Oxide Electrolysis Cell-Methanation Reactor for Renewable Energy Storage", *Applied Energy*, 215 2018 371 - 383
4. X.Y. Wu, M. Uddi, A.F. Ghoniem, "Enhancing co-production of H_2 and syngas via water splitting and POM on surface-modified oxygen permeable membranes", *AIChE Journal*, 62 (12) 2016, 4427 - 4435, (invited contribution to AIChE Journal as "Best paper" initiative)
5. X.Y. Wu, L. Chang, M. Uddi, P. Kirchen, A.F. Ghoniem, "Toward enhanced hydrogen generation from water using oxygen permeating LCF membranes", *Physical Chemistry Chemical Physics*, 17 2015 10093-10107
6. Huang, X. Y. Wu, Z. Wu, W. Li, H. T. Zhu, B. Sunden, "Experimental Study on Heat Transfer of Nanofluids in a Vertical tube at Supercritical Pressures", *International Communications in Heat and Mass Transfer*, 63 2015 54-61

542035 Fueling Our Future with Membrane Technology:Clean Energy Conversion and Process Intensification

Abstract:

Fueling our Future with Membrane Technology:

Clean Energy Conversion and Process Intensification

Simona Liguori – Department of Chemical and Biological Engineering, Colorado School of Mines, Golden, CO

Research Assistant Professor – *Colorado School of Mines* 2016 – present

Physical Science Research Associate – *Stanford University* 2014 – 2016

Post Doc – *Institute on Membrane Technology – ITM-CNR* 2012 – 2014

<>Motivation and **Research Interests:**

In the view of a significant reduction of greenhouse gases emissions amidst fossil fuel exploitation, the use of sustainable energy sources is a mandatory option as well as the use of alternative fuels for energy conversion. In this context a strong interest has been addressed towards the possible development of a global 'hydrogen economy', by considering hydrogen as the energy carrier of the future and as reliable fuel source for transportation and stationary system. Among all of the different technologies to produce green and sustainable hydrogen, inorganic membrane reactors are a promising alternative to enable distributed and centralized production and separation of hydrogen, respectively.

Hydrogen can be generated using feedstock from renewable sources, such as biogas, biomethanol, bioethanol, and bioglycerol or from fossil fuel such as natural gas. In both cases, two streams are produced by performing the reforming reaction in the membrane reactor. One stream contains high-purity hydrogen, which can be directly fed to a PEM fuel cell or internal combustion engine, while the other stream is mainly composed of CO_2, which can be easily separated and stored. An evident outcome of using a membrane reactor is emissions-free energy generation.

Inorganic membrane reactors are a clear example of process intensification and their utilization provides several other benefits when applied to chemical processes. Specifically, they are able to reduce the equipment footprint, energy consumption, and environmental impact of manufacturing processes. For these reasons, they may also be applied to NH_3 production. They allow for performing the process at milder operating conditions than the conventional Haber-Bosch process and to produce ammonia on-site eliminating the need for the U.S. import of foreign ammonia. Specifically, for the first time it has been demonstrated that N_2 permeates through metallic membranes by solution-diffusion and, by using the hydrogen as a sweep gas, it is feasible to produce NH_3 at a lower operating pressure.

This new method to produce NH_3 is highly significant as the current process is the second most energy-intensive chemical manufacturing process in the US and worldwide due to the harsh operating conditions.

In general, inorganic membranes have the potential to play a critical role in the area of clean energy conversion and the study of different materials through alloying is the key to reaching high membrane reactor performance.

In summary, my research is focused on:

- Hydrogen production by renewable sources or natural gas at lower operating conditions,
- NH_3 production via an alternate route from the conventional Haber-Bosch process,
- Membrane manufacturing and materials-alloying investigations to produce high-performing membranes,
- the direct capture of CO_2 from a residue stream before it is released into the atmosphere.

<>Teaching Experience and **Teaching Interests:**

Sharing knowledge and communicating effectively is very important to me. I have served as a Teaching Assistant (TA) for Inorganic Chemistry courses for two years in Italy. Teaching is a very complex activity. It involves planning for learning, organizing materials, prioritizing ideas, interacting with students, learning to "monitor and adjust", "differentiate instruction" for students of various abilities, and learning how to accomplish goals that sometimes seem to be mutually exclusive, all while keeping "control" of a class and meeting the expectations of parents, administrators, and peers. Anyhow, I am always enthusiastic and passionate when I have to share my knowledge with student and when they are engaged during my lesson. I have had the opportunity to serve as mentor for three undergraduate students in Italy and two PhD student at Stanford and currently mentor three PhD student at the Colorado School of Mines.

Courses that I would be interested in teaching are fluid mechanics, transport phenomena, thermodynamics, energy conversion processes, and chemical kinetics.

Research Experience

I have been working on several fields on chemical engineering: reactor analysis and chemical kinetics, heat/mass transfer, and reaction/diffusion through catalysts and membranes.

During my PhD, my research focused on experimental and simulation analysis of reforming reactions of bio-fuels in both membrane and fixed-bed reactors for hydrogen production. Since completing my PhD, I have continued my research in membrane science and technology in my position as a Post-Doctoral Research Fellow at the Institute on Membrane Technology from 2012 to 2014 and as a Physical Science Research Associate at Stanford University from 2014 to 2016. I have been conducting research as a Research Assistant Professor with the Colorado School of Mines since June 2016 continuing my focus on membrane science and clean energy conversion processes.

Future Direction

As a faculty member I would like to continue applying membrane science and technology to different areas of engineering, as well as industrial applications. In particular, I would like to merge the knowledge that I obtained during my PhD and post-doctoral position: chemical engineering, and environmental and applied science. I believe that the synergy between the two fields can make feasible the reduction the greenhouse gases, above all CO_2, by producing H_2 and NH_3 by means of an alternative device including a membrane reactor.

Although I only started working with the Colorado School of Mines in 2016, I have hit the ground running in terms of improving my knowledge base associated membrane separations. I have had the opportunity while at Mines to continue my collaborations with Prof. Jennifer Wilcox in addition to beginning collaborations with membrane expert, Professor Doug Way. I am currently working on the evaluation of the performance of a membrane reactor to produce NH_3 at ambient pressure and by using a layer of catalyst deposited on the membrane surface (H_2 permeates through the membrane and N_2 is used a sweep-gas); NH_3 production in a membrane reactor at low operating conditions compared to the Haber-Bosch process (*opposite case*: N_2 permeates through the membrane and H_2 is used a sweep-gas); H_2 production and separation in Pd-alloys membrane reactor from reforming reactions; writing proposal for NSF, DOE and ARPA-E related to the H_2 production and separation in membrane reactor technology from renewable and natural gas reforming, respectively; writing scientific articles and chapters.

Selected Publications:

1. **S. Liguori**, J. Wilcox, Membrane considerations and plant design for post-combustion CO_2 capture, Ch. 14, In Current Trends and Future Developments on (Bio-) Membranes, A. Basile, E. Favvas (Eds.), (2018) Elsevier, ISBN 9780128136454

2. B. Anzelmo, J. Wilcox**, S. Liguori (Corresponding Author)**, "Hydrogen production via natural gas steam reforming in a Pd-Au membrane reactor. Reaction temperature and GHSV effects and long-term reaction", 2018, *submitted to Journal of Membrane Science*

3. B. Anzelmo, J. Wilcox**, S. Liguori (Corresponding Author)**, "Hydrogen production via natural gas steam reforming in a Pd-Au membrane reactor. Comparison between methane and natural gas steam reforming reactions", 2018, *submitted to Journal of Membrane Science*

4. B. Anzelmo, J. Wilcox**, S. Liguori (Corresponding Author)**, "Natural gas steam reforming reaction at low temperature and pressure conditions for hydrogen production via Pd/PSS membrane reactor", *Journal of Membrane Science*, 522 (**2017**) 343-350.

5. **S. Liguori**, J. Wilcox, Silica membranes application for Carbon Dioxide separation, Ch. 11, In Current Trends and Future Developments on (Bio-) Membranes, A. Basile, K. Ghasemzadeh (Eds.), (2017) Elsevier, pp. 265-294, ISBN 9780444638663

6. **S. Liguori**, J. Wilcox, Membrane considerations and plant design for post-combustion CO_2 capture, Ch. 14, In Current Trends and Future Developments on (Bio-) Membranes, A. Basile, E. Favvas (Eds.), (2018) Elsevier, ISBN 9780128136454

7. **S. Liguori**, K. Lee, J. Wilcox, " Innovative N_2-selective metallic membranes for a higher-efficiency of CO_2 capture" 2018, *submitted to Journal of Membrane Science*

8. J. Wilcox, P.C. Psarras, **S. Liguori**, "Assessment of reasonable opportunities for direct air capture", Environmental Research Letters 12 (**2017**) 065001. (**It was invited**)

9. A. Iulianelli, **S. Liguori**, J. Wilcox, A. Basile, "Advances on methane steam reforming to produce hydrogen through membrane reactors technology: a review", *Catalysis Reviews Science & Engineering*, 58 (**2016**) 1-35.

10. A. Iulianelli, **S. Liguori**, Y. Huang, A Basile, "Model biogas steam reforming in a thin Pd-supported membrane reactor to generate clean hydrogen for fuel cells", *Journal of Power Sources*, 273 (**2015**) 25-32.

11. **S. Liguori**, A. Iulianelli, F. Dalena, P. Pinacci, F. Drago, M. Broglia, Y. Huang, A Basile, "Performance and long-term stability of Pd/PSS and Pd/Al_2O_3 membranes for hydrogen separation", *Membranes*, 4 (**2014**) 143-162.

12. **S. Liguori**, M. Yuan, K. Lee, B. Anzelmo, N. Buggy, T. Fuerst, D. Way, S. Paglieri, J. Wilcox, "Opportunities and Challenges for Hydrogen Production/Separation via Metallic Membranes", 2018, in preparation, *Energy & Environmental Science*. (**It is invited**).

534478 Applications of Functional Fiber-Based Materials in Energy and Engineering Fields

Abstract:

Research Interests:

My research interests in various areas of materials science and engineering including but not limited to synthesis of polymers, fiber-based nanomaterials, and carbon-based materials *via* ***various spinning technologies*** (e.g., wet-spinning, melt-spinning, electrospinning, etc.) and ***casting approaches*** for their applications in the fields of energy, filtration, and advanced fiber composites manufacturing. The motivation behind my research stems from solutions to real-world problems.

Teaching Interests:

Passion and preparation are the key factors that I focus on in my teaching. I concentrate my efforts on developing, improving, and teaching courses related to materials and chemical engineering. In the classroom, I am driven by a passion to inspire and nurture students, and make the learning process an interactive, enjoyable, and rewarding experience for students.

535078 Nanoscale Solid State Electrolyte Synthesized through Atomic Layer Deposition for Interfacial Engineering and All-Solid-State Batteries

Abstract:

Research Interests:

A. New Materials Development and Processing â€“ *for Novel Manufacturing, Flexible Electronics and Solid State Energy Storage*

A1. Development of Novel atomic layer deposition (ALD) Chemistry of Solid State Electrolyte

A2. Flexible Hybrid Inorganic/Organic Materials via Vapor Phase ALD/MLD Synthesis

A3. Rational Designs toward High Performance, 3D Solid State Battery

A4. Interfacial Modification for Stable Solid-Solid Interfaces

B: Advanced Material Properties Exploration - Promoting and Suppressing Solid-State Phase Transformation under Interfacial Controls â€“ *from fundamental thermodynamics to ion-configurable solid-state device*

B1. Phases Competitions under Kinetic Constraints

B2. Application: Solid-State, Ion-Configurable Devices

C: Next-Generation Energy Storage and Sustainability:

C1. Rational Materials Desig, Engineering and Processing of Protection layers for Advanced Layered Electrode Materials

C2. Metal Anodes Protection for Next-Generation Battery

Teaching Interests:

Thermodynamics, Kinetics, Solid-state physics, Analytic Electrochemistry, Materials for Energy and Environmental Sustainability, Micro-processing and Engineering of Advanced Materials, Nanotechnology Characterization

540466 Energy Storage in Clathrate Hydrates – Recent Advancements in Solidified Natural Gas (SNG) Technology

Abstract:

Energy storage in clathrate hydrates – Recent advancements in Solidified Natural Gas (SNG) technology

Hari Prakash Veluswamy

Department of Chemical and Biomolecular Engineering, National University of Singapore, Singapore 117 585

SNG (solidified natural gas) technology is a promising and potential technology for long-term and large-scale storage of methane (natural gas) in hydrates due to the plethora of benefits it offers in comparison to conventional methods of natural gas (NG) storage. The technology has the potential to store multi-fold volumes of natural gas in compact hydrate crystals offering the safest and environment friendly mode of NG storage. Energy recovery from SNG is very easy by simple thermal stimulation (low waste heat is sufficient) with the possibility of complete recovery of the stored gas. Few key existing challenges that impede the commercial deployment of SNG include the need for severe operating conditions (low temperature and high pressure requirement for hydrate formation), slow kinetics of hydrate formation and the stringent storage conditions (storage temperature of -20 deg C at atmospheric pressure based on the anomalous 'self-preservation' effect). Recently, we demonstrated the rapid hydrate formation kinetics in presence of methane and 5.6 mol% tetrahydrofuran (THF) in a simple unstirred reactor configuration at moderate pressures and near ambient temperature conditions. The feasibility of scale-up potential of this approach was also documented in our study. Molecular level characterization studies have revealed the characteristic two-step mixed hydrate formation process and a plausible mechanism for the rapid hydrate growth was formulated. Hydrates formed in presence of THF are very stable, thus storage is possible at temperatures even at 0 deg C at atmospheric pressures without depending on the 'self-preservation' effect.

PhD work:

My PhD thesis investigation was on energy storage in clathrate hydrates. This involved the research in storing energy rich gases like hydrogen and methane in the form of hydrates. My initial research focused on examining the challenging aspect of storing hydrogen in the form of hydrates. I have worked on different domains of hydrate research that include hydrate phase equilibrium studies, investigation of macroscopic kinetics of hydrate formation/dissociation and examining the morphology of hydrate crystals during formation and dissociation of hydrates. Next part of my PhD research involved storing methane via clathrate hydrates. I found out an effective method to form mixed methane hydrates rapidly at moderate conditions (closer to ambient temperature and low pressures) in a simple unstirred reactor configuration. Feasibility to scale up this method without significant reduction in the methane storage capacity was demonstrated through my research. The experimental data generated from my PhD research played a significant role in the award of a prestigious multi-million external grant by Energy Markets Authority (EMA), Singapore. This project had also won the highly commended research project award presented by Institution of Chemical Engineers (IChemE) Singapore in 2016.

Post-doctoral work:

Since the completion of my doctoral study, I am working as a lead in the project for developing the 'Solidified Natural Gas (SNG) Technology for Energy Storage to Strengthen Energy Resilience in Singapore' granted by EMA. The objective of this project is to develop a prototype to demonstrate the feasibility of hydrate based SNG technology for storing natural gas in the form of hydrate pellets. I investigate the storage and gas recovery aspects from hydrate pellets for the effective integration of SNG technology. I am also working on the design and fabrication of different units that aid in the realization of the SNG technology. I am also learning hydrate characterization techniques like Differential Scanning Calorimetry and In-situ Raman spectroscopy to understand the hydrate formation at molecular level.

Research Interests:

- Gas hydrates
    - Energy storage
    - Energy recovery from hydrates
    - Gas separation
- Flue gas desulphurization in power plants
- Integrated gasification combined cycle (IGCC)

Teaching Interests:

Unit operations of Chemical Engineering, Chemical process technology and calculations, Heat transfer operations and electives like Environmental engineering, Corrosion engineering, Petroleum refining and petrochemicals are the courses that I might be interested in teaching. During my PhD tenure, I have served as a grader for Design project and Heat & Mass transfer modules offered in the Department of Chemical and Biomolecular Engineering of National University of Singapore (NUS). I have served as an able mentor to seventeen undergraduate students during their final year project in our laboratory during my research tenure at NUS. I am proud to state that three of my mentees received Innovation and Research Merit Awards by NUS and five of them had received American Institute of Chemical Engineers (AIChE) - Singapore Local Section (SLS) award for their project work.

Awards:

- **Highly commended Young Chemical Engineer in Research** by IChemE Singapore in 2016
- **Honorable mention** in recognition of contributions to research during graduate study by American Institute of Chemical Engineers (AIChE) Singalore Local Section (SLS) 2016
- **2015 Applied Energy Award for highly cited review paper**
- **President's Graduate Fellowship (PGF)** for the year 2014-15 by NUS

Selected Publications:

1. Veluswamy, H. P.; Kumar, A.; Seo, Y.; Lee, J. D.; Linga, P., A review of solidified natural gas (SNG) technology for gas storage via clathrate hydrates., 2018. *Applied Energy*, 216, 262-285.
2. Veluswamy, H. P., Wong, A. J. H., Babu, P., Kumar, R., Kulprathipanja, S., Rangsunvigit, P., & Linga, P., Rapid methane hydrate formation to develop a cost effective large scale energy storage system., 2016. *Chemical Engineering Journal*, 290, 161-173. **[Highly cited paper]**
3. Veluswamy, H.P., Kumar, R., Linga, P., Hydrogen storage in clathrate hydrates: current state of the art and future directions., 2014. *Applied Energy*, 122, 112-132. **[Highly cited paper]**
4. Veluswamy, H. P., Linga, P., Macroscopic kinetics of hydrate formation of mixed hydrates of hydrogen/tetrahydrofuran for hydrogen storage., 2013. *International Journal of Hydrogen Energy*, 38,

4587-4596.

Contact details: Tel: (65) 6601-1604; e-mail: chehpv@nus.edu.sg

540500 Clathrate Hydrates for Sustainable Development

Abstract:

Clathrate Hydrates for Sustainable development

Ponnivalavan Babu

Department of Chemical and Biomolecular Engineering, National University of Singapore, **Singapore, Singapore 117 585**

Keywords: gas hydrates; desalination; clathrate process; seawater, cold energy, LNG

*corresponding author: e-mail: valavan@nus.edu.sg; Tel: (65) 8693-7709; Fax: (65) 6779-1936.

Research Interests:

Energy and Water are key resources for sustainable and smart nation development. Climate change is the most significant challenge to achieving sustainable development. Water, Energy and Climate change are intricately interlinked. Therefore there is a need for developing innovative technologies to address these challenges in a holistic way. My research focus is on Energy, Water and Climate change pertaining to clathrate hydrate as a technology enabler.

Clathrate hydrate or gas hydrate are ice-like crystalline solid compounds, consisting of a small molecule surrounded by a cage of water molecules. In nature, natural gas hydrate exists in permafrost regions and marine sediments and it is considered as a huge potential energy resource for the future. Selective fractionation of gas components and high storage capacity in gas hydrates have enabled a wide range of applications including seawater desalination, natural gas storage, and transport, CO2 capture and sequestration, cold storage and refrigeration.

My Ph.D. thesis title was "Hydrate based gas separation (HBGS) technology for pre-combustion capture of Carbon dioxide". While most of the research groups employed traditional stirred tank reactors, I employed an innovative fixed bed reactor with new porous media and demonstrated improved process efficiency. Further, a major finding that emanated from my Ph.D. work is that when propane is present as co-guest, it has the ability to draw dispersed water above the porous media towards the gas phase to form hydrate. Based on this mechanism, I proposed a clathrate hydrate based desalination (HyDesal) process utilizing LNG cold energy and simultaneous HBGS – HyDesal process.

Currently, I am developing an innovative technology, HyDesal to desalinate seawater utilizing LNG cold energy. HyDesal process was studied for last 70 years but never commercialized due to slow kinetics; difficulty in separation of hydrate crystals from brine solution, higher energy cost involved due to refrigeration. My innovative approach/reactor design will address all the major challenges that were associated with the HyDesal process. Based on the innovative design, I have built a start of the art prototype to demonstrate the HyDesal process which can strengthen the water-energy nexus. A patent application on the innovative reactor designs is pending (allocated application number PCT/SG2018/050083). This work is of strategic interest for any country since both energy and water demand can be met using an environmentally benign process.

I intend to extend this technology for simultaneous CO2 capture and desalination utilizing LNG cold energy and for wastewater treatment. My experience and experimental know-how to enhance the hydrate formation kinetics would enable me to design and make operational an innovative reactor for simultaneous CO2 capture and desalination and wastewater treatment.

Another area of my research interest is the development of next-generation high-efficiency thermal energy carriers for district cooling. District cooling is an energy efficient, low carbon and affordable solution for meeting the rising cooling demand. Hydrate slurry is a promising cooling medium for air conditioning. I intend to evaluate the techno-economic feasibility of hydrates as thermal energy carriers for next-generation sustainable district cooling systems.

Teaching Interests:

I believe that learning needs to be student-centered and students need to be equal partners in the learning process. The teacher's role should be to use his/her expertise on lesson organization and its clarity, to hand the necessary resources to the students, promote self-learning, build positive classroom climate and use different teaching techniques to reach every student in the classroom. The teacher should make an effort to kindle student's interest in the subject and stimulate their critical thinking and apply the concepts to real-world problems. Teacher-student interactions are the key to quality classrooms and should be bidirectional. Continuous feedback will help to improve teaching effectiveness.

My interest in teaching started during my undergraduate degree at Crescent Engineering College, Chennai, India. During those days, I enjoyed sharing my knowledge and explaining the new topics to my fellow classmates on Chemical reaction Engineering, Fluid Mechanics and Polymer Chemistry. My confidence in teaching was boosted when my fellow students often approached me for clarification and explanations regarding various concepts. During my doctoral degree in National University of Singapore, Singapore, I worked as teaching assistant in CN2125 Heat and Mass Transfer, CN3102 Chemical Engineering Laboratory II and CN4123R Final Year Design Project. This experience as teaching assistant provided me with greater insight into the challenges of teaching in a multicultural environment. Feedback from the students helped me strengthen my teaching abilities. My interactions with the students as a teaching assistant gave me an insight into the student's interest, motivations and their point of view.

Having a strong background in Chemical Engineering and Polymer Technology, I would be interested to teach Chemical Reaction Engineering, Separation Processes, Fluid Mechanics, Polymer Chemistry and Polymer Engineering. However, I would like to teach rest of the courses if required by the department. Apart from teaching, I would also embrace the opportunity to mentor students for their projects. As a research advisor, I would teach my students the necessary technical skills to operate safely and effectively in a laboratory and also providing them the freedom to think, learn from mistakes and develop as an independent researcher.

Selected Publications:

1. **Babu, P.;** Kumar, R.; Linga, P. Pre-combustion capture of carbon dioxide in a fixed bed reactor using the clathrate hydrate process. *Energy* 2013, 50 (1), 364-373. "***Highly cited paper***" by Essential Science Indicators of Thomson Reuters since September, 2013; Top 25 "**Most Cited Paper**" in Energy journal (top 1% among 7000+ publications in the Journal since 2013). Highlighted by Elsevier in a Virtual Special Issue on ***Chemistry and Materials for Energy***; "***h5 core paper***" Google Scholar Metrics in 2016 & 2017.

2. **Babu, P.;** Kumar, R.; Linga, P. Medium pressure hydrate based gas separation (HBGS) process for pre-combustion capture of carbon dioxide employing a novel fixed bed reactor. *International Journal of Greenhouse Gas Control* 2013, 17, 206-214. **"*Highly cited paper*"** by Essential Science Indicators of Thomson Reuters since July, 2014; Top 25 "***Most Cited Paper***" in International Journal of Greenhouse Gas Control (top 1% among 1600+ publications in the Journal); "***h5 core paper***" Google Scholar Metrics in 2017.

3. **Babu, P.;** Kumar, R.; Linga, P. Unusual behavior of propane as a co-guest during hydrate formation in silica

sand: Potential application to seawater desalination and carbon dioxide capture. *Chemical Engineering Science* 2014, 117, 342-351.

4. **Babu, P.**; Linga, P.; Kumar, R.; Englezos, P. A Review of the Hydrate Based Gas Separation (HBGS) Process for Carbon Dioxide Pre-Combustion Capture. *Energy* 2015, 85, 261-279. "***Highly cited paper***" by Essential Science Indicators of Thomson Reuters; Top 25 "***Most Cited Paper***" in Energy journal (top 1% among 7000+ publications in the Journal since 2013); "***h5 core paper***" Google Scholar Metrics in 2017.

5. Chong, Z. R.; Yang, S. H. B.; **Babu, P.**; Linga, P.; Li, X.-S. Review of natural gas hydrates as an energy resource: Prospects and Challenges. *Applied Energy* 2016, 162, 1633-1652. "***Highly cited paper***" by Essential Science Indicators of Thomson Reuters; 2016 ***Applied Energy Award***; Top 25 ***"Most Cited Paper"*** Award (2013-2018); "***h5 core paper***" Google Scholar Metrics in 2017.

540604 Modeling the UV/H_2O_2 Oxidation of Trace Organic Compounds in a Continuous-Flow Reactor with Reflective Walls

Abstract:

Research Interests:

During the last 10 years I have been involved in research related to wastewater treatment. The combination of my previous experiences have made me passionate about this field. My concern about water pollution is genuine; moreover, this branch of research is a global priority in order to ensure a sustainable future. Being born in an overpopulated city with many environmental concerns where potable water is not always available created in me the desire to contribute to the solution of environmental problems and water scarcity. Currently, my research is mainly focused on understanding the mechanisms of degradation of contaminants of emerging concern in wastewater, in order to develop mathematical models that predict the behavior of these contaminants in different natural and engineered systems. Understanding these mechanisms will allow the design of more efficient means of treatment. As a researcher I would like to continue my research in the combination of treatment technologies for the degradation of emerging contaminants, developing methods for the design of treatment trains for wastewater potable reuse.

Teaching Interests:

As an instructor, I am committed to inclusive excellence and active learning in my classes. I understand that, besides being up to date in the advances in my field of research, I need to keep finding strategies to facilitate the learning process of diverse students. I am willing to adapt my teaching style to include a variety of ways of participation, so that students with different learning styles, different capabilities and personal and academic backgrounds can engage in my classes and learn while I learn from them too.

I have previously taught Chemical Engineering laboratories at the undergraduate level, specifically Environmental analysis, Kinetics and reactors, Transport phenomena, Unit operations and Thermodynamics. I have also taught General Chemistry laboratory and assisted the lecture in an active learning environment. I would like to continue teaching classes related to Chemical and Environmental Engineering as a lecture or in laboratory settings.

Abstract:

The occurrence and fate of trace organic compounds (TOrCs) in the aquatic environment are important issues in environmental chemistry. Advanced oxidation processes (AOPs) can be used to remove a wide range of TOrCs by means of reactive species among which hydroxyl radical is prevalent. The UV photolysis of hydrogen peroxide (UV/H_2O_2) produces hydroxyl radicals and is a well-established AOP technology used in full-scale treatment of wastewater effluents.

Understanding of the UV/H_2O_2 AOP process is based on simulation of physicochemical aspects of the process. Recent UV/H_2O_2 process models include consideration of hydroxyl radical scavenging by compounds in complex water matrices and by organic intermediates produced from the oxidation of the parent targets.

Continuous-flow AOPs are commonly used for the destruction of TOrCs that survive conventional wastewater treatment processes. The UV/H_2O_2 process is applied to large-scale operations using tubular reactors that contain one or more UV lamps. A common configuration consists of a single concentric lamp located at the center of a pipe.

Poor absorbance by hydrogen peroxide in the UV range contributes to energy demand in UV/H_2O_2 processes. In poor light-absorbing matrices, a significant portion of the energy is wasted. This inefficiency can be remedied by reflecting unabsorbed light into the reactive mixture. In this work, two novel flow-through UV reactors with a concentric lamp and a reflective surface are used to investigate the destruction of trace organics in wastewater. A kinetic model for the oxidation of *p*-cresol by hydrogen peroxide photolysis developed previously was modified to account for annular flow hydrodynamics, surface light reflection and the full mechanism of *p*-cresol oxidation by hydroxyl radicals. In addition, a mechanism for the degradation of caffeine, bisphenol A and carbamazepine is proposed and calibrated.

The model includes elementary reactions of the UV/H_2O_2 system; direct UV photolysis of the target; and reactions of hydroxyl radical with the target, H_2O_2 and reaction intermediates. The model incorporates UV light reflection from the reactor walls, as well as the hydrodynamics of the annular flow. The model accurately predicts the destruction of the target compounds in a wide range of experimental conditions. Experimental and theoretical results demonstrate that wall provides a significant enhancement of the rate of conversion of the target for high transmittance solutions.

541630 Construction of Ultrasonic / Magnetic Combined Reactor for Rapid Clarification of Turbid Metamorphic Diesel Oil

Abstract:

Research Interests:

The combination of Ultrasonic treatment and Magnetic field applied to diesel oil can alter its characteristics in terms of forces that hold the hydrocarbons together. Consequently, a magnetization reactor was designed combined with ultrasonic treatment to improve the quality and change the property of turbid metamorphic diesel fuel.

Teaching Interests:

Based on the previous study of this experimental work, it was found that 50 mL of modified turbid diesel oil was pretreated with 20 kHz and 80 W power ultrasonic at room temperature for 2 min, and then clarified by 60 mT magnetic field under the temperature of 50 °C, and 800 r/min stirring speed for 10 min. The magnetic field distribution was simulated by ANSYS software and methylene blue dye channel ultrasonic sound field test were utilized to obtain design parameters of the combined magnetic/acoustic treatment of turbid diesel oil.

541687 Fundamental Discovery and Materials Design for Energy Storage

Abstract:

Yuzhang Li - Ph.D. `18 (Stanford University)

-

Research Interests:

Electrochemical energy conversion and storage is critical for vehicle electrification and storing clean energy from intermittent sources (i.e. wind, solar) on the grid. Towards these goals, my work as a doctoral student has encompassed the broad spectrum of stages necessary for technological innovation: (1) developing new tools to discover fundamental understandings, (2) designing and synthesizing unique materials to address intrinsic failure modes, and (3) patenting the technology for real world applications.

As a doctoral student, I have developed [1] and patented [2] a novel graphene cage materials design approach that enables silicon (Si) to cycle as a lithium-ion battery anode with over four times the theoretical capacity of commercial graphite (372 mAh g^{-1}). Using *in situ* transmission electron microscopy (TEM), I showed that the graphene cage is electrically conductive and mechanically robust enough to overcome the dramatic failure modes (i.e. particle rupture, unstable interface) of Si during battery operation. This work is important because it demonstrated for the first time that fractured Si particles could still be recharged in a battery, something that the battery community believed to be previously impossible. Furthermore, the concept of direct graphene growth on each Si particle shows that simple physical mixing of two materials may be insufficient to achieve the desired properties of a composite materials system.

Looking beyond Si, I established strategies to investigate fundamental aspects of pure lithium (Li) metal, which can potentially enable batteries with energy density approaching that of gasoline. Previous techniques were unable to study Li metal at the nanoscale due to its extreme reactivity to the ambient environment and sensitivity to the electron beam. Using environmental [3] and cryogenic [4] TEM techniques that I developed, my research revealed the nanoscale corrosion behavior and detailed crystallography of Li metal dendrites. The new insights guided the design of a protective nitride layer that enabled improved cycling stability of Li metal. My results represent a simple but telling demonstration of the practical applications that can be initiated by fundamental studies on reactive battery materials. These newly developed techniques open up many excellent opportunities for future scientific discovery and invention.

My future research seeks to develop fundamental understandings of the working mechanisms and failure modes in electrochemical devices at both the macroscopic and atomic scale. Guided by these insights, I will design, synthesize, and characterize materials systems that can overcome intrinsic challenges and enable devices for practical application. Using the advanced characterization tools and chemistry expertise developed in my doctoral studies, I intend to further study the atomic structure of electrocatalysts at reaction conditions to understand their behavior during operation [5].

Selected Publications:

[1]: **Y. Li**, K. Yan, H.W. Lee, Z. Lu, N. Liu, Y. Cui. ***Nature Energy*** 1 (2016): 15029.

[2]: **Y. Li**, K. Yan, Z. Lu, Y. Cui. ***US Patent Application No***. 15/178,366. (2016).

[3]: **Y. Li**, Y. Li, Y. Sun, B. Butz, K. Yan, A.L. Koh, J. Zhao, A. Pei, Y. Cui. ***Nano Letters*** 17.8 (2017): 5171-5178.

[4]: **Y. Li**, Y. Li, A. Pei, K. Yan, Y. Sun, C.L. Wu, L.M. Joubert, R. Chin, A.L. Koh, Y. Yu, J. Perrino, S. Chu, Y. Cui. ***Science*** 358.6362 (2017): 506-510.

[5]: H. Wang, S. Xu, C. Tsai, **Y. Li**, C. Liu, J. Zhao, Y. Liu, H. Yuan, F. Abild-Pedersen, F.B. Prinz, J.K. Norskov, Y. Cui. ***Science*** 354.6315 (2016):1031-1036.

Teaching Interests:

My teaching audiences have spanned middle school students to post-doctoral scholars. These experiences have trained my ability to clearly explain difficult concepts in a simple or detailed manner. Coming from a chemical engineering undergraduate background, I am excited to teach a variety of courses at the undergraduate and graduate levels. In particular, I am interested in teaching materials chemistry, transport phenomena (heat, mass, momentum), and atomic structure and arrangements. Creating a great educational experience for young scientists and researchers is critical for the continued advancement of science and it is with this view that I will approach teaching as a future faculty member.

Proposal Experience:

- Office of Energy Efficiency and Renewable Energy Materials Research Program (2016)
- Department of Energy Batteries for Advanced Transportation Technologies (2016)
- Intelligence Community Postdoctoral Research Fellowship (2018)

541821 Reinforced Anion Exchange Membrane (AEM) Separators Based on Triblock Copolymers for Electrode-Decoupled Redox Flow Batteries (RFBs)

Abstract:

Reinforced anion exchange membrane (AEM) Separators Based on Triblock Copolymers for Electrode-decoupled redox flow batteries (RFBs)

Shrihari Sankarasubramanian

Postdoctoral research associate

Department of Energy, Environmental and Chemical Engineering, Washington University in St. Louis, St. Louis, MO

Research motivation

Energy consumption is one of the barometers of economic and human well-being and the equitable development of humankind requires sustainable energy equity. Achieving this overarching goal requires us to address issues such as the intermittency of ecologically non-disruptive renewable energy technologies (solar and wind), decarbonize the transportation sector, and find alternative chemical pathways for large scale, energy intensive industrial processes. *I aim to help address these key issues by leveraging my training and experience in multi-scale theoretical and experimental investigations of electrochemical systems.*

Postdoctoral project – *"Reinforced anion exchange membrane (AEM) Separators Based on Triblock Copolymers for Electrode-decoupled redox flow batteries (RFBs)"*

Supervisor: Prof. Vijay K. Ramani, Department of Energy, Environmental and Chemical Engineering, Washington University in St. Louis

PhD dissertation – *"Investigation of the oxygen reduction reaction at the lithium-oxygen cell cathode"*

Supervisor: Prof. Jai Prakash, Department of Chemical and Biological Engineering, Illinois Institute of Technology

Research Interests:

My research journey began with my undergraduate work on continuum scale modeling of the capacity fade mechanisms in lithium-ion batteries with Prof. Balaji Krishnamurthy. An internship at Tata Power Solar, where I assisted in formulating a new silicon texturing bath composition (resulting in a ~1% improvement in final cell efficiency), whetted my appetite for translational research.

During my PhD research with Prof. Prakash at Illinois Institute of Technology, I: 1) evaluated novel lithium-ion cell chemistries for use in sub-zero temperatures (in conjunction with Hydro-Quebec), 2) performed the first ever accelerating rate calorimetry study of the thermal safety of lithium-sulfur cells (again with Hydro-Quebec), and 3) performed a multi-year theoretical and experimental investigation of catalysts, solvents and kinetics of the oxygen reduction reaction (ORR) in lithium-oxygen cells in collaboration with Toyota Motors. Over the course of these projects, I extensively employed density functional theory (DFT) simulations, continuum scale kinetic modeling, and classical electrochemical and electroanalytical experimental techniques.

As an integral part of my postdoctoral research project, I am investigating anion exchange membranes for use in electrode-decoupled redox flow batteries (RFBs). Herein, I am applying my knowledge of electrochemistry and transport phenomena to understand the transport of ions across a functionalized membrane. The final goal of this project, funded by ARPA-E, is to develop selective membranes that allow for the deployment of low-cost, electrode-decoupled RFBs such as the Fe-Cr system.

Future research directions

I plan to leverage my training and experience in multi-scale theoretical and experimental investigations of electrochemical systems to carry out transformative, translational research into next generation, beyond lithium-ion batteries, grid scale energy storage using redox flow batteries, and the electro-synthesis of bulk and fine chemicals.

Project 1: *Metal- (metal superoxide) batteries* – Lithium-oxygen batteries have been the holy grail of next generation battery systems due to their unmatched theoretical specific energy (3458 Wh kg^{-1}). Despite years of extensive work on catalysts, redox mediators and ever-more-stable electrolytes, a practical lithium-oxygen battery has not been realized. As part of this project I will examine sodium- and potassium- oxygen batteries whose reduction products are superoxides instead of the peroxides, allowing me to circumvent the columbic efficiency and degradation problems of lithium-oxygen batteries. I will combine insights from electrochemical kinetics studies with oxygen impermeable membranes and water tolerant ionic liquids to demonstrate long life alkali metal- (metal superoxide) batteries. *The end goal will be the development of a true metal-ambient air battery.*

Project 2: *Low-cost, grid scale energy storage* – The adoption of intermittent renewable energy sources calls for the large-scale deployment of energy storage solutions to maintain grid reliability. The DOE estimates that $100/kWh is the cost at which storage solutions become cost competitive. This cost target precludes both the Li-ion battery and the all-vanadium redox-flow battery. I will investigate alternative, low cost, earth abundant chemistries for redox-flow batteries. The use of catalytic inhibitors to suppress the hydrogen evolution and oxygen evolution reactions, in conjunction with electrolyte systems that increase the nominal redox potential of the redox couple of choice, will pave the way for >2V RFBs utilizing couples such as Cr-Ce and provide an unparalleled cost advantage compared to existing technologies. Further, I will develop highly perm-selective, high anion transference number anion-exchange membranes for this system both in-house and in collaboration with polymer scientists and material chemists.

Project 3: *Electro-synthetic alternatives to energy intensive chemical production –*

Ammonia production – Ammonia is produced to the tune of 140 million tons a year via the energy intensive Haber-Bosch process with a massive carbon footprint. Biomimetic or electro-catalytic ammonia production is a low CO_2 alternative to this process that will allow for low capital cost and distributed production of ammonia at the demand site. Further, the elimination of steam reforming to produce hydrogen and the avoidance of transportation costs and associated emissions makes this an extremely tantalizing proposition. I will examine metal, metal oxide, carbide, nitride and alloy catalysts for electrochemical ammonia synthesis from water and air. Combining DFT screening of catalyst candidates (nitrides, carbides, Pt and Pd alloys) with robust half-cell and device-level testing, I will identify candidate catalysts at the top of the activity pyramid, evaluate their nitrogen reduction kinetics and design, test and optimize larger-scale cells to produce ammonia.

CO_2 capture, concentration and conversion (C3)- Any realistic timeline of a transition to renewables envisions significant dependence on fossil fuels in the near-term. Thus, it is imperative that efficient CO_2 capture technologies

be developed. In collaboration with other groups working on CO_2 capture, I will develop alkali metal-O_2/CO_2 batteries for direct conversion of CO_2 to electricity. An intriguing idea that will be investigated is a combined N_2 electro-reduction/ alkali metal-O_2/CO_2 system wherein the ammonia produced is used to capture and concentrate the CO_2 before use in the battery.

Teaching Interests:

The experience I have gained assisting in the teaching of 9 courses across the spectrum of ChE, from the sophomore level to the advanced PhD level, has prepared me well to serve as an instructor teaching the traditional chemical engineering core. I have served as Assistant to instructor and Support to instructor respectively for courses on Electrochemical Engineering and Energy Conversion and Storage at Washington University, where I have been actively involved in class planning and laboratory sessions in addition to traditional teaching assistant roles. During my PhD, I served as a Teaching assistant for 7 courses including thermodynamics, transport phenomena, undergraduate laboratory and electives in energy technology. The experience of assisting in the teaching of thermodynamics as an introductory first course, an intermediate course for 1st year master's students and as an advanced course for PhD students provided invaluable insights into the nuances of balancing depth and clarity without sacrificing technical accuracy. The mentor-mentee relationships I formed with high school, undergraduate and graduate students in the lab redoubled and renewed my commitment to excellence in teaching and training. I look forward to bringing this passion to my future role as a ChE faculty member.

Invention and technology translation (1 patent issued, 3 disclosed)

I have always wanted to see my work changing society in a reasonable timeframe. My PhD work on Pd and Pt alloy catalysts for lithium-oxygen batteries with Toyota resulted in a patent and is the basis of their continued research in this field. My on-going postdoctoral work on reinforced membranes for flow batteries has already resulted in a $50,000 Leadership in Entrepreneurial Acceleration Program (LEAP) grant from Washington University to serve as a pre-seed fund for a start-up. My future independent research career will maintain that same strong practical, inventive and translational emphasis.

Selected Publications (10 published, 2 under review and 4 under preparation)

1. **S. Sankarasubramanian,** and B. Krishnamurthy, A Capacity Fade Model for Lithium ion batteries including kinetics and diffusion, *Electrochim. Acta* **70**, 248-254 (2012).

2. **S. Sankarasubramanian*,** N. Singh, F. Mizuno, J. Prakash, *Ab initio* investigation of the Oxygen Reduction Reaction activity on noble metal (Pt, Au, Pd), Pt_3M (M=Fe, Co, Ni, Cu) and Pd_3M (M=Fe, Co, Ni, Cu) alloy surfaces, for Li-O_2 cells, *J. Power Sources,* **319**, 202-209 (2016).

3. **S. Sankarasubramanian*,** J. Seo, F. Mizuno, N. Singh, J. Prakash, Rotating ring-disc electrode investigation of the aprotic superoxide radical electrochemistry on multi-crystalline surfaces and correlation with Density functional theory modeling – implications for Lithium-air cells, *J. Electrochem. Soc.*, **163**(10), A2377-A2384 (2016).

4. **S. Sankarasubramanian*,** J. Seo, F. Mizuno, N. Singh, K. Takechi, J. Prakash, Enhancement of oxygen reduction reaction rate by addition of water to an oxidatively stable ionic liquid electrolyte for lithium-air cells, *Electrochem. Commun.,* **73,** 55-57 (2016).

5. **S. Sankarasubramanian*,** J. Seo, F. Mizuno, N. Singh, J. Prakash, Elucidating the Oxygen reduction reaction kinetics and the origins of the anomalous Tafel behavior at the Li-O_2 cell cathode, *J. Phys. Chem. C*, **121,** 4789-4798 (2017).

6. Y. Zhang, J. Parrondo, **S. Sankarasubramanian,** V. Ramani, Detection of Reactive Oxygen Species in AEM Fuel Cells using *in situ* Fluorescence Spectroscopy, *ChemSusChem*, **10**, 3056-3062 (2017).

544353 Rational Design of Novel Catalysts for Energy Applications

Abstract:

Research Interests:

As CO_2 emissions increase, efforts much be put forth to reduce the impact on climate change and ocean acidification. An interesting and attractive alternative approach would be the conversion of CO_2 into chemicals and fuels, which has been intensively pursued for renewable and sustainable energy.[1] However, due to the negative adiabatic electron affinity and large ionization potential, the CO_2 molecule is chemically inert, thus making the conversion difficult under normal conditions. Heterogeneous catalysts, with high stability, superior efficiency and low cost are urgently needed to facilitate the problems concerning the processes of CO_2 conversion and utilization.

Research Experience:

My research career path has been a blend of many fields: chemical engineering, materials science and heterogeneous catalysis. My graduate research was devoted to the synthesis of the following high-surface area bimetallic catalysts for energy applications: Fe-Ce solid solution for hydrogen production via ethanol steam reforming[2] and Cu-Cr catalysts for hydrogenolysis of glycerol.[3-4] During my PhD study and postdoctoral training at the ETH Zurich, I focused on catalytic conversion of biomass to value-added chemicals over zeolite and zeolite-based catalysts.[5-7] The catalysts were fully characterized by state-of-the-art techniques, including two-dimensional solid-state nuclear magnetic resonance spectroscopy (2D SS NMR), electron microscopy, X-ray absorption spectroscopy.[8-10] The dual role of zeolites and zeolite-based catalysts was proposed via both stabilization of the intermediates and catalytic transformation of the intermediates to the desired products, e.g. aromatic hydrocarbons. We demonstrated that by choosing the proper transition metals, such as iron, and fine tuning the acidity and morphology of zeolites, the selectivity to the desired products was controllable.[7] My current research at University of Rochester includes rationally designing and developing a fundamental understanding of highly active and selective zeolite-based catalysts for CO_2 hydrogenation into light olefins.

Successful Proposals:

1. Contributing author to an industry grant: Phenol and aromatics from lignin, funded by SABIC Americas, 2015-2017, Switzerland. (~ $400,000 for two years)
2. Contributing author to the National Research Program "Resource Wood" (NRP 66), funded by Swiss National Science Foundation, 2011-2013, Switzerland. (~ $200,000 for two years)

Future Research Directions:

As faculty member, I plan to continue my research on design and synthesis of novel catalysts for CO_2 conversion to chemicals and fuels. My vision is to combine the extensive knowledge obtained during my academic training: catalyst design, synthesis, characterization, and performance evaluation, to develop selective catalysts for reduction of carbon oxides. Iron catalysts, typically used for Fischer-Tropsch synthesis (FTS), have shown the greatest potential in converting CO_2 to value-added hydrocarbons;[11] however, the selectivity to specific products, for example light olefins, is generally low. Zeolites, due to their unique structure and controllable morphology and acidity, have shown excellent activity, shape selectivity and stability for energy applications, e.g. biomass catalytic conversion to fuels and chemicals.[5-7, 9] Therefore, the addition of zeolites to iron catalysts, to synthesize hybrid catalysts e.g. Fe-ZSM-5, is essential to fine tune the product distribution to the more desirable light olefins.

A second approach is to design a multifunctional catalyst that first converts CO_2 to CO, the main component in syngas; then CO (or syngas) is converted to hydrocarbons over the same catalyst via FTS. The design of such a multifunctional catalyst is extremely challenging because thermodynamics dictates that the conversion of CO_2 to CO is slightly endothermic and FTS is exothermic, which requires that the catalyst be operated under two different temperature regimes. The ideal catalyst should be composed of three components: 1) an active component for the conversion of CO_2 to CO; 2) a FTS catalyst that converts the produced CO to hydrocarbons; and 3) a component which controls the hydrocarbon selectivity. Although CO_2 is chemically inert, it could be converted into CO with high selectivity over transition metal carbides at relative low temperature, e.g. Molybdenum carbide (Mo_2C).[12] For the FTS component, iron and transition metal oxides have showed promising performance,[13-14] while zeolites can control the hydrocarbon selectivity. Therefore, a catalyst that combines Mo_2C, Fe, and a zeolite (Mo_2C/Fe/zeolite) could be an initial candidate. Upon further research studying the effect of varying each component, the knowledge gained will provide fundamental understanding of desirable catalysts of CO_2 conversion to chemicals and fuels.

Teaching Interests:

Courses that I would be interested in teaching are an introductory course in chemical engineering principles and surface science. The introductory course could be heat and mass transfer or catalytic reaction engineering; and the surface science course covers the different characterization techniques used in heterogeneous catalysis, from basic principles to applications. I have taught the chemical engineering laboratory course at the undergraduate and graduate level, and the surface science course as a teaching assistant to graduate students. Through my participation as a teaching assistant, I have developed confidence and an interest in teaching and look forward to the opportunity to both teach assigned classes and develop my own classes in chemical engineering. And lastly, I have extensive experience mentoring graduate students in my group during my doctoral and postdoctoral research (1 PhD, 2 third-year PhD candidates, 5 master students).

References:

1. Kondratenko, E. V.; Mul, G.; Baltrusaitis, J.; Larrazabal, G. O.; Perez-Ramirez, J., Status and perspectives of CO2 conversion into fuels and chemicals by catalytic, photocatalytic and electrocatalytic processes. *Energy Environ. Sci.* **2013,** *6* (11), 3112-3135.
2. Ma, Z.; Xiao, Z.; van Bokhoven, J. A.; Liang, C., A non-alkoxide sol–gel route to highly active and selective Cu-Cr catalysts for glycerol conversion. *J. Mater. Chem.* **2010,** *20*, 755-760.
3. Liang, C.; Ma, Z.; Lin, H.; Ding, L.; Qiu, J.; Frandsen, W.; Su, D., Template preparation of nanoscale CexFe1-xO2 solid solutions and their catalytic properties for ethanol steam reforming. *J. Mater. Chem.* **2009,** *19* (10), 1417-1424.
4. Liang, C.; Ma, Z.; Ding, L.; Qiu, J., Template Preparation of Highly Active and Selective Cu–Cr Catalysts with High Surface Area for Glycerol Hydrogenolysis. *Catal. Lett.* **2009,** *130*, 169-176.
5. Ma, Z.; Troussard, E.; van Bokhoven, J. A., Controlling the selectivity to chemicals from lignin via catalytic fast pyrolysis. *Appl. Catal., A.* **2012,** *423–424* (0), 130-136.
6. Ma, Z.; Ghosh, A.; Asthana, N.; van Bokhoven, J., Optimization of the Reaction Conditions for

Catalytic Fast Pyrolysis of Pretreated Lignin over Zeolite for the Production of Phenol. *ChemCatChem* **2017,** *9* (6), 954-961.
7. Ma, Z.; Custodis, V.; van Bokhoven, J. A., Selective deoxygenation of lignin during catalytic fast pyrolysis. *Catal. Sci. Technol.* **2014,** *4* (3), 766-772.
8. Li, T.; Ma, Z.; Krumeich, F.; Knorpp, A. J.; Pinar, A. B.; van Bokhoven, J. A., Properties modification of nano-sized hollow zeolite crystals by desilication. *ChemNanoMat* **2018**, doi: 10.1002/cnma.201800225.
9. Ma, Z.; Ghosh, A.; Asthana, N.; van Bokhoven, J., Visualization of structural changes during deactivation and regeneration of FAU zeolite for catalytic fast pyrolysis of lignin using NMR and electron microscopy techniques. *ChemCatChem* **2018**, doi: 10.1002/cctc.201800670.
10. Ma, Z.; van Bokhoven, J. A., Deactivation and Regeneration of H-USY Zeolite during Lignin Catalytic Fast Pyrolysis. *ChemCatChem* **2012,** *4* (12), 2036-2044.
11. Abelló, S.; Montané, D., Exploring Iron-based Multifunctional Catalysts for Fischer–Tropsch Synthesis: A Review. *ChemSusChem* **2011,** *4* (11), 1538-1556.
12. Porosoff, M. D.; Yang, X.; Boscoboinik, J. A.; Chen, J. G., Molybdenum Carbide as Alternative Catalysts to Precious Metals for Highly Selective Reduction of CO2 to CO. *Angew. Chem. Int. Ed.* **2014,** *126* (26), 6823-6827.
13. Jiao, F.; Li, J.; Pan, X.; Xiao, J.; Li, H.; Ma, H.; Wei, M.; Pan, Y.; Zhou, Z.; Li, M.; Miao, S.; Li, J.; Zhu, Y.; Xiao, D.; He, T.; Yang, J.; Qi, F.; Fu, Q.; Bao, X., Selective conversion of syngas to light olefins. *Science* **2016,** *351* (6277), 1065-1068.
14. Torres Galvis, H. M.; Bitter, J. H.; Khare, C. B.; Ruitenbeek, M.; Dugulan, A. I.; de Jong, K. P., Supported Iron Nanoparticles as Catalysts for Sustainable Production of Lower Olefins. *Science* **2012,** *335* (6070), 835-838.

544552 Sustainable Production of Renewable Specialty Chemicals and Fuels from the Catalytic Conversion of Lignocellulosic Biomass

Abstract:

Lignocellulosic biomass is one of the most abundant carbon sources in the world and has the potential of becoming the main source for the renewable production of valuable chemicals such as pharmaceuticals, fuels and specialty chemicals. Biomass can be obtained from agricultural residues, wood processing, and food industry wastes, among others. The use of these continuously produced wastes is a good approach for the reduction in the production costs of high valuable chemicals. In order to obtain a selective and efficient production of valuable chemicals, it is crucial to select and understand the reaction conditions, solvent and catalyst. Hence, it is important the development and understanding of the physicochemical properties of the catalysts used for the conversion of biomass and its derivatives, such as, cellulose, hemicellulose, and lignin, among others. Studying the effect of the reaction media (solvent and mixture of solvents) on the catalysts, reactants and products during the reaction helps to elucidate the catalytic performance of the materials. This in combination with existing knowledge can be used to optimize the current and new thermo catalytic processes for the production of high valuable chemicals from biomass.

Research Interests:

During my academic career, I had the opportunity to work on different projects related to heterogeneous catalysis. My research experience in this field started during my master degree with the synthesis and study of catalysts for the conversion CO_2 to fuels. We used multifunctional catalysts to improve the production of dimethyl ether from the hydrogenation of CO_2. During my PhD, we studied the catalytic fractionation of agricultural biomass residues to glucose, xylose and lignin using sulfuric acid in polar aprotic solvents. At the same time, we were also studying the catalytic conversion of cellulose and levoglucosan to levoglucosenone using acid functionalized mesoporous silica. While working in these projects, I gained experience in materials synthesis and characterization, and acquired expertise in the evaluation and analysis of catalytic reactions. Using the skills acquired during my academic research career, I want to expand my research work in the catalytic production of renewable specialty chemicals, fuels and pharmaceuticals from lignocellulosic biomass wastes and greenhouse gases. At the same time, I want to develop novel solid catalysts for the production of commodity chemicals and investigate the solvent effects on the chemical properties of the catalyst during the reaction. In addition, I would like to perform kinetic studies for the determination of complex reaction routes in combination with in-situ spectroscopy techniques to monitor reactions and support kinetics studies. I am also interested in the production of bio-oil and its catalytic cracking for the production of organic liquid products.

Teaching Interests:

During my graduate academic experience, I had the opportunity to be teacher assistant of the Chemical Engineering Unit Operations Laboratory at University of Puerto Rico, which includes topics related to mass transfer operation, reaction and kinetics analysis. Also, I had the opportunity to guide approximately 12 undergraduate and graduate students in their research projects. Moreover, during this year I started to co-advise a graduate student from the Chemical Engineering Department at University of Puerto Rico, which was my undergraduate student. My teaching philosophy is based in my experience as a student and a researcher, not only in the academia but also from the industry point view. From the academia, I learned how fundamental is the theoretical background in order to understand and find a solution to a problem. From the industry, I learned techniques on how to apply in a practical way what I learned in the academia. Having experienced both the academia and the industry, has helped me to have a wide vision on what engineering and science students need and how to help them to prepare themselves in any of the professional routes they decide to take: industry or academia. I believe that education has to be student-centered and that students need to be involved in active learning, thereby practicing what they learn in real situations and gaining hands-on experience. Furthermore, I want to include young high school students in my research projects to impact, shape and give them the essentials skills to become successful college students and researchers.

Keywords: Catalysis, carbon dioxide, synthesis, characterization, reactions, kinetics, biomass, cellulose, and lignin.

542458 Control of Slip at the Fluid-Surface Interface Using Molecular Additives

Abstract:

Research Interests:

The fluid behavior in a confined geometry is dictated by the boundary conditions at the fluid-wall interface. From the macroscopic point of view, fluid is immobile at a solid wall that is known as the no-slip boundary conditions. This assumption shows shortcomings at microscopic length scales, where the interaction between the wall and the fluid plays an important role in determining the fluid motion. At a rough wall, fluid velocity is the same as the wall velocity (i.e. stick condition), while at a smooth wall, the fluid velocity is larger than the wall velocity that is called slip condition. The slip between fluid/surface and fluid/fluid interfaces occurs in various applications such as polymer processing, and microfluidic and nanofluidic devices. To create wall slip, short hydrocarbons with functional groups, such as fatty amides, are used as additives in polypropylene processing. It is been reported in the literature that a very low concentration (500 ppm) of these molecules is required for the surface modification. A possible scenario for this phenomenon is that these small molecules migrate to the vicinity of the wall, and form an interphase (boundary layer) that has a low cohesive energy with the polymer phase. The underlying mechanisms of this process are obscure, and I intend to focus on the experimental and computational design of molecular additives that can alter the slippage of large molecules, such as polymers, in confined geometry. I will be using atomistic and coarse-grained simulations techniques, and experiments to characterize the mechanisms of slip at the fluid-surface and fluid-fluid interphases of polymeric systems that are discussed briefly:

Molecular Design of slip modifiers: In order to understand the mechanisms of the slip and effect of different chemical groups in the slip modifiers (additives), atomistically-detailed molecular dynamics (MD) simulations will be used. MD simulations will provide the details of microstructure and local dynamics of different components in the interphase of the polymer and boundary layer.

Coarse-grained model for polymerâ€"boundary layerâ€"wall: Given that the timescale of the atomistic simulations is very limited (<0.1 Î¼s), the shear rates in the atomistic simulations is above 10^7 s^{-1}, which is extremely larger than accessible shear rates observed in experiments. To tackle this problem, a coarse-grained model will be developed to simulate the rheology of polymer melts in the presence of small molecules in a confined geometry. The interaction between small molecules and polymer melt will be estimated from the atomistic simulations. Results will be used to determine a model for the slip velocity as a function of the shear stress in these systems.

Interaction between polymer and additive molecules: To validate the coarse-grained model results, a surface force apparatus (SFA) will be used to quantify the cohesive energy between the small molecules and the polymer melt. These results will be used to enhance the accuracy of the atomistic simulations at the interphase of the boundary layer and polymer phase. Furthermore, using rheometry experiments, the rheology of the polymer under shear flow will be determined and compared with the numerical data obtained from the coarse-grained models.

Teaching Interests:

I am excited to teach any core course in the undergraduate curriculum of chemical engineering, and I want to develop graduate elective courses on simulation. I will develop an interactive teaching method for students by using a computer-based teaching method. Using programming languages such as MATLAB or POLYMATH, students will learn how to apply different computational tools to solve not only problems that they face in the class, but also various intricate modeling problems that emerge in chemical engineering.

In particular, I look forward to teaching numerical analysis, process control and design courses that involve mathematical modeling and simulation. An advanced project-based course that covers molecular and Monte-Carlo simulations with a focus on complex fluids, such as ILs and biomolecules will be created as an elective course.

Research Background:

I have been fortunate to work on various research topics during my graduate studies in the Khare research lab at Texas Tech University. I started my graduate research by investigating the structural and rheological properties of dilute polymer solutions of different chain architectures. Later on, I created rigorous atomistically-detailed models for determining the volumetric, structural, and rheological properties of neat and polymer modified asphalts. I showed that master curve can be constructed from simulation values of viscoelastic properties, thus this increases the range of time and frequency scales accessible to simulations by a few decades and facilitates a comparison with experiments. The simulations were able to capture the sub-glass transition temperature relaxation of the model structures that has been reported in experimental studies of glass former liquids. I have also closely collaborated with an experimental group in order to develop novel polymeric materials that can be utilized as the membrane in pervaporation applications. During the latter project, I demonstrated that the dynamics of solvent molecules (water and ethanol) are closely correlated with dynamics of cross-linked polymer regardless of the solvent concentration and this dynamic coupling could be used as an additional factor in designing polymeric materials for pervaporation based separations. In addition, in a collaborative study and, we showed that MD simulations can capture the experimental trend of the rheology of imidazolium-based ILs and is able to connect the macroscopic properties to the microstructure of ILs.

During my post-doctoral appointment in the Bonnecaze research group at the University of Texas (UT) at Austin, I am developing models and simulations of magnetically-driven flows of dilute suspensions of magnetite particles. In this study, we aim to model and simulate the drug delivery to a fully occluded blood vessel using a magneto-rheological fluid that is caused by a rotary magnetic field with a gradient. I also study the rheological properties of soft particle glasses. In particular, I found that the polydispersed soft particles glasses, which are jammed beyond the random close packing volume fraction, can form crystalline structures in the presence of shear flow. In addition, I found that the macroscopic properties, such as shear stress, normal stress differences, and long-time diffusion coefficient of the SPGs can be extracted from the microscopic dynamics of the particles. I have also created atomistic models to study the interface of calcite and water in the presence of oil and surfactant molecules. This project is of interest to oil recovery applications in which the interfacial properties of a reservoir play a crucial role in determining the efficiency of the process. By varying the surfactant coverage, the evolution of surfactant morphologies near the calcite surface is being studied. The current study is being extended to simulate oil-surfactant-calcite interactions to investigate mechanisms of surfactant-induced wettability alteration. Furthermore, in a collaborative MRSEC project at the Center for Dynamics and Control of Materials at UT, we have created two-dimensional network models of germanium nanowire to study the mechanical properties under tension, compression, and shear. In addition, I have been involved to develop a Brownian dynamics simulations tool to study the dynamics and rheology of colloidal gels in shear flow.

In summary, I will apply these skills (atomistic and coarse-grained MD simulations of polymers/ionic liquids, microscale simulations of suspensions, and continuum mechanics) to other problems in the field of engineering to understand the mechanisms underlying different phenomena that occur at nano- and micro-length scales.

542496 Engineering Non-Equilibrium Materials with Controllable Spatiotemporal Patterns: Oscillator Networks and Active Suspensions

Abstract:

Research Interests:

Creating spatial and temporal structure from inert or homogeneous media is central to the function of living material but is also a generic behavior of far from equilibrium systems. Studying model active systems will lay the groundwork for engineering spontaneously organizing materials that exhibit desired and robust dynamical attractors. Such materials will have novel means by which to process information, spontaneously create control signals, or execute mechanical tasks, all by consuming chemical energy. My research group will study the fundamental processes of pattern formation in chemical oscillator networks and active suspension with nematic ordering. In the oscillator system, these structures take the form of patterns of temporal synchrony characterized by the relative phase at which connecting units oscillate. The gallop or trot of a quadruped, for example, are two alternate forms of synchrony orchestrated by its nervous system. In the active liquid crystal system, active stress creates mobile topological defects that generate fluid flow and whose trajectories exhibit both temporally periodic and chaotic dynamics. At my poster, I will present both my current research on these systems as well as my plans for developing novel non-equilibrium materials by bringing together a group that integrates both experiment and theoretical investigation.

Teaching Interests:

I would be happy contributing to courses in fluid dynamics, transport, and soft matter where my experience modeling micro- and nano-scale fluid dynamics, multiphase flows, electro-kinetic flows, and transport phenomena will enrich the curriculum. I would also like to create my own special topics courses for graduate students based on my research interests in active, low Reynolds number flows, reaction-diffusion systems, and nonlinear dynamics. To develop these courses I will use elements from the following texts: Pattern Formation and Dynamics in Non-equilibrium Systems (Cross and Greenside), Nonlinear Dynamics and Chaos (Strogatz), Low Reynolds Number Hydrodynamics (Happen and Brenner), and Dissipative Structures and Chaos (Kuramoto).

542248 Complex Interfacial Dynamics, Deformation-Based Microrheology, and Beyond

Abstract:

Advances in visualization and manipulation over length scales ranging from the molecular to the macroscopic have ushered a need and relevance in connecting the structure, flow, and rheology of soft materials with applications in materials design and micro-scale/biological transport. I intend to look at these critical questions through the lens of microhydrodynamics, while providing the soft matter community with insights into designing new experiments, materials, and measurement techniques.

As a doctoral student with David Saintillan at the *University of Illinois at Urbana-Champaign* and later at the *University of California, San Diego*, I developed a theoretical and numerical framework to study the hydrodynamics of elastic filaments such as stiff biopolymers and fibers. I applied this model to illustrate unique instabilities and shape transitions of elastic biopolymers in microfluidic devices, and describe, for the first time, the effect of particle flexibility on suspension stability.

More recently, as a postdoctoral researcher with Todd Squires at the *University of California, Santa Barbara*, I have focused on the dynamics of complex surfactant-laden interfaces, such as in membranes, industrial coatings, and biological interfaces like lung alveoli. My theoretical work has revealed qualitatively new features such as non-intuitive pair interactions between interfacial probes, spurring new directions of experimental investigation with far-reaching consequences in the measurement and interpretation of interfacial rheology.

Recent theoretical work and its broader consequences in '2D suspensions' will be the focus of my oral presentation at this meeting.

Selected Publications:

H. Manikantan & T. M. Squires, *Proceedings of the Royal Society A* 473, 0346 (2017).
H. Manikantan & T. M. Squires, *Physical Review Fluids* 2, 023301 (2017).
H. Manikantan & D. Saintillan, *Physics of Fluids* 28, 013303 (2016).
H. Manikantan & D. Saintillan, *Physical Review E* 92, 041002 (2015).
H. Manikantan, L. Li, S. E. Spagnolie & D. Saintillan, *Journal of Fluid Mechanics* 756, 735-764 (2014).
L. Li, **H. Manikantan**, D. Saintillan & S. E. Spagnolie, *Journal of Fluid Mechanics* 735, 705-736 (2013).
H. Manikantan & D. Saintillan, *Physics of Fluids* 25, 073603 (2013).

Research Interests:

Moving forward, my research will involve mathematical modeling, numerical simulations, and lab experiments in close collaboration with other researchers to understand complex soft systems. A primary area of interest will be fluid-fluid interfaces populated with surfactants, proteins, particles, or macromolecules. The subtle interplay of surface rheology, Marangoni stresses, adsorption/desorption, and 2D phase behavior makes complex interfaces as rich a field of exploration and relevance as traditional 3D fluids. Teasing apart these different phenomena in real examples using analytical theory will be a central theme of my work. These studies will also advise new avenues and model problems for collaborative experiments and ultimately guide better prediction and design. Armed with these fundamental insights, my vision is to explore more involved and applied problems such as the collective motion and assembly of surface aggregates, lipid domains, and surface-attached probes.

Another broad avenue of research will be the fluid-structure interaction of deformable inclusions, and its interplay with bulk rheology. Elongated elastic particles such as actin and microtubules occur in nature as a way to provide locomotive and structural properties to cellular systems. Internal degrees of freedom of particles embedded in a medium can relax stresses, and in turn contribute to the bulk behavior of the material. I will focus on elucidating the impact of shapes and deflections of suspended elastic particles on effective rheology, and developing a deformation-based mathematical framework for nonlinear microrheology with exciting experimental opportunities. I will also explore related themes in the hydrodynamic 'training' and 'healing' of biopolymers for cellular functions, and in assembling artificial micro-structures.

Teaching Interests:

I am committed to being an effective and engaging teacher, and have been recognized with teaching awards and certifications from both the *University of Illinois at Urbana-Champaign* and the *University of California, San Diego*. I strive to provide a solid foundation in the mathematical and physical principles, while connecting concepts to real applications. I also make it a point to encourage participation and communication, while continually collecting and acting on student feedback. I have been a teaching assistant in four undergraduate and two graduate classes, through which I have practiced and implemented my teaching strategies, with student ratings that reflect the success of my methods. I have a particular interest in teaching foundational undergraduate and graduate courses in transport phenomena, biophysics, and mathematical methods, and specialized courses in soft matter, colloids and suspensions, and microhydrodynamics. Beyond college-level teaching, I have also independently designed and delivered two high-school-level courses, with excellent feedback on both.

542105 Fluid Dynamics at Different Length Scales in Confinements

Abstract:

Research Interests:

My work seeks to elucidate the multi-scale dynamics of fluids in environmental water systems. To this end, my research agenda entails the study of multi-scale flow and interaction with solid bodies in the context of 3 different phenomena: flow dissipation, sedimentation and contamination.

I will do experimental measurement of turbulent flow at large scales past obstacles and investigate sedimentation regimes at different Re and for different sizes and shapes of sediment particles. I will also study the infiltration of water in soil surfaces and exchange with subsurface water. I will measure the dispersion and transport of particulates in different forms from small organisms to nanoparticles and nutrients in porous material such as soil.

Teaching Interests:

Facing a generation of students with abundant access to knowledge via nontraditional means, instructors bear a great responsibility for channeling their curiosity and teaching them the skills necessary to make use of this knowledge. In this context, my main teaching objective is to provide students the scientific and analytical tools they will need to understand problems and work out solutions in their future lives and careers.

I have served as both a Lecturer at Azad University, Iran, and a Teaching Assistant at Wesleyan University and mentored a large number of graduate and undergraduate students in research during my postdoc. Given this experience and my training, I am prepared to teach both a broad range of undergraduate courses in Chemical engineering, including Statistics, as well as more advanced undergraduate or graduate courses in my area of specialty, notably Continuum Mechanics, and Thermodynamics. Finally, I would also be excited to develop a new undergraduate and graduate courses on immiscible and two phase flow and interfacial phenomenon.

540633 Research on the Vertical Falling Film Behavior in the Scrubbing-Cooling Tube

Abstract:

Research Interests: Chemical Engineering, Gasification, Fluid flow, Heat and mass transfer

Teaching Interests: Principle of transport process

Research on the vertical falling film behavior in the scrubbing-cooling tube

Yifei Wang*, Xin Peng, Liucheng Yan , Guangsuo Yu, Fuchen Wang

Key Laboratory of Coal Gasification and Energy Chemical Engineering of the Ministry of Education, East China University of Science and Technology, Shanghai 200237, PR China

Being a part of the opposed muti-burner (OMB) gasifier, the scrubbing-cooling chamber is used to cool, humidify and scrub the high temperature raw syngas. The chamber mainly consists of a scrubbing-cooling ring, a scrubbing-cooling tube, a liquid pool and a bubble bed. The cooling water flows through the ring and forms vertical liquid film inside the tube. The liquid film protects the wall of the tube from the high temperature syngas and slag. Most of the falling film studies are carried out in micro-scale tube with small liquid Reynolds number. The research here focuses on the study of both of the velocity and thickness distribution of falling liquid film under relatively large Reynolds number which is closed to actual industrial condition.

An Ultrasound Doppler Velocimetry was employed to measure the velocity and thickness distribution of falling liquid film and the liquid Reynolds number ranged from 1.0¡Á10^{4} to 3.1¡Á10^{4}. A 3D numerical model of the scrubbing-cooling tube was established and the simulation results were in good agreement with the experimental data as shown in Fig. 1. The average liquid film thickness became largest at the 0 circumferential position while a part of the liquid was separated from the local liquid film and dripped in the gas core when Reynolds number was larger than 1.5¡Á10^{4} as shown in Fig. 2-3. In the lower section of the tube, the circumferential liquid film thickness distributed more evenly. With the increase of Reynolds number, both of the thickness and velocity of global liquid film increased and the thickness distribution became unevener. Under the effect of gravity, the liquid film velocity increased with the increasing flow distance at small Reynolds numbers while decreasing at large Reynolds numbers due to the resistance.

Keywords: falling film, film velocity and thickness, numerical simulation, large Reynolds numbers, Ultrasound Doppler Velocimetry

Fig. 1 Comparison of liquid film velocity distribution between simulated and experimental data

Fig. 2 Liquid film thickness distribution at different Reynolds numbers

Fig. 3 Phase contour graph of the outlet of the tube

1. Name of poster session you would like to submit to: **Fluid Mechanics**

2. Abstract Title: **Research of the vertical falling film behavior in the scrubbing-cooling tube**

2. Presenting author and all co-authors, with affiliations and emails for each:

Yifei Wang, East China University of Science and Technology, wangyf@ecust.edu.cn

Xin Peng, East China University of Science and Technology, Y10150127@mail.ecust.edu.cn

Liucheng Yan, East China University of Science and Technology, 413733589@qq.com

Guangsuo Yu, East China University of Science and Technology, gsyu@ecust.edu.cn

Fuchen Wang, East China University of Science and Technology, wfch@ecust.edu.cn

3. Abstract: **As seen in Page 1 and 2.**

4. Keywords: **falling film, film velocity and thickness, numerical simulation, large Reynolds numbers, Ultrasound Doppler Velocimetry**

---

* Corresponding Author: Tel: +86 21 6425 2522. Fax: +86 21 6425 1312. E-mail: wangyf@ecust.edu.cn

541674 Fluid Mechanics of Two-Phase Flows: Concentrated Suspension of Non-Spherical and Deformable Particles

Abstract:

Research Interests:

Multiphase flows are present in many industrial, environmental, and biomedical applications including the flow of crude oil, the dispersion of smokestack plumes, and the movement of cells in the human body. The presence of the second phase has a marked effect on the fluid mechanical behavior of the mixture. The current understanding of multiphase flows is considerably less developed than their single-phase counterparts, and many phenomena still need to be studied. The focus of my research program is the investigation of the behavior of multiphase flows and the study of the underlying physics involved at different flow conditions. In particular, I am interested in the experimental study of suspensions in Newtonian and non-Newtonian fluids, mixtures with high solid concentrations, and flows of deformable and non-spherical particles.

My complementary research experience in pilot-scale equipment, microfluidics, and multiphase flows uniquely enables me to develop a research program at the macro and micro scale with applications ranging from biomedicine to environmental and industrial processes. Specific areas of interest include the characterization of blood flow, and its response to disease, which may aid in the development of mechanical valves used in heart surgery; measurements of the radial distribution and flow of red blood cells in capillaries, for the development of point-of-care diagnostic tools for diseases like malaria and sickle cell; and the determination of two-phase coupling of deformable particles in dense pipe flow for the development of mathematical models for fiber flow in industrial plants. The overarching goal is to provide further insight into the interaction between phases and how this affects the flow mechanics in different regimes. To develop this research program, I will leverage my work to date that includes solid-liquid and liquid-liquid flows in a wide variety of experimental configurations including turbulent pipe flows, fluidized beds, and droplet-based microfluidics.

Teaching Interests:

I have been inspired by great professors throughout my scientific education. It is my goal to encourage students in that same way. I have been involved in teaching and mentoring since my undergraduate years. During my graduate studies, I engaged in recitation responsibilities in addition to the required duties for teaching assistants. I have also taken formal training to teach in the STEM disciplines, including a course on "College Science Teaching" at University of Michigan, and the yearlong course "Institute for Development on Engineering Faculty" at University of Florida. Throughout my teaching and mentoring endeavors, I plan to incorporate inclusive teaching techniques to promote an effective learning environment. I am comfortable teaching any undergraduate course in the Chemical Engineering curriculum, and I am especially interested in the fluid and heat transport, transport phenomena, and introduction to chemical engineering courses. In addition, I hope to develop specialty courses on particulate flow, data analysis, and multiphase flow at both graduate and undergraduate level.

541822 Active Soft Matters and Soft Interfaces

Abstract:

Research Interests:

My main research interest is in biological active soft material focusing on 1) colloidal motion near and on active biological interfaces and 2) hydrodynamic interaction of active colloids with soft surfaces. I have spent my research career advancing state of art techniques to image colloids such as bacteria in 3-d and to quantitatively determine the mechanical properties of complex and heterogeneous soft matters such as living cells. In addition to successfully developing state of art techniques, I am very interested in implementing these techniques to explain the physics of many complex active systems. Previously, implementing a robust 3-d particle tracking method using Digital Holographic Microscopy, I have shown that the hydrodynamic forces induced by the solid surfaces and ow shears strongly alters the ability of the wild-type *Escherichia coli* to tumble. I have also studied how the hydrodynamic and physiochemical boundary condition at interfaces alter bacterial motility. In this line of research, I have focused on the motion of *Pseudomonas* cells on the oil-water interfaces and their encounter phenomena with oil droplets. I am also very interested in the motion of colloids over membranes of active soft matters such as living cells. It has been well stablished that the mechanical structure of the cells and their mechanosensing ability play important roles in their biological functions. Tracking the 3-d orientation and motion of gold nanorods on the plasma membrane using state of art Dark Field Microscope, I am trying to measure its viscoelastic response, bending modulus, and stiffness. Throughout my research carrier I have planned my future research to address two main issues.

I. Physical interaction of bacteria with planktonic and sessile life style with bio-inspired interfaces:

In this research area, I aim at fundamental understanding of the relation between complex mechanical structure of the surfaces that covers the internal organs of the human body and bacterial behavior near and on these surfaces. Mucous membranes of digestive, respiratory, and urogenital tissues which establish the rst barrier to limit the invasion by bacteria, have a complex structures with diverse length scales. Furthermore, the mucosal environment in proximity of these membranes have a complex viscoelastic characteristic. These complex membrane structure and the viscoelasticity of mucus can strongly aect bacterial locomotion, surface attachment, and biolm formation. My main goal in this line of research is to study these interactions using well controlled lab on a chip platforms and advanced imaging techniques.

II. Rheological characterization of active biological interfaces:

Many biological soft materials are far from equilibrium. Many of these active materials are uids or deformable solids with complex viscoelastic responses. This complexity is partially due to their heterogeneous and convoluted structures and partially is due to their activity. Therefore, wide range of mechanisms with wide range of length scales need to be explored to explain mechanical behavior of active soft materials. Over the years, many instruments and techniques have been developed to study soft materials. For example, active and passive microrheology, atomic force measurement (AFM), rheometry, or micro pipet aspiration have been utilized to study rheology of living cells and have provided treasured knowledge in cell mechanics. My research in this eld aims to extend active and passive microrheology to nanometer length scales and microseconds time scale to explore active phenomena closer to molecular motor levels.

Teaching Interests:

I strongly believe that engineering students should learn how to encounter and internalize new concepts in their academic lives to be prepared to face dierent challenges and problems in their professional lives as researchers or engineers. My teaching philosophy therefore is while learning new course topics, students should also practice to teach themselves. While I was a M.Sc. student in Aerospace Engineering and also during my Ph.D program in Mechanical Engineering, I had to learn great deal about microbiology to perform my research on bacterial locomotion. Later on as a postdoc researcher in the Chemical Engineering Department, I had too teach myself cell biology. These experiences showed me the value of the self teaching. I intend to engage the students in the teaching process by assigning smaller projects to them. Students will need to present the way they approach the problem and teach what they learned to their classmates.

Following my teaching philosophy, I plan to engage research in my teaching. Performing experiments, analyzing data, or surveying literature can help students to get involved in advanced topics and higher education experiences. Having research experiences can form a strong learning platform for students. Exploring research environment shows students how they can employ course materials in solving research problems. Although performing specic research project in a short period of time could be challenging I have great experiences in the Research Experience for Undergraduate (REU) program. Through Summer REU Programs, I successfully trained 6 undergraduate fellows to perform experiments, analyze data, interpret the result and present the research during a short period of time.

I have established a solid background in Chemical and Mechanical Engineering and in the eld of Biophysics. At this interface between engineering and biophysics, my research leverages many disciplines including advanced imagining, cell mechanics, and advanced rheology. These experiences will allow me to teach courses like Transports, Fluid Mechanics, Biotechnology, and BioFluid. I am also very interested to develop new interdisciplinary courses to contribute to the department curriculum.

541841 Experimental and Numerical Studies on the Micromixing Process in Novel Reactors with Multiphase System

Abstract:

Research Interests:

As a PhD candidate, my goal is to become a professor who devotes himself to research and education purpose in related chemical engineering fields. My current research including: novel reactor based on high gravity technology such as rotating packed bed (RPB) and rotor-stator reactor (RSR), numerical simulation of multiphase system and reactive flow, micromixing process and turbulence model.

Rotating Packed Bed (RPB) is an apparatus consisting mainly of a packed rotator and a fixed casing. The basic principle of RPB is to create a high-gravity environment via the action of centrifugal force as so called "Higee". The fluids going through the packing of RPB are spread or split into very fine droplets, ligaments, and thin films by the strong shear, resulting in a significant intensification of micromixing and mass transfer between the fluid elements. The rate of mass transfer between gas and liquid in RPB is 1–3 orders of magnitude larger than that in a conventional packed bed, allowing a dramatic reduction of reaction time. Since RPB was invented, it has been successfully applied in polymerization, absorption, distillation and preparation of nanomaterials due to its high mass transfer and micromixing efficiency. Rotor-stator reactor (RSR) is also a novel Higee device which is evolved from the RPB but employs an internal of rotor-rings and stator-rings in lieu of the packing in an RPB. Due to the unique structure, RSR can not only significantly intensify micromixing and mass transfer like RPB, but also exhibits advantages over RPB, such as better liquid distribution, less fouling, no gas cut-short, and so on. RSRs have been widely studied for their performance in the intensification of mass transfer and micromixing as well as for the preparation of ultrafine particles and emulsion.

In addition, studying the micromixing controlled process is also the main content in my PhD projects. Micromixing performance of reactors is the controlling factor for fast reactions in polymerization, pharmaceutical and crystallization applications. In recently years, combined with computational fluid dynamics (CFD), the micromixing process in reactors could be investigated comprehensively. Due to the conflict between the scale of the micromixing and the grid size of Finite Difference Method (FDM), it's difficult to building a proper micromixing algorithm. Therefore, numerical simulation of practical applications related to micromixing processes is still challenging.

CFD technique has been widely used in academic researches and industrial applications when it comes to fluid flow. Turbulence modeling as a fundamental part of CFD has great effect on the flow characteristic and mass transfer process in numerical simulation. Due to the computational cost and grid requirement of Direct Numerical Simulation (DNS) and Large Eddy Simulation (LES), many one- and two-equation turbulence models have been developed based on Reynolds-Averaged Navier-Stokes (RANS) and Boussinesq approximation. However, certain deficiencies in the validation accuracy of the models still exist, especially in complex flow conditions, such as strong rotation and multiphase flow (typical flow characteristic in RPB). Proper turbulence model is needed to increase the accuracy of flow simulations for mildly separated flows, flows with rotation and curvature effects, and flows with surface roughness.

Current PhD Dissertation:

"Experimental and numerical studies on the micromixing process in novel reactors with viscous fluids"

Research Experience:

Under the supervision of Prof. Chen Jian-Feng (Academician of Chinese Academy of Engineering, Beijing University of Chemical Technology), studying the micromixing controlled process, practical applications such as polymerization and alkylation process in RPB and microreactors, experimentally and numerically.

Under the supervision of Prof. Ramesh K. Agarwal (William Palm Professor of Engineering, Washington University in St. Louis), I finished a Joint doctoral program. I studied a new turbulence model-Wray-Agarwal model, including rotation/curvature and surface roughness corrections, and implemented it

Teaching Interests:

As mentioned above, my goal is to become a professor for research and education purpose in related chemical engineering fields. I was a teaching assistant in the courses of "Chemical Engineering Principles", "Transport phenomena" and "Chemical Reaction Engineering". Besides, based on the my research background of chemical engineering and CFD, I became proficient in numerical calculation of multiphase flow, reactive flow and high performance computing.

Future Direction:

As mentioned in "Reaction engineering: Status and future challenges (Dudukovic M P. Chemical Engineering Science, 2010, 65(1): 3-11.)", reactor miniaturization provides opportunities for scale-up in parallel, and quantification of multiphase flow fields reduces the risk of scale-up. Novel reactor plays a significant role in process intensification. My future direction will focus on the design and optimization of Higee reactors and microreactors in multiphase system using multiscale method and visualizaiton technology, including Finite Difference Method, Lattice Boltzmann Method or even Molecular Simulation combined with validation experiments.

Besides, with the rapid development of numerical method, new models and algorithms are needed (such as mass transfer model and turbulence model) for complex flow simulation in order to obtain better accuracy and computational efficiency, which is my future direction as well.

Selected Publications:

Ouyang, Y., Xiang, Y., Zou, H. K., Chu, G. W., & Chen, J. F. (2017). Flow characteristics and micromixing modeling in a microporous tube-in-tube microchannel reactor by CFD. *Chemical Engineering Journal*, 321, 533-545.

Ouyang, Y., Zou, H. K., Gao, X. Y., Chu, G. W., Xiang, Y., & Chen, J. F. (2018). Computational fluid dynamics modeling of viscous liquid flow characteristics and end effect in rotating packed bed. *Chemical Engineering and Processing: Process Intensification*, 123, 185-194.

Ouyang, Y., Wang, S., Xiang, Y., Zhao, Z., Wang, J., & Shao, L. (2018). CFD analyses of liquid flow characteristics in a rotor-stator reactor. *Chemical Engineering Research and Design*, 134, 186-197.

Gao, X. Y., Chu, G. W., **Ouyang, Y.**, Zou, H. K., Luo, Y., Xiang, Y., & Chen, J. F. (2017). Gas Flow Characteristics in a Rotating Packed Bed by Particle Image Velocimetry Measurement. *Industrial & Engineering Chemistry Research*, 56(48), 14350-14361.

Li, W. L., **Ouyang, Y.**, Gao, X. Y., Wang, C. Y., Lei, S., & Xiang, Y. (2018). CFD analysis of gas-liquid flow characteristics in a microporous tube-in-tube microchannel reactor. *Computers & Fluids.*

Presented at conferences:

Ouyang, Y., Xiang, Y., Gao, X. Y., Zou, H. K., Chu, G. W., & Chen, J. F. Micromixing efficiency in a rotating packed bed with non-Newtonian fluids. in: *The 17th Congress Of The Asian Pacific Confederation Of Chemical Engineering*

Ouyang, Y., Xiang, Y., Gao, X. Y., Zou, H. K., Chu, G. W., & Chen, J. F. Micromixing efficiency optimization of premixed liquid distributor of rotating packed bed by CFD. in: *The 8th Global Chinese Chemical Engineers Symposium*

543209 Multiscale Computation of Miscroscale Fluid Dynamics in Porous Materials

Abstract:

Research Interests:

A grand challenge of our generation is the procurement of clean sustainable energy and water, while minimizing environmental footprints. In meeting this challenge, *porous materials occupy a very central role.* As geologic rocks, they are the very medium for geothermal energy, CO_2 sequestration, unconventional hydrocarbon, and groundwater remediation. They also serve as key components in Li-batteries and fuel cells. *The scientific focus of my research is to understand and develop capabilities to predict and control multiphase fluid dynamics in porous media at the micron scale and to translate this knowledge to large-scale engineering systems.* My approach involves: (i) characterizing porous materials via multiscale imaging techniques that combine microscopy, spectroscopy, and optics, (ii) simulating fluid dynamics at the microscale to optimize design and operational controls, and (iii) scaling microscopic predictions to large engineering systems via multiscale multiphysics computation. I focus my attention on short- as well as long-term solutions to key problems facing our energy and water security, such as: CO_2 sequestration, geothermal energy, oil/gas recovery (shale), ground water, and fuel cells.

Postdoctoral Projects:

- *"Multiscale computation of multiphase fluid dynamics in porous materials for rapid fluid-structure optimization"*
- *"High-resolution imaging of hydro-thermo-mechanical properties of geologic porous materials (e.g., shales)"*

Advisor: Prof. Hamdi Tchelepi, Energy Resources Engineering Department, Stanford University, CA

PhD Dissertation:

- Title: "Modeling single-phase flow and solute transport across scales"

  Focus:
  1) Implications of CO2-mineral-water interactions on leakage in geologic carbon storage
  2) Multiscale methods for bridging microscopic physics to macroscopic observations

  Advisor: Prof. Matthew Balhoff, Petroleum & Geosystems Engineering Department, University of Texas at Austin, TX

Research Experience:

My career is a mixture of chemical and petroleum engineering, computational physics, applied mathematics, and hydrology. However, the question I have always deeply cared about is our energy, water, and environmental security and the role porous materials play to provide it. My undergraduate training was in chemical and petroleum engineering. But my graduate research drove me heavily into applied mathematics, computational physics, and high performance computing. I developed new mathematical methods for modeling reactive flow and transport in porous media at a variety of scales (micron to meter). Simultaneously, I collaborated closely with microfluidic experimentalists to validate/enrich my models. In my postdoc, I developed multiscale techniques to accelerate multiphase flow simulations, which enabled fluid-structure optimizations in a variety of applications (electrodes, membranes, oil recovery). I also developed experimental imaging techniques for mapping hydro-thermo-mechanical properties of geomaterials (e.g., oil/gas shales). This involved training in hyperspectral imaging, microscopy (SEM, FIB, fluorescence), FTIR/Raman spectroscopy, and optical spectrophotometry to name a few. My modeling background gave a unique way of approaching experimentation, in which I could identify data most useful to modelers. The combination of experiments and modeling also enabled me to identify important opportunities for research and allowed me to communicate and collaborate with several colleagues across various universities and national laboratories (PNNL, LBNL, Sandia).

Teaching Experience:

Philosophy. Students graduating from universities today face the biggest global challenges in the water, energy, and environmental sectors since the beginning of human history. My contention is that quantitative literacy in basic sciences as well as creativity is what equips them to solve these problems in the future, which should be given equal weight in their education.

University of Texas at Austin. I was a teaching assistant for the undergraduate Numerical Methods and Programming course (60 students). I designed and graded homework, held in-class programming sessions, and gave review lectures. I always looked forward to the office hours where I could interact and help some of the more reserved and struggling students, who refrained from asking questions in class. I received an evaluation score of 4.2/5, among highest for the course. I was also an undergraduate research mentor of three summer interns to develop a computer model for non-Newtonian flow in porous media. I taught programming, non-Newtonian fluid mechanics, and encouraged collaboration and discussions to promote creativity and teamwork. In the end, the project finished one month ahead of schedule and one of the students joined our group.

Stanford University. I co-instructed the graduate Multiphase Flow in Porous Media course (22 students) taught by my postdoctoral advisor. I prepared lecture notes, taught in-class, and designed homework. Even though no formal evaluation was conducted, both times I received very positive feedback from several students and my advisor.

Teaching Interests/Future Direction:
As a faculty I will apply my expertise in porous media science to develop transformational solutions to provide sustainable energy and water for future generations. Among other directions, I am interested in the following areas:

Fuel cells and battery optimization. Fuel cells and batteries are very promising technologies for harnessing renewable energy. They can also be used for in-situ conversion of fossil fuels to cleaner hydrogen. They consist of several porous components. Device-level optimizations ensure membrane-electrode-assemblies that have maximal power output (minimal transport/ohmic losses). Experimental and modeling techniques (μ-tomography, domain decomposition) that incorporate microstructural effects allows optimizing design and operational controls.

CO_2 storage & enhanced oil recovery. While we transition from fossil energy to renewables, it is important to continue providing energy for a growing global population while simultaneously managing carbon. Geologic carbon storage coupled to enhanced oil recovery is an attractive option. The problems: (a) safety issues regarding CO_2 leakage, and (b) poor displacement of heavy hydrocarbons. Fluids with optimal rheological/interfacial properties can be identified through microscopic fluid dynamics simulations followed by their synthesis in the lab. This leads to custom-designed fluids for a given rock/reservoir type, cutting-down on conventional trial-and-error experimentation. The outcome is of interest to oil companies and regulatory agencies.

Sustainable shale gas development. Two biggest concerns in shale gas development are: (a) how to minimize water usage, and (b) how to increase production without inducing seismicity. While natural gas remains the cleanest form of fossil energy, these public concerns must be addressed to avoid irreversible environmental damage. The high heterogeneity of these geomaterials further complicates analysis. Micro/nanofluidic experiments and modeling coupled to image-based characterization (microscopy/spectroscopy) of shales hold immense potential for increased sustainability.

I am also very interested in designing fouling-resistant membranes, implications of biomass growth at river-aquifer interfaces on contaminant transport and climate change, increasing the efficiency of geothermal energy production and storage, drug delivery in vascular networks (some of my PhD work focused on flow and transport in general graph-based networks).

Selected Publications:

1. Mehmani, Y., Tchelepi, H.A., "Multiscale computation of pore-scale fluid dynamics: two-phase flow." Journal of Computational Physics, (in prep).
2. Mehmani, Y., Tchelepi, H.A., "Multiscale computation of pore-scale fluid dynamics: single-phase flow." Journal of Computational Physics, (in review).
3. Mehmani, Y., Tchelepi, H.A., "Minimum requirements for predictive pore-network modeling of solute transport in micromodels." Advances in Water Resources, (2017).
4. Mehmani, Y., Burnham, A.K., Vanden Berg, M.D., Tchelepi, H.A., "Quantification of organic content in shales via near-infrared imaging: Green River Formation." Fuel, (2017).
5. Mehmani, Y., Burnham, A.K., Tchelepi, H.A., "From optics to upscaled thermal conductivity: Green River oil shale." Fuel, (2016).
6. Mehmani, Y., Burnham, A.K., Vanden Berg, M.D., Gelin, F., Tchelepi, H., "Quantification of kerogen content in organic-rich shales from optical photographs." Fuel, (2016).
7. Mehmani, Y., Balhoff, M.T., "Eulerian network modeling of longitudinal dispersion." Water Resources Research, (2015).
8. Mehmani, Y., Balhoff, M.T., "Mesoscale and hybrid models of fluid flow and solute transport." Reviews in Mineralogy and Geochemistry, (2015).
9. Mehmani, A., Mehmani, Y., Prodanović, M., Balhoff, M.T. "A forward analysis on the applicability of tracer breakthrough profiles in revealing the pore structure of tight gas sandstone and carbonate rocks." Water Resources Research, (2015).
10. Yang, X., Mehmani, Y., Perkins, W.A., Pasquali, A., et al. "Intercomparison of 3D pore-scale flow and solute transport simulation methods", Advances in Water Resources, (2015).
11. Mehmani, Y., Balhoff, M.T., "Generalized semi-analytical solution of advection-diffusion-reaction in finite and semi-infinite cylindrical ducts." Int. Journal of Heat and Mass Transfer, (2014).
12. Mehmani, Y., Oostrom, M., Balhoff, M.T., "A streamline splitting pore-network approach for computationally inexpensive and accurate simulation of species transport in porous media," Water Resources Research, (2014).
13. Oostrom, M., Mehmani, Y., Romero-Gomez, P., Tang, Y., Liu, H., Yoon, H., Kang, Q., et al. "Pore-scale and continuum simulations of solute transport micromodel benchmark experiments." Computational Geosciences, (2014).
14. Altman, S.J., Aminzadeh-Goharrizi, B., ... Mehmani, Y., ... (alphabetical order) "Chemical and hydrodynamic mechanisms for long-term geological carbon storage." The Journal of Physical Chemistry C, (2014).
15. Mehmani, Y., Balhoff, M.T., "Bridging from pore to continuum: a hybrid mortar domain decomposition framework for subsurface flow and transport." SIAM Journal of Multiscale Modeling and Simulation, (2014).
16. Mehmani, Y., Sun, T., Balhoff, M.T., Eichhubl, P., Bryant, S., "Multiblock pore-scale modeling and upscaling of reactive transport: application to carbon sequestration." Transport in Porous Media, (2012).
17. Sun, T., Mehmani, Y., Balhoff, M.T. "Hybrid multiscale modeling through direct substitution of pore-scale models into near-well reservoir simulators." Energy & Fuels, (2012).
18. Sun, T., Mehmani, Y., Bhagmane, J., Balhoff, M.T., "Pore to continuum upscaling of permeability in heterogeneous porous media using mortars." Int. Journal of Oil, Gas and Coal Technology, (2012).
19. Balhoff, M., Sanchez-Rivera, D., Kwok, A., Mehmani, Y., Prodanović, M., "Numerical algorithms for network modeling of yield stress and other non-Newtonian fluids in porous media." Transport in porous media, (2012).

542026 From Liquid Crystalline Solutions to Functional Materials

Abstract:

Postdoctoral Project: *Nanomaterial Building Blocks for Functional Materials,* Kavli Energy NanoScience Institute, University of California, Berkeley (Advisor: Professor A. Paul Alivisatos).

PhD Dissertation: *Morphology of Carbon Nanotube Liquid Crystalline Solutions: Insights into Tactoids and Columnar Phase*, Department of Chemical and Biomolecular Engineering, Rice University (Advisor: Professor Matteo Pasquali)

Research Interests:

Anisotropic particles such as carbon nanotubes are known to self-assemble into ordered phases at high concentrations due to entropic reasons. My PhD research in Professor Matteo Pasqualiâ€™s laboratory was focused on studying the phase behavior and morphology of carbon nanotube liquid crystalline solutions. I studied the formation of spindle-shaped nematic droplets (known as tactoids) coexisting with the isotropic phase and developed a method for measuring the line tension of lyotropic system of rod-like particles. Furthermore, I developed the necessary techniques to use SANS and SAXS to characterize higher order liquid crystalline phases formed in length-polydisperse solutions of carbon nanotubes in chlorosulfonic acid to study the lateral ordering of the particles in solution, which is essential for fluid-phase processing of macroscopic materials such as fibers. The alignment of carbon nanotubes within the fiber structure provides unprecedented mechanical and electrical properties in the macroscale. This enables us to use these fibers as the backbone material for fabricating functional fibers with sensing and actuating properties. As a postdoctoral researcher in Professor Paul Alivisatosâ€™s laboratory, I focused on colloidal synthesis of nanoparticles with unique optoelectronic properties. At the individual particle level, these nanoscale materials can serve as sensitive sensors yet once integrated with other materials can further form into active devices of unconventional functionalities at the micro- and macro-scale. Leveraging these unique properties across a spectrum of length-scales leads to the development of emerging concepts in science and technology. Example applications can range from pico-newton biological sensors to smart fibers and fabric that offers new capabilities including sensing, energy generation and storage, light emission, localized data processing and communication. As an independent PI, I aim to study the colloidal synthesis of nanoparticles with sensing properties, active assembly of such nanoparticles into ordered phases, and ultimately fabricating macroscale functional materials using nanomaterial building blocks to bridge the fundamental research and applications for developing functional active materials.

Teaching Experience: During my Ph.D., I served two semesters as a teaching assistant (TA) for the transport phenomena I (CHBE 401) and chemical engineering lab (CHBE 433) courses. I was awarded the 2012 best teaching assistant award by the department of Chemical and Biomolecular Engineering at Rice. On my third year, I was selected to be the Deanâ€™s teaching assistant which is a position offered to one graduate student in each class to work closely with a professor on a preferred course. I chose to work with one of our professors on the thermodynamics (CHBE 411) course. This position offered me the opportunity to interact more with the students, teaching as a guest lecturer in some classes, and designing homework and exam problems. In addition to my teaching experiences, I volunteered several times to teach one-day courses on STEM topics to young underrepresented girls in local science festivals in Houston, TX. Throughout my PhD, I also had the opportunity to mentor undergraduate (both through Rice and NSF-REU program) and high school students. I had the privilege to mentor an outstanding undergraduate student who will be joining as a graduate student at the University of Pennsylvania this coming fall.

Teaching Interests: My teaching interests include the courses in the area of nanotechnology. In addition, I am also interested in teaching fluid dynamics, colloidal and interfacial phenomena, rheology, and thermodynamics.

542430 Modeling across Disparate Spatiotemporal Scales – Enabling Answers to Grand Engineering Challenges

Abstract:

Research Interests:

My braoder research interest is to conduct theoretical/simulation studies to develop a fundamental understanding about transport phenomena at the solid/metal surface or interface relevant to energy and electronics applications. My training in graduate and postdoctoral research is in developing surface mass transport models across disparate spatiotemporal scales and will serve as the foundation for pursuing my future research interests. Surface transport has ubiquitous applications in Chemical Engineering, from thin film growth to heterogeneous catalysis. Alongside traditional fields there are emerging areas of interdisciplinary nature, like the plasma-material interface in a fusion reactor where surface transport plays an important role and the underlying physics is mostly unexplored. The structural and thermomechanical response of the plasma-facing materials that are exposed to the harsh plasma environment inside the nuclear reactor and damaged by the high fluxes of helium atoms produced by nuclear fusion – is one of the major limitations toward realizing nuclear fusion. Considering its future potential and current impediments in realizing that potential, nuclear fusion is widely recognized as one of the grand challenges for engineering and applied science in the 21st century. Another grand challenge in the field of energy application is to improve the solar cell efficiency for making solar energy economically viable; an idea for enhancing efficiency, which is worth exploring, involves developments of nanocrystalline active photovoltaic materials and integrating these nanostructures into a system that can transmit the energy into a circuit. Transferring core Chemical Engineering knowledge to another field to address such grand challenges is an exciting research proposition for me.

Theory, computation, and experiment taken together to probe the structure, chemistry, and function of new and complex materials is a comprehensive approach to address such a grand problem. However, connecting an atomic-scale (fusion reaction) or quantum-scale phenomenon (photon absorption and emission of charge carriers in photovoltaics) to experimentally/operationally relevant spatial and temporal scales will require O(300 Million years) wallclock time of atomic-scale simulations, using current state-of-the-art supercomputing facilities. And modeling/simulation challenges will increase by several folds if we make the models realistically accurate by taking materials heterogeneity and imperfection into account. Systematic development of atomistically-informed multiscale mass and energy transport model(s) can bridge the gap between atomistic modeling and experimental data, and my research will focus precisely on the development of such an approach.

Teaching Interests: I am comfortable with teaching any of the core Chemical Engineering undergraduate and graduate courses, including courses of interdisciplinary nature. The only exception is courses which require substantial background in bioengineering or the life sciences. I will particularly be interested in teaching transport phenomena and fluid mechanics to graduate students. In addition to core courses, based on my academic and research experience, I would like to design two specially structured courses for first-year graduate students, one on *Mathematical and Statistical Methods for Chemical Engineering (including design of experiments)*: an introductory graduate-level course to support the needs of students interested in experimental research, and another one on *Computational Methods in Materials Science*: an introductory graduate-level course to prepare engineering students for computational research in the interdisciplinary arena of materials science and engineering.

Postdoctoral Project: *"Modeling of Plasma-Surface Interactions and Radiation Damage Dynamics*;" Advisor: Prof. Brian D. Wirth, University of Tennessee Knoxville and Oak Ridge National Laboratory.

PhD Dissertation: *"Analysis of External-Field-Driven Surface Stabilization and Patterning*;" Advisor: Prof. Dimitrios Maroudas, University of Massachusetts, Amherst.

Research Experience:

My academic and research career path has traversed through many fields of Chemical Engineering and covered the entire spectrum of simulation scales, from the atomic level to the chemical plant scale. Nevertheless, my graduate and postdoctoral research have been focused on development of multiscale models, hierarchically parametrized coarse-grained models, using atomistic information through molecular-dynamics (MD) simulations and/or first-principles density functional theory (DFT) calculations, which are experimentally validated, for surface mass transport in diverse systems of interest and applications. For example, a part of my Ph.D. thesis research was focused on developing a model to facilitate external-field-driven complex nanopattern formation strategies starting from single-layer islands of a conducting material on crystalline conducting substrates, including current-driven formation of nanowires. In my postdoctoral research, I focus on developing an atomistically informed model to predict the structural and morphological evolution of plasma-facing material components in the fusion reactor. My current projects involve extensive collaborations with research institutes and national laboratories including close collaborations with experimentalists from Oak Ridge and Sandia National Laboratories. Also, as a part of a DOE-funded Scientific Discovery through Advance Computing (SciDAC) project, I have acquired proficiency in high-performance computer simulations in leadership computing facilities. Before joining the UMass Amherst Chemical Engineering Ph.D. Program, I have worked on a collaborative research project with Honeywell UOP to develop system identification and model predictive control strategies for an industrial fluidized bed polymerization reactor.

Teaching Experience:

I have extensive teaching experience, mostly in the capacity of teaching assistant (TA) for undergraduate and graduate level courses, including some laboratory courses. I have TAed Chemical Engineering Analysis (graduate level), Chemical Reactor Engineering (graduate level), Introduction to Statistics and Design of Experiments (undergraduate level) with regular responsibilities of holding office hours, grading homework assigfnments, and preparing solution sets for assignments and exams. In addition to these, I have guest-lectured a graduate-level Chemical Reactor Engineering course at UMass Amherst (I was nominated for the departmental TA award by the students). Moreover, for Introduction to Statistics and Design of Experiments offered to ~650 freshman students, I had special responsibility as the head TA to host regular training sessions for 32 other TAs for the course. Finally, I have mentored graduate students in the research groups where I conducted my Ph.D. research and currently conducting my postdoctoral research.

Publications:

1. **D. Dasgupta**, R. Kolasinski, L. Du, D. Maroudas, and B. D. Wirth, "Modeling of nano-fuzz formation in helium-ion-irradiated tungsten," (Manuscript in preparation).
2. **D. Dasgupta**, A. Kumar, and D. Maroudas, " Complex pattern formation from current-driven dynamics of single-layer homoepitaxial islands on crystalline conducting substrates," *Surface Science* **669**, 25-33 (2018).
3. A. Kumar, **D. Dasgupta**, and D. Maroudas, "Complex pattern formation from current-driven dynamics of single-layer homoepitaxial islands on crystalline conducting substrates," *Physical Review Applied* **8**, Article No. 014035, 14 pages (2017).
4. A. Kumar, **D. Dasgupta**, and D. Maroudas, "Surface nanopattern formation due to current-induced

homoepitaxial nanowire edge instability," *Applied Physics Letters* **109**, Article No. 113106, 4 pages (2016).

5. A. Kumar, **D. Dasgupta**, C. Dimitrakopoulos, and D. Maroudas, "Current-driven nanowire formation on surfaces of crystalline conducting substrates," *Applied Physics Letters* **108**, Article No. 193109, 5 pages (2016).
6. L. Du, **D. Dasgupta**, and D. Maroudas, "Weakly nonlinear theory of secondary rippling instability in surfaces of stressed solids," *Journal of Applied Physics* **118**, Article No. 035303, 14 pages (2015).
7. L. Du, **D. Dasgupta**, and D. Maroudas, "Stabilization of the surface morphology of stressed solids using simultaneously applied electric fields and thermal gradients," *Journal of Applied Physics* **116**, Article No. 173501, 12 pages (2014).
8. L. Du, **D. Dasgupta**, and D. Maroudas, "Stabilization of the surface morphology of stressed solids using thermal gradients," *Applied Physics Letters* **104**, Article No. 181901, 4 pages (2014).
9. **D. Dasgupta** and D. Maroudas, "Surface nanopatterning from current-driven assembly of single-layer epitaxial islands," *Applied Physics Letters* **103**, Article No. 181602, 4 pages (2013).
10. **D. Dasgupta**, G. I. Sfyris, and D. Maroudas, "Current-driven morphological evolution of single-layer epitaxial islands on elastic substrates," *Surface Science* **618**, L1-L5 (2013).

    - This work was highlighted by the editor and a perspective article on this work was written by Prof. Yannis G. Kevrekidis in the same issue of the journal.

11. G.I. Sfyris, **D. Dasgupta**, and D. Maroudas, "The effect of a thermal gradient on the electromigration–driven surface morphological stabilization of an epitaxial thin film on a compliant substrate," *Journal of Applied Physics* **114**, Article No. 023503, 13 pages (2013).
12. **D. Dasgupta**, G. I. Sfyris, M. R. Gungor, and D. Maroudas, "Electromigration- driven complex dynamics of void surfaces in stressed metallic thin films under a general biaxial mechanical loading," *Journal of Applied Physics* **112**, Article No. 083523, 10 pages (2012).
13. **D. Dasgupta**, G. I. Sfyris, M. R. Gungor, and D. Maroudas, "Surface morphological stabilization of stressed crystalline solids by simultaneous action of applied electric and thermal fields," *Applied Physics Letters* **100**, Article No. 141902, 4 pages (2012).
14. **D. Dasgupta** and S. C. Patwardhan, "NMPC of a continuous fermenter using Wiener-Hammerstein model developed from irregularly sampled multi-rate data," *IFAC Proceedings Volumes* **43**(5), 637-642 (2010).
15. U. Sarkar, **D. Dasgupta**, T. Bhattacharya, S. Pal, and T. Chakroborty, "Dynamic simulation of activated sludge based wastewater treatment processes: Case studies with Titagarh Sewage Treatment Plant, India," *Desalination* **252**(1-3), 120-126 (2010).
16. S. Pal, U. Sarkar, and **D. Dasgupta**, "Dynamic simulation of secondary treatment processes using trickling filters in a sewage treatment works in Howrah, West Bengal, India," *Desalination* **253**(1-3), 135-140 (2010).

542454 Engineering Nanoscale Materials and Interfaces for Sustainable Energy and Chemical Processes

Abstract:

Research Interests:

Interfaces are critical elements of many technologies and chemical processes, particularly in the areas of clean energy and efficient separations. My proposed research focuses on the rational, molecular level design of functional interfaces and materials by **bringing together cutting-edge tools and concepts from interface science, electrochemistry, and nanotechnology**. These studies will enable technologies to address critical needs, from energy efficient water purification, to next generation batteries based on earth-abundant chemistries.

For decades, colloid and interface science has been driven by equilibrium experiments and ideas. Yet, many important chemical processes, like electrochemical energy conversion, inherently operate under far-from-equilibrium conditions, where equilibrium intuition breaks down. These processes are dynamic, controlled by cooperative molecular interactions which span many nanometers in distance, and often hinge on fleeting metastable states. The spatiotemporal scales on which these processes operate **remain inaccessible to molecular simulation, and few experimental techniques can directly measure dynamic, molecular level processes at interfaces**. This presents a major challenge for targeted design of surfaces and nanoscale materials.

I propose to address this challenge by **leveraging my perspective and expertise in molecular force spectroscopy, electrochemistry, and advanced microscopy to measure and control dynamic processes at non-equilibrium interfaces.** I am particularly excited to explore how ionic and electronic transport can be controlled at solid-liquid and liquid-liquid interfaces using dynamic electric and magnetic fields.

My proposed research program will address the following aims:

1. Engineering confined solid-liquid interfaces for energy storage and conversion.

   Understanding how nanoscale confinement influences electrochemical reaction dynamics, barriers, and pathways will enable advanced electrolytes and interfaces for next generation batteries.

2. Creating electroactive nanomaterials to achieve energy efficient separations.

   Studying how dynamic electric fields drive molecular transport, adsorption, and charge transfer in nanoporous materials will enable efficient CO_2 adsorption, water purification, and enantiomeric separations.

3. Controlling the dynamics of soft materials and interfaces with magnetic fields.

   Magnetic interactions are conventionally neglected in colloid science. Designing magnetic soft materials to exhibit field dependent interface energies will permit magneto-responsive emulsions for separations, phase transfer catalysis, and water treatment.

Research Experience:

I apply quantitative experiments and systematic tuning of molecular structures to establish design principles for materials and interfaces. During my Ph.D., **I developed deep expertise in colloid and interface science, electrochemistry, and materials synthesis.** In my primary project, I combined surface forces and electrochemical methods to discover that more than 99.99% of the ions that comprise ionic liquids behave as neutral pairs (1). This contrasted prior expectations, which suggested that most ions were freely dissociated. From these results, we proposed a new way of envisioning concentrated electrolytes to guide the design of advanced ionic liquids and battery electrolytes (1–3). I used analogous approaches to assess the role of electrostatics in mussel protein adhesion. We found that dynamic ionic bonds are critical for marine bio-adhesion and unveiled new stable functionalities to expand the toolbox for designing underwater adhesives (4).

During my postdoctoral tenure, **I broadened my expertise towards nanomaterials synthesis, nanofabrication, and advanced microscopy.** At Stanford, I studied diamond nanomaterials for single-molecule fluorescence. A poor understanding of nucleation has been a major hurdle facing the synthesis of high quality diamond for many applications. By mapping the nucleation landscape using molecular seeds, we found that diamond nucleation during chemical vapor deposition is a non-classical process with a barrier that is four orders of magnitude lower than prior classical estimates (5). We also developed methods to pattern fluorescent color center crystal defects in diamond lattices with nanoscale precision. Both thrusts address key challenges to enable using diamond in molecular sensing, bio-imaging, and quantum computing.

Selected Publications:

1. **Gebbie MA,** Valtiner M, Banquy X, Fox ET, Henderson WA, Israelachvili JN. Ionic liquids behave as dilute electrolyte solutions. *PNAS* 110, 9674–9679 (2013). **DOE Basic Energy Sciences Research Highlight, June 2013**
2. **Gebbie MA**, Dobbs HA, Valtiner M, Israelachvili JN. Long-range electrostatic screening in ionic liquids. *PNAS* 112, 7432–7437 (2015).
3. **Gebbie* MA**, Smith AM, Dobbs HA, Lee AA, Warr GG, Banquy X, Valtiner M, Rutland* MW, Israelachvili* JN, Perkin* S, Atkin* R. Long range electrostatic forces in ionic liquids. *Chem. Commun.* 53, 1214–1224 (2017). *** - Co-corresponding authorship, Invited Feature Review**
4. **Gebbie* MA**, Wei* W, Schrader AM, Cristiani T, Waite JH, Israelachvili JN. Tuning underwater adhesion with cation-π interactions. *Nature Chemistry* 9, 473–479 (2017). * - Equal contribution, **Highlighted in Chemical & Engineering News, February 2017**
5. **Gebbie MA**, Ishiwata H, McQuade PM, Petrak V, Taylor A, Dahl JE, Carlson RMK, Fokin AA, Schreiner PR, Shen ZX, Nesladek M, Melosh NA. Experimental measurement of the diamond nucleation landscape reveals classical and non-classical features. *In revision.*

Teaching Interests:

I have broad interests in teaching core chemical engineering courses, especially transport, fluids, thermodynamics, interfaces, and polymers. I am also excited to develop a graduate course on applying concepts from colloid and interface engineering to sustainability and energy science.

542583 Designing Functional Soft Materials Using Anisotropic Fluids

Abstract:

Research Interests:

My broad research interest is to develop functional materials guided by fundamental approaches from soft-matter physics, interfacial phenomena and statistical mechanics. Of particular importance will be the studies on anisotropic fluids in the context of self-assembly, formation/dynamics of topological defects, flow and interfacial phenomenon. The immediate research themes that I intend to pursue are: 1) Developing new sensor platforms for detection and mechanistic studies of bio-entities including viruses, bacteria, erythrocytes and other biomolecules 2) Achieving directed active transport of colloids and biological cargo using anisotropic fluids and 3) Control and functionalization of topological defects in confined ordered phases as building blocks for complex materials. In the longer term, I intend to lead a collaborative research group advancing our understanding of fundamental physical and biological mechanisms involved in complex soft-material systems that are either- a) selectively responsive to a range of targeted species or b) engineered via mesoscopic building blocks with anisometric interactions resulting in unique properties.

Teaching Interests:

My teaching philosophy regardless of the particular subject matter will be guided by methods to maximize student involvement in the learning process and enabling students to effectively communicate science. I have received training at Georgia Tech as a teaching assistant (TA) on courses including physical chemistry of polymer solutions and a laboratory course on fundamental concepts of materials for which I received the outstanding teaching assistant award. I am comfortable teaching transport phenomena, thermodynamics, reaction kinetics, polymer physics, and colloidal/ interfacial phenomena. I intend to develop a course on scaling concepts for soft matter physics.

Selected Publications (out of 11):

1) **K. Nayani**, P. Rai, N. Bao, H. Yu, M. Mavrikakis, R. J. Twieg, N. L. Abbott. Liquid Crystals with Interfacial Ordering that Enhances Responsiveness to Chemical Targets. *Advanced Materials*, 1706707 (2018).

2) R. Mangal, **K. Nayani**, Y. K. Kim, E. Bukusoglu, U. M Córdova-Figueroa, N. L. Abbott. Active Janus Particles at Interfaces of Liquid Crystals. *Langmuir*, 33, 10917 (2017).

3) **K. Nayani**, J. Fu, R. Chang, J. O. Park, M. Srinivasarao. Using chiral tactoids as optical probes to study the aggregation behavior of chromonics. *PNAS*, 114, 3826 (2017).

4) **K. Nayani**, R. Chang, J. Fu, P. Ellis, A. F. Nieves, J. O. Park, M. Srinivasarao. Spontaneous emergence of chirality in achiral lyotropic liquid crystals confined to cylinders. *Nature Communications*, 6, 8067 (2015).

542319 Deep Learning in Dynamic and Complex Systems

Abstract:

Transport Phenomena at Microscopic Scales and their Effects on Macroscopic Scale Processes

1) Research Interests:

Research Interests

Understanding transport phenomena is important to many processes both living and non-living. New technology in fabricating nano- and micro-structures allows us to control the transport process from the nano- and micro-levels, and therefore the whole process at the macroscopic scale. This strong connection between what happens at the microscopic scale and the macroscopic scale opens many opportunities for scientists in the transport phenomena field to make current processes more efficient, create new processes, and gain new fundamental knowledge. Understanding what happens at microscopic scales will help us reach the next step of more advanced technology. I want to focus my academic career on studying the transport phenomena at microscopic scales and their effects on macroscopic scale processes.

Research Experience

During my Ph.D. research, in collaboration with NASA, I studied the interfacial forces and fluid dynamics in a wickless heat pipe in microgravity at both macroscopic and microscopic scales. On Earth, gravity dominates over Marangoni force. But in microgravity, Marangoni and capillary forces control fluid flow in the heat pipe. At the macroscopic scale, the strong Marangoni flow created a flooded region instead of dry-out near the heater end of our heat pipe on the ISS. At the microscopic scale, we found that the large Marangoni force in the region near the heater end generated instability in the liquid film on the wall surface which led to condensation at an anomalously high temperature. We developed a numerical model to understand this phenomenon and found that oscillation can enhance net evaporative heat transfer if a short period of condensation exists within each oscillation cycle. The effects of capillary and Marangoni forces continued to amaze us, both in their microscopic dynamics and their macroscopic impact on heat pipe performance. Analyzing the interferometry pattern, we discovered a fluid flow pattern inside the heat pipe that mimics "rip currents" along shore lines. This rip current pattern helps to release the heat supplied to the heater end. To reduce the effect of the Marangoni force, we added a second component to the working fluid and increased the heat transfer coefficient significantly. This knowledge about interfacial phenomena in microgravity is extremely important to understanding fluid mechanics within microsystems on Earth where the effect of gravity is small.

My Ph.D. training gave me a solid background in transport processes with a focus on interfacial phenomena and thin film. It taught me the importance of combining experimental work, analytical and theoretical analysis, and simulation to solve problems and gain new understanding. I am a scientist who is always on the lookout for opportunities to expand my skills and knowledge, and I'm excited about unexpected results because that is where we discover new phenomena.

To expand my knowledge in transport processes I joined a team of thermal engineers at Corning Inc. in 2017. I got the opportunity to do both fundamental research and process development work. On the research side, I'm leading a mid-sized project to study the heat transfer process during glass forming. On the development side, I work in a large team of thermal management experts to study, develop, and improve full-size production processes. My experience in industry gave me a practical view of transport phenomena and showed me how fundamental knowledge gained from academic research is used in real production. I also learned the importance of using simulation to guide experimental work. These experiences show how the transport phenomena at microscopic scales affects full-size processes and how important it is to understand what happens at microscopic scales.

Future Direction

In the next step of my career, I will focus on understanding transport phenomena at both microscopic and macroscopic scales in different directions. I will incorporate both experimental and simulation work in my research.

The first direction is thermal management. With the continuing increase in heat dissipation of computer chips and power electronics, the need for more efficient electronic cooling systems is critical. Oscillating heat pipes (OHP) have the advantage over traditional heat pipes in that the flows of liquid and vapor are separated. The oscillatory flow of liquid and vapor plugs also offers an additional mechanism for heat transfer. However, the inherent instability of the OHP keeps it from being used widely. Also, only less than 10% of heat transfer in an OHP is from latent heat. I'd like to develop micro-structures inside the OHP so that the heat transfer process can be manipulated and improved at the microscopic scale, leading to better performance at the macroscopic scale. The knowledge gained from this research can be used to improve many other processes including other forms of heat pipe, boiling heat transfer, and transport processes in thin films.

The second direction is microfluidic bio-separation. Separation and accurate identification of highly valuable biomarkers that are present at micro-and nano-levels becomes critical for disease diagnostics. Traditional separation devices require large sample volumes, are relatively time-consuming, or limited to only several separation techniques. Microfluidic devices are becoming promising alternatives that can overcome these shortcomings of traditional devices. The efficiency of the microfluidic devices depends strongly on the nano- and micro-structures of the stationary phase materials and the ligands immobilized. This is a perfect niche to utilize my background in chemistry, biochemistry, and transport processes. The knowledge gained in this direction will also complement my work in thermal management.

Beyond these directions, I continue looking for other systems where the transport phenomena at microscopic scales affect the macroscopic processes.

2) Teaching Interests:

Everyone has a unique, inner creativity and the desire to better themselves. I believe that the job of a teacher is to create an environment that fosters this creativity and desire. The role of teacher requires that we guide each student in the direction that is right for that student. We should make sure students have a strong fundamental understanding of the subject and know how to apply it to solve the real world problems they'll face in their careers. Technology changes at the speed of light. The ability to self-motivate, independently learn new material, and apply it to the problem at hand is more important than ever.

I come from a family of teachers, where I've been taught to see the unique path that every student will take. I've been building my teaching philosophy with this idea at the forefront, and I've developed methods to help me follow it. RPI gave me a chance to learn what works, and what doesn't, so I could refine my methods and hone my skills. At RPI, I was a teaching assistant for five courses in the Chemical Engineering Department, the Biology Department, and the Mathematical Sciences Department. I mentored seven undergraduate and graduate students from RPI and three students visiting from other universities. They each went down a different road in life. Some work in different industries and some are pursuing their Ph.D. programs at Ivy League schools. Each of these students had a passion for how they wanted to be successful, and enabling students to accomplish their goals by following their paths was one of the most rewarding experiences I had as a mentor.

3) Proposal Writing and Funding Management Experience

- Leading and managing funding for a mid-sized research project at Corning Inc.
- Writing a proposal with Prof. Joel Plawsky to form collaboration between RPI and Corning Inc.

4) Industry Experience

- Corporate Thermal Engineer, Corning Inc., Corning, NY (2017 – present).
- Research project: Leading a mid-sized research project on studying the heat transfer process during glass forming.

o Conducted literature search to develop initial scope and methods.

o Formed a team of engineers from different departments to develop decision analysis and project timeline.

o Working with the team to design, build, and test the equipment and run experiments.

o Building analytical model in COMSOL to analyze results and create final model.

- Development project: Working on one of the largest and most visible projects in the company.

o Working with top researchers and engineers in the field to develop new production processes for a critical customer.

o Developing large scale thermal management techniques to achieve desired material properties.

o Conducting experiments and proposing solutions for process improvement.

o Collaborating with people from different backgrounds and skill sets from the plant, division, and corporate engineering teams.

5) Ph.D. Dissertation

"The Constrained Vapor Bubble heat pipe: interfacial forces at both macroscopic and microscopic scales".

Under the supervision of Prof. Joel L. Plawsky and Prof. Peter C. Wayner, Jr.

- 2018 Karen and Lester Gerhardt Best Thesis Prize.
- 2018 William N. Gill Prize for Excellence in Dissertation Research.

6) Publications

("*" denotes equal contribution.)

1. **Nguyen, T.**, Plawsky, J., Kundan, A., Wayner Jr., P., Chao, D., Sicker, R. *Rip currents in microgravity. (Submitted to PNAS).*
2. **Nguyen, T.**, Yu, J., Plawsky, J., Wayner Jr., P., Chao, D., Sicker, R. *Spontaneously oscillating menisci: Maximizing evaporative heat transfer by inducing condensation,* Int. J. Thermal Sciences, 128 (2018) 137-148.
3. Plawsky*, J., **and Nguyen***, T. *Wickless heat pipes in microgravity,* Physics Today, 70, 9, 82 (2017). **Invited article.**

4. Kundan*, A., **and Nguyen*, T.**, Wayner Jr., P., Plawsky, J., Chao, D., Sicker, R.*Condensation on highly superheated surfaces: wickless heat pipe operation in microgravity*, Phys. Rev. Lett., 118 (2017) 094501.**PRL Editors' Suggestion. Featured in PHYS.ORG.**
5. **Nguyen, T.**, Kundan, A., Wayner Jr., P., Plawsky, J., Chao, D., Sicker, R.*Experimental study of the heated contact line region for a pure fluid and binary fluid mixture in microgravity*, J. Colloid Interface Sci.,488 (2017) 48-60.
6. **Nguyen, T.**, Kundan, A., Wayner Jr., P., Plawsky, J., Chao, D., Sicker, R. *The effect of an ideal fluid mixture on the evaporator performance of a heat pipe in microgravity*, Int. J. Heat Mass Transfer, 95 (2016) 765-772.
7. **Nguyen, T.**, Kundan, A., Wayner Jr., P., Plawsky, J., Chao, D., Sicker, R. *Effects of cooling temperature on heat pipe evaporator performance using an ideal fluid mixture in microgravity*, Exp. Therm. Fluid Sci., 75 (2016) 108-117.
8. Yu, J., **Nguyen, T.**, Plawsky, J., Wayner Jr., P., Chao, D., Sicker, R. The effect of cooling temperature on the interfacial forces in a wickless heat pipe in microgravity. *(In preparation).*
9. Kundan, A., **Nguyen, T.**, Wayner Jr., P., Plawsky, J., Chao, D., Sicker, R. *Arresting the phenomenon of heater flooding in a wickless heat pipe in microgravity*, Int. J. of Multiphase Flow, 82 (2016) 65-73.
10. Popov, Yu., Mokhov, V., **Nguyen, T.**, *Direct alkylation of amines by alcohols with catalysis by nickel and cobalt nanoparticles*, Inter-university collection of scientific articles, Higher Attestation Committee (HAC) of the Ministry of Education and Science of the Russian Federation (MESRF), Apr. 2011.
11. Mokhov, V., Popov, Yu., **Nguyen, T.**, *The reaction of alkyl halides with 2, 4-pentanedione catalyzed by copper nanoparticles*, Inter-university collection of scientific articles, HAC of MESRF, Apr. 2011.

541909 Laboratory of Interfaces, Flow and Electrokinetics (LIFE)

Abstract:

Research Interests:

Interfaces, by definition, are responsible for heterogeneity in materials that in turn gives rise to critical properties like surface tension, wetting, colloid-colloid interactions and many others. My group will study interfacial phenomena of liquid-liquid, liquid-solid and liquid-liquid-solid interfaces coupled with multiphase flow and ion transport. The overarching goal of my group is to utilize the expertise in interfacial science for applications in personal care, food science, enhanced oil recovery, and energy storage. More specifically, I will focus on the following research directions:

(1) Advanced nanoemulsions: Nanoemulsions are nano-sized droplets of oil suspended in water or vice-versa. Due to their small size, nanoemulsions provide an efficient way to deliver hydrophobic nutrients, be it orally or topically, by dissolving the nutrients in the oil phase. In my group, we will develop techniques that are specifically tailored to create double nanoemulsions, i.e. nanodroplets inside nanodroplets (oil-in-water-in-oil and water-in-oil-in-water nanoemulsions). Furthermore, we will design energy friendly routes to create pickering nanoemulsions, i.e. nanoemulsions stabilized by particles. The pickering emulsions are attractive since they do not rely on harsh surfactants and instead utilize the particles for stabilizing emulsions. Our efforts will advance food science and personal care research through the formulation of ready-to-use therapeutic food as well as functional skin care products. To this end, we plan to collaborate with expert nutritionists and dermatologists to develop relevant and stable formulations.

Prior work: During my doctoral research under Prof. Patrick S. Doyle and Prof. T. Alan Hatton at MIT, I proposed and experimentally validated new scaling relations to predict average droplet size of nanoemulsions. I also built a general platform to prepare nanoemulsions using energy friendly routes and demonstrated their utility in nano-hydrogel synthesis and in efficient hydrophobic drug dissolution.

(2) Flow in porous media: Despite the fact that the crude oil sources are rapidly decreasing worldwide, it is estimated that only 20 â€“ 40% oil is recovered from the oil fields. Therefore, recovery of the trapped oil from the oil fields is crucial for our mounting energy needs. In my group, we will study the evolution of oil-water interface past different geometric features such as array of different-shaped obstacles and/or bifurcations. Specifically, we will explore the effects of wetting and relative viscosity by developing invasion-percolation algorithms and multiphase-flow simulators. In addition to the oil recovery applications, these results will also of value in micro-reactor design, flow inside lungs, and for trapping / releasing drops inside microfluidic devices. We will also collaborate with leading scientific groups that study flow in porous media experimentally to validate and strengthen our theoretical capabilities.

Prior work: During my doctoral research, I examined formation of liquid pockets during immiscible liquid-liquid displacement over an obstacle / patterned wall where I theoretically predicted the criterion for entrapment. The theoretical predictions were validated through microfluidic experiments performed by a collaborator.

(3) Electrokinetics for energy and the environment: Electrochemical capacitors, or supercapacitors, are an active area of research since they are useful in high power applications such as in trains and elevators. My group will develop theoretical and experimental tools to investigate the effects of electrolyte choices (including flowable electrodes) on the charging and discharging of supercapacitors. We will also study electrokinetic remediation, i.e. the process where toxic waste is removed by applying a voltage, with an emphasis on the effect of surfactants on ion transport.

Prior work: In my post-doctoral position with Prof. Howard Stone at Princeton University, I developed general relations for charge storage and capacitance for asymmetric electrolytes near a charged surface. I am currently conducting experiments to validate these results for use in supercapacitor applications. Furthermore, I am also investigating the utility of bi-disperse carbon suspensions for flowable supercapacitors.

Teaching Interests:

I am passionate about teaching and have had several teaching responsibilities such as instructor for special topics in graduate fluid mechanics (Princeton), instructor for undergraduate fluid mechanics (MIT), teaching assistant for undergraduate heat and mass transfer (MIT), and instructor of high school physics and mathematics (iitjeelectures.com). My philosophy while teaching a course is to provide students with multiple learning opportunities. I like to engage them through short conceptual quizzes, supplementary video lectures to clarify recurring questions (sample from my course at MIT here: http://bit.ly/2tN87UX), and in general promoting an interactive class environment. I am prepared to teach the following courses in future: (a) undergraduate / graduate fluid mechanics (b) undergraduate / graduate heat and mass transfer and (c) undergraduate reaction engineering.

Selected publications:

- Badruddoza, A.Z M,* Gupta, A.*, Myerson, A.S., Trout, B.L., and Doyle, P.S. (2018), *Advanced Therapeutics*, 2018
- Gupta, A.*, Lee, H.*, & Doyle, P. S. (2017), *Physical Review Fluids*, 2, 094007.
- Gupta, A.*, Badruddoza, A. Z. M*, & Doyle, P. S. (2017), *Langmuir*, 33, 7118-7123.
- Lee, H.*, Gupta, A.*, Hatton, T. A., & Doyle, P. S. (2017), *Physical Review Applied*, *7*(4), 044013.
- Gupta, A., Narsimhan, V., Hatton, T. A., & Doyle, P. S. (2016), *Langmuir*, *32*(44), 11551-11559.
- Gupta, A., Eral, H. B., Hatton, T. A., & Doyle, P. S. (2016), *Soft Matter*, *12*(11), 2826-2841
- Gupta, A., Eral, H. B., Hatton, T. A., & Doyle, P. S. (2016), *Soft Matter*, *12*(5), 1452-1458.
- Gupta, A., & Roy, S. (2013), *Chemical engineering journal*, *225*, 818-836.

* denotes equal contribution

542103 Achieving Next-Level Transport with Soft Matter and Interfaces

Abstract:

Research Interests:

We often rely on the use of gradients, such as thermal gradients for power generation or concentration gradients for chemical processing. Our command of transport behavior, however, is quite rudimentary when compared to the sophisticated transport found in the natural world (e.g. ion pumping against concentration gradients in cell membranes, aerosol production at the ocean-air interface, water harvesting by beetles). Inspired by nature, I intend to advance our understanding of transport by studying gradients in the presence of colloids, polymers, interfaces, biomatter, and active matter.

My goal as a professor is to achieve next-level transport capabilities across temperature, concentration, and momentum gradients by focusing on three areas:

- Molecular engineering of interfaces using surfactants[1,2] and nanoparticles[3] for the development new drug delivery methods
- Phase-change heat transfer[1] and storage[4] for power generation/distribution and water harvesting
- Mass transport and fluid mechanics through complex porous media[5] and soft materials[6] to enhance groundwater remediation

I will take a multifaceted approach to study fundamental problems: a combination of novel experimentation with a particular focus on optical techniques and computer simulations at multiple scales. This provides both a top-down and bottom-up interpretation of physical phenomena, allowing us to develop rich physical insights. These insights will ultimately help us to better inform and develop technologies in the biotech, energy, and environmental spaces.

Selected References:

[1] **H.J. Cho**, J.P. Mizerak, E.N. Wang, "Turning bubbles on and off during boiling using charged surfactants," *Nature Communications*, 2015.

[2] **H.J. Cho**, V. Sresht, E.N. Wang, "Predicting Surface Tensions of Surfactant Solutions from Statistical Mechanics," *Langmuir*, 2018.

[3] C. Tian, J. Feng, **H.J. Cho**, S.S. Datta, R.K. Prud'homme, "Adsorption and denaturation of structured polymeric nanoparticles at an interface," in review.

[4] H. Kim, **H.J. Cho**, S. Narayanan, S. Yang, H. Furukawa, S. Schiffres, X. Li, Y. Zhang, J. Jiang, O.M. Yaghi, E.N. Wang, "Characterization of adsorption enthalpy of novel water-stable zeolites and metal-organic frameworks," *Scientific Reports*, 2016.

[5] H.K. Mutha, **H.J. Cho**, M. Hashempour, B.L. Wardle, C.V. Thompson, E.N. Wang, "Salt rejection in flow-between capacitive deionization devices," *Desalination*, 2018.

[6] **H.J. Cho**, M.P. Howard, N.B. Lu, S.S. Datta, "Cracking and healing of hydrogel suspensions," in preparation.

Teaching Interests:

Teaching is a core reason why I seek a faculty position. In the past, I have found great satisfaction mentoring numerous students from the high-school to PhD levels, guest lecturing, and attending teaching/mentoring workshops. In order to successfully inspire students and equip them for research or the workplace, a key aspect of my teaching approach will be to promote more interaction beyond the traditional lecture and to incorporate more open-ended assignments. This way, I can cater to different learning styles and instill a rich physical intuition of the subject matter. I would like students to invoke questions like, "what is happening physically?" rather than "which equation do I need to use to 'plug and chug?'" By incorporating discussions, writing assignments, and projects, I believe students will develop communication skills, metacognition, and creativity.

Classes that I would be naturally inclined to teach would be thermodynamics, physical chemistry, heat and mass transfer, and fluid mechanics. I would also be interested in developing courses in molecular simulation and imaging/microscopy techniques.

532281 Application of Gas Hydrate Slurry Relative Viscosity Models for an Advanced Hydrate Management Strategy

Abstract:

Research Abstract

Clathrate hydrates (or gas hydrates) are crystalline structures that trap small gas molecules such as methane, ethane and carbon dioxide in a three-dimensional hydrogen-bonded lattice of water cages. Gas hydrates usually form under high pressure and low temperature conditions. As such, they are considered a nuisance by the oil/gas industry as gas hydrates can plug flowlines. Over several decades, the oil/gas industry has injected Thermodynamic Hydrate Inhibitors (THIs), such as methanol and monoethylene glycol, to completely prevent the formation of gas hydrates in flowlines. However, this method has been proven to be uneconomical especially for maturing fields. In the mid-2000s, there was a paradigm shift in dealing with gas hydrates in flowlines. A new approach called "hydrate management" has been explored, whereby the formation of gas hydrates in flowlines is allowed but its properties are altered by injecting a small amount of Low Dosage Hydrate Inhibitors (LDHIs). The formation of gas hydrates in flowlines changes the flow properties of the fluids, as well as the flow pattern and pressure drop of the system. As a result, advanced knowledge of the rheological properties of gas hydrates slurries is necessary for safe and economic operation of oil/gas flowlines. We have recently developed a model (Majid-Wu-Koh model) for predicting the relative viscosity of a gas hydrate slurry [1]. This new model has been shown to be able to predict the relative viscosity of hydrate slurries for oil-dominated system across various conditions relatively well. The model has also been used for the prediction of pressure drops in large scale flowloop investigations. Limitations of the current relative viscosity model and areas for improvement have been identified. Finally, potential application of the knowledge of rheological properties of hydrate slurries to aid in advancing hydrate management in flowlines will be discussed. Based on the current success of the proposed model, it currently being implemented to industry software. The objective of this work, is to predict the pressure drop of oil/gas flowlines.

References

[1] A. A. A. Majid, D. T. Wu, C. A. Koh, "A Perspective of Rheological Studies of Gas Hydrate Slurry Properties," *Engineering,* 2018 (Accepted)

Research Interests: My primary research focused on understanding the transportability of gas hydrate particles in flowlines. In my post-doctoral and doctoral work, I have conducted both lab and large scale research. I conducted several experiments using lab scale high-pressure rheometer to investigate the viscosity change in the presence of gas hydrates. With the knowledge obtained from this investigation, I conducted investigations using a large/industrial scale flowloop in order to confirm my lab results can be translated to large scale facilities. In addition, in my doctoral work, I also conducted research in gas hydrate technology. In this research, I apply my knowledge on gas hydrates properties for engineering application: sea water desalination and natural gas transportation. My future research plan is to focus on understanding the transportability of gas hydrates in flowlines for different conditions: water- and gas-dominated system. This is an interesting area of research since there is no model for predicting the pressure drop for this system. Additionally, I would like to focus my research on thermodynamic and kinetic properties of gas hydrates. It can be said that the thermodynamic properties of gas hydrates are well understood but currently there is limited knowledge on the kinetic properties. This is a crucial knowledge for gas hydrates to be applied for engineering applications.

Teaching Interests:

Beyond my research work, I am also fortunate to obtain various teaching experiences. My first college-level teaching experience was during my senior year where I was an Instructional Aids for Separation Processes (junior level course) as well as a tutor for Chemical Engineering Reaction Kinetics (junior level course). While holding these positions, I was responsible for holding additional lectures for students. These additional lectures focused on students' critical thinking in problem solving. During my doctoral years, I was also a Teaching Assistant (TA)for Advanced Thermodynamics (graduate level course). From my teaching experience, I learned that it is necessary for the instructor to promote students interest in the subject. As such, as a TA, I always try to relate the materials that I teach in class with my research as well as real world applications. For this reason, my teaching interests are in Thermodynamics and Transport Phenomenon courses as these courses are closely related to my area of research. During my undergraduate years, I found out that these two courses are the most challenging courses. Therefore, I hope that by focusing on these courses, I can help students see the connection between their classroom knowledge with real world applications. I believe by doing this, it promotes interest in students.

542323 From Training in Polymer Physics to Developing Nonwovens for Advanced Applications

Abstract:

From Training in Polymer Physics to Developing Nonwovens for Advanced Applications

Behzad Nazari

Scientist at H&V and Research Affiliate/postdoc at MIT

PhD, PE Chemical Engineering, Polymer Scientist, Innovator and Educator

Research Interests:

As a scientist at Hollingsworth & Vose Company in Massachusetts, I get to use polymer physics and chemistry to develop nonwovens for air filtration applications. Throughout my training as a chemical engineer, filtration was always kept at a distance, but now it has become a realization for me how exciting it is to modify fibrous media in a way to be able to capture dust and to purify air. This particularly becomes an art when one needs to do it with minimized energy (low pressure drop across the media) for the most penetrating particles. Spinning processes and nonwovens are now definitely areas that I will be active in for the years to come.

I have taken my interest in controlling polymer physics and its effect on processing and final properties of the product to my current postdoctoral/research affiliate position at MIT with Prof. Greg Rutledge, where we seek to utilize structure-property relationship of polymer configuration and its molecular orientation to significantly alter the mechanical properties of polymeric fibers. For instance, by improving the molecular orientation and degree of crystallinity from gel-electrospinning ultra high molecular weight polyethylene, our group has successfully fabricated ultrathin fibers with high stiffness that rivals the commercial protective vest such as Spectra. By defining a fundamental connection between the predicted nanostructure and its functionality, we can expand much further on the capabilities of electrospun nanofibers to significantly impact the polymer processing and its application as a whole.

A second project I have been working on at MIT is embedding a shear thickening fluid within a nonwoven structure for protection applications. By using sub-micron fibers, capillary effects dominate the gravitational forces. This makes it difficult for the fluid to be extruded out of the system under its weight. Via capillary flow porometry and rheological measurements on the fluids, we now understand that the yield stress of the fluids play a significant role in keeping them inside the nonwoven pores.

Training students and researchers on classic topics of conformation and molecular dimensions of polymer chains remains challenging. As a postdoc at Penn State with my supervisor, Prof. Ralph Colby, I was able to assemble a 17-student team of superb undergraduate researchers spanning across different aspects of polymer science. These undergraduate students worked on the following projects:

1. Flow-induced crystallization of PEEK, nylon 66, iPP, and PET.
2. Rheology of ionic liquids/cellulose/chitosan solutions.
3. Investigation of polyolefin-based interpenetrating polymer networks for collecting oil spills.
4. Renewable lignocellulose-based foams via Pickering emulsion templates.

To provide these undergraduates with opportunities for continuing research experience, I submitted two separate proposals to the Fall 2017 University Park Research Experience for Undergraduates Funding Program through the College of Engineering at Penn State. These grants were successful in enabling two of our students to work on a project that aims to develop biomass-based films.

Through my Penn State experience, I came to realize that newcomers to the polymer field are easily attracted to studying relationships between structure and function in polymeric systems. In particular, our newly purchased rheometers made it easier for us to have our students experience and understand how conformational ordering under shear in a molten polymer can significantly change its crystallization rate and resulting crystal morphology.

My interest in "green materials" builds upon my Ph.D. work, undertaken at UMaine under the guidance of Prof. Doug Bousfield, which elucidated the interactions between rheology of cellulose nanofibers (CNFs) suspensions and their processability. I focused on the development of new applications for CNFs as an alternative platform for petroleum-based products. That work contributed to the feasibility of utilizing waste biomass, and therefore has huge environmental benefits. As the PI, I wrote a proposal entitled "Renewable lignocellulose-based foams via Pickering emulsion templates" to USDA's Fiscal Year 2017 Biomass Research and Development.

Teaching Interests:

During 2011, I was a fulltime teacher of Chemical Reaction Engineering, Numerical Methods, Unit Operations (+Lab), Aspen Plus Workshop, Heat Transfer (+Lab), and Fluid Mechanics (+Lab). At UMaine, I TAed in Biological Transport Phenomena. More recently at Penn State, I taught Polymer Rheology and Processing. At MIT, I also colectured in Physical Chemistry of Polymers. In the future, I would welcome teaching these courses, but I would also like to teach Polymer Physics, Chemical Engineering Thermodynamics and Transport Phenomena. In 2015, I published a textbook entitled "Numerical Methods for Process Engineers" with Shiraz University Press. I believe effective teaching empowers the students' ability to think in a critical way and create their own knowledge, and this ultimately benefits the research community. I also think other legitimate purposes of teachers include the need to make a useful contribution to society by helping induct younger human beings into the society's thrive process.

542415 Continuous Technology Platforms Enabled By Molecular Design of Disperse Multiphase Soft Matter

Abstract:

Research Interests:

Soft materials encompass a wide range of materials, such as polymers, colloids, emulsions, gels, whose properties arise from an interplay of processes across multiple scales and phases. Conquering the major challenges, especially in food, pharmaceuticals and healthcare in the 21st century requires novel technology platforms that can provide unprecedented properties and extraordinary functions, capable of exquisitely interfacing between engineering, chemistry, medicine and biology. The discovery and development of novel processing technologies not only bring about improved performance, but also enable innovative approaches of material manufacturing that meet economic, regulatory, and societal constraints. My vision for training a new generation of technology leaders is strongly motivated by the burgeoning need for better multi-scale understanding the structure - dynamic function relation of soft materials and also to develop novel processes for materials chemistry that enable production of such materials ranging in scale from nanocomposites to hierarchically structured polymer *particles* and *gels.*

As a faculty candidate, I am poised to leverage my expertise in designing and developing novel bottom-up approaches for soft materials with tunable architectures together with my experience in the use of innovative approaches at the interface of chemical engineering, polymer and materials science, and emulsion and interface science to have immediate impact in teaching, scholarship, professional service, and economic development. For example, in my teaching, I emphasize fundamental principles and relate underlying physics and chemistry to application. I provide the molecular basis for precise control of thermodynamic stability and molecular transport, as well as the tools for understanding structure-function relationships. My research group will be a recognized leader in (i) Developing technology platforms for scaleable, high-throughput processes for multiphasic soft matter systems; (ii) Design next generation intelligent hydrogel encapsulation systems for applications in biopharmaceutics and healthcare; and (iii) Establishing structure/function paradigms in polymeric materials/nanocomposites for manufacturing, formulation, and separation of food, pharmaceutical, and consumer products;

My poster will highlight the key aspects of some of my past and current professional development, including teaching experience (an advance undergraduate/graduate course and an industrial short course), grant proposal writing (high throughput emulsification and advanced characterization, advanced nutraceutical formulation, and biopharmaceutical stabilization), and professional service, along with an outline of future projects.

Teaching Interests:

Advanced separation and mass transfer, transport phenomena, process design, colloids and interface science

Selected publications (26 total, 14 first author, 1250 citations, H-index 13):

Abu Zayed Md Badruddoza, P. Douglas Godfrin, Allan S. Myerson, Bernhardt L. Trout, Patrick S. Doyle, Core-shell composite hydrogels for the controlled nanocrystal formation and release of hydrophobic soluble active pharmaceutical ingredients, *Advanced Healthcare Materials*,5 (2016) 1960-1968.

Ankur Gupta,Abu Zayed Md Badruddoza, Patrick S. Doyle, A general route for nanoemulsions synthesis using low energy methods at constant temperature, *Langmuir*, 33 (28) (2017)7118-7123.

Abu Zayed Md Badruddoza, Ankur Gupta,Patrick S. Doyle, Low energy nanoemulsions as templates for the formulation of hydrophobic drugs, *Advanced Therapeutics,* 1 (1) (2018) 1700020.

Abu Zayed Md. Badruddoza, Bikash Bhattarai, Rominder Suri, Environmentally friendly Î²-cyclodextrin â€“ ionic liquid polyurethanes modified magnetic sorbent for the removal of PFOA, PFOS and Cr(VI) from water, *ACS Sustainable Chemistry & Engineering,* 5 (10) (2017) 9223-9232.

Lilian C Hsiao, Abu Zayed Md Badruddoza, Li-Chiun Cheng, Patrick S. Doyle, 3D Printing of Self-Assembling Thermoresponsive Nanoemulsions into Hierarchical Mesoporous Hydrogel Scaffolds, *Soft Matter*, 13 (2017) 921-929.

Arpad Istvan Toldy, Abu Zayed M. Badruddoza, Lu Zheng, T. Alan Hatton, Rudiyanto Gunawan, Raj Rajagopalan, Saif A. Khan, Spherical Crystallization of Glycine from Monodisperse Microfluidic Emulsions, *Crystal Growth & Design,* 12 (2012) 3977-3782.

Reno A. L. Leon, Wan Wai Yew, Abu Zayed Md. Badruddoza, T. Alan Hatton, Saif A. Khan, Simultaneous spherical crystallization and co-formulation of drug(s) and excipient from microfluidic double emulsions, *Crystal Growth & Design,* 14, (2014) 140-146.

542505 Self-Assembly, Elasticity, and Rheology of Soft Materials

Abstract:

Research Interests:

Amorphous materials – including emulsions, foams, gels, pastes, glasses, etc. – have been, and may well remain, some of the most technologically and industrially relevant soft materials. The somewhat fuzzy boundary between fluid- and solid-like behaviors of these materials is largely what makes them so practically useful, but their intrinsic structural complexity has severely limited our ability to understand and prescribe how macroscopic behavior arises from microscopic properties. During my Ph.D. studies in Prof. David Weitz's lab at Harvard University I focused primarily on understanding the elasticity and solidification of amorphous materials. There I discovered that the liquid and solid regimes of quiescent, compressed emulsions are separated by a thermally induced "unjamming" instability of the solid: the solid falls apart when the thermal agitation of the droplets that compose it are strong enough to locally and continually yield it. The properties of emulsions that make this instability possible are also shared by foams and many pastes, and understanding how material failure via unjaming may affect the properties of these and other materials will require developing new, non-invasive experimental and analytical tools to study them. Thus far I have worked on integrating light scattering, MRI, microscopy, rheology, and computational tools to better understand this transition in emulsions, microgel suspensions, and colloidal glasses. More broadly, understanding and controlling the elasticity and flow behavior of soft materials is a long-standing fundamental problem with a myriad practical applications, and I am excited to continue my work in this area.

Introducing various kinds of microscopic order into soft materials is another, very powerful way to control the mechanical, rheological, electrical, and optical properties of these materials; however, despite an incredible diversity of bottom-up and top-down approaches that accomplish this in a laboratory environment, there is still a need for scalable methods that can organize colloids composed of inexpensive ingredients. Most recently, as a postdoctoral associate in Prof. Paul Chaikin's lab at New York University, I have focused on the self-assembly and frustrated crystallization of particles adsorbed on curved surfaces. However, I have also begun work on a new technology that will enable scalable colloidal self-assembly by *in situ* capacitive deionization of colloidal suspensions, and which I expect will become a major focus of my upcoming research. The process depends only on surface charging – a nearly universal property of colloidal particles – to induce ordering, using supercapacitive electrodes immersed in the fluid to modulate the ionic strength of the suspension: a process that can be combined with ordered templates for epitaxial assembly. Thorough deionization can dramatically affect the optical properties of colloidal suspensions, and precisely modulating the salinity of a suspension of oppositely charged particles can induce ordering and crystallization. While I am still working to catch up with state-of-the-art electrode fabrication techniques, these processes are mature and reliable enough to make it possible to print large-scale, locally-addressable electrodes. Combining them with large-scale, perfectly ordered templates, will make it possible to fabricate large-scale structured colloids.

Teaching Interests:

A person's ability to work well as part of a team is often a crucial part of a successful career. As a TA for graduate and undergraduate classes, as well as a student myself, I have found that encouraging students to work in groups helps the class cover more advanced material and helps students sharpen the basic social skills required to work well with others. I look forward to working with students to incorporate more collaborative and peer learning approaches to the classroom, and to help them become successful scientists and engineers.

I am also interested in helping more people understand how amazing science and engineering can be. I have found that explaining what makes the soft materials that surround us – foods, toiletries, plastics, digital displays, etc. – interesting and special is a concrete and appealing way to introduce people to physical and chemical concepts that they might otherwise have found to be alien or abstract. I would welcome the opportunity to develop a cross-disciplinary class aimed at reaching out to people who might not have though science or engineering was for them, and would be especially willing to work with high-school science teachers to help them broaden and enrich their lesson plans: I believe helping students see science and engineering as an integral part of their lives early on will spark their curiosity and make them more likely to become scientists and engineers themselves.

542518 New Frontiers in Materials Chemistry for Sustainable Energy Technologies

Abstract:

Andrew B. Wong, Postdoctoral Scholar, Stanford University, Thomas Jaramillo Group

Research Interests:

Development of new materials has driven society's progress. The goal of a sustainable energy future necessitates continued breakthroughs in materials development. I will explore new material chemistries that will become enabling technologies relevant to clean energy and to environmental sustainability by developing new catalytic and semiconducting materials. I will also explore novel catalytic systems. Particular reactions of interest include capture and utilization of CO_2, which is a vital area to develop in order create technologies for carbon negative emissions.

Research Experience:

Postdoctoral Scholar, Stanford University with Prof. Thomas Jaramillo, 2017 - Present

- Synthesis of novel 2D PdPt 'nanoclams' for electrochemical CO_2 Reduction (In Preparation)
- Study of the fundamental properties of PdPt alloys for CO_2 Reduction (In Preparation)
- Integration of microorganisms with electrochemical systems for overall CO_2 Reduction
- MOCVD of InGaN for PEC applications

Graduate Student Researcher, University of California, Berkeley, with Prof. Peidong Yang, 2011 - 2016

- Investigated light-induced halide migration in halide perovskite semiconductor materials and developed new cation exchange method to control halide migration (*JPCL, under revision,* 2018)
- Developed new synthetic methods for lead-free halide perovskite semiconductors and developed new methods to improve the stability of this material (*Nano Letters,* 2018)
- Discovered the first ever example of a crystalline material with ultralow thermal conductivity but high electrical conductivity (*PNAS,* 2017)
- Co-originated study to photosensitize non-photosynthetic microbes for the first time by bio-synthesis of CdS nanoparticles for novel cyborgian microbes (*Science*, 2016)

- Investigated atomically-thin two-dimensional hybrid perovskites, revealing novel optical and structural properties in ultrathin hybrid perovskites (*Science*, 2015)
- Demonstrated first hybrid perovskite nanorod array while demonstrating a novel anion exchange reaction to tune composition and maintain morphology, independently developed synthesis of $CH_3NH_3PbI_3$ nanostructures (*Nano Letters*, 2015)
- Developed cation exchange synthesis and fabrication strategy for CdS–Cu_2S core–shell nanorod array solar cells (*Nano Letters*, 2015)

Joint Undergraduate/Masters Student, University of Chicago, with Prof. Laurens Mets, 2009 - 2011

- Investigated hydrogen production using microorganisms

Selected Successful Proposals:

- DOE Basic Energy Sciences, FWP: 'Physical Chemistry of Inorganic Nanostructures'
- LAM Research Graduate Student Fellowship
- Department of Energy Office of Science Graduate Fellowship, Finalist, 2012

Teaching Interests:

In addition to my research career, I have had extensive opportunities to teach courses on diverse topics. Before beginning my PhD, I served as a teaching assistant for PHSC 13400: Global Warming, Understanding the Forecast for two quarters at the University of Chicago. In this course, I mentored non-science majors to understand scientific concepts in order to approach scientific problems.

As a PhD student, I served as a graduate student instructor for:

- CHEM 1A: General Chemistry I
- CHEM 3B: Organic Chemistry II
- CHEM 4A: Honors General Chemistry I

While teaching these courses, I encouraged students to work through problems by using probing questions. I also utilized active learning techniques while lecturing. In recognition of my teaching effectiveness, I was awarded an Outstanding Graduate Student Award after teaching CHEM 3B: Organic Chemistry, despite the fact that my research focus was in physical chemistry and inorganic nanostructures.

Additionally, during the final year of my PhD, I co-instructed and co-wrote the course materials for CHEM 98/198: Space: An Applied Role-Playing Game, which I co-designed with a fellow graduate student. In this course, we used a series of fun, imaginary scenarios set in outer space to teach students how to critically understand research claims, justify their own scientific claims using quantitative reasoning, and solve scientific problems in groups under time constraints in situations with no single "correct" answer. Although the course was by design highly unconventional, it was a highly unique and valuable teaching experience. Through course evaluations, students expressed that they felt that this unconventional teaching method allowed them to improve their scientific problem solving skills significantly as well as improve their ability to assimilate new scientific knowledge.

Future Plans:

My career goal is become a principle investigator at an academic institution that is at the forefront of developing sustainable energy technologies. I see my future research group as being based around: (1) the synthesis and utilization of novel semiconductor and catalytic materials and (2) the exploration of new, unconventional catalytic systems. I have a particular interest in CO_2 capture and CO_2 catalysis given the pressing need for carbon negative technologies. Consequently, I see an ideal academic institution as one that has potential for synergistic and collaborative research in semiconducting materials, catalytic materials, and applied systems involving the biotic-abiotic interface. Collaborations with theorists will have the potential to be mutually beneficial in these areas.

In the long term, I see success stories from my group's research as advancing a virtuous cycle based on: (1) applying knowledge to create new systems, (2) experimentally characterizing these systems, (3) interpreting these results, and (4) creating new knowledge. This virtuous cycle of knowledge creation can drive new innovation towards a more sustainable future.

Another critical part of this endeavor of knowledge creation is to mentor the students, who will perform the work. A major focus of mine will be mentorship of my students through research, so that they will join the next generation of scientists and engineers. I will seek out academic, governmental, and industrial collaborations that will create an interdisciplinary environment and enable students to pursue career trajectories that will match with their career goals while providing opportunities to enter the workforce as professional researchers.

Selected Publications:

(* Denotes equal contribution, papers in each section listed in chronological order)

A. Synthesis and Fundamental Properties of Semiconducting Halide Perovskites

1. C. G. Bischak*, **A. B. Wong***, E. Lin, D. T. Limmer, P. Yang, N. S. Ginsberg, Photoinduced phase separation in single-crystalline mixed halide perovskites, *JPCL*, (*Under Revision*), 2018.

2. **A. B. Wong***, Y. Bekenstein*, J. Kang, C. Kley, D. Kim, N. A. Gibson, D. Zhang, Y. Yu, S. R. Leone, L. W. Wang, A. P. Alivisatos, P. Yang, Strongly Quantum Confined $CsSnI_3$ Nanoplates: Lessons for Improved Stability, *Nano Letters*, **18**, 2060–2066, 2018.

3. W. Lee*, H. Li*, **A. B. Wong***, D. Zhang, M. Lai, Y. Yu, Q. Kong, E. Lin, J. J Urban, J. C. Grossman, P. Yang, Ultralow Thermal Conductivity in All-Inorganic Halide Perovskites, *PNAS*, **114**, 8693–8697, 2017.

4. L. Dou*, **A. B. Wong***, Y. Yu*, M. Lai, N. Kornienko, S. W. Eaton, A. Fu, C. G. Bischak, J. Ma, T. Ding, N. S. Ginsberg, L.-W. Wang, A. P. Alivisatos, P. Yang, Atomically Thin Two-dimensional Organic-inorganic Hybrid Perovskites, *Science*, **349**, 1518–1521, 2015.

B. Synthesis and Fabrication of Novel Nanostructured Solution Processable Optoelectronics

1. **A. B. Wong**, M. Lai, S. W. Eaton, Y. Yu, E. Lin, L. Dou, A. Fu, P. Yang, Growth and Anion Exchange Conversion of $CH_3NH_3PbX_3$ Nanorod Arrays for Light-Emitting Diodes, *Nano Lett.*, **15**, 5519–5524, 2015.

2. **A. B. Wong**, S. Brittman, Y. Yu, N. P. Dasgupta, P. Yang, Core–Shell CdS–Cu_2S Nanorod Array Solar Cells, *Nano Lett.*, **15**, 4096–4101, 2015.

C. Nanostructured Electrodes for Renewable Energy

1. **A. B. Wong**, C. Hahn, A. M. Spormann, T. F. Jaramillo, Synthesis and CO_2 reduction activity of Quasi-2D PdPt Nanoclams, (*In Preparation*).

2. L. Wang, S. A. Nitopi, **A. B. Wong**, J. Snider, A. C. Nielander, C. G. Morales-Guio, M. Orazov, D. C. Higgins, C. Hahn, T F. Jaramillo, Engineering Effective Strategies for Reducing Carbon Monoxide into Liquid Fuels by Copper Catalysts, *ACS Catalysis*, (*Under Revision*), 2018.

3. L. Zhang*, C. Liu*, **A. B. Wong**, J. Resasco, P. Yang, MoS_2-wrapped silicon nanowires for photoelectrochemical water reduction, *Nano Research, 8, 281–287, 2015.*

4. L. Zhang, K. Liu, **A. B. Wong**, J. Kim, X. Hong, C. Liu, T. Cao, S. G. Louie, F. Wang, P. Yang, Three-Dimensional Spirals of Atomic Layered MoS_2, *Nano Lett.*, **14**, 6418–6423, 2014.

D. Integration of Nanoscale Electrochemical and Biological Systems for Renewable Energy

1. **A. B. Wong**, F. Kracke, A. D. Antoniuk-Pablant, C. Hahn, A. M. Spormann, T. F. Jaramillo, PdPt for Stepwise CO_2 Reduction in Tandem with Microbial Communities, (*In Preparation*)

2. K. Sakimoto, **A. B. Wong**, P. Yang, Self-photosensitization of non-photosynthetic bacteria for artificial photosynthesis, *Science*, **351**, 74–77, 2016.

542522 Design and Fabricate Functional Materials for Biological and Energy Applications

Abstract:

Research Interests:

Materials have always been the foundation of human civilization that we associate Ages with them. My research seeks to promote the progress of materials science and apply its knowledge to advance health and to address energy related challenges. Designing and fabricating cutting-edge, unconventional functional materials to solve specific biological questions and exploit applications in energy fields will form the core of my future research. In particular, my research objectives will focus on three areas: 1) Designing nanomaterials to investigate fundamental biological mechanism; 2) Designing novel stimuli-responsive systems for controlled release; 3) Developing new strategy to fabricate functional materials for energy applications. I envision my research emerging at the interface of chemistry, physics, biology, energy and materials science.

During my research, I have also built wide but close collaborations and connections with both academic communities (universities and institutes in Asia, Europe and US) and industrial societies (both large companies, such as BASF, Moderna, and start-ups), which not only broaden my research horizon, but give me opportunities to identify the urgent problems and needs in industrial world, so that I can actively solve practical problems and benefit our society.

Research Experience:

My research has involved many fields of science and engineering. My research starts with molecular design, combines chemical and physical methods, as well as microfluidic technology, to fabricate advanced materials from sub-nm scale (ultrathin carbon nanofiber, polymer of intrinsic microporosity) to nanoscale (mesoporous materials, silicon nanowires, graphene, photonic crystals), to microscale (microgels, microcapsules), and to bulk (hydrogel), and applies them for certain applications in biomedical fields.

My formal research training started in Chemistry Departments at both Tsinghua University and Purdue University, in which I have studied organic synthesis, self-assembly, nanomaterials synthesis and biomedical applications of nanomaterials. After completing my PhD, I joined the School of Engineering at Harvard University working on microfluidics for producing functional materials, such as stimuli-responsive microcapsules for controlled release and bio-sensing applications.

Synthesis of Nanomaterials: With my chemistry background for molecular design and organic synthesis, I fabricated nanomaterials through rationally designing and synthesizing molecular building blocks with different functional groups, followed by investigating their self-assembly behavior, and finally using these self-assembled structures as templates and/or precursors to controllably prepare highly ordered mesoporous materials, ultrathin carbon nanofiber and pure single layer graphene. Besides nanomaterials fabrication, I also synthesized ionic liquids and incorporated them into 3D photonic crystal structure to develop chemical sensors with capability for naked-eye detection.

Bio-applications of Nanomaterials: My PhD research mainly focused on biomedical applications of nanomaterials. I used chemical vapor deposition methods to controllably prepare silicon nanowires (SiNWs) and graphene. By visualizing SiNWs with their intrinsic non-linear optical (NLO) signals, I investigated the impact of size and surface functionalities of SiNWs on their cellular binding and internalization. In addition, for the first time, I demonstrated that graphene protects metals from corrosion in biological environment and reduces immune responses toward metallic devices. Moreover, I worked with a collaborator to develop a new NLO imaging method based on highly sensitive transient absorption to characterize graphene and graphene oxide in bulk and in biological systems.

Microfluidics for Functional Materials Production: As a postdoc in Prof. David Weitz's group at Harvard University, I am engaged to produce novel functional materials using microfluidic techonolgy. By using glass capillary microfluidic devices, I efficiently encapsulated a variety of enzymes and antibodies within microcapsules with different shell structures. These encapsulated bio-actives can be well maintained for long-term storage without losing their biological activities and can be controllably released upon applying osmotic pressure changes. This encapsulation technique shows great potential in protection and delivery of enzymes and antibodies. Additionally, I prepared biosensors by encapsulating quantum dots within microcapsules for glucose detection. This study opened the possibilities of developing implantable biosensors with high sensitivity.

In addition to microfluidics, I also opened a new research direction in my post-doc lab by developing a versatile strategy to fabricate 3D porous conductive collectors using various inorganic, organic and metallic substrates, as lithium battery anodes. I demonstrate the performance, such as battery life and efficiency, of lithium batteries with these anodes is as good as the best reported results in the literature.

Future Direction:

As a research faculty member, I would like to continue the philosophy of designing and fabricating functional materials based on chemistry and microfluidics to explore and solve challenges in biology and energy fields. I will contribute my efforts to prepare functional materials with hierarchical structures and novel properties to meet requirements for studying both fundamental and practical problems.

One interesting direction is to further explore biomedical applications of nanomaterials, particularly, graphene. Due to its biocompatibility, excellent mechanic and electric properties, graphene is an ideal candidate to study cellular biomechanics and differentiation behavior of neuron stem cells under an external electrical field.

I am also anticipated to fabricate "smart" microcapsules for specific release. In comparison to the release triggered by osmotic pressure, the release of encapsulated actives triggered by specific chemical or biological species can be precisely controlled. The specific release can be achieved by introducing functional groups into shell materials, which can be degraded by adding specific additives. This precisely controlled specific release will have broad applications in both industry and biomedical fields.

Additionally, I am interested in designing functional materials for developing high-performance lithium batteries. A major problem related to lithium batteries is the formation of lithium dendrites. I want to develop new strategies to design novel anode structures and explore unique interfacial layer materials to suppress the formation of lithium dendrites to enhance the electrochemical performance of lithium batteries.

In addition to the above directions, due to the interdisciplinary nature of my research work, I look forward to establishing a strong and active collaboration with fellow faculties in science and engineering to explore new opportunities at the interface of chemistry, biology and materials science.

Teaching Interests:

Teaching is a very important part in academia, since teaching is not only transferring knowledge, but also a process of learning and inspiring. I am always eager to involve myself in teaching. My teaching experience came from teaching in classes as a teaching-assistant, and mentoring undergraduate and graduate students in the lab. During my PhD study at Purdue University, I was repeatedly chose by my department to teach several courses including both lectures and laboratory for students from Chemistry and Chemical Engineering Departments. My duties included lecturing undergrads, experiment demo, holding office hours, preparing course materials and exams. In addition, I mentored undergraduate students as a part of their undergraduate thesis, and one of these undergraduates received the undergraduate research award and graduated in Purdue Chemistry Honors Program with the Highest Distinction. I have also been mentoring more than 10 undergraduate and graduate students to conduct research programs in my labs at both Purdue University and Harvard University.

With my education and research background, I am well prepared to teach a broad range of courses in chemistry, chemical engineering and materials science at both undergraduate and graduate levels. I also envision instructing and encouraging students and postdocs to deliberate about questions and to conduct independent research studies in my lab.

Selective Peer-Reviewed Publications

-**27** peer-reviewed journal articles (**12** first-authored or corresponding-authored papers, H-index (Google Scholar, July 1, 2018): 14), **2** patents, **1** book chapter.

First-Author or Corresponding-Author Articles (†co-first author, *corresponding author)

W. Zhang†, L. Qu†, H. Pei, Z. Wu, J. Didier, D. Ingber, D. Weitz, Controlled Fabrication of Inhomogeneous Microcapsules for Biologics Release. (In Preparation).

H. Zhao, R. Ding, X. Zhao, Y. Li, L. Qu, H. Pei, L. Yildirimer, Z. Wu, **W. Zhang***, Graphene-based Nanomaterials for Drug/Gene Delivery, Bioimaging and Tissue Engineering. *Drug Discovery Today*, 2017, 22, 1302-1317. (Invited Review).

W. Zhang†, A. Abbaspourrad†, D. Chen, E. Campbell, H. Zhao, Y. Li, Q. Li, D. Weitz, Osmotic Pressure Triggered Rapid Release of Encapsulated Enzymes with Enhanced Activity. *Adv. Funct. Mater.*, In print, (Featured as inside front cover).

Y. Li†, **W. Zhang†**, et al., Rapid Assembly of Large Scale Transparent Circuit Arrays Using PDMS Nanofilm Shaped Coffee Ring. *Adv. Funct. Mater.*, 2017, 27, 1606045.

X. Xie†, **W. Zhang†**, A. Abbaspourrad†, et al., Microfluidic Fabrication of Colloidal Nanomaterials-Encapsulated Microcapsules for Biomolecular sensing. *Nano Lett.*, 2017, 17, 2015-2020.

J. Li†, **W. Zhang†**, et al., Highly sensitive transient absorption imaging of graphene and graphene oxide in living cells and circulating blood. *Sci. Rep.*, 2015, 5, 12394.

W. Zhang, S. Lee, et al., Use of graphene as protection film in Biological Environments. *Sci. Rep.*, 2014, 4, 4097.

W. Zhang, L. Tong, C. Yang, Cellular Binding and Internalization of Functionalized Silicon Nanowires. *Nano Lett.*, 2012, 12, 1002-1006.

W. Zhang, J. Cui, et al., A strategy of Producing Pure-Layer Graphene Sheets Based on a Confined Self-Assembly Approach. *Angew. Chem.*, 2009, 121, 5978-5982.

- Selected as **Hot Paper** by *Angewandte Chemie*
- Reported by *C&EN*, Latest News, "Graphene Via Self-Assembly". July 8 (2009)
- Reported by *Nature China*, Research Highlight, "Material chemistry: Casting graphene", DOI:10.1038/nchina.2009.147, July 22 (2009)

Reported by *NPG Asia Materials*, Research Highlight, "Graphene: Confined approach", DOI:10.1038/asiamat.2009.22, October 27 (2009)

W. Zhang, J. Cui, et al., Pyrrole containing ionic liquid as tecton for construction of ordered mesoporous silica with aligned polypyrrole nanowires in channels. *J. Mater. Chem.*, 2009, 19, 3962-3970.

W. Zhang, J. Cui, et al., Confined Self-Assembly Approach to Produce Ultrathin Carbon Nanofibers. *Langmuir*, 2009, 25, 8235-8239.

W. Zhang, Y. Li, et al., Electrochemical Polymerization of Imidazolum-Ionic Liquids Bearing a Pyrrole Moiety. *J. Polym. Sci. A: Polym. Chem.*, 2008, 46, 4151-4161.

542528 Radical-Bridged Dinuclear, Trinuclear and Metallacyclic Lanthanide Molecular Magnets

Abstract:

Research Interests:

Single molecule magnets (SMMs) are compounds that exhibit slow relaxation of magnetization. These molecules are approximately 1 nm in size. This size is approximately 100 times smaller than current nanoparticle-sized permanent magnets and represent the next generation of magnetic data storage materials. The quality of an SMM can be measured by the thermal energy barrier to reversal of magnetization (U_{eff}) and the magnetic blocking temperature (T_B). In lanthanide compounds, the magnetic anisotropy of the Ln^{3+} ions gives rise to high thermal energy barriers. Despite recent advances in mononuclear lanthanide SMMs that have resulted in astonishing thermal barriers of > 1000 K, the highest blocking temperature recorded for SMMs has only reached 20 K. The lack of high blocking temperatures in SMMs primarily results from quantum tunneling of magnetization (QTM), which serves to circumvent the thermal barrier by providing alternative relaxation pathways. QTM can be mitigated by using ligands that promote strong magnetic coupling between metal centers in an SMM. In lanthanide complexes, the valence 4f-orbitals are contracted, making it difficult for them to achieve high coupling constants.

Utilization of open-shell ligands to bridge lanthanide ions has recently gained traction as a method for improving the magnetic properties SMMs. Polylanthanide complexes bridged by closed shell ligands (e.g. CN^-, CH_3COO^-, OH^-, O^{2-}) typically display very weak magnetic superexchange interactions between the lanthanide centers. As a result of this weak superexchange, the lanthanides in such complexes typically exhibit single-ion effects, rather than acting as a coupled unit. In contrast, open-shell bridging ligands engender direct exchange interactions with the lanthanide ions, leading to higher degrees of magnetic coupling and improved SMM behavior.

In this poster, I will describe my PhD research in the field of transition metal metal-metal bonded complexes as well as my postdoctoral research into the synthesis and magnetic characterization of the aforementioned radical-bridged SMMs. This postdoctoral research has led to a recent successful NSF proposal. I will also detail my plans to merge these two fields in my independent career through the synthesis of metal-metal bonded molecular magnets. Such compounds will leverage direct exchange coupling mechanisms to produce superior molecular magnets.

Teaching Interests:

In graduate school at UW-Madison, I had ample Teaching Assistant experience for four semesters. In my TAship, I led discussion sections twice a week in which I planned and taught lessons to the students to supplement what they learned in the corresponding lectures for the course. For two semesters, I was awarded an Honored Instructor recognition for my teaching.

In my postdoctoral research, I planned and delivered a short workshop detailing the problems encountered in the course of solving and refining small-molecule X-ray crystallographic structures.

Selected Publications:

1. <u>Dolinar, B. S.</u>; Alexandropoulos, D. I.; Vignesh, K. R.; James, T.; Dunbar, K. R., Lanthanide Triangles Supported by Radical Bridging Ligands. *Am. Chem. Soc.*, **2018**, *140*, 908-911.
2. <u>Dolinar, B. S.</u>; Gómez-Coca, S.; Alexandropoulos, D. I.; Dunbar, K. R., An Air Stable Radical-Bridged Dysprosium Single Molecule Magnet and Its Neutral Counterpart: Redox Switching of Magnetic Relaxation Dynamics. *Commun.* **2017**, *53*, 2283-2286.
3. <u>Dolinar, B. S.</u>; Kozimor, S. A.; Berry, J. F., $K_3[Mo_2(SNO5)_4Cl]_3[Mo_2(SNO5)_4]$: The First Example of a Heterometallic Extended Metal Atom Node (HEMAN). *Dalton Trans.* **2016**, *45*, 17602-17605.
4. <u>Dolinar, B. S.</u>; Berry, J. F., Influence of Lewis acid charge and proximity in MoMo···M linear chain compounds with M = Na^+, Ca^{2+}, Sr^{2+}, and Y^{3+}. *Polyhedron*, **2016**, *103*, 71-78.
5. <u>Dolinar, B. S.</u>; Berry, J. F., Electronic tuning of Mo_2(thioamidate)$_4$ complexes through π-system substituents and *cis/trans Dalton Trans.* **2014**, *43*, 6165-6176.
6. <u>Dolinar, B. S.</u>; Berry, J. F., Lewis Acid Enhanced Axial Ligation of $[Mo_2]^{4+}$ *Inorg. Chem.* **2013**, *52*, 4658-4667.

542550 Soft Materials and Bio-Integrated Devices: From Complex Colloidal Systems to Skin/Brain-Interfaced Biosensors

Abstract:

Research Interests:

My primary research interests lie in colloids, interfacial science, and bio-integrated electronic devices for solving grand challenges in energy-efficient separations and human healthcare. **My doctoral research focused on colloids, polymer, and interfacial science**. I made two major contributions to the colloid community. My first accomplishment is the discovery of a new class of soft materials called capillary foams. I studied its stabilization mechanism, the effect of system parameters, and its potential applications in energy-efficient separations. This research has been reported by many press releases, such as Science Daily, NSF News, and World Industrial Reporter. My second contribution is that I answered a question that exists in the colloid community for a few decades. Over one hundred years ago, people found that colloidal particles with suitable wettability can adsorb to (and stabilize) the fluid-fluid interfaces. However, there is a long ongoing debate about whether the adsorption of non-amphiphilic particles in the interface can change the interfacial tension. I answered this question by conducting systematic experimental and theoretical studies. My studies show that particles do reduce the interfacial tension, which depends on the packing density and contact angle of particles in the interface. This insight is important for understanding and controlling the assembly of non-amphiphilic nanoparticles at fluid–fluid interfaces, which is crucial for applications from food technology to oil recovery.

My postdoc training focuses on skin/brain-interfaced devices for human health applications. Chronic kidney disease (CKD) is a condition where kidneys are damaged and cannot filter blood as well as healthy ones. 30 million or 15% of US adults are estimated to have CKD but most of them are not aware of it at an early stage of the disease. Early detection and treatment can often keep chronic kidney diseases from getting worse. Therefore, a novel method for painless and stress-free kidney health monitoring is highly desirable. I developed a wearable sweat sensor that can capture and analyze creatinine and urea in sweat for the early detection of CKD. Another significant direction I am working on is the design and fabrication of new tools for neuroscience. Diseases related to the central and peripheral nervous systems, such as depression, anxiety, addiction, pain, Alzheimer's, and amyotrophic lateral sclerosis, affect the life quality of millions of people worldwide. To understand the function of neural networks and the causes of these mental health problems, I designed and fabricated injectable/implantable optofluidic devices for brain and peripheral nerve interfaces. This technology has been transferred from the research bench to commercialization by a startup company Neurolux (http://neurolux.org/).

Postdoctoral Advisor: Professor John A. Rogers, Northwestern University

Ph.D. Advisors: Professor Sven H. Behrens and Professor J. Carson Meredith, Georgia Tech

Teaching Interests:

My teaching philosophy is to develop a supportive environment for my students, where they can develop their interests in both fundamentals and a broader impact of a given subject. I also intend to equip my students with an effective way of learning and collaboration that they can benefit from for their lifetime. '**Big picture**': I believe that courses that focus on both impact and core principals will better motivate students from various backgrounds, interests, and career goals. As a teacher, it is my responsibility to convey the importance of Chemical Engineering topics and my passion for them, so that students see the value and potential impact of their work. I will accomplish this by relating course materials to real-life problems. I will begin each semester by asking students why they register for my course and tailoring examples to the students' interests. These activities can be incorporated into large classes without sacrificing effectiveness. **Effective way of learning:** In this rapidly changing world, the knowledge in science and engineering is expanding exponentially. Therefore, instead of solely subject-oriented teaching, it is critical to stimulate students' curiosity to explore different fields and teach them learning skills that they can apply to various learning processes. **Collaboration:** Learning to work together to solve problems is a crucial skill for any engineering career. To promote this, I will present problems and probe questions pertaining to lecture concepts for students to work on and discuss together during class. Collaboration makes engineering more accessible to a broader range of students by exposing them to a range of approaches to problems and by reducing competition. While competition can help some types of students thrive, highly competitive classrooms have been shown to turn students away from engineering. As a graduate student at Georgia Tech, I experienced an environment where students worked together to solve homework problems and make sure everyone in their study group understand the material.

542230 Porous, Conductive Crystals: Expanding the 2D Materials Library with Metal-Organic Frameworks (MOFs)

Abstract:

Research Interests:

My graduate research with Prof. Charles Lieber at Harvard University focused on the gas phase synthesis of inorganic, semiconductor nanowires, and specifically on exploiting previously unrealized, synthetic opportunities for growth on one-dimensional materials with applications in energy conversion and photonic devices. As first author on the project, I discovered and provided a mechanistic understanding for the 'Plateau-Rayleigh crystal growth' (*Nature Nanotechnology* **10**, 345), which is unique to growth on one-dimensional substrates at the nanoscale. I extended the PR crystal growth to nanowire heterostructures (*Nano Letters* **16**, 4713), and to growth on nanofabricated top-down Si substrates for photonic applications (in prep.). I also have significant experience with and co-led the electrical and optical characterization of photovoltaic devices fabricated from synthetically tuned nanowires (*Nano Letters* **12**, 4971).

For my postdoctoral work, I sought out the opportunity to work with Prof. Mircea Dinca at MIT on conductive metal-organic frameworks (MOFs), which are hybrid inorganic/organic porous, crystalline materials that have demonstrated promise for energy storage, electrocatalysis and chemiresistive sensing. One of the biggest bottlenecks to understanding these new materials is an inability to grow sufficiently large crystals. I have improved our understanding for how these materials grow and, consequently, obtained much larger crystals for the porous MOF that currently has the highest conductivity to date. And by using nanofabrication techniques, I have fabricated, for the first time, single crystal electrical devices for both reported and new 2D MOFs.

My future research goals seek to develop materials that can efficiently store and convert energy, can detect molecules with high selectivity and sensitivity, and generate new platforms for fundamental transport studies. I think significant opportunities lie at the interface of porosity, chemistry and conductive, crystalline materials. My lab will pursue an understanding of how porous crystals conduct, and how interactions within these pores can influence their electrical and optical properties.

Teaching Interests:

I have extensive experience teaching undergraduate students and mentoring graduate students. For two years, I was a TA for Prof. Lieber's "Nanoscience and nanotechnology" course and was asked to lecture for him in his absence.

Publications

1. T. Ozel, B.A. Zhang, R. Gao, **R. W. Day**, C. M. Lieber, and D. G. Nocera, "Electrochemical deposition of conformal and functional layers on high aspect ratio silicon micro/nanowires," *Nano Letters,* **17**, 4502-4507 (2017).

2. Y.-S. No, R. Gao, M. Mankin, **R. W. Day**, H.-G., Park, and C. M. Lieber, "Encoding active device elements at nanowire tips," *Nano Letters,* **16**, 4713-4719 (2016).

3. **R. W. Day**, M. N. Mankin, and C. M. Lieber, "Plateau-Rayleigh crystal growth of nanowire heterostructures: Strain-modified surface chemistry and morphological control in 1, 2 and 3 dimensions," *Nano Letters,* **16**, 2830-2836 (2016).

4. **R. W. Day***, M. N. Mankin*, R. Gao, Y-S. No, S-K. Kim, D. Bell, H-G. Park and C. M. Lieber, "Plateau-Rayleigh crystal growth of periodic shells on one-dimensional substrates," *Nature Nanotechnology* **10**, 345-352 (2015).

5. M. N. Mankin, **R. W. Day**, R. Gao, Y-S. No, S-K. Kim, A. A. McClelland, D. C. Bell, H-G. Park, and C. M. Lieber, "Facet-selective epitaxy of compound semiconductors on faceted silicon nanowires," *Nano Letters* **15**, 4776-4782 (2015).

6. T. J. Kempa, S.-K. Kim, **R. W. Day**, H.-G. Park, D. Nocera and C. M. Lieber, "Facet-selective growth on nanowires yields multi-component nanostructures and photonic devices," *Journal of the American Chemical Society* **135**, 18354-18357 (2013).

7. T. J. Kempa, **R. W. Day**, S.-K. Kim, H.-G. Park and C. M. Lieber, "Semiconductor nanowires: A platform for exploring limits and concepts for nano-enabled solar cells," *Energy & Environmental Science* **6**, 719-733 (2013).

8. S.-K. Kim*, **R. W. Day***, J. F. Cahoon*, T. J. Kempa, K.-D. Song, H.-G. Park and C. M. Lieber, "Tuning light absorption in core/shell silicon nanowire photovoltaic devices through morphological design," *Nano Letters* **12**, 4971-4976 (2012).

9. T. J. Kempa, J. F. Cahoon, S.-K. Kim, **R. W. Day,** D. C. Bell, H.-G. Park and C. M. Lieber, "Coaxial multishell nanowires with high-quality electronic interfaces and tunable optical cavities for ultrathin photovoltaics," *Proc. Natl. Acad. Sci. USA* **109**, 1407-1412 (2012).

542336 Sheikhi Laboratory for Sustainable Soft Matter and Active Interfaces

Abstract:

Research Interests:

How can natural resources be engineered at the micro- and nanoscale to address the current limitations of synthetic, often hazardous materials? How may Nature inspire us with its sophisticated structure-property relationships to overcome everyday life challenges? These two key questions propel us to seek solutions for persistent bottlenecks in environment, water, healthcare, and food sectors by combining sustainability with the fundamentals of soft materials and interfaces. Our research focuses primarily on micro- and nanoengineering natural or semi-natural materials to develop advanced soft matter, including colloidal particles and macromolecules, as well as active interfaces with tailored structure-property relationships. We are particularly excited about the interdisciplinary research in the intersection between materials science, chemical engineering, environmental engineering, and bioengineering wherein biopolymers, such as cellulose and proteins as well abundant minerals play key roles in designing functional materials with unique properties. The overarching theme of our Lab is reflected in the following three platforms:

(1) Nanoengineered colloidal systems

Colloids, aggregates of molecules that form particles with a size between 1 nm-1 μm, distributed in a fluid, include dispersion, gels, foams, and emulsions. These materials are the main pillars of everyday life, contributing to a broad spectrum of applications and extraordinary phenomena. Our research endeavors to take control over the structure and properties of colloidal systems and ultimately use them for water treatment, environmental remediation, biosensing, energy storage, and smart foods. To this end, our effort is geared towards harnessing the potential of cellulose, the most abundant biopolymer in the world, by nanoengineering the building blocks of trees. We will develop a library of nanocelluloses with tailored physical and surface properties, providing stable colloidal dispersion (e.g., in extreme ionic strengths and biofluids), multifunctional hydrogels, ultralight aerogels, and robust films. Parallel efforts will be devoted to taking advantage of structural heterogeneity of nanoclays in developing adaptive nanocomposites.

Selected publications:

- **Sheikhi**, A. Kakkar, and T.G.M. van de Ven. Nanoengineering colloidal and polymeric celluloses for threshold scale inhibition: towards universal biomass-based crystal modification. Materials Horizons, 5:248-255. 2018.+Featured on the inside back cover.
- **Sheikhi** and T.G.M. van de Ven. Colloidal starch and cellulose nanocrystals unite to improve the mechanical properties of paper: From enhanced coatings to reinforced nanocomposites. ACS Applied Nano Materials, 1:1841-1852. 2018.
- **Sheikhi**, S. Afewerki, R. Oklu, A.K. Gaharwar, and A. Khademhosseini. Effect of ionic strength on

shear-thinning nanoclay-polymer composite hydrogels. Biomaterials Science, Accepted. 2018.+Featured on the outside back cover.

- **Sheikhi**, H. Yang, P.J. Carreau, and T.G.M. van de Ven. Colloidal nanotoolbox for molecularly regulated polymerization: chemorheology over 6 decades of viscoelasticity. Materials Horizons, 4:1165-1170. 2017.
- **Sheikhi** and T.G.M. van de Ven. Squishy nanotraps: hybrid cellulose nanocrystal-zirconium metallogels for controlled trapping of biomacromolecules. Chemical Communications, 53:8747-8750. 2017.
- **Sheikhi** and T.G.M. van de Ven. Colloidal aspects of Janus-like hairy cellulose nanocrystalloids. Invited Review Article to the Special Issue on Nanocelluloses. Current Opinion in Colloid & Interface Science, 29:21-31. 2017.

(2) Biomimicry for sustainable material design

Nature has mastered on-demand, highly engineered responsive architectures to minimize the material demand while maximizing the performance and functionality. We build on this invaluable lesson by carefully exploring natural systems and mimicking their structures. We have two main approaches here: (i) taking advantage of naturally-derived nanomaterials to design hybrid systems with inherited inspiration from all-natural systems, and (ii) achieving biomimicry through gaining a precise control over the molecular architecture of designer dendritic macromolecules, which are typically highly symmetric and monodisperse. These compounds enable elucidating the role of functional groups as well as, often neglected, structural backbones in designing biomimetic materials. We envision that understanding these principles may have direct implications in developing advanced materials and devices for water technology and biomedical applications.

Selected publications:

- **Sheikhi**, A. Kakkar, and T.G.M. van de Ven. Biomimetic scale-resistant polymer nanocomposites: Towards universal additive-free scale inhibition. Journal of Materials Chemistry A, 6:10189-10195. 2018.
- **Sheikhi**, A. Kakkar, and T.G.M. van de Ven. A leaf out of Nature's book: hairy nanocelluloses for bioinspired mineralization. Crystal Growth & Design, 16:46274634. 2016.
- **Sheikhi**, S.L. Mejlsoe, N. Li, E. Bomal, T.G.M. van de Ven, and A. Kakkar. Biomimetic dendrimers for mineralization: Rare fibrous amorphous calcium carbonate (ACC) and branch-and-bud ACC-vaterite polymorphs. Submitted. 2018.

(3) Microfluidic-enabled biomaterials

Reducing the scale of common "bulk" soft materials, such as hydrogels may open a new horizon for next-generation biomaterials with exceptional properties that would otherwise be implausible to achieve. An example is hydrogel scaffolds for three-dimensional biological cell culture that have interdependent porosity and stiffness, imposing inevitable limitations when highly stiff substrates with large pores are demanded. We are interested in revisiting biomaterial design by breaking down bulk platforms into modular compartments and rebuild them in a bottom-up approach. We aim at building biomaterials, such as polysaccharide- and protein-based hydrogels, using annealable microparticles, produced in common microfluidic platforms, e.g., a flow focusing device. Directed self-assembly of these soft building blocks will provide microporous scaffolds, which may enable engineering spatiotemporal cell-cell and cell-matrix interactions for a variety of applications, including organoid development for drug evaluation and screening.

Selected publications:

- **Sheikhi**, J. de Rutte, R. Haghniaz, O. Akouissi, M. Rasti, K. Suthiwanich, H. Mary, D. di Carlo, and A. Khademhosseini. Annealable gelatin methacryloyl microbeads: A versatile microenvironment with orthogonal stiffness and porosity. To be submitted. 2018.

Teaching Interests:

Amir Sheikhi has long been passionately involved in active teaching, initiated from numerous Teaching Assistantships in the undergraduate core courses as well as the specialized graduate-level curriculum of Chemical Engineering, encompassing Chemical Manufacturing Processes (Freshman), Fluid Mechanics (Sophomore), Heat and Mass Transfer (Sophomore), Kinetics and Chemical Reactions Engineering (Junior), Process Simulation and Modeling (Senior), Process Control (Senior), Material Surfaces: A Biomimetic Approach (Senior/Graduate), and Biochemical Engineering (Senior). In addition, he has developed and instructed lectures in Advanced Computer-Aided Process Simulation (Senior/Graduate), Advanced Materials (Graduate), and Colloid Chemistry (Graduate). Amir believes that the quality of undergraduate teaching directly shapes the next-generation scientists, and pursues three main goals in undergraduate teaching: (i) to facilitate active learning in which students are highly engaged in the process of learning through cognitive-affective-sensory domains within the Bloom's taxonomy, (ii) to bridge the gap between undergraduate coursework and graduate research by integrating real-life and laboratory experiences in the lectures, and (iii) to develop new, highly multidisciplinary curricula for extending the vision of young scholars beyond their major degree, eternalizing a built-in "curiosity". Amir has also been active and passionate about chemical education, hands-on demonstrations, and academic life advising.

Selected publications:

- **Sheikhi**, N. Boonsukha, O. Akouissi, S.-K. Hu, J. Hayashi, and A. Khademhosseini. Assembling an affordable miniature microscope (mini-microscope). Submitted to the Journal of Chemical Education. 2018.
- **Sheikhi**, Learning to be more than my publications: Thriving in academic life through a silence of publications, Submitted to *Science*, 2018.
- **Sheikhi**, H. Yang, Md. N. Alam, T.G.M. van de Ven. Highly stable, functional hairy nanoparticles and biopolymers from wood fibers: Towards sustainable nanotechnology. Journal of Visualized Experiments, 113:e54133. 2016.

542352 Programmable 3D Transformation of Smart Soft Materials

Abstract:

Research Interests:

My research interests lie in the intersection of soft matter, photonics, microfluidics and nanostructure engineering for development of smart soft materials and platforms. From this interdisciplinary approach, my independent research aims to exploit the fascinating interplay between material composition, geometry and prescribed deformation, and create novel structured materials and actuation devices with unprecedented properties and functions, including tunable optics, flexible electronics and bio-inspired transformable robotics. The early stage of my research program will span mainly two areas below.

1. **Taming instabilities toward multi-functionality for reconfigurable active matter**

In the course of evolution, nature has developed strategies to create materials with exquisite adaptability and optimized efficiency. Learning from nature, people have developed stimuli-responsive engineering polymers, known as a class of smart materials. Yet, there remains extensive untapped knowledge of the ways in which to harness functional, dynamic, and adaptable materials into the diverse integrated systems. To establish a rational strategy to control instabilities of soft polymeric materials, understanding of their phase behavior and active interactions with their surrounding is critical. **My first research goal is to investigate dynamic behavior/mechanism of the responsive materials in terms of structure and physical property triggered by mechanical strain or different external field conditions by using various microscopy techniques.** This knowledge should help resolve significant challenges of materials integration into complex and robust 3D structures. I believe this research regarding reconfigurable metamaterials can address broader technological and societal challenges, including those in the areas of renewable energy harvesting, water purification, drug delivery, etc.

2. **Development of a new 4D printing based on photon upconversion and flow lithography**

Two-photon polymerization (TPP), also known as direct laser writing, has been recognized as a promising technique for the fabrication of arbitrary 3D sub-micron features with high-order shapes and functions in photosensitive materials without photomasks. Typically, it requires an ultrashort pulsed laser to trigger non-linear two-photon absorption only in the focal spot volume, so called voxel. Unlike other 3D printing technologies, thus, it allows for the highest resolution or layer thickness with a previously unavailable freedom of design. Although TPP is a powerful and attractive technique, the major bottleneck in unleashing its potential lies in point-by-point writing with high power pulsed laser and long processing times unacceptable for practical applications. **Here, I am interested in triplet fusion upconversion-assisted polymerization which allows for use of a non-coherent, continuous wave light at ultra-low excitation power as a pumping source to generate high-energy photons.** Besides photochemistry, my interests include methods **to develop high-throughput microfabrication** by combining of flow lithography using digital micromirror device. I believe this approach provides a short processing time, offering possibility of mass production for high-fidelity freeform microstructures with programmed internal heterogeneity. If successful, I am utilizing them to synthesize novel functional building blocks like anisotropic, shape-deformable colloids for the assembled complex structures that dynamically respond to energy input from their surroundings.

Research Experiences:

For my postdoctoral research at the University of Massachusetts Amherst (Prof. Ryan C. Hayward, Polymer Science & Engineering), I am currently working on precise control of mechanical instabilities in responsive polymer patterns. I am particularly interested in programming robustly self-folding action in rigid origami without bifurcation of vertices which leads to undesired misfolding states. Using mathematical prediction, I have designed a new structural transformation of multilayer polymer films that can be dynamically reconfigured to change their shapes and properties on demand. I also have developed the active vertices to apply a driving force that avoids such bifurcation from a flat unfolded state using constrained deformation by external stimuli such as pH, temperature and light.

During my PhD at Georgia Institute of Technology (Prof. Elsa Reichmanis, Chemical Engineering), I focused on optofluidic encapsulation of triplet fusion photon upconversion. I investigated photon management in intermolecular energy transfer between metastable states and diffusion-mediated delayed fluorescence in polymer thin films and microcapsules. I had been leading the development of photo-induced interfacial polymerization and one-step emulsification of multi-phase fluids with microfluidic devices for high quantum efficiency and photochemical stability. Further, I demonstrated a photonic shell containing cholesteric liquid crystals with planar alignment to amplify spontaneous upconverted emission yield and tailor optical properties with tunable bandgaps. I believe the upconversion capsules have a wide application in photonics such as photonic ink, microlaser, volumetric display, solar energy harvesting, and photocatalytic water treatment.

Early on in my career at KAIST (Prof. Seung-Man Yang, Chemical Engineering), I was involved in research on creation of nanostructures with free-defects over a large area by multi-coherent light without a photomask (i.e. holographic interference lithography, HIL). I developed a new type of hydrogel photoresist that undergoes viscoelastic deformation and produced 3D photonic crystals with tunable structural color. I also worked on hierarchical porous electrodes by colloidal templating and HIL for photovoltaic cells and batteries.

Teaching Interests:

During my teaching career development, I enjoyed serving as a teaching assistant for undergraduate courses in "Chemical Process Safety" and "Unit Operations Lab" at Georgia Tech. Chemical Process Safety requires students to tackle real safety problems with chemical engineering knowledge. As a TA, I prepared to guide their quizzes and projects with many aspects such as toxicology, release & dispersion models and case studies. For the course of UO Lab, I was assigned as a TA of the fluidized bed to give students a brief introduction and standard operation procedures to run a series of experiments. I evaluated their knowledge from their reports so I also learned how to clarify concepts to help them understand. These experiences have built my confidence and an interest in teaching. Besides, I am excited to teach a wide range of the stem courses in Chemical Engineering such as "Chemical Reaction Engineering", "Mass, Momentum and Energy Transport", and "Thermodynamics". From my multi-disciplinary research experience, I am also confident in teaching specific courses of "Nano-chemical Technology", "Photochemistry and Photophysics", "Fluidic Mechanics", "Capillarity and Wetting", and "Polymer Chemistry". I look forward to the opportunity to not only teach existing courses but also work to develop new ones that I can contribute to. Also, I have a lot of experience in mentoring undergraduate students from SURE and REU (NSF MRSEC), and graduate students to develop and trim their own creative ideas. Based on these teaching and mentoring experience, I have learned about the roles and responsibilities of the principal investigator regarding how to communicate better with my students and encourage them to proceed.

Selected Publications:

1. **Kang,-H.**; Kim, S.-H.; Fernandez-Nieves, A.; Reichmanis, E., Amplified Photon Upconversion by Photonic Shell of Cholesteric Liquid Crystals, *Journal of American Chemical Society* **2017**, 139, 5708.
2. **Kang, J.-H.**; Lee, S. S.; Guerrero, J.; Fernandez-Neives, A.; Kim, S.-H.; Reichmanis, E., Ultrathin Double-Shell Capsules for High Performance Photon Upconversion, *Advanced Materials* **2017**, 29, 1606830.
3. **Kang, J.-H.**; Reichmanis, E., Low-Threshold Photon Upconversion Capsules Obtained by Photoinduced Interfacial Polymerization, *Angewandte Chemie International Edition* **2012**, 51, 11841.
4. Kim, Y.; Cho, C.-Y.; **Kang,-H.**; Cho, Y.-S.; Moon, J. H., Synthesis of Porous Carbon Balls from Spherical Colloidal Crystal Templates, *Langmuir* **2012**, 28, 10543.
5. Kim, H.-N.*; **Kang**, **-H.***; Jin, W.-M.; Moon, J. H., Surface Modification of 2D/3D SU-8 Patterns with a Swelling-Deswelling Method, *Soft Matter* **2011**, 7, 2989. (*equally contributed)
6. Shin, J.-H.; **Kang**, **-H.**; Jin, W.-M.; Park, J. H.; Cho, Y.-S.; Moon, J. H., Facile Synthesis of TiO2 Inverse Opal Electrodes for Dye-Sensitized Solar Cells, *Langmuir* **2011**, 27, 856.
7. Jin, W.-M.; **Kang**, **-H.**; Moon, J. H., Fabrication of 3D Copper Oxide Structure by Holographic Lithography for Photoelectrochemical Electrodes, *ACS Applied Materials and Interfaces* **2010**, 2, 2982.
8. **Kang**, **-H.**; Moon, J. H.; Lee, S.-K.; Park, S.-G.; Jang, S. G., Yang, S.; Yang, S.-M., Thermoresponsive Hydrogel Photonic Crystals by Three-Dimensional Holographic Lithography, *Advanced Materials* **2008**, 20, 3061.
9. Ha, S.-J.; **Kang, J.-H.**; Choi, D.H.; Nam, S.K.; Reichmanis, E.; Moon, J.H., Upconversion-assisted Dual-band Luminescence Solar Concentrator Coupled for High Power Conversion Efficiency Photovoltaic Systems, *ACS Phtonics* **2018**, in revision.

542361 Harnessing Flow-Microstructure Interactions Towards Improved Soft Materials Manufacturing and Processing

Abstract:

Research Interests:

My interests are in soft matter, transport phenomena, rheology and rheo-optics in a variety of engineering and scientific problems, including emulsions, polymer blends, oil recovery, polyelectrolytes, macromolecules of different kind, gels, foams, biomaterials, printing techniques and flow chemistry at the micro-scale. All of these studies have in common the goal of taming flow-microstructure interactions towards functionalities, or, alternatively said, how to manipulate the microstructure in order to get tailored mechanical and flow properties of materials and fluids.

In recent years I have been using many different experimental techniques (Optical Microscopy, Confocal-Laser-Scanning Microscopy, Bulk Rheology, Interfacial Rheology, Lipid Membranes Electro-formation, Microfluidics, Flow-Visualization) to address how flow affects microstructural properties of complex fluids and materials spanning from suspensions of biocompatible polymeric fibers, vesicles, emulsions, gels, foams, surfactant solutions, crystals, aromatic amines, composites of polymers plus particles, polyelectrolyte solutions etc., to understand how these insights can be transferred towards improved materials/chemicals manufacturing and printing processes.

- *Biocompatible Polymeric Gels and Fiber Suspensions*

For example, suspensions of flexible fibers are known to be a shear thinning material (resistance to flow decreases by increasing flow strength). However, we demonstrated for the first time that by using ultra-flexible long microfibers, a hydrophilic gel made of a jammed network of fibers can be obtained [1]. This is a noticeable case where the resistance to flow is irreversibly increased by increasing flow strength or applying a constant strain rate for a long enough time. Simply said, one can make a gel by shaking the suspension. We showed these phenomena to be purely topological and to be associated with mechanical entanglements of the fibers, without the requirement of any chemical cross-linking or chemical interactions. The microfibers used in the experiments were hydrophilic, biocompatible and loosely suspended in water, hence the hydrogel can be simply obtained by injection of the microfiber suspension through a syringe needle and is potentially suitable for biomedical applications. This research was awarded the cover page of the "Proceedings of the National Academy of Sciences" in 2017.

- *Inkjet and 3D Printing*

In another study, we identified the relevant variables for optimal printing within a novel laser-induced printing technique by using polyelectrolyte solutions as a model fluid and dimensionless parameters involving dynamic variables of fluid flow and viscoelasticity of the fluid [2]. Shear and extensional rheometrical measurements were conducted. The use of dimensionless analysis allowed to predict the printing outcome and matched the experimental results. A similar dimensional analysis has been conducted to optimize the printing pattern in 3D-embedded printing, a technique where an ink is deposited in a viscoplastic matrix (a composite made of polymers and particles) to permit manufacturing of morphologically complex 3D objects [3]. Here, the experimental campaign has been conducted coupling rheometrical measurements and particle image velocimetry (PIV).

- *Polyelectrolyte Solutions*

Another result of my work concerns the effect of salt type on polyelectrolyte solution rheology [4]. Recently, we demonstrated that for a given polyelectrolyte and salt concentration, different salts affect elasticity of the solution depending on the salt counterion ionic radius. This feature can be used to manipulate the solution relaxation time by an order of magnitude, i.e. from tens of seconds to less than a second.

- *Emulsions, Immiscible Fluids and Interfaces*

Concerning emulsions, we combined rheometry, organic electrochemical transistors, optical microscopy, confocal laser scanning microscopy, cryo-SEM and small angle neutron scattering in flow (Rheo-SANS) to understand the structure of viscoelastic depletion-flocculated emulsion where surfactant self-assembly is relevant and droplet attractive interactions are at play [5,6,7]. We showed for the first time flow-induced structural re-arrangements of the droplets involving droplet coalescence, droplet cluster disruption, droplet breakage and consecutive flow-alignment in strings until the droplets become nanometric in size. The droplet chaining is purely of elastic origin and does not depend on confinement. The typical percolated structure of the emulsion has been also associated with increased conductivity and the peculiar organic electrochemical transistor (OECT) response [8]. We were first to use OECT to "read" emulsion microstructure.

In a separate project, we used the capillary flow of the emulsion to estimate droplet interfacial tension using a simple method that exploit the linear shear rate profile close to channel walls [9]. Interfacial dilational rheology of the surfactant covered droplet has been also performed.

We also focused, in another study, on development of improved emulsion design characterizing the interfacial physical chemistry of the droplets when mixed surfactants are used [10]. Mixed surfactants are key to achieve extremely low interfacial tension values (close to zero) as well as minimize the amount of amphiphile requested to stabilize the emulsion, being the latter one of the most impairing factor regarding the use of emulsion in enhanced oil recovery. We reviewed methods of emulsion, microemulsion and nanoemulsion preparation by phase inversion highlighting avenues for possible improvements [11] and we wrote a book chapter reviewing methods and perspectives within phase inversion emulsification [12]. In another review, we summarized findings on emulsion flow in porous media for enhanced oil recovery and oil-contaminated aquifer remediation and reported several novel perspectives for future work both on the side of emulsion improved design and on the relevance of flow-microstructure interactions to obtain better oil recovery performance [13].

- *Micro-scale Reactive Flows, Crystals and Fouling*

We were the first to realize the synthesis of an important pharmaceutical intermediate compound in a continuous flow stainless steel micro/millireactor [14]. This simple small-sized reactor can be easily scaled-up by running several reactors in parallel, and was used to characterize the reaction kinetics. Building on these results, an industrial pilot plant has been built through a collaboration of many academic groups and a pharmaceutical company [15]. The chemical reaction is based on the most popular catalytic reaction used in pharmaceuticals, known as Buchwald-Hartwig cross-coupling, i.e. an organometallic synthesis to produce aromatic amines, and using an ad-hoc developed Pd-based catalyst. Clogging issues of the reactor where then characterized [16] and the manipulation of fluid flow has been used to tailor microfluidic crystals production [16].

- *Future Research Directions*

Flow can be used as a microstructural designer to provide soft materials and chemicals with specific functionalities. Likewise, these materials are often harnessed in continuous operations/processing and the effect of flow on their microstructure and functionality is a key-step towards materials development. As a faculty, I aim to exploit various flow conditions and physico-chemical interactions to design new materials with tailored properties and understand how to control these parameters to achieve a unique realization in many applications such as flow in porous media and materials for biomedical applications.

Teaching Interests:

I am interested in teaching courses broadly focused on fluid mechanics, non-Newtonian fluid mechanics, rheology, transport phenomena, thermodynamics, polymer physics, physical chemistry (with a focus on colloids and interfaces), soft materials mechanics (both biomaterials and non) and chemical reaction engineering at both the undergraduate and graduate levels.

[1] Perazzo, A., Nunes, J. K., Guido, S., & Stone, H. A. (2017). Flow-induced gelation of microfiber suspensions. Proceedings of the National Academy of Sciences, 114(41), E8557-E8564.

[2] Turkoz, E., Perazzo, A., Kim, H., Stone, H. A., & Arnold, C. B. (2018). Impulsively Induced Jets from Viscoelastic Films for High-Resolution Printing. Physical review letters, 120(7), 074501.

[3] Grosskopf, A., Truby, R., Kim, H., Perazzo, A., Lewis, J. A., & Stone, H. A. (2018). Viscoplastic Matrix Materials for Embedded 3D Printing. ACS applied materials & interfaces. DOI: 10.1021/acsami.7b19818

[4] Turkoz, E., Perazzo, A.*, Arnold, C. B., Stone, H. A., (2018). Salt Type and Concentration affect the Viscoelasticity of Polyelectrolyte Solutions, Applied Physics Letters, 112, 203701, *co-first author

[5] Preziosi, V., Perazzo, A.*, Tomaiuolo, G., Pipich, V., Danino, D., Paduano, L., & Guido, S. (2017). Flow-induced nanostructuring of gelled emulsions. Soft matter, 13(34), 5696-5703. *co-first author

[6] Preziosi, V., Perazzo, A., Tomaiuolo, G., & Guido, S. (2018). Flow-switchable morphology of concentrated emulsions. Chemical Engineering and Processing-Process Intensification, 125, 275-279.

[7] Preziosi, V., Perazzo, A., Tomaiuolo, G., & Guido, S. (2018). The effect of flow on viscoelastic emulsion microstructure. The European Physical Journal E, 41(3), 45.

[8] Preziosi, V., Barra, M., Perazzo, A., Tarabella, G., Romeo, A., Marasso, S. L., ... & Guido, S. (2017). Monitoring emulsion microstructure by using organic electrochemical transistors. Journal of Materials Chemistry C, 5(8), 2056-2065.

[9] D'Apolito, R., Perazzo, A., Preziosi, V., D'Antuono, M., Tomaiuolo, G., Miller, R. and Guido, S., (2018). Measuring interfacial tension of emulsions in situ by microfluidics. Langmuir, 34 (17), 4991-4997

[10] Posocco, P., Perazzo, A.*, Preziosi, V., Laurini, E., Pricl, S., & Guido, S. (2016). Interfacial tension of oil/water emulsions with mixed non-ionic surfactants: comparison between experiments and molecular simulations. RSC Advances, 6(6), 4723-4729.*corresponding author

[11] Perazzo, A., Preziosi, V., & Guido, S. (2015). Phase inversion emulsification: Current understanding and applications. Advances in colloid and interface science, 222, 581-599.

[12] Perazzo, A., & Preziosi, V. (2018). Catastrophic Phase Inversion Techniques for Nanoemulsification. In Nanoemulsions (pp. 53-76), Elsevier-Academic Press.

[13] Perazzo, A., Tomaiuolo, G., Preziosi, V., & Guido, S. (2018). Emulsions in porous media: From single droplet behavior to

applications for oil recovery. Advances in colloid and interface science, 256, 326-339

[14] Perazzo, A., Tomaiuolo, G., Sicignano, L., Toscano, G., Meadows, R. E., Nolan, S. P., & Guido, S. (2015). A microfluidic approach for flexible and efficient operation of a cross-coupling reactive flow. RSC Advances, 5(78), 63786-63792.

[15] Falß, S., Tomaiuolo, G., Perazzo, A., Hodgson, P., Yaseneva, P., Zakrzewski, J., ... & Meadows, R. E. (2016). A continuous process for Buchwald–Hartwig amination at micro-, lab-, and mesoscale using a novel reactor concept. Organic Process Research & Development, 20(2), 558-567.

[16] Sicignano, L., Tomaiuolo, G., Perazzo, A., Nolan, S. P., Maffettone, P. L., & Guido, S. (2018). The effect of shear flow on microreactor clogging. Chemical Engineering Journal, 341, 639-647.

542031 Engineering Transport in Microporous Materials for Next-Generation Energy Technologies

Abstract:

Research Interests: I aim to establish new technologies that can solve pressing challenges in developing renewable energy and sustainable resources. To accomplish this, I will lead investigations into fundamental gas, liquid, and ion interactions in nanoengineered porous materials, focusing on host-guest thermodynamics and kinetics. Through my academic training in fields that span chemical engineering, inorganic chemistry, and polymers chemistry, I have acquired a unique perspective in understanding and developing cutting-edge materials; my areas of interest include two newer generation material classes: cooperative materials and microporous polymer nanocomposites. Cooperative materials, which can be synthesized by grafting oligomers into microporous hosts, are a new class of microporous materials that possess unexplored transport behavior. Microporous polymer nanocomposites, which can selectively permeate guest species, have the potential to revolutionize chemical process technology by replacing decades old distillation and absorption-based processes. Collaborating with faculty across academic disciplines, including chemistry and materials science, will enable me to more effectively find solutions to problems in energy and resources.

Teaching Interests: Universities are the only institutions that believe that cultivating intellectual curiosity can lead to world-changing ideas. This is why professors, who are uniquely positioned to foster intellectual growth through instructing and mentoring students, hold such noble positions in society. My own curiosity for and joy of sharing fundamental concepts in chemical engineering drove me to pursue advanced study and is now the driving force in my pursuit for a career in academia. I initially began teaching during my time as an undergraduate as a Teaching Assistant for core chemical engineering curricula, including process engineering, kinetics, and mass transport & separations. I was awarded the Koch Fellowship for Excellence in Undergraduate Teaching. It was during my undergraduate teaching experience that I had my initial foray into scientific research, and it was through these experiences together that I realized the vast personal and intellectual growth that is facilitated by their joint pursuit. In graduate school, I served as a Graduate Student Instructor (GSI) for graduate-level statistical thermodynamics. Additionally, as the GSI of a lower level survey of energy technologies course for chemical engineers, I found myself inspired by the passion that undergraduates found in chemical engineering. As an Assistant Professor, I would be happy to teach any core undergraduate course, with a particular fondness for thermodynamics. For graduate education, I would be thrilled to again teach statistical thermodynamics, or introduce a new course on the properties and applications of polymer membranes and microporous adsorbents.

Graduate Research, Advisor: Prof. Jeffrey Long, UC Berkeley (2013-2018)**:** My graduate work focused on the understanding and development of metal–organic frameworks and the design of metal–organic framework/polymer composites. In Prof. Jeffrey Long's group at UC Berkeley, my research was at the interface of inorganic chemistry and chemical engineering; here I saw a unique opportunity to understand the transport of gases and thermodynamics of gas-adsorbent interactions within metal–organic frameworks. This led to insights into advanced materials for olefin/paraffin separation, CO_2 capture, and natural gas processing. I developed a novel class of adsorption-selective membrane materials that overcome fundamental limitations of polymer membranes by compositing metal–organic framework nanoparticles with high-performance polyimides. Going forward, I sought to explore materials with an even greater ability to separate olefin/paraffin mixtures. I demonstrated the record performance of a metal–organic framework featuring coordinatively unsaturated metal sites for ethylene/ethane and propylene/propane mixtures. Finally, I introduced these materials as a solution to a long-standing challenge in alternative ethylene production by enabling the oxidative coupling of methane process. This work required a fundamental understanding of metal-gas interactions and was enabled by the tuning of metal site electronic structure.

Postdoctoral Research, Advisor: Prof. Yi Cui, Stanford University (March 2018-Present): While my graduate work focused primarily on understanding gas transport and thermodynamics in microporous hosts, I then went on to expand my scope by pursuing my postdoctoral studies on ion transport in solid polymer electrolytes under the supervision of Prof. Yi Cui at Stanford University. Here, I have studied polymers of intrinsic microporosity in order to understand the implications of salt transport and reactivity on Li-S battery performance.

Presentations at AIChE 2018:
[1] *Adsorption-Enhanced, Plasticization Resistant Composite Membranes Using Metal–Organic Framework Nanocrystals* (Session: Mixed-Matrix Materials for Gas Separation)
[2] *Highly Selective, High-Capacity Metal–Organic Frameworks for Olefin Production* (Session: MOFs, COFs, and Porous Polymer Materials: Characterization and Application)
[3] *Engineering Ion Transport in Microporous Polymer Separators for Li-S Batteries* (Session: Polymers in Batteries)

Selected Publications (out of 16):

[1] Mason, J. A.; Oktawiec, J.; Taylor, M. K.; Hudson, M. R.; Rodriguez, J.; **Bachman, J. E.**; Gonzalez, M. I.; Cervellino, A.; Guagliardi, A.; Brown, C. M.; Llewellyn, P. L.; Masciocchi, N.; Long, J. R., "Methane Storage in Flexible Metal–Organic Frameworks with Intrinsic Thermal Management" *Nature* 2015, *527*, 357–361.

[2] **Bachman, J. E.**; Smith, Z. P.; Li, T.; Xu, T.; Long, J. R., "Enhanced Ethylene Separation and Plasticization Resistance in Polymer Membranes Incorporating Metal–Organic Framework Nanocrystals" *Nature Mater.* 2016, *15*, 845–849.

[3] **Bachman, J. E.**; Long, J. R., "Plasticization-Resistance Ni_2(dobdc)/Polyimide Composite Membranes for the Removal of CO_2 from Natural Gas" *Energy Environ. Sci.* 2016, *9*, 2031–2036.

[4] Shete, M.; Kumar, P.; **Bachman, J. E.**; Ma, X.; Smith, Z. P.; Xu, W.; Mkhoyan, A.; Long, J. R.; Tsapatsis, M., "On the Direct Synthesis of Copper 1,4-Benzenedicarboxylate (CuBDC) MOF Nanosheets and their Performance in Mixed-Matrix Membranes" *J. Membr. Sci.* 2017, *549*, 312–320.

[5] **Bachman, J.E.**; Kapelewski, M. T.; Reed, D. A.; Gonzolez, M. I.; Long, J. R., "M_2(*m*-dobdc) (M = Mn, Fe, Co, Ni) Metal–Organic Frameworks as Highly-Selective, High-Capacity Adsorbents for Olefin/Paraffin Separations" *J. Am. Chem. Soc.* 2017, *139*, 15363–15370.

[6] Maserati, L.; Meckler, S. M.; **Bachman, J. E.**; Long, J. R.; Helms, B. A., "Diamine-Appended Mg_2(dobpdc) Nanorods as Phase-Change Fillers in Mixed-Matrix Membranes for Efficient CO_2/N_2 Separations" *Nano Lett.* 2017, *17*, 6828-6832.

[7] Smith, Z. P.; **Bachman, J. E.**; Li, T.; Gludovatz, B.; Kusuma, V. A.; Xu, T.; Hopkinson, D. P.; Ritchie, R. O.; Long, J. R., "Increasing M_2(dobdc) Loading in Selective Mixed-Matrix Membranes: A Rubber Toughening Approach" *Chem. Mater.* 2018, *30*, 1484-1495.

[8] **Bachman, J. E.**; Reed, D. A.; Kapelewski, M. T.; Chachra, G.; Jonnavittula, D.; Radielli, G.; Long, J. R., "Enabling Alternative Ethylene Production through its Selective Adsorption in Mn_2(*m*-dobdc)" *Energy*

Environ. Sci. 2018, Advance Article.

[9] Meckler, S. M.; **Bachman, J. E.**; Robertson, B. P.; Zhu, C.; Long, J. R.; Helms, B. A. "Thermally Rearranged Polymer Membranes Containing Tröger's Base Units with Exceptional Performance for Air Separations" *Agnew. Chem. Int. Ed.*, 2018, *130*, 5006-5010.

542039 Engineering Complex Polymer Materials with Tailored Chemistry, Morphology, and Functionality

Abstract:

Research Interests:

My research program meets the ever-present challenge in the field of materials engineering to identify innovative methods for producing materials that address a complex performance profile. I use polymers as the platform from which I build functional materials, and my research aims to have full design authority over their ultimate chemistry and structure so that I can engineer several desired properties including flexibility, mechanical robustness, wettability, biocompatibility, response to external stimuli, etc. To enable this design, I study the fundamental driving forces that dictate polymer architecture, conformation, heterogeneity, and micro-structure during formation. The over-arching goal of my program is to use fundamental studies to identify versatile and improved methods towards material designs.

Heterogeneous polymer materials, such as blends or composites are often used to address the need for bulk performance or properties that cannot be achieved using a single monomer building block. To address the need for more efficient techniques to create heterogeneous materials, my graduate research studied the dynamic process of *in situ* phase separation during the formation of polymer networks. Specifically, I demonstrated that by understanding the simultaneous yet non-equivalent changes in thermodynamics and kinetics during photopolymerization, it is possible to form well-structured, heterogeneous materials with broad differentials in local properties despite rapid network formation. Furthermore, with this approach I synthesized network materials with utility and performance superior to their homogeneous, non-structured analogues. I also developed polymerization protocols that simultaneously conformed to processing constraints, yet also optimized heterogeneous phase structure in order to attain target property profiles. An example of this work being the demonstration of significant interfacial stress reduction in phase separated acrylic networks (a class of materials that are often employed as dental restoratives, which typically fail due to polymerization stress) based on heterogeneous phase structure and the rate of network formation.

Interfacial materials represent another class of complex materials, as for many critical surface properties (wettability, adhesion, etc.) both the surface chemistry and three-dimensional topography play an important role. In my post-doctoral research, both at the Université de Nice (Nice, France) and currently at Northwestern University, I have developed bio-inspired, structured interfaces using polymerization techniques. I synthesized thiophene-based monomers and studied the effects of variations in structure and polymerization protocol on the resultant *in situ* surface polymerization of these moieties via electrochemical deposition. With these studies, I created interfaces with robust features such as spherical agglomerates, nano-folds, and nano-tubes which all formed *in situ* and without any templating. I have also explored the use of heterogeneous photopolymerizations as an approach to tune surface structure and chemistry for interfacial materials. Using my expertise in heterogeneous network polymerizations, I developed coating materials where the formation of topographical surface features was driven by local differentials in polymerization stress. These studies have resulted in the rapid fabrication of interfaces with desirable properties such as superhydrophobicity and strong adhesion. My continued work on this topic explores the influence of polymer chemistry, structure, and architecture on interfacial properties when developing coating materials.

Future research directions:

My independent career research will combine my skills to develop advanced materials addressing needs from varied fields including: coatings, adhesives, and biomaterials. Initial work in the Szczepanski group will address three specific aims: (1) manipulating heterogeneous network structure as a means to prevent hydrolytic degradation in aqueous environments, (2) engineering of biomimetic interfaces with controlled functionality, (3) incorporation of stimuli-responsive moieties in polymer coating materials to create robust, switchable interfaces.

Selected Publications (15 total, 10 first author)

1. **C.R. Szczepanski*,** T. Darmanin, G. Godeau, F. Guittard "Template-less Patterning of Polymeric Interfaces for Controlled Wettability via in situ Heterogeneous Photopolymerizations" *Macromol. Chem. Phys.***2018**. ***corresponding author**
2. **C.R. Szczepanski,** I. M'Jid, T. Darmanin, G. Godeau, F. Guittard "A template-free approach to nanotube-decorated polymer surfaces using 3,4-phenylenedioxythiophene (PhEDOT) monomers." *J. Mater. Chem. A* **2016**, *4*, 17308-17323
3. **C.R. Szczepanski,** T. Darmanin, F. Guittard "Spontaneous, Phase-Separation Induced Surface Roughness: A Novel Method to Design Parahydrophobic Polymer Coatings with Rose petal-like morphology." *ACS Appl. Mater. Interf.* **2016**, *8*, 3063-3071.
4. **C.R. Szczepanski,** J.W. Stansbury "Modification of linear prepolymers to tailor heterogeneous network formation through photo-initiated polymerization-induced phase separation." *Polymer* **2015**, *70*, 8-18.
5. **C.R. Szczepanski,** J.W. Stansbury JW. "Stress reduction in phase-separated, cross-linked networks: influence of phase structure and kinetics of reaction." *J. Appl. Polym. Sci.* **2014**, *131*, 40879.
6. **C.R. Szczepanski,** C.S. Pfeifer, J.W. Stansbury "A new approach to network heterogeneity: Polymerization Induced Phase Separation in photo-initiated, free-radical methacrylic systems" *Polymer* **2012**, *53*, 4694-4701.

Teaching Interests:

When educating and training the future generation of engineers, it is crucial to emphasize that each course is not designed solely to introduce a narrowly focused set of analytical skills, but also to improve a student's overall aptitude at solving complex problems that draw on a variety of subject matters. The fundamental core subjects of chemical engineering are not topics that exist in isolated circumstances but must be called upon throughout one's professional career. I have established a teaching philosophy that is centered around three essential goals for my students: (1) that they successfully acquire a new set of skills to add to their engineering toolbox, (2) that they gain an understanding as to how the techniques they learn can be applied to a variety of problems especially outside the context of class, and (3) that they improve their overall problem-solving approach during the course. To ensure that these goals are being met, I organize my lectures such that both the student's comprehension and my own teaching are being assessed throughout the course, using active learning techniques to engage all students.

Up to this point in my career, my teaching appointments have been diverse in that I have worked with students at various stages of their educational careers, giving me valuable insight as to how to best engage students based on their educational experiences. As a graduate teaching assistant at the University of Colorado, I taught laboratory sections and developed recitation materials for General Chemistry (first year undergraduates, 2010) and Process Controls (fourth year undergraduates, 2012). In 2014 I was a Teaching Fellow for my graduate department, where my responsibilities included preparing course materials for 10 independent General Chemistry lectures. In my current role at Northwestern, I have been a guest lecture for an introductory polymers course, which consists of upper level undergraduate as well as graduate students. Beyond these classroom experiences, I also served as the Lead Graduate Teacher for my graduate department for one academic year, during which I was responsible for training graduate students in engineering pedagogy and effective teaching techniques. Because of my interest on developing and improving problem-solving approaches, I am most interested in teaching foundational Chemical Engineering courses such as: Material & Energy Balances, Thermodynamics, and Transport Phenomena. Given my research expertise I would also be interested in developing and teaching courses related to polymer science, polymer engineering, photopolymerizations, materials characterization techniques, as well as bio-inspired approaches to materials engineering.

516127 Multifunctional Soft-Nano Interfaces for Energy, Environment, and Healthcare

Abstract:

Research Interests:

My research interests are in micro/nano fabrication of functional materials, colloids and interfaces of soft nanostructures, self and directed assembly, nano/micro-electronics, microfluidics, 3D printing, photovoltaics, polymer thin-films, carbon nanomaterials, carbon composites, soft elastomers, liquid metals, flexible electronics, stretchable transistors, carbon MEMS/NEMS in security, health, energy and environmental applications.

Teaching Interests:

In addition to research, I would also like to teach Chemical Engineering, Materials Engineering and Physics based on my interdisciplinary academic background and experiences.

I would like to help with the countryâ€™s emerging problems in healthcare, agriculture and security sectors to benefit a broad spectrum of society while using minimal resources. Multifunctional nanointerfaces can be used to develop sensor and detection devices such as biosensors, explosives trace detectors, mechanical-stress sensors, wastewater management and energy storage owing to their nanoscopic surface properties. During my doctoral study in Chemical Engineering at Indian Institute of Technology Kanpur, I have performed research on synthesis of various nanostructured materials and their application towards electrochemical biosensing, waste water treatment, Li-ion batteries and supercapacitors. I have used electrospun carbon nanofibers and nanocomposites with metal and metal-oxide nanophase to fabricate sensors for enzymatic detection of cholesterol, breast cancer, food toxin and soil nutrient. Catalytic and photocatalytic properties of those nanostructures were used to degrade industrial and agricultural wastes. The pseudo capacitance of carbon nanomaterials was integrated with metal oxides to store electrochemical energy in supercapacitors and Li-ion batteries. At present in my postdoctoral research here at NC State University, I am focusing on flexible electronics such as soft stretchable transistors, strain and pressure sensors. I have used microfluidic and 3D printing approaches to pattern liquid metal alloy on soft elastomers to create stretchable conducting electrodes. I am also working with Eastman Chemical Company, USA on automotive coating formation and impact resistive coating on glass windshields.

529797 Building Hierarchical Materials for Energy and Catalysis

Abstract:

The growth of hierarchical materials from solution/vapor is an important challenge for developing new and improved materials and materials synthesis pathways for energy and catalysis. One way to address this challenge is understanding crystal growth mechanisms via directly probing the crystal growth and using the detected information to develop predictive modeling capabilities for new materials. Therefore I aim to understand crystal growth mechanisms *in situ* via directly measuring forces at mutually oriented particle-particle interfaces and imaging the nucleation and crystal growth using various techniques, such as in situ liquid phase scanning/transmission electron microscopy and homemade high temperature high pressure atomic force microscopy. Based on the understanding the nucleation and crystal growth mechanisms, I will design and synthesize novel hierarchical materials using in energy and catalysis.

Research Interests: Crystal growth, Material synthesis, Nanofabrication, Energy and Catalysis

Teaching Interests: I would like to teach courses related to Thermodynamics, Kinetics, Heat Transfer, Nanofabrication, and Nanoscience. I am also willing to develop new courses based on my research experiences and interests such as electron microscopy and scanning probe microscopy.

541699 Machine Learning and Data-Enabled Design and Discovery of Nano and Soft Materials

Abstract:

Research Interests: High-throughput materials design, Energy storage materials, Polymer nanocomposites, Superlattice, Self-healing materials, Self-assembly, Phase transition and free energy calculation, and Machine learning

Teaching Interests: Thermodynamics, Transport phenomena, Polymer Physics, Numerical Methods and Molecular modeling

Soft materials are highly correlated many body systems with complex structures and dynamics spanning a wide range of length and time scales. Their relaxation processes involve non-equilibrium phenomena such as vitrification, jamming and semi-crystallization, which are highly process-dependent and for which no comprehensive theoretical framework and understanding exist. A significant advancement in the understanding of these processes in soft materials are vital for their successful implementation in future electronics, medicine and energy devices. The current understanding of soft materials is limited by many factors. First, atomic level interactions in many soft materials, for a wide range of physicochemical conditions, is not well-established. Second, unlike biomolecules where informatics-based design has been accelerated by the availability of large, open, and homogenous structure/property relation databases such as the protein data bank, similar data in soft materials tend to be sparse, heterogenous, and often outright unavailable. Third, due to a general lack of rapid, parallelizable techniques for measuring materials properties, building such databases de novo for a specific problem is often cost-prohibitive. Furthermore, many critical design problems, such as the search for highly flexible barrier materials, target materials properties well outside the range of what have been discovered so far.

The overarching goal of the proposed research program is to address these limitations by combining molecular modeling, machine leaning and evolutionary computing. By combing these tools, I plan to develop more accurate and transferable force fields for a wide range of soft materials, including polymers, small organic molecules, colloids; discover their hitherto unknown structure-property relationship; and accelerate materials design. By integrating data science and first principle calculations, my goal is to address some of the open questions of soft matters â€" such as the non-universal growth of relaxation time upon cooling a material, the interrelationship between ion conductivity, modulus and glass formation behavior of a polymeric ionic liquids, and the universal relationship between the governing forces and structural order parameter of nanoparticles assembly within a polymer matrix. Addressing these questions are crucial for many future technologies, such as safer battery with high energy density and longer durability; capsule that prevents drugs from unwanted release and degradation caused by temperature variation; membrane that pulls salts out of sea water rapidly; flexible smart phone that can be wrapped around humanâ€™s wrist; and a coating that prevents roads and bridges from corrosion. These limitless possibilities can only be realized by efficient screening of the vast configurational space of soft materials, and investigate the above mentioned open questions. The vision of the proposed program is, therefore, to address these challenging problems and establish new materials design rules by building large amount of accurate homogeneous structure property data; use artificial intelligence (AI) and machine learning to understand their interrelationship; and employ evolutionary algorithms for discovering new structures and properties. This way, the proposed research program aims to push the frontier of materials research towards more predictive designing and phenomenological modeling.

With successful implementation of this program, I hope to study complex physical science problems and attract external funding. I am confident that this research program will lead to the development of useful design rules for flexible electronics; polyelectrolytes for energy storage application; and nanostructure materials for gas storage, water purification and many other applications. In this program, students will be trained in statistical mechanics, wide range of machine learning approaches and artificial intelligence and their applications in chemical engineering and materials science.

Selected Publications (Total: 14, Under review: 3, Under preparation: 4)

1. **Patra T K**, Zhang F, Schulman D; Chan H, Cherukara M, Terrones M; Das S, Narayanan B, Sankaranarayanan S KRS, Defect dynamics in 2D materials probed by combining machine learning, molecular simulation and high- resolution microscopy, Accepted for publication, *ACS Nano* (2018)
2. **Patra T K**, Meenakshisundaram V, Hung J, and Simmons D S, Neural network biased genetic algorithm for materials design: Evolutionary algorithms that learn, *ACS Combinatorial Science* **19**, 96 (2017)
3. **Patra T K** and Singh J K, Polymer directed aggregation and dispersion of anisotropic nanoparticles, *Soft Matter* **10,** 1823 (2014)
4. **Patra T K** and Singh J K, Coarse-grained molecular dynamics simulations of nanoparticle-polymer melts: Dispersion vs. Agglomeration, *Journal of Chemical Physics* **138**, 144901 (2013)
5. **Patra T K**, Hens A, Singh JK, Vapor-liquid phase coexistence and transport properties of two dimensional oligomers, *Journal of Chemical Physics* **137**, 084701 (2012)

541815 Sustainable Materials for Separations and Catalysis

Abstract:

Research Interests: A long-standing challenge in materials science is the development of functional materials with the ability to separate and convert chemicals in the same material system. Traditionally, these processes are separated into several multi-step operations and an improvement in materials resulting in a reduction of the necessary number of steps would provide considerable benefit beyond that offered by the production of value added products. Inorganic/organic hybrid materials such as MOFs have unique properties including flexibility, hierarchical structure, and tunable chemistries that make them interesting candidates for health, separations, and catalysis applications. In such systems, the MOF architecture strongly affects material functionality. However, synthesizing these materials in functional forms is challenging due to their inherent instability. My research aims to directly address this challenge by developing new approaches to synthesize MOF hybrid materials with precise control over stability, structure, and functionality to elucidate the relationship between molecular-scale properties and macroscopic material performance, with the overarching goal of enabling better understanding of structure and properties of MOF hybrid materials.

Successful Proposals: Oak Ridge National Laboratory, CNMS (2016); Brookhaven National Laboratory, CFN (2018)

Postdoctoral Projects: *Activity and Stability Design Principles of Non-Noble Metal Catalysts to Promote Oxygen Reduction and Oxygen Evolution Reaction Kinetics* and *Active and Sinter Resistant Transition Metal Nitride/Carbide Nanomaterials for Methane Dry Reforming*, Department of Chemical Engineering, Massachusetts Institute of Technology (Advisor: Professor Yuriy Román-Leshkov).

Ph.D. Dissertation: *Acid and Base Gas Exposure and Solvent Effects on Metal-Organic Framework (MOF) Structural and Gas Adsorption Properties*, Department of Chemical & Biomolecular Engineering, Georgia Institute of Technology (Advisor: Professor Krista S. Walton)

Research Experience: My research focuses on the development of new classes of inorganic and inorganic/organic hybrid materials by studying and engineering the effects of synthesis and acid and base gases on material structure and properties. In my Ph.D. research, I studied the adsorption of acid and base gases on metal-organic frameworks. Specifically, I studied the effects of acid gas degradation on metal-organic frameworks to develop an understanding of adsorbed species-property relationships and determine the acidic species involved in the degradation mechanism through a combined experimental and computational approach. Using this approach, I synthesized a variety of metal-organic frameworks and characterized their material properties during acid gas adsorption through the use of a variety of characterization techniques including in situ diffuse reflectance infrared Fourier transform spectroscopy.[2-4] These studies have increased the understanding of the interactions between acidic and basic species and MOF materials.

In my postdoctoral research, I developed synthetic methods to produce unsupported transition metal carbide nanoparticles and monolayer phase-pure carbide nanosheets for catalytic applications. Using these materials I studied the effects of oxidation on monolayer carbides and the effect of morphology on the activity and stability of carbide materials for dry reforming of methane using a variety of characterization techniques, including powder X-ray diffraction, transmission electron microscopy, scanning transmission electron microscopy, and Raman spectroscopy. This work holds the potential to advance the fundamental understanding of carbide materials, impacting the design of novel functional catalysts for a variety of applications in thermal and electrocatalysis.

Future Research Directions: As an independent investigator, I will study materials for reactive adsorption and separation. Initial investigations in the Mounfield Group will consist of three aims (1) synthesis and investigation of inorganic/organic hybrid materials for adsorption of specific chemicals, (2) synthesis and investigation of catalysts for the conversion of these chemicals to value added products, and (3) preparation of reactive adsorbents for separation and conversion for industrial application. This work will require a multidisciplinary approach, combining my Ph.D. work in synthesis and separations with my postdoctoral work in synthesis and catalysis.

Teaching Experience: During my Ph.D., I served two semesters as a teaching assistant (TA) for Chemical Process Principles (ChBE 2100) and one semester for Graduate Mass Transport (ChBE 6260). In both semesters of serving for the introductory undergraduate course, I was selected to lecture several classes, which afforded me the opportunity to lecture weekly and prepare problem sets. Furthermore, I taught several lectures for the graduate course as well. In addition to my teaching experiences in my Ph.D., while studying at University of Florida, I led the development of an undergraduate laboratory course utilizing fuel cells to illustrate principles of general chemistry. During my graduate and postdoctoral career, I have had the privilege to mentor several outstanding undergraduate and graduate students, who continued their careers at places such as, ExxonMobil, Dow, and Tesla and graduate research in Chemical Engineering at UPenn.

Teaching Interests: My teaching interests include the central courses in Chemical Engineering â€“ Mass Transport, Transport Phenomena, and Kinetics and Reactor Design. In addition, I am also interested in teaching Thermodynamics, Math Modeling and courses on materials characterization techniques.

Selected Publications:

1. **Mounfield, WP**; Garg, A; Shao-Horn, Y; RomÃ¡n-Leshkov, Y. Electrochemical Oxygen Reduction for the Production of Hydrogen Peroxide. *Chem.* 4, **2017**, 18-19.
2. **Mounfield, WP**; Han, C; Pang, SH; Tumuluri, U; Jiao, Y; Bhattacharyya, S; Dutzer, MR; Nair, S; Wu, Z; Lively, RP; Sholl, DS; Walton, KS. Synergistic Effects of Water and SO_2 on Degradation of MIL-125 in the Presence of Acid Gases. *Phys. Chem. C*, 120, **2016**, 27230-27240.
3. **Mounfield, WP**; Taborga Claure, M; Jones, CW; Walton, KS. Synergistic Effect of Mixed Oxide on the Adsorption of Ammonia with Metal-Organic Frameworks. *Eng. Chem. Research.* 55, **2016**, 6492-6500. **ACS Editorâ€™s Choice Paper**
4. **Mounfield, WP**; Tumuluri, U; Jiao, Y; Li, M; Dai, S; Wu, Z; Walton, KS. Role of Defects and Metal Coordination on Adsorption of Acid Gases in MOFs and Metal Oxides: An In Situ IR Spectroscopic Study. *Meso. Mater.* 227, **2016**, 65-75.
5. **Mounfield, WP** and Walton, KS. Effect of Synthesis Solvent on Breathing Behavior of MIL-53(Al). *Colloid Interface Sci.* 447, **2015**, 33-39.

543951 Task-Specific Functional Porous Materials: From Academic Laboratory to the Commercial Marketplace

Abstract:

Professional Experience:

06/17 – Present Senior Postdoctoral Researcher, National Energy Technology Laboratory (NETL), Pittsburgh, PA, USA

11/15 – 06/17 Visiting Assistant Professor (Research Scholar), Pacific Northwest National Laboratory (PNNL), Richland, WA, USA

01/15 – 06/17 Assistant Professor of Chemistry, Faculty of Science, Alexandria University (AU), Alexandria, Egypt

05/15 – 11/15 Department Coordinator for the National Authority of Quality Assurance and Accreditation of Education, AU, Egypt

09/14-03/15 Graduate Intern, Alternate Sponsored Fellowship (ASF), PNNL, USA

01/11 – 12/14 Research and Teaching Assistant, USF, Tampa, FL, USA

11/12 – 12/14 Graduate Mentor, USF, Tampa, FL, USA

08/05 -12/10 Lecturer, Faculty of Science, AU, Alexandria, Egypt.

Research Interests: My research focuses on the discovery and synthesis of such novel functional materials as metal organic frameworks (MOFs), covalent organic frameworks (COFs), porous aromatic frameworks (PAFs), porous polymers, and porous organic cages. Currently, my emphasis is on mixed-matrix membranes, nanocomposites, and hollow fibers. Porous materials are all around us, playing significant roles in many aspects of our daily lives. Their "made-to-order" design means their applications are nearly boundless where the characteristic features of their pores in terms of size, geometry and chemical functionality can determine the ways in which they can be employed. The key step in unlocking the functionality of these materials is transferring them from the lab to the marketplace, but their dispersive nature and the labor involved in processing them has impeded their exploitation. In this proposed research, I discuss the design and synthesis of task-specific porous materials that can be engineered in the form of mixed-matrix membranes, nanocomposites, or hollow fibers based on the needs of the market. The combination of porous materials with other compounds, such as organic polymers, nanoparticles, or magnetic molecules, can generate further hybrid materials with great potential for applications ranging from gas storage and separations to catalysis, water purification, energy storage, light harvesting, and beyond. I draw particular attention to the fact that the research projects that I anticipate involving the fabrication of novel end-use materials for real-world applications including environmental remediation, heterogeneous catalysis, and alternative energy applications—are highly interdisciplinary in nature.

Postdoctoral Project-NETL: "Mixed-Matrix Membranes for Post-Combustion Carbon Capture."

Advisor: Dr. David Hopkinson

AU-PNNL Joint Project: "Metal-organic Frameworks for the Selective Removal of TcO_4^-."

PI: Dr. Praveen Thallapally

PhD Dissertation: "Crystal Engineering of Functional Metal-Organic Material Platforms for Gas Storage and Separation Applications."

Advisor: Professor Michael Zaworotko

Teaching Experience:

Summer 2016 Assistant Professor of Chemistry, Faculty of Science, AU, Egypt

C 101-General Chemistry

C 243-Fundumentals of Organic Chemistry

04/15–11/15 Assistant Professor of Chemistry, Faculty of Science, Alexandria University (AU), Egypt

C 101-General Chemistry

C 311-Instrumental Analysis I

C 331-Chemistry of Transition elements and coordination chemistry

C 356- Surface Chemistry and Catalysis

C 243-Practical Physical Chemistry (Coordinator)

01/11–09/14 Graduate Teaching Assistant, USF, USA

CHM 2045L-General Chemistry I Lab (Summer A 2012)

CHM 2046L-General Chemistry II Lab (Spring 2011, Summer B 2011, Fall 2011, Spring 2012, Summer B 2012)

CHM 2210L-Organic Chemistry I Lab (Fall 2013, Fall 2014)

CHM 2211L-Organic Chemistry II Lab (Spring 2013, Spring 2014, Summer 2014)

CHM 3610L-Inorganic Chemistry Lab (Fall 2012)

2011–13 CHM 3610- Inorganic Chemistry Lecture (Fall 2011, Fall 2012, Fall 2013) on behalf of Professor Michael Zaworotko

09/06–12/10 Lecturer, Faculty of Science, AU, Egypt

C 101L-General Chemistry Lab (Fall 2006, Fall 2007, Fall 2008, Fall 2010)

C 211L-Fundamentals of Analytical Chemistry Lab (Spring 2006, Spring 2007, Spring 2010)

0403701L-Advanced Analytical Chemistry Lab (Fall 2009)

C 311L-Instrumental Analysis Lab (Spring 2008, Spring 2009, Spring 2010)

09/05-12/10 Chemistry Tutor, Faculty of Science, AU, Egypt

C 101-General Chemistry

C 211-Fundamentals of Analytical Chemistry

C 243- Fundamentals of Organic chemistry

C 311-Instrumental Analysis I

C 331-Chemistry of Transition elements and coordination chemistry

C 356- Surface Chemistry and Catalysis

C 411- Instrumental analysis II

Teaching Interests: I have more than a decade of experience teaching general, organic, inorganic, analytical, and physical chemistry. I have also had the benefit of working in a variety of educational environments, at various universities in different countries, and at national laboratories. The bottom line is that I can teach just about any chemistry/chemical engineering/materials science course in just about any context. I am, however, particularly excited about the prospect of teaching general chemistry, organic chemistry, inorganic chemistry, thermodynamics, kinetics, introduction to chemical engineering, energy and mass transport and materials science. I would also greatly appreciate the opportunity to create a new graduate course that focuses specifically on my area of expertise, functional porous materials.

Mentorship Experience:

I have consistently during my time as a graduate student, visiting assistant professor, and postdoctoral researcher supervised and directed the research of undergraduate students and graduate interns. These students have worked on the synthesis and characterization of porous metal-organic frameworks and membranes for gas storage and separation applications. In recognition of these efforts, which included supporting the students' presentation of their work at conferences and in high-profile publications, I received a mentor certificate and award from USF. My CV includes further details regarding my mentoring experience.

Selected Publications:

Elsaidi, S.; Sinnwell, M.; Banerjee, D.; Devaraj, A.; Kukkadapu, R.; Droubay, T.; Nie, Z.; Kovarik, L.; Manandhar, , S.; Nandasiri, M.; McGrail, B.; Thallapally, P.; Vijayakumar, M. **Reduced Magnetism in Core-Shell Magnetite@MOF Composites**. *Nano Letters* **2017**, *17*, 6968-6973.

Elsaidi, S. K.; Mohamed, M. H.; Simon, C. M.; Braun, E.; Pham, T.; Forrest, K.; Xu, W.; Space, B.; Zaworotko, M. J.; Thallapally, P. K. **Effect of Ring Rotation upon Gas Adsorption in SIFSIX-3- M (M = Fe, Ni) Pillared Square Grid Networks**. *Chemical Science* **2017**, *8*, 2373-2380.

Elsaidi, S. K.; Mohamed, M. H.; Loring, J.S.; McGrail, B. P.; Thallapally, P. K. **Covalent Coordination Frameworks: A New Route for Synthesis and Expansion of Functional Porous Materials**. *ACS Applied Materials & Interfaces* **2016**, *8*, 28424.

Elsaidi, S. K.; Mohamed, M. H.; ; Schaef, H.; Kumar, A.; Lusi, M.; Pham, T.; Forrest, K.; Space, B.; Xu, W.; Halder, G.; Liu, J.; Thallapally, P. K.; Zaworotko, M. J. **Hydrophobic Pillared Square Grids for Selective Removal of CO_2 from Simulated Flue Gas**. *Chemical Communications* **2015**, *51*, 16872.

Elsaidi, S. K.; Mohamed, M. H.; Wojtas, L.; Chanthapally, A.; Pham, T.; Space, B.; Vittal, J. J.; Zaworotko,

M. J, **Putting the squeeze on CH_4 and CO_2 through control over interpenetration in diamondoid nets**. *Journal of the American Chemical Society*, **2014,** *136*, 5072.

Reviews:

Elsaidi, S. K.; Mohamed, M. H.; Banerjee, D.; Thallapally, P. K. **Flexibility in Metal Organic Frameworks: A Fundamental Understanding**. *Coordination Chemistry Reviews* **2018**, *358*, 125-152.

Patents:

Michael J. Zaworotko, Mona H. Mohamed, and Sameh K. Elsaidi, **Metal-Organic Materials (MOMs) for Adsorption of Polarizable Gases and Methods of Using MOMs**, US 9676807.

544253 Engineering Soft Materials with Different Length Scales for Diversity Applications

Abstract:

Research Interests:

My research is engaged in the field of soft materials for a variety of application. To this end, I use my background in polymer chemistry and materials science to engineer soft materials, often times using microfluidics technique. These soft materials are engineered for use in diverse applications and fields such as cell encapsulation in biology, drug encapsulation in pharmacology, and additive encapsulation in petroleum engineering.

In the future, my research pursues both fundamental and applied science of soft materials. The initial step of my research would be design and synthesis of novel functional soft responsive materials for 1) the fundamental science of therapeutic encapsulation and transport process; 2) Biomaterials for stem cell regulation, 3) flow control by hydrogel for green chemistry. This all leads to the development of new materials as well technologies for a variety of soft materials application.

Teaching Interests:

All the researchers and scientists have the responsibility to guide students who have enormous curiosity in the science. As several rich years of research, I have plenty of teaching experience with students who have very diverse background. Based on these experience, I know how to communicate with students and have developed my own teaching strategy: to give students an opportunity to do hands-on science; to encourage them to explore beyond the disciplinary boundaries of the field; and to interact with student to ensure that the student understands the basic and fundamental concept.

I have prepared to teach both broad undergraduate and graduate courses in Chemistry, Polymer Chemistry and Physics, and Soft Matter. In additional to the fundamental course, I also dedicate to creating new courses related to topic from my own research, notably *Modern Microfluidic Techniques*, *Therapeutic Delivery and Regenerative Medicine*, and *Stem Cell Therapy*.

545753 Advanced Deposition and Characterization of Thin Films for Electronics and Sustainable Energy

Abstract:

Advances in the deposition of thin films have enabled a wide range of technologies that have changed the way we live. These technologies include integrated circuits (e.g. personal computers and mobile electronics), light emitting diodes (e.g. lighting and displays), optical and protective coatings, and solar cells. In the future, thin films will continue to enable the development of new technologies, especially in electronics and sustainable energy. Two important aspects in the deposition of thin films are (1) control of the deposited material in terms of chemical composition, crystallinity, nanostructure, and morphology, and (2) regulation of the energy levels and charge transport within the deposited materials and at their interfaces.

I am interested in advanced methods for depositing thin films with simultaneous control of the chemical composition and morphology in order to obtain desired functionalities. This includes vapor phase and liquid phase methods such as reactive ballistic deposition, glancing angle deposition, electrochemical anodization, and electrodeposition. Combining these methods allows for the synthesis of multicomponent thin films that are either dense or nanostructured with conformal layering. The deposited materials can be used in microelectronic, nanoelectronic, photovoltaic, electrochemical, and photoelectrochemical devices. I am interested in emerging characterization techniques that quantify the energy levels and charge transport within these materials and at their interfaces. I plan to characterize photovoltaic, electrochemical, and photoelectrochemical thin films using new scanning probe measurement techniques and a consistent laser light source that is variable in wavelength and photon flux.

Research Interests:

I am currently a postdoctoral scholar in the group of Prof. Roel van de Krol in the Institute for Solar Fuels at Helmholtz-Zentrum Berlin. My postdoctoral work focuses on the development of complex metal oxide semiconductors and photoelectrode materials for solar water splitting. I completed my Ph.D. in Chemical Engineering at the University of Texas at Austin in the group of Prof. C. Buddie Mullins where I worked on a wide range of projects with additional guidance from Prof. Allen J. Bard. I synthesized mixed metal oxides (e.g. doped $BiVO_4$ and $CuBi_2O_4$) and metal carbides using a variety of methods including vacuum deposition, electrochemical and hydrothermal/solvothermal synthesis, and piezoelectric dispensing. I characterized them using surface science, material science, electrochemistry, and photoelectrochemistry techniques.

Selected Publications:

1. Fuxian Wang, Wilman Septina, Abdelkrim Chemseddine, Fatwa F. Abdi, Dennis Friedrich, Peter Bogdanoff, Roel van de Krol, S. David Tilley, and **Sean P. Berglund**, Gradient Self-Doped $CuBi_2O_4$ with Highly Improved Charge Separation," *Am. Chem. Soc.* 139, 15094-15103 (2017).

   http://dx.doi.org/10.1021/jacs.7b07847
2. Fuxian Wang, Abdelkrim Chemseddine, Fatwa Abdi, Roel van de Krol, and **Sean P. Berglund**, "Spray pyrolysis of $CuBi_2O_4$ photocathodes: improved solution chemistry for highly homogeneous thin films," *Mater. Chem. A* 5, 12838-12847 (2017). http://dx.doi.org/10.1039/C7TA03009F
3. Fatwa Abdi, **Sean P. Berglund**, "Recent developments in complex metal oxide photoelectrodes," *Phys. D: Appl. Phys.* 50, 193002 (22pp) (2017). http://dx.doi.org/10.1088/1361-6463/aa6738
4. **Sean P. Berglund**, Fatwa F. Abdi, Peter Bogdanoff, Abdelkrim Chemseddine, Dennis Friedrich, and Roel van de Krol, "Comprehensive Evaluation of $CuBi_2O_4$ as a Photocathode Material for Photoelectrochemical Water Splitting," *Mater.* 28, 4231-4242 (2016). http://dx.doi.org/10.1021/acs.chemmater.6b00830
5. **Sean P. Berglund**, Huichao He, William D. Chemelewski, Hugo Celio, Andrei Dolocan, and C. Buddie Mullins. "p-Si/W_2C and p-Si/W_2C/Pt Photocathodes for the Hydrogen Evolution Reaction," *Am. Chem. Soc.* 136, 1535-1544 (2014). http://dx.doi.org/10.1021/ja411604k
6. **Sean P. Berglund**, Heung Chan Lee, Paul D. Núñez, Allen J. Bard, and C. Buddie Mullins, "Screening of transition and post-transition metals to incorporate into copper oxide and copper bismuth oxide for photoelectrochemical hydrogen evolution," *Chem. Chem. Phys.* 15, 4554-4565 (2013). http://dx.doi.org/10.1039/C3CP50540E

Teaching Interests:

Teachers and professors have been among the most influential people in my life. I have always admired their commitment to uncovering the truth, mastering specific subjects, and helping others to learn and grow. Teachers play an extremely important role in preparing future generations to think critically and solve new challenges. It would be a great privilege for me to contribute to society as a professor.

My personal journey towards a career as a professor and researcher has been less direct than most. Before studying engineering I studied other fields and after obtaining my B.S. degree in chemical engineering I spent 4.5 years working as a process engineer in the semiconductor industry before returning to academia to obtain my Ph.D. I believe that my experiences have given me a valuable and unique perspective that I can share with students.

As a professor I would like to help students to master fundamental concepts and think rationally about new concepts. In my teaching methods I will emphasize active problem solving to achieve deep learning of science and engineering fundamentals. I will also apply problem solving to non-ideal situations that might be encountered outside of academia. I am interested in teaching a variety of chemical engineering courses from essential core courses to specialty topics such as electrochemistry, photoelectrochemistry, and sustainable energy.

Course Experience:

Chemical Engineering Materials, Transport Phenomena (Heat Transfer, Mass Transfer, and Fluid Mechanics), Introduction to Chemical Practices (General Chemistry Lab), *University of Texas at Austin*, 2009-2013

Q-Team (Research-Based Learning), *Humboldt University*, 2016

Directly Supervised Students:

Two Ph.D. student theses, *Technical University Berlin*, 2016-Present

One M.S. student thesis and one undergraduate student researcher, *Hochschule für Technik und Wirtschaft Berlin*, 2015-2016

Four undergraduate researchers, *University of Texas at Austin*, 2009-2013

545960 Colloidal Templating of Model Mesostructured Surfaces for Electrochemistry, Optics, and Sensing

Abstract:

Research Interests: Mesostructured surfaces such as colloid-based porous materials have been investigated extensively as optical materials, sensors, catalysts, and electrodes, allowing them to be used to help solve global problems in energy, healthcare, and security [1]. These applications are largely enabled by structure on the micron and sub-micron scale (i.e., mesoscale), which controls or enhances physical effects such as light scattering, capillary forces, and mass transport. Many unanswered questions remain regarding these fundamental structure-property relationships, which could be addressed using well-defined mesostructures as a model system. My research uses synthetic control over mesostructured surfaces to investigate their structure-dependent properties and ultimately design new multifunctional materials.

Research Experience: My doctoral research at Harvard probed the self-assembly of colloids into mesoscopically ordered, porous structures known as inverse opals. My studies of colloidal assemblies included their fabrication (using the molecular state of precursors to achieve new structures) and their optical [2, 3] and wetting [4] properties. The overall theme was structure-property relationships, especially how our synthetic control over the structure allowed us to control the materials' properties. As a postdoc at MIT, I have gained expertise in electrochemistry, working on several topics related to electrochemical CO2 capture and conversion [5], developing materials where the molecular, nanoscale, and mesoscale structure influenced their activity.

Future Research Directions: As an independent scientist, I plan to work at the interface of my previous research areas by using colloidal templating to make model systems for investigating different physical effects at mesostructured interfaces. In particular, my lab will develop methodologies to make novel hierarchical mesomaterials with well-controlled properties, and then use them as model systems for electrochemical, optical, and sensing applications. The Phillips Lab's advanced understanding of both the synthesis and properties of hierarchical materials will lead to improved manufacturing processes, more efficient carbon capture and conversion, and better pollution detection, ultimately opening the door to additional future applications in energy, healthcare, and security such as fuel cells, drug delivery, and optical displays.

Teaching Experience and Teaching Interests: I have taught a wide range of courses, including introductory and advanced undergraduate and graduate classes in chemistry and engineering. In particular, I was a teaching fellow (~TA) for an introductory chemistry course, an introductory food science/soft matter course (for which I received the White Prize for Excellence in Teaching), a quantum chemistry course, a graduate teaching course, and an advanced course on the chemistry of materials (for which I received a Certificate of Distinction in teaching). I would be interested in teaching similar courses in the future, as well as classes related to materials synthesis, nanotechnology, and colloid chemistry. Beyond formal teaching, I also have extensive undergraduate advising experience, as well as experience mentoring undergraduate and masters students in the lab. I have volunteered at a number of science outreach activities, and started a new outreach program for high school students.

Selected Publications:

1. K. R. Phillips, G. England, S. Sunny, E. Shirman, T. Shirman, N. Vogel, and J. Aizenberg. "A Colloidoscope of Colloid-Based Porous Materials and Their Uses." Chem. Soc. Rev., 2016, 45, 281-322. *cover article
2. K. R. Phillips*, T. Shirman*, E. Shirman, A. V. Shneidman, T. Kay, and J. Aizenberg. "Nanocrystalline Precursors for the Co-Assembly of Crack-Free Transition Metal Oxide Inverse Opals." Adv. Mater. 2018, 30, 1706329. (*equal contribution)
3. K. R. Phillips, N. Vogel, Y. Hu, M. Kolle, C. C. Perry, and J. Aizenberg. "Tunable Anisotropy in Inverse Opals: Mechanism and Emerging Optical Properties." Chem. Mater. 2014, 26, 1622-1628.
4. K. R. Phillips, N. Vogel, I. B. Burgess, C. C. Perry, and J. Aizenberg. "Directional Wetting in Anisotropic Inverse Opals." Langmuir, 2014, 30, 7615-7620.
5. K. R. Phillips, Y. Katayama, J. Hwang, and Y. Shao-Horn. "Sulfide-derived copper for electrochemical conversion of CO2 to formic acid." J. Phys. Chem. Lett., 2018, 9, 4407-4412.

542504 Nanoengineering Materials with Atomic Specificity for Catalysis and Energy Applications

Abstract:

Research Interests:

Catalysts that transform low carbon feedstocks into energy, fuels, and chemicals are at the core of future sustainable energy solutions. Advances in first principles techniques such as density functional theory (DFT) have, in turn, enabled the *in-silico* screening of a surprisingly large ensemble of catalytic architectures. Candidate materials are rapidly screened by directly mapping rates and selectivity onto one or two descriptors which are usually energies of atomic adsorbates. This mapping is facilitated by linear scaling relations connecting energies of large reaction intermediates to atomic descriptors. Despite transforming catalyst screening from a primarily Edisonian endeavor to one guided by design principles, current approaches are limited by three key features. First, descriptor-based screening while routinely applied to metallic systems, remains largely elusive for structurally and mechanistically complex materials like oxides, metal/oxide interfaces and two-dimensional materials. Second, despite revealing targeted descriptor values that the best catalysts possess, a direct mapping to the local morphology and composition around a binding site for materials beyond monometallic systems is currently lacking. Finally, catalytic processes are inherently dynamic and do not retain idealized structures under reaction conditions, which is often assumed in screening studies. Addressing these compelling questions will further unlock the promise of computational driven catalyst design.

My graduate research at Purdue University with Prof. Jeffrey Greeley focused on identifying design principles for reducible oxides and bifunctional metal/oxide interfaces. Using methanol oxidation on reducible MoO
reaction, I examined the interplay between catalytic reactions and surface redox processes on reaction kinetics. Building on this analysis, reactivity trends for doped MoO_3 were established. Periodic trends uncovered a relationship between electronic and morphological features of defect sites that can ultimately be exploited to rationally engineer reducible oxides. Design principles for bifunctional metal/oxide interfaces were investigated using linear scaling relations. In contrast to metals, alloys, oxides and zeolites, scaling laws at bifunctional interfaces deviate from strict bond order conservation constraints, permitting greater flexibility in engineering bifunctional catalysts. These deviations together with their physical origin are explored using electrostatic interactions at the interface.

Over the past year, my postdoctoral research at Stanford with Dr. Frank Abild-Pedersen within the SUNCAT-Nørskov group has focused on creating a site-specific mapping of adsorption energies to local morphological features of bimetallic alloys. With minimal inputs from DFT, energies of both adsorbates and surface metal atoms are expeditiously predicted with atomic resolution, across a large combinatorial space of structure and composition. Using site specific energies of adsorbates and metal atoms, we can directly screen for active site architectures that not only display high rates/selectivity but are also thermodynamically feasible under reaction conditions.

Future plans for my independent research group will be focused on nanoengineering multicomponent oxides and two-dimensional heterostructures that contain cooperative catalytic functions (acid, base, and redox sites). Targeted catalytic applications will include synthesizing platform chemicals (alkenes, alcohols, and aldehydes) using raw materials derived from both shale gas and biomass-based sources. The first phase of this agenda will identify functional forms explicitly linking energy descriptors to morphological features of active sites. Predicted descriptor values will then be integrated within multiscale models simulating catalytic cycles on inherently dynamics active sites that are characteristic to these materials. These studies will ultimately open pathways towards reverse engineering realistic active site motifs displaying targeted rates and selectivity on fundamentally new classes of catalytic materials.

Teaching Interests:

My teaching interests include core chemical engineering courses like mass and energy balances, thermodynamics, separations, transport phenomena, chemical reaction engineering, and chemical engineering laboratory. Depending on departmental needs, I plan on developing a mezzanine level course on computational materials design and an advanced graduate elective on computational modeling of catalytic processes.

I believe that effective teachers enthusiastically engage students, clearly communicate complex concepts to diverse audiences, emphasize continuous learning, and ultimately inspire the next generation of students. Within an engineering curriculum, it is important to deconstruct complex ideas by building scaffolds, and continuously reinforce linkages between course content, the broader curriculum, and industrial applications. This can be greatly facilitated by proven pedagogical advances like active and blended learning strategies. To mold myself into an effective educator, I enrolled in a course preparing postdocs for faculty careers and I am currently participating in the *postdoc teaching certificate* program at Stanford. My prior teaching experiences include preparing and delivering lectures in recitations and conducting laboratory sessions in core undergraduate courses of *chemical reaction engineering* (junior year) and *chemical engineering laboratory* (senior year). I also served as a guest lecturer in *advanced modeling for catalysis studies* (advanced graduate elective). I have mentored two high school students (ACS SEED program), one undergraduate honors student, one graduate student, and peer-mentored one postdoc in research projects involving *ab initio* simulations of heterogeneous catalysts. I strongly endorse STEM outreach programs which tend to propel non-traditional students into STEM fields and I have been involved in several such activities during graduate school.

542421 Strategic Advancement of Targeted Nanomedicines: Intelligent Bio-Nanoengineering Using Molecular Imaging in 3D and *In Vivo* tumor Models

Abstract:

Abstract ID: 542421

Girgis Obaid, PhD, Postdoctoral Research Fellow, Wellman Center for Photomedicine, Massachusetts General Hospital, Harvard Medical School

Title

Strategic Advancement of Targeted Nanomedicines: Intelligent Bio-Nanoengineering Using Molecular Imaging in 3D and *In Vivo* tumor Models

Email

girgisobaid@gmail.com

Abstract Text

Research Interests:

I am a biochemist with strong doctoral training in nanochemistry and ligand targeted photodynamic therapy of cancer. My current Postdoctoral training in the laboratory of Dr. Tayyaba Hasan is funded by my NIH:NCI K99-R00 award and focuses on image-guided ligand targeted nanotherapeutic combination regimens.

With the clinical emergence of targeted nanomedicines, a concurrent emergence of skepticisms has arisen regarding the true capacity for ligand directed nanotechnologies to engage with their tumor cell targets *in vivo*. This skepticism has created a growing knowledge gap between nanotechnologists developing cutting edge ligand targeted nanotherapeutics, with their various permutations and utilities, and the broadly inconsistent *in vivo* responses. It is becoming increasingly evident that the complexity of multiplexed bio-nanoconstructs requires detailed and rigorous tuning, where the combination of multiple nanoconstruct properties must be evaluated together. Such an approach is insufficient when 2D cellular binding specificity is used as the primary metric for success, and it becomes imperative that specificity to 3D tumor tissue and *in vivo* molecular specificity is validated.

In light of this unmet need, I am driven to establishing a collaborative fully-funded K99-R00 NIH sponsored independent program that is dedicated to bridging the knowledge gap between fundamental nanoengineering of targeted nanomedicines and true molecular binding and biological responses to combination-based targeted therapies for cancer. Utilizing my consistent evidence of cross-disciplinary expertise, and my dedicated postdoctoral training in clinically-relevant targeted nanomedicines and molecular imaging in both complex 3D models and *in vivo*, I aspire to establish a unique program. This program will specifically serve to enrich the understanding of fundamentals in nanotechnology, to provide well-thought out and validated workflows of *in vivo* testing in order to expedite the translation of safer, more specific targeted nanomedicines. Alleviating this bottleneck in the literature that centers on the skepticism with regards to the reported inconsistencies will provide invaluable centralized information that will serve to forge both industry and academic partnerships with my future host institution. This is consistent with my current research interests that include partnerships with two industry corporations and with four national esteemed academic institutions.

Publications *(see bottom)*

Education

Ph.D. University of East Anglia, United Kingdom (Advisor: Professor David Russell)

B.Sc. Biochemistry (Hons) University of East Anglia, United Kingdom

Funding

1. Molecular imaging-directed bioengineering of nanoconjugates as adaptable tumor targeting platforms

NIH/NCI 1K99CA215301

Role on Project: PI

02/01/18 – 02/01/23

2. A multifunctional drug delivery system for image-guided combination of PDT and anti-angiogenic agent in an *in vivo g*lioblastoma model

Bullock-Wellman Fellowship

Role on Project: PI

01/01/14 – 01/01/15

Peer Review

Independent Editor for Journals: *Scientific Reports, RSC Advances, PLoS ONE, Photochemistry and Photobiology, Photochemical and Photobiological Sciences, Photodiagnosis and Photodynamic Therapy, Lasers in Surgery and Medicine*

Teaching Interests:

Having been the lucky recipient of outstanding teaching and mentorship throughout my entire academic path, I am passionate about empowering the next generation of students to question, observe, evaluate and educate others, themselves. obseGiven my background in biochemistry, nanotechnology and clinical nanomediciens, I am driven to pursue a cross-disciplinary Tenure Track Teaching position at departments including Chemical Engineering, Pharmacy, Pharmacology, Bioengineering, Drug Discovery and Medicinal Chemistry.

I have given lectures on “Nanomaterials in the Human Body: *Behavior, Interactions and Customization*” for a BMED 4200 – Biomaterials and Tissue Engineering class at Wentworth Institute of Technology and on “Light activated nanomedicines” at the Harvard-MIT (HST) Summer Institute in Biomedical Optics. I have also been an active member of the Mass General Postdoc Association (MGPA) for five years and have founded a highly successful initiative to bridge communication between academia and industry. I have mentored over 10 scientists at training levels varying from undergraduate, doctoral, medical, postdoctoral and clinical fellowships. My long-standing academic philosophy that I teach my mentees is to observe with critical diligence to evaluate from every angle and to question every assumption made.

Publications

Obaid, G., Samkoe, K., Mallidi, S., Tichauer, K., Bano, S., Park, Y., Silber, Z., Jing, W., Pogue, B. and Hasan., T. Paired-Nanotracer Imaging quantifies in vivo nanoconstruct binding to EGFR in glioblastoma and enhances the specificity of imaging for fluorescence guided surgery (2018) *Pending submission*

Obaid, G., Jin, W., Bano, S., Kessel, D. and Hasan., T. Nanolipid formulations of benzoporphyrin derivative: Exploring the dependence of nanoconstruct photophysics and photochemistry on their therapeutic index in ovarian cancer cells (2018) *Under review*

Obaid, G., Bano, S., Mallidi, S., Kuriakose, J., Broekgaarden, M., Silber, Z., Bulin,A-L., Wang, Y., Mai, Z.,Simeone, D. and Hasan., T. 2018. Chemically tuned NIR-light activated photoimmuno-nanoconstructs using copper-free click chemistry exhibit efficient penetration and selective destruction in heterocellular 3D tumor models (2018) *Under review*

Rizvi, I.*, Obaid, G*., Bano, S., Hasan, T. and Kessel, D., (2018) Photodynamic therapy: Promoting in vitro efficacy of photodynamic therapy by liposomal formulations of a photosensitizing agent. Lasers in surgery and medicine. doi: 10.1002/lsm.22813. https://www.ncbi.nlm.nih.gov/pubmed/29527710

Obaid, G., Spring, B.Q., Bano, S. and Hasan, T., (2017) Activatable clinical fluorophore-quencher antibody pairs as dual molecular probes for the enhanced specificity of image-guided surgery. Journal of biomedical optics, 22(12), p.121607. https://www.ncbi.nlm.nih.gov/pubmed/28853247

Mallidi, S., Anbil, S., Bulin, A. L., Obaid, G., Ichikawa, M., & Hasan, T. (2016). Beyond the Barriers of Light Penetration: Strategies, Perspectives and Possibilities for Photodynamic Therapy. Theranostics, 6(13), 2458. https://www.ncbi.nlm.nih.gov/pmc/articles/PMC5118607/

Obaid, G., Broekgaarden, M., Bulin, A. L., Huang, H. C., Kuriakose, J., Liu, J., & Hasan, T. (2016). Photonanomedicine: a convergence of photodynamic therapy and nanotechnology. Nanoscale, 8(25), 12471-12503. http://pubs.rsc.org/en/content/articlelanding/2016/nr/c5nr08691d#!divAbstract

Obaid, G., Chambrier, I., Cook, M.J. and Russell, D.A. Cancer targeting with biomolecules: A comparative study of photodynamic therapy efficacy using antibody or lectin conjugated phthalocyanine-PEG gold nanoparticles, Photochemical and Photobiological Sciences 2015, 14 (4), 737-747. DOI: 10.1039/c4pp00312h http://www.ncbi.nlm.nih.gov/pubmed/25604735

Obaid, G., Chambrier, I., Cook, M.J. and Russell, D.A. Targeting the Oncofetal Thomsen–Friedenreich Disaccharide Using Jacalin-PEG Phthalocyanine Gold Nanoparticles for Photodynamic Cancer Therapy, Angewandte Chemie International Edition 2012, 51 (25), 6158-6162. DOI: 10.1002/anie.201201468 http://onlinelibrary.wiley.com/doi/10.1002/anie.201201468/abstract

542431 Utilizing Nano- and Micro-Particles for Safe and Efficient Gene and Drug Delivery

Abstract:

Research Interests: My research expertise centers around cell culture techniques. I have been trained in cell culture for human embryonic stem cells, particularly with the aim of differentiating them into pancreatic progenitor cells. I have also investigated the toxicity of nanomaterials in several cell lines including 3T3 fibroblasts and A549 adenocarcinoma. I have studied the nanotoxicity of nanomaterials fabricated in the lab (nickel on silicon dioxide), commercial particles (silicon dioxide) and industrially relevant particles generated by flame spray pyrolysis by collaborators.

In addition to studying the nanotoxicity of commercial silicon dioxide nanoparticles, I have also investigated the bovine serum albumin (BSA) protein corona on these particles. Utilizing a hybrid of chemistry and cellular engineering techniques, I was able to quantitatively assess the secondary protein structure and the protein adsorption onto the particles. The main focus of this project was to investigate the pH effects on the protein corona, which was done as part of a larger investigation within the Grassian laboratory. This project was further expanded into an investigation of ion effects on the protein corona as ions are a large component of many biological fluids. Each of these projects included mentorship of full-time undergraduate researchers whose contributions to the projects and publications were exceptionally valuable.

Currently, my investigations focus on characterizing particles, delivering particles to cells *in vitro* as well as animals to assess how particles either enhance or mitigate certain diseases. The formulations are developed based on pharmaceutical knowledge as well as empirical evidence. Initial formulations are largely dependent on what has previously been shown successful by Salem laboratory members, however, improvements are constantly underway. These projects span many different diseases including gene delivery to treat cancer, interactions of vaccine and engineered nanomaterials in allergic asthma, and neural inhibitors to aide in recovery following heart failure events.

Throughout graduate school, I have received several fellowships and scholarships. Prior to entering graduate school, I was awarded the National GEM Consortium Associated Fellowship. Along with my application materials, I was nominated for and received a Dean's Graduate Research Fellowship through the Graduate College at the University of Iowa, which provides full funding support for the first and fifth years, as well as every summer. I was also awarded the Alfred P. Sloan Foundation Minority PhD Scholarship, which provides external scholarship funding to be used in any manner that supports graduate research and productivity. This award is specific to students who are members of underrepresented groups and who are considering academia as a career path. After the completion of my M.S. thesis, I was also awarded a two-semester fellowship from the University of Iowa Graduate College. Additionally, my research was featured in the Dare to Discover Campaign sponsored by the Office of Research and Economic Development along with 62 other undergraduate and graduate students at the University of Iowa in the 2018 campaign.

<u>M.S. Projects</u>:

The influence of pH on the bovine serum albumin corona about commercially available silicon dioxide nanoparticles

Givens, B. E.; Diklich, N. D.; Fiegel, J.; Grassian, V. H., "Adsorption of bovine serum albumin on silicon dioxide nanoparticles: Impact of pH on nanoparticle-protein interactions" 2017. Biointerphases. 12 (1), DOI: 10.1116/1.4982598.

Givens, B. E.; Xu, Z.; Fiegel, J.; Grassian, V. H., "Bovine serum albumin adsorption on SiO_2 and TiO_2 nanoparticle surfaces at circumneutral and acidic pH: A tale of two nano-bio surface interactions." 2016. Journal of Colloid and Interface Science. 493, 334-341.

The influence of ions on the bovine serum albumin corona about commercially available silicon dioxide nanoparticles

Givens, B. E.; Wilson, E.; Fiegel, J., "The effect of solution properties on protein corona formation." 2018. Langmuir. *In preparation.*

Ph.D. Projects:

Nanotoxicity of engineered nanomaterials at occupationally-relevant exposure levels

Morris, A. S.; Givens, B. E.; Salem, A. K., "Serum effects on copper oxide nanoparticle colloidal behavior and toxicity in BEAS-2B cells" 2018. ACS Nano. *In preparation.*

Tuning particle properties for desired sustained-release kinetics

Terry, T. L.; Givens, B. E.; Rodgers, V. G. J.; Salem, A. K., "Tunable properties of poly-DL-lactide-poly(ethylene glycol) porous microparticles for sustained release of polyethylenimine-DNA polyplexes." 2018. AAPS PharmSciTech. *Invited Manuscript.*

Non-viral gene delivery decreases over-expression of tumor-linked protein

Givens, B.E.; Devor, E. J.; Warrier, A.; Salem, A. K., "Non-viral nanoplex delivery of CRISPR-Cas9 can knockdown PLAC1 in endometrial cancer cells." 2018. Biomaterials, *In preparation.*

Teaching Interests: I realized that I wanted to enter academia midway through my undergraduate studies and therefore took two teaching assistantships through other departments in areas of interest to me. I served as a teaching assistant and grading support for Introduction to Shakespeare in the English Literature department. Most valuably, this opportunity gave me additional practice with written communication. The following year, I was one of 16 teaching assistants for Introduction to Biology, and with three other teaching assistants, I planned a discussion session each week. Working as part of a team of students that were all interested in the same topic, but not necessarily majoring within the sponsoring department provided valuable experience with effectively communicating to individuals outside of your chosen field. I was also able to solidify that teaching was something I was thoroughly interested in and wanted to pursue as a career.

In graduate school, I have served as the teaching assistant and laboratory instructor for Process Dynamics and Controls. I chose this course because it was a course that I really enjoyed when I took it, and because it contained material that I was not able to practice with my research project. I felt that it kept the areas of my brain active that may otherwise not have been. I also taught a GRE preparation course through the University of Iowa Graduate College's Summer Research Opportunity Program (SROP), specifically for the quantitative section. Not only has this improved my ability to deliver information at the front of the classroom, it has also made me more passionate about delivering material in a way that holds students' interests. I believe in creating a safe space and an active learning environment for students, particularly with smaller class sizes.

In my future, I hope to instruct courses such as drug delivery and transport phenomena because they pertain to my research. My favorite course at both the undergraduate and graduate levels was thermodynamics, and I would enjoy teaching this course at either or both levels. My philosophy is to take the scary connotation out of the course and deliver the information in such a way that it is informative and relevant.

542432 Engineering Optical Nanomaterials to Probe Brain Chemistry

Abstract:

Research Interests:

Unlocking our understanding of neurological function requires methods that can probe chemical activity at the smallest and fastest time scales. The emergence of optical nanomaterials provides a unique toolset to develop multifunctional optical probes to explore neurochemistry at the nanoscale and millisecond timescale where neurotransmission occurs. I am interested in exploring new methods to engineer hybrid nanomaterials that incorporate covalent and non-covalent attachment of small molecules and polymers to develop fluorescent chemical sensors to probe biological systems.

My recent work focuses on using infrared-fluorescent nanosensors composed of single-walled carbon nanotubes to image the release and re-uptake of dopamine in brain tissue. Dopamine is a neuromodulatory neurotransmitter involved in neural circuits controlling motion, learning, and motivation. Understanding the role of dopamine as a neuromodulator, and the role its dysregulation plays in disease, remains elusive in neuroscience and psychiatric medicine. With custom-built infrared microscopy and spectroscopy equipment I designed (Del Bonis-O'Donnell *et al. Adv. Funct. Mater.* 2017), and neuromodulator nanosensor technology I developed (Beyene *et al. bioRxiv*, 2018; Del Bonis-O'Donnell *et al. ACS Chem Neurosci.* 2018), I have created a platform to image dopamine, and other catecholamines, with unprecedented spatial and temporal resolution required to study endogenous neuromodulation. This approach promises to enable non-invasive, in vivo imaging of neurochemistry deep in the living brain.

My future independent work aims to combine my nanosensor and imaging techniques to probe fundamental biological questions related to neurotransmission. Specifically, the merger of nanosensors with deep-tissue infrared imaging promises to reveal new details into the mechanisms of neurochemical dysregulation involved in neurodegenerative diseases, such as Parkinson's disease (*DoD PRP Young Investigator Award*, Awarded 2018). Towards this goal, I will construct microscopy systems that will enable the cooperative use of infrared neuromodulator nanosensors with existing tools for neuroimaging, including calcium imaging and optogenetic stimulation, to create a new 'neurochemical dimension' in our neuroscience interrogative toolkit to address the unknowns of brain function.

Additionally, I am enthusiastic to explore new directions in nanomaterial functionalization to create new optical probes for biosensing. Borrowing strategies from combinatorial chemistry, molecular biology, and biomaterials, I will expand the toolkit for biochemical sensing using both carbon nanomaterials and noble metal nanoclusters to visualize the dynamics of new biochemical targets critical to biological function, including additional neurotransmitters, hormones and regulatory proteins.

Select References:

1. JT Del Bonis-O'Donnell, *et al.* Dual Near-Infrared Two-Photon Microscopy for Deep-Tissue Dopamine Nanosensor Imaging, *Funct. Mater.*, 2017, **27**, 1702112.
2. JT Del Bonis-O'Donnell *et al.* DNA-Stabilized Silver Nanoclusters as Specific, Ratiometric Fluorescent Dopamine Sensors. *ACS Chem. Neurosci.*, 2018, acschemneuro.7b00444.
3. A Beyene, K Delevich, JT Del Bonis-O'Donnell *et al.* Imaging Striatal Dopamine Release Using a Non-Genetically Encoded Near-Infrared Fluorescent Catecholamine Nanosensor. *BioRxiv*, 2018.

Teaching Interests:

I am excited to teach topics including transport, thermodynamics, statistical mechanics, and biophysics, with a personal emphasis on integrating hands-on, discovery-based laboratory curriculum with a foundational theoretical background that can be translated to the lecture hall. I intend to leverage my academic background (B.S. Mathematics-Physics, Ph.D. Mechanical Engineering with emphasis in thermofluids and microelectromechanical systems, Postdoc in Chemical and Biomolecular Engineering) and previous experience as a mentor both in the classroom and in the laboratory to provide guidance for undergraduate and graduate students to grow into independent thinkers, scientists and engineers. Previous experience working in laboratory settings comprised of diverse technical and cultural backgrounds will aid me in growing and managing a research group that excels at tackling interdisciplinary problems at the intersections of engineering, physics, chemistry and biology.

542459 Reprogramming Tumor-Clearing Macrophages with Nanotherapeutics

Abstract:

Research Interests:

Macrophages (MÏ†s) are key immune effectors that infiltrate into the tumor in high numbers. They undergo a switch from an activated (M1) tumoricidal state to an immunosuppressive (M2) phenotype within the immunosuppressive tumor milieu, which facilitates tumor growth and metastasis. The design and manipulation of nanoparticle platforms for drug delivery is an increasingly prominent area of interest for cancer treatment; however, while many drug delivery platforms focusing on targeting macrophages have shown efficacious pre-clinical results in certain cancer types, this has not translated to a significant number of FDA approved, clinically applied nanoparticle systems. This is especially true for solid tumors such as glioblastoma, of which immune suppressive cells such as macrophages prevented effector T cells from penetrating into the solid tumor. Therefore, there is great opportunity to bring macrophage-targeting nanotherapeutics for the treatment of glioblastomas, with results that can be translated to other solid tumors. *My research goals focus on using nanoparticle as biophysical probes to study the behaviors of macrophage within the tumor. This knowledge will then be used to better design and implement therapeutic nanoparticle platforms in clinically relevant models of solid tumors.* My laboratory will be built on multidisciplinary work that involves a team of experts in engineering, nanomedicine, materials science, and cancer immunology. It will provide training for engineering students, cancer immunology students, and clinical fellows.

I am uniquely suited to establish a laboratory that integrates training in engineering nanoparticle platforms with an understanding of nanoparticle formulation, pharmacokinetics, and cancer-immunology. During my doctoral work with Dr. Kannan Rangaramanujam, I developed dendrimer-based targeted therapy for brain tumor and CNS inflammation. a now patented technology. These works built the guiding principles for designing nanomedicine that can efficiently target CNS diseases. Throughout my graduate studies, I led collaborations with faculty and students in the departments of neuroscience, neurology, and oncology. My postdoctoral training with Dr. Matthias Stephan and Dr. Eric Holland has built my foundation in cancer immunology, leading to experience in understanding the role of macrophages in cancer development, as well as the development of clinically relevant animal models of cancers. Based on my combined doctoral and postdoctoral work, I was recently awarded the American Brain Tumor Association Basic Research Fellowship Award.

Teaching Interests: My laboratory will provide training for engineering students, cancer immunology students, and clinical fellows.

Education

Ph.D., 2016, Johns Hopkins University, Materials Sciences and Engineering

Selected Publications

- **Zhang F**, Stephan SB, Ene CI, Smith TT, Holland E, Stephan MT. Nanoparticles that reshape the tumor milieu create a therapeutic window for effective T cell therapy in solid malignancies. Cancer research. 2018. canres. 0306.2018
- **Zhang F***, Mastorakos P*, Mishra M, Mangraviti A, Hwang L, Zhou J, Hanes J, Brem H, Olivi A, Tyler B, Kannan RM. PAMAM Dendrimer Biodistribution in Glioblastoma and Intrinsic Targeting of Tumor Associated Macrophages. Biomaterials. 2015. 52:507-16.
- **Zhang F**, Magruder T, Lin YA, Crawford TC, Grimm J, Sciortino C, Wilson MA, Blue M, Kannan S, Johnston M, Baumgartner W, Kannan RM. Generation-6 hydroxyl PAMAM dendrimers for improved delivery of therapeutics to the brain in a large animal brain injury model. J Control Release. 2017. 249:173-182.
- **Zhang F**, Nance E, Jasty V, Mishra MK, Kambhampati SP, Zhang Z, Kannan RM, Kannan S. Surface functionality affects the biodistribution and targeting to neuroinflammation for intra-amniotically delivered dendrimers. J Control Release. 2016. 10;237:61-70.

542495 Complex Nano-Architectures from Self-Assembly and Surface-Confined Chemistry for Energy Storage and Beyond

Abstract:

Research Interests:

Self-assembly of amphiphilic molecules such as surfactants and block copolymers has enabled the fabrication of ordered morphologies with single to 10s of nanometers dimensions, including ordered three-dimensionally (3D) co-continuous network structures. The utilization of such self-assembling molecules to structure-direct inorganic, organic, and polymeric materials allowed the fabrication of nano- and mesoporous functional materials with structure-based performance enhancements in numerous applications, including energy storage and conversion. While the range of nanostructured materials between the sub-nanometer to mesoscale (<50 nm) is well explored, there exists a gap in characteristic length scales of structured and porous materials in the range of 100s of nanometers. This range is of particular interest to obtain well-defined architectures and assemblies of multiple functional materials layered on top and confined within each other for exploring solid interfacial phenomena and to obtain radically new device concepts and designs, such as folding all device components into a single hybrid architecture. In my research program, I aim to develop methods to obtain 3D structured functional materials and multicomponent architectures with pore sizes and characteristic dimensions beyond the mesoscale in the 100s of nanometers. To achieve such hybrid architectures and devices, my efforts will initially focus on three distinct directions:

(1) *Self-assembled 3D structures beyond the mesoscale (100-1000 nm)* through solution assembly, ultra-large self-assembling amphiphiles, and controlled phase-separation. In particular, initial focus will be put on the use of block copolymers as the interfacially active, amphiphilic molecule in microemulsions, lyotropic liquid crystals, bicontinuous interfacially jammed emulsion gels, and for the combination of macro- and microphase separation. Furthermore, I aim to develop and synthesize a new class of ultra-large soft amphiphiles such as colloid-like molecular Janus architectures. Since the parameter space of self-assembly and structuring methods is very large, we will employ droplet-based microfluidics to screen for optimal conditions and compositions using small amounts of materials at fast rates.

(2) *Multilayered three-dimensionally inderdigitated nanostructures from surface confined syntheses methods* on nanopore surfaces for rationally designed nano-integrated architectures. There exists a lack of syntheses and deposition techniques that can conformally coat nanoporous but macroscopic materials with high uniformity. In addition to gas-phase deposition processes, I aim to develop electrodeposition methods of inorganics and polymers, as well as layer-by-layer syntheses using surface sol-gel (SSG) and surface-initiated step-growth polymerizations, respectively, on high-surface area nanostructures. These efforts will be complimented with microfluidic reactors for fast screening of synthesis conditions.

(3) *Three-dimensionally integrated battery architectures ("3D Batteries").* Our initial focus for applying multilayered 3D nanoarchitectures is the development of so-called 3D batteries, in which the anode and

cathode are interpenetrating each other in three dimensions on the nano- or microscale. The integrated electrodes are fully separated by a 3D continuous solid electrolyte layer to prevent electrical contact between the electrodes while keeping electrode distances to a minimum for fast ion transport. The small distance between the electrodes conceptually allows very fast charging and discharging of the batteries, while the lack of porosity due to the integrated architecture is volumetrically efficient and enables high energy densities.

Another focus of my research program will be the development of novel dynamically responsive materials. We will utilize thiol-ene chemistry-derived polymer networks as ultra-durable hosts to attach dynamically responsive molecular moieties. These dynamic networks will also be synthesized in 3D nanotemplates to study the effect of nanostructuring in combination with molecular functionality and architecture on the dynamic bulk behavior and responsive properties. One particular class of material we will study are liquid crystal elastomers (LCEs), including the development of new LCE monomers from inexpensive precursors in simple syntheses and the alignment of LCE orientation in nanostructures and on liquid surfaces and interfaces.

Research Experience:

My research program will build on my experience in block copolymer structure direction of functional inorganic materials for energy applications that I obtained during my PhD under the supervision of Prof. Uli Wiesner at Cornell University, and expertise in emulsion templating of dynamically responsive polymeric materials from droplet microfluidics that I obtained during my current postdoctoral research position with Prof. David Weitz at Harvard University.

During my PhD, we developed and fabricated a trifunctional three-dimensionally integrated nanohybrid, a so-called 3D battery, featuring an ordered 3D networked (gyroidal) mesoporous carbon anode, coated with an ultra-thin polymer electrolyte, and backfilled with an inorganic-polymer cathode composite. We employed a combination of molecular self-assembly, electropolymerization, and nano-confined synthesis to obtain a final monolithic solid-state 3D architecture with each layer possessing a thickness below 20 nm. The synthesized 3D nanohybrid with macroscopic overall dimensions exhibited battery-like properties [1]. This first demonstration of a functional 3D battery with nanoscale dimensions and interdigitation was based on and only possible due to our development of highly stable gyroidal mesoporous carbon monoliths with an ultra-large pore size of 40 nm and high temperature stability up to 1600 ºC [2]. We employed these ordered mesoporous carbons to demonstrate their most influential structural and chemical properties on the electrochemical performance as sulfur hosts in lithium-sulfur batteries [3]. We further showed the limitations of atomic layer deposition (ALD) on coating nanoporous monolithic materials with macroscopic dimensions, with a maximum deposition depth of 10 microns into 40 nm 3D pores [4]. Furthermore, we synthesized hierarchically macro- and mesoporous inorganic structures with graded macroporosity using polymer phase separation on multiple length scales [5,6].

In my postdoctoral research in the soft condensed matter lab of Prof. David Weitz, I pioneered the synthesis of dynamically responsive microcapsules employing complex drops from microfluidic fabrication. In particular, I developed functional monomer systems with a controlled hydrophobic-to-hydrophilic transition for microfluidic emulsion processing. The highly uniform and tunable microcapsules reversibly change their size, shape, and permeability in response to stimuli such as pH, ionic environment, and light for controlled capture, trap, and release of cargo [7].

Teaching Interests:

During my education and research career I have been part of a variety departments giving me a diverse background in chemistry, materials science and engineering, and applied physics. I am interested and prepared to teach organic, solid state, polymer, physical and general chemistry, electrochemistry and thermodynamics on the introductory level, and polymer chemistry, polymer physics, materials chemistry, and materials characterization at an advanced undergraduate and graduate level. Furthermore, I am very interested in teaching interdisciplinary courses of interest to a wide variety of students such as nanomaterials, nanocharacterization, surface deposition and characterization techniques, electrochemical devices and materials, and materials phase behavior at any level.

During my undergraduate education in Germany, I learned to appreciate the importance of taking the classroom material into the lab to put it to practical use. In particular, the application of physical and chemical principles in experimental setups and projects that allow the students to explore and discover the utility and applicability of these concepts, but also to realize shortfalls and limits of certain models, are crucial to understand when and how to use them. Hence, I would teach most courses in combination with small projects where the students can apply analytical methods using a combination of characterization tools materials of interest to the students. This will give students an appreciation of the physical concepts, as well as the importance of exhaustive characterization with complementary techniques. In particular in polymer science, most courses are focused on the synthesis and the fundamental physics of polymers. I will develop a course that introduces the breadth of polymer characterization tools such as size exclusion chromatography, light scattering, rheology, X-ray scattering, osmolarity, and electron microscopy in a combined picture, and introduce them to students as complementary characterization tools in practical use.

Selected Publications:

[1] **J. G. Werner**, G. G. Rodríguez-Calero, H. D. Abruña, U. Wiesner, "Block copolymer derived 3-D interpenetrating multifunctional gyroidal nanohybrids for electrical energy storage" *Energy Environ. Sci.* **11**, 1261-1270 (2018).

[2] **J. G. Werner**, T. N. Hoheisel, U. Wiesner, "Synthesis and Characterization of Gyroidal Mesoporous Carbons and Carbon Monoliths with Tunable Ultralarge Pore Size", *ACS Nano* **8** (1), 731-743 (2014).

[3] **J. G. Werner**, S. S. Johnson, V. Vijay, U. Wiesner, "Carbon-Sulfur Composites from Cylindrical and Gyroidal Mesoporous Carbons with Tunable Properties in Lithium-Sulfur Batteries", *Chem. Mater.* **27**, 3349-3357 (2015).

[4] **J. G. Werner**, M. Scherer, U. Steiner, U. Wiesner, "Gyroidal Mesoporous Multifunctional Nanocomposites via Atomic Layer Deposition", *Nanoscale* **6**, 8736-8742 (2014).

[5] S. A. Hesse,* **J. G. Werner,*** U. Wiesner, "One-pot Synthesis of Hierarchically Macro and Mesoporous Carbon Materials with Graded Porosity", *Macroletters* **4**, 477-482 (2015). (*Contributed equally)

[6] Y. Gu, **J. G. Werner**, R. M. Dorin, S. W. Robbins, U. Wiesner, "Graded Porous Inorganic Materials Derived from Self-Assembled Block Copolymer Templates", *Nanoscale* **7**, 5826-5834 (2015).

[7] **J. G. Werner**, David A. Weitz *et al.*, "Dynamic Microcapsules with Rapid and Reversible Permeability Switching " *Adv. Funct. Mater.* **2018**, in review; „Hydrogel Microcapsules with Dynamic pH-responsive Properties from Methacrylic Anhydride" *Macromolecules* **2018**, in review.

542512 Continuous Manufacturing of Ultrathin Electronic/Optoelectronic Devices with Colloidal Nanocrystals

Abstract:

Research Interests:

The pervasiveness of electronic devices in our everyday life is increasing, heading towards a ubiquitous environment involving cutting-edge technology in fields of artificial intelligence, the â€œInternet of Things,â€ autonomous vehicles, etc. In particular, wireless communication between the human body and objects is an emerging multi-disciplinary field that finds numerous applications from many diverse research areas. Merging electronic devices with our human body can be realized through the development of wearable/implantable ultrathin electronic devices which conform to the soft human tissues. It is expected that the wearable/implantable technology will lead to electronic skin displays, real-time health monitoring systems, brain-computer interfaces, and many others. Colloidal semiconductor nanocrystals have drawn considerable interest as attractive building blocks for flexible and printable electronic/optoelectronic devices. A critical barrier to commercialization, however, is the toxicity associated with usage of heavy metals (*e.g.* Pb or Cd), which makes RoHS compliant nanocrystals (*e.g.* CISeS, CZTS, Si, Ge etc.) particularly attractive for potentially biocompatible electronic devices. Hence, understanding the charge carrier transport characteristics of non-toxic semiconductor nanocrystals thin films and their practical application are crucial to their implementation in wearable/implantable electronic/optoelectronic technologies.

I have previously demonstrated the first switchable-logic devices (CMOS inverter, and NAND or NOR gate devices) using engineered *non-toxic* colloidal semiconductor nanocrystals. To expand on this, starting from the functionalized nanocrystals, I seek to develop ultrathin printable logic-gate and optoelectronic devices on flexible substrate for achieving wearable/implantable electronic devices. This research will involve: 1) synthesis of solution-processed non-toxic semiconductor nanocrystals; 2) exploration of suitable surface ligands for biocompatible electronic devices; and 3) design and fabrication of ultrathin logic-gate and optoelectronic device architectures. In particular, engineering the surface ligands of tailored nanocrystals strongly influences the carrier transport polarity (unipolar *p*- or *n*-type, or ambipolar) of electronic devices, which are basic characteristic for fabricating efficient logic-gate devices and other devices such as light-emitting field-effect transistors (FETs). These devices are highly applicable to wearable/implantable ultrathin electronic gadgets. Therefore, understanding the ligand chemistry for non-toxic semiconductor nanocrystals will be crucial for fabricating high-performance electronic/optoelectronic devices. Recently, inorganic ligands such as halide, chalcogenide and metal-chalcogenide complex ions have been extensively used to produce strong electronic coupling between nanocrystals, and tune the transport polarity. My research group will employ various classes of atomic inorganic ligands to non-toxic semiconductor nanocrystals for achieving high performance logic-gate and opto-electronic devices on the ultrathin substrates. The versatile tunability of transport polarity of nanocrystal film with applying various inorganic ligands is expected to enable inexpensive and high-throughput solution-based processing of ultrathin electronic/optoelectronic devices in ambient environment. In particular, this significant simplification of integration processes implies the feasibility of high profile fabrication of various device geometries.

In order to pursue such ambitious lines of research, significant financial resources will be needed. All funding opportunities available to me will be pursued, starting with National Science Foundation (NSF) fellowship applications for all grad students with a reasonable chance of success. In the beginning phases, I will also apply for early career grants such as the ACS PRF Doctoral New Investigator Grant and the NSF Early Career Award. Furthermore, national research foundation of South Korea (my home country) opens the chance to make international collaboration with exceptional research groups. I will tackle the main funding agencies with my previous collaborator in Seoul National University, KAIST, Samsung, and so on.

Teaching Interests:

As a teaching assistant, I helped students at undergraduate courses of Chemical Engineering Experiments and Transport Phenomena in both Ajou University and Seoul National University. In addition, I mentored undergraduate and graduate students in research when I was senior PhD graduate students and post-doctoral associate. I was also invited to give research talks several universities and high school.

I have three important key words as mentoring philosophy. 1) Socratic elenchus. As mentor, I lead a mentee to find their own answer by themselves. For this great endeavor, I asked and got answers to stimulate critical thinking. 2) Open mind. In research society, somebody was really unwilling to show their results because of a fear of being scooped. But I emphasized we are studying to make better world, not to get a fund or good papers. Open mind always makes scientific revolution in various research fields. 3) Beyond the ivory tower. As an engineer, I lead mentee to link our knowledge from text book and our real life. This kind of linkage made the learner feel fun of studying and motivate their creativity.

542535 Toward Next Generation of Colloidal 2D Nanomaterials:Â Liquid-Phase Characterization, Modification, and Controlled Assembly

Abstract:

Research Interests:

In last few years, 2D nanomaterials have been exploited for their unique properties to prepare electronic devices, sensors, membranes, and energy storage devices. Most advances in this field are based on the fundamental understanding of the physics and properties of these materials in solid state. Meanwhile, little is known about their structure-property relationship in the liquid phase. Colloidal 2D nanomaterials produced by liquid-phase exfoliation methods are polydispersed in size and aspect ratio, surface chemistry, electronic structure, and reactivity. This polydispersity is the main challenge in understanding the interactions of nanosheets. Developing new platforms for minimizing the polydispersity while maintaining the production yield of liquid-phase exfoliation methods is critical for advancing this research field. Such platforms equip us with fundamental understanding and control over the intersheet interactions as well as the interactions with their surrounding environment. Monodispersed colloidal nanosheets with controlled surface chemistry present opportunities for developing novel chemical and biological sensing platforms, catalysts with promoted activity, and 3D self-assembled macrostructures, holding the potential to revolutionize biomedical, structural composites, and energy storage and conversion technologies.

Currently, in Prof. Michael Stranoâ€™s lab at MIT, I am investigating the corona-phase of the surfactant-stabilized nanosheets and their size distribution using single particle tracking techniques, as well as their potential applications in biomedical devices and photocatalysis.My research in Prof. Micah Greenâ€™s lab at Texas A&M University has been focused on graphene-based colloidal dispersions and their self-assembly. Combining my experimental knowledge with the particle-based computational tools such as Brownian dynamics and using techniques from statistical mechanics will allow me to precisely predict and quantify polydispersity of the nanosheets. I will use novel size analysis techniques developed during my postdoctoral studies, optical and spectroscopic methods, and controlled deformation with simultaneous measurement of microscopic stresses to study the interfacial energy and interactions of 2D nanomaterials as polydispersed colloidal systems. Using the knowledge gained through these studies, it is feasible to separate and sort the nanosheets based on their size, surface chemistry, and reactivity. Next, I will apply functionalization and surface treatments as means to tune the assembly of the sorted 2D nanomaterials into desired 2D and 3D structures for biosensing, targeted drug delivery, photocatalysis, and energy storage applications. Additionally, I will focus on process development for scalable preparation of these finely tuned nanosheets to make them available for industrial applications.

Teaching Interests:

As a new faculty, I will use my experiences as graduate teaching fellow and teaching assistant to guide the undergraduate students in gaining a fundamental understanding of the key concepts of chemical engineering. I expect all my students to apply these concepts to a wide variety of physical systems based on problem categorization. I will teach my students to tackle the smaller components of problems through defining the system in terms of proper mathematical models and then using key techniques to achieve a solution. This problem-solving algorithm enables the students to approach any engineering problem in the real-life and industrial applications. In regard to graduate education, my goal is to empower students to become independent academic researchers. I will focus my efforts on not only imparting scientific knowledge, but also teaching the methodology of scientific research and independent thinking. My ultimate goal is to produce researchers who think in a multidisciplinary fashion and can generate creative solutions for challenging and complicated situations. While my teaching experience allows me to feel comfortable with teaching any of the core courses in the chemical engineering undergraduate curriculum, my specific teaching interests include transport, numerical methods in chemical engineering, and reaction engineering. My teaching interests for elective and graduate level courses closely match my research interests, and I would like to teach or develop courses in the areas of Advanced Functional Materials, and Colloids and Interfacial Phenomena. In particular, the latter course will cover an overview of the fundamental topics in colloids including the colloidal interaction at atomic and molecular level, interfacial forces, and self-assembly of the colloidal particles.

542163 A Comprehensive Study of Photocatalytic Degradation of Methylene Blue By ZnO Nanoparticles and Its Nano-Composites with Ag an C_3N_4 Under UV Light

Abstract:

A Comprehensive study of photocatalytic degradation of Methylene Blue By ZnO Nano particles and its Nano composites with Ag and C_3N_4 Under UV Light.

Sadia Ata[1], Samina Ghafoor[1], Ifra Mirza[1], Qurat-ul-Ayne[1]

[1]Institute of Chemistry, University of the Punjab, Lahore, Pakistan.

sad_ziam@yahoo.com[1], saminaghafoor0@gmail.com,

Abstract:

We report a simple and convenient method for the synthesis of Ag/ZnO and Ag/ZnO/g-C_3N_4 heterostructurenanorods by applying simple routes. Initially, pure ZnOnanorods have been synthesized by hydrothermal method followed by Ag deposition on ZnO surface by photodeposition. Structure, crystallinity, and morphology have been assessed by X-ray diffraction, UV-Vis spectroscopy, Fourier transform Infrared, and scanning electron microscopy techniques. The influences of the deposited Ag nanoparticles on the surface of ZnO and subsequent composite with g-C_3N_4 have been emphasized by applying the as synthesized nanostructure in dye degradation under illumination of UV light. The basic motivation behind this work is to find a superior photocatalyst, which can efficiently degrade the organic pollutants from water. Photocatalytic performances of bare ZnO, Ag/ZnO, g-C_3N_4/ZnO and Ag/ZnO/g-C_3N_4 have been studied thoroughly. Photodegradation results demonstrated that the incorporation of silver and graphitic carbon nitride nanoparticles significantly increases the catalytic efficiency by promoting the photogenerated charge carrier separation. The main advantage of the proposed Ag/ZnO/g-C_3N_4 semiconductor is that it delays the recombination process of the electronâ€"hole pairs generated by the photon absorption, which in lieu increases the photocatalytic efficiency. It is a challenging issue to fabricate stable photocatalysts which is efficient upto 96% after 5th time recyclic study.

Research Interests

Material synthesis at nano scale acting as a sorbent, functionalization studies on the Polymeric skeleton, Water quality, Environmental issues, Disinfection by products, Haloacetic acids , Ploy aromatic hydrocarbon, Pesticides,

Trihalomethane, Modeling (predictive and time series), Method development, Monitoring of Environmental pollutants, Heavy metal studies, Removal of heavy metals, Method development based on instrumental technique such as GC- ECD, GC-MS, HPLC, Solid phase extraction (SPE), Solid phase microextraction (SPME) etc.

Teaching Interest

Analytical Chemistry, Inorganic Chemistry, Instrumental Techniques

Image:

Figure:1 Photocatalytic degradation of methylene blue by Ag/g-C_3N_4/ZnO composite under UV light.

Research Interests:

Material synthesis at nano scale acting as a sorbent, functionalization studies on the Polymeric skeleton, Water quality, Environmental issues, Disinfection by products, Haloacetic acids , Ploy aromatic hydrocarbon, Pesticides, Trihalomethane, Modeling (predictive and time series), Method development, Monitoring of Environmental pollutants, Heavy metal studies, Removal of heavy metals, Method development based on instrumental technique such as GC- ECD, GC-MS, HPLC, Solid phase extraction (SPE), Solid phase microextraction (SPME) etc.

Teaching Interests: Analytical Chemistry, Inorganic Chemistry, Instrumental Techniques

542224 Novel Nanomaterials for Chemical and Life Sciences

Abstract:

Brief bio:

I am a Research Assistant Professor in the Chemical and Life Science Engineering Department at Virginia Commonwealth University (VCU), Richmond, VA and I am running a full research lab which I inherited from Professor (Emeritus) McHugh. Before joining VCU, I worked at a global pharmaceutical company as process R&D engineer for two years, where I was responsible for active pharmaceutical ingredients process development, optimization, scale-up, and technology transfer. I have a broad range of skills that I acquired through my experiences both in academia and industry that make me confident that I can contribute to any institution's research and teaching.

Research Interests:

My research work is at the interface of material science, nanotechnology and thermodynamics to design and develop novel multifunctional materials which find applications in wide range of industries such as pharma, automotive, and oil & gas. This research helps to create highly efficient products and processes that uses state-of-the-art experimental techniques coupled with robust mathematical models. For example, combining polymer science with supercritical fluid technology allows to create nano-porous bio aerogels for drug delivery or helps to design and formulate novel automotive lubricants.

Research Experience:

My research experience spans across many areas such as pharma crystallization, nano materials for drug delivery, and polymer solution thermodynamics. Working for two years at industry, I learned how to perform cutting-edge applied product and process research. During post-doctoral research, I have had an excellent opportunity to work on an Office of Naval Research sponsored project in collaboration with Dr. Kevin Ward (VCU Medical School), where I developed novel bio-aerogels for drug delivery, hemostats, and tissue engineering. Currently, I am running an active research program in collaboration with a chemical company on high pressure phase behavior and thermophysical properties of polymer + hydrocarbon mixtures.

Projects/proposals:

- Multi-year project with Afton Chemical Corporation, "Phase behavior and thermodynamic properties of polymer + solvent mixtures," $98,500/year (01/2015 to 12/2018)
- One active proposal, as co-PI, submitted to DoE NETL – decision pending

Teaching Interests:

Throughout my academic career, I have had many opportunities to be involved in tutoring, mentoring.

preparing examinations, and supervising laboratory courses in polymer engineering, material science, unit operations, and thermodynamics. I firmly believe that I can integrate research concepts into both basic and advanced chemical engineering courses. My education and research experience has been in core chemical engineering and pharmaceuticals. Hence, I am confident that I can teach any of the undergraduate/graduate chemical engineering courses.

Future Research:

As a faculty, I would like to continue the research on aerogels to further understand the structure-property relationships at the fundamental level to better design and engineer with tailored properties. More specifically, I intend to use my expertise of material science and engineering combining with supercritical fluid technology to create unique bio-aerogels for controlled drug release applications. The other area that I will focus on polymer solution thermodynamics, the ultimate goal of this research is, to design, develop, and evaluate alternative, cost-effective hyperbranched/star polymeric technologies for the application of petroleum additive products with improved performance characteristics especially for automate and enhanced oil recovery. My ultimate goal is to use all of this experimental knowledge and data to validate and develop simple, easy to use models for practical applications for example in computational fluid dynamics.

Education:

Ph.D. Chemical Engineering, University of Erlangen, Erlangen, Germany - Mar 2009

M.S. Polymer Science & Engineering, Technical University of Berlin, Berlin, Germany - Feb 2005

B.Tech. Chemical Engineering, Osmania University, Hyderabad, India - April 2002

Selected publications:

Mallepally R.R., Bamgbade B.A., Rowane A.J., Newkirk M.S., and McHugh M.A., "Fluid Properties at High Pressures and Temperatures: Experimental and Modelling Challenges" (in review) Journal of Supercritical Fluids (2017)

Mallepally R.R., Marin MA., and McHugh M.A., "CO_2-assisted synthesis of silk fibroin hydrogels and aerogels," Acta Biomaterialia, **10**, 4419-4424 (2014)

542257 Design of Functional Nanomaterials for Energy Applications Using Flow Reactors

Abstract:

Design of Functional Nanomaterials for Energy Applications Using Flow Reactors

Research Interests: The design of functional nanoscale materials with predictable physicochemical properties, which can act as building blocks for the development of efficient and renewable energy sources, is essential toward environmentally sustainable societies. Semiconducting nanomaterials, including traditional quantum dots, perovskite nanocrystals, upconversion and conjugated polymeric nanoparticles, have attracted increasing attention owing to the unique photo-electroluminescent properties, which make them ideal candidates for energy, biological and environmental related applications. In such systems, size, morphology, surface chemistry and photophysical properties can dramatically alter material functionality. Conventional flask-based synthetic approaches seem incapable of providing direct answers for the nucleation and growth mechanisms due to the slow reagent mixing and heat transfer and at the same time the reproducible synthesis of particles, in a continuous fashion, with well-controlled properties remains a challenge. My research aims to tackle these challenges by developing flow-based synthetic and surface engineering techniques for the end-to-end continuous formation of complex colloidal semiconducting nanomaterials while combining optical and structural characterization techniques for the tailor-made design of existing and novel compositions.

Research Experience: To date, my experimental research has focused on understanding nanocrystal reaction mechanisms particularly for PbS and PbSe quantum dots while developing methodologies for the better design of perovskite nanocrystals for their further implementation in optoelectronic devices. In particular, I have focused on the development and application of high-throughput segmented-flow, microfluidic platforms with incorporated in-situ optical detection systems for efficient and fast screening of reaction conditions and a better understanding of mechanisms governing nanocrystal synthesis. Such automated microfluidic configurations allowed for a detailed monitoring of nucleation and growth of various quantum dots on a millisecond time-scale and at high temperatures. In addition, I have developed real-time detection modalities, including steady-state and time-resolved fluorescence and absorbance techniques, for the parametric screening and optimization of the synthesis of all-inorganic and organic-inorganic perovskite nanocrystals aiming for enhancing their structural and photophysical stability for their further implementation in optoelectronic and photovoltaic devices. Part of my research work contributed to the development of multinary perovskite compositions which exhibited phase- and chemical stability for the fabrication of efficient NIR nanocrystal-based light emitting diodes.

My postdoctoral work at MIT revolves around the development of miniature continuous stir-tanked reactors for the facile multistage synthesis and upscaling of core-shell nanocrystal morphologies with precise control of their shell thickness. Such reactor configurations overcome limitations associated with the controlled multi-addition of reagents in robust devices that can operate at high temperatures while advancing the continuous and on-demand manufacturing of colloidal non-toxic III-V based nanostructures. Besides that, my current research focuses on the development of single-molecule detection modalites, based on solution photon correlation Fourier spectroscopy, which can be coupled to flow systems aiming for understanding the nanocrystal photophysics at the single particle level. Such implementation of single-molecule optics with flow reactors could provide unprecedented precision in the synthesis of existing and novel nanocrystal compositions while shifting the current paradigm of reaction optimization from ensemble to single-dot level.

Future Research Directions: Establishing my own research group, I envision to develop a framework that revolves around three major research areas, addressing questions of fundamental and practical relevance: (1) Use of flow reaction systems for the precise formation and mechanistic investigation of quantum confined nano-heterostructures with distinctive properties, (2) surface engineering/characterization and purification technologies for the end-to-end production of functionalized nanomaterials, (3) high-thoughput formation and characterization of semiconducting polymer nanoparticles. The proposed research will combine concepts from both my Ph.D work in quantum dot synthesis and postdoctoral work in the development of multistage synthetic processes and single-molecule characterization.

Teaching Interests: During my Ph.D, I served two semesters as a teaching assistant for the Biomicofluidic Engineering course, while designing a laboratory exercise for graduate students for the microfluidic synthesis of quantum dots. During the past five years I have been fortunate to have supervised a number of graduate and undergraduate students on conducting experimental and analytical work. While conducting my PhD and postdoctoral work I was the direct supervisor of seven graduate and two undergraduate students conducting their master and bachelor theses. After working with all these students in an educational and focused direction I have come to confirm how working effectively and closely as a team has a large impact on the quality of research conducted. I am ready to implement this pedagogical knowledge to teach a range of courses from the chemical engineering curriculum both at the undergraduate and graduate level, including chemical reaction engineering, heat and mass transfer and process design. In addition, my research interests drive me toward the development of advanced classes on microfluidic engineering and materials characterization techniques.

542398 Taking the Lab to the Field: Performing Real-Time Environmental and Diagnostic Monitoring

Abstract:

Research Interests:

The ability to receive real-time information is a necessity in today's society. To a diabetic, the use of a continuous glucose monitor can help them be proactive rather than reactive toward an abnormal blood sugar reading. In an occupational setting, understanding the way the body is reacting to the surrounding environment provides a more comprehensive picture than knowing solely what is in the air. By leveraging transport phenomena and varying sensor techniques, real-time monitoring can be brought to the field.

My Ph.D. work at Virginia Commonwealth University under Dr. Nastassja Lewinski has focused on the design and use of a portable *in vitro* exposure system for aerosol sampling. Starting with a commonly used system by industrial hygienists, I designed and characterized a device for exposure at the air-liquid interface. I compared the application of our device with a similar exposure system designed by my colleagues at the Centre for Research and Technology – Hellas as a Fulbright student in Greece. I have applied our device to a number of scenarios including enhanced diesel exhaust using a metallic nano-based fuel additive, aerosols from metal 3D printers, and aerosols generated by consumer products. Additionally, our device incorporates an electrochemical enzyme-based sensor for real-time response monitoring, allowing users to better understand the effects of their environment. This work has been partially supported by a Dissertation Assistantship Award granted by the Graduate School at Virginia Commonwealth University.

My research career has included the experimental and computational investigation of electrochemical thermodynamics of prismatic lithium-ion batteries, the evaluation of toxicological effects of nanomaterials *in vitro*, and the design and evaluation of enzymatic biosensors. I believe my combination of skills will allow for the design of electrochemical biosensors and their varied applications, particularly in point-of-care diagnostics.

Building on this foundation, my future research plans include development of electrochemical sensors for rapid and quantitative monitoring and diagnosis and use of these sensors to determine human impact of nanomaterials. I am interested in leveraging various electrochemical reactions to bring diagnostic techniques to the field of occupational health and safety as well as the medical field. Within the medical field, there is a need for a long term (lasting over a month time period) continuous glucose sensor that provides similar sensitivity to current monitoring methods (short term implantable and self-monitoring blood glucose sensors). I am particularly interested in the development of electrochemical sensors for monitoring dehydration through electrolytes and for heavy metals in drinking water. Through the incorporation of nanostructures, improvement in the sensitivity of current sensing platforms can be achieved due to the dramatic increase in surface area. This work involves a combination of: 1) material synthesis, 2) electrochemical methods, and 3) cell biology techniques. These methods used in my research can be applied in a wide-class of complex systems including pharmaceuticals, biofuel production, and catalytic chemical processes.

Teaching Interests:

I believe that the role as an educator is to catalyze the learning process. My wish for students is that they will recognize tools available to them for creative and successful problem solving as they gain understanding of the material. My teaching philosophy includes the following three strategies: 1) Teach in a simple and direct manner, 2) Emphasize relations to real-world scenarios, and 3) Treat students at different levels fairly and with patience. I am eager and excited to continue in undergraduate and graduate education. I have experience conducting lectures for Material Balances and Energy Balances in addition to having the pleasure of assisting with these classes at multiple universities. I am interested in teaching the major courses in chemical engineering, especially Fluid Dynamics, Heat and Mass Transfer, Separation Process, and Chemical Kinetics. Additional courses I am interested in include Electrochemistry, Nanotechnology, and Environmental Health and Safety. In addition to classroom settings, students will have hands-on research experience under my tutelage. By participating in my laboratory, students will learn applications including how electrochemical kinetics and mass transfer apply to reactions occurring on a biosensor, how the fluid flow within an exposure system can affect the observed biological outcome, and how concepts covered in curriculum behave at the nanoscale or microscale.

Publications:

1. **Secondo LE**, Wygal NJ, Lewinski NA. "Design and Characterization of a New, Portable *In Vitro* Exposure Cassette with Real-Time Monitoring for Aerosol Measurements". Annual AIChE Meeting 2018.
2. Lewinski NA, **Secondo LE**, Ferri JK. "Enabling Real-Time Hazard Assessment at the Workplace." *Global Congress on Process Safety*. (2018). (Citations: 0)
3. Lewinski NA, Avrutin V, Izadi T, **Secondo LE**, Ullah, MB, Ozgur U, Morkoc H, Topsakal E. "Influence of ZnO Thin Film Crystallinity on In Vitro Biocompatibility." *Toxicology Research* (2018). DOI: 10.1039/c8tx00061a (Citations: 0)
4. **Secondo LE**, Liu N, Lewinski NA. "Methodological Considerations When Conducting In Vitro, Air-Liquid Interface Exposures to Engineered Nanoparticle Aerosols." *Critical Reviews in Toxicology* DOI: 10.1080/10408444.2016.1223015. (2016). (Citations: 8)
5. Oh K, Siegel JB, **Secondo LE**, Kim S, Samad N, Qin J, Anderson D, Garikipati K, Knobloch A, Epureanu B, Monroe CW, Stefanopoulou A. "Rate Dependence of Swelling in Lithium-Ion Cells." *Journal Power Sources*. 267, 197. (2014). (Citations: 55)

542403 Iron Oxide Nanoparticles Inhibit Metastasis and Tumor Growth in Lung

Abstract:

Research Interests:

My research goals are directed toward developing Clinical Translational nanoscale technologies with particular emphasis on developing nanoscale biomaterials and immunoengineering systems for cancer immunotherapy, drug delivery and molecular/cellular imaging. My goal is to link the fields of nanotechnology, immunotherapy, cellular biology, and medical imaging towards more efficient and accurate diagnosis, personalized therapies and ultimately improving treatment outcomes and patient quality of life.

Teaching Interests:

My teaching interests lie in the interface of nanotechnology, cancer immunotherapy, cell biology, and medical imaging.

Iron oxide nanoparticles inhibit metastasis and tumor growth in lung

Saeid Zanganeh[1], Morteza Mahmoudi[2]

[1]Sloan Kettering Institute for Cancer Research, New York, New York 10065, USA

[2]Center for Nanomedicine, Department of Anesthesiology, Brigham and Women's Hospital, Harvard Medical School, Boston, MA 02115, United States

Introduction:

We recently revealed a hidden therapeutic effect of the iron oxide nanoparticles (ferumoxytol) on the breast cancer tumors. Here, we show the immunotherapeutic effect of these nanoparticles on the growth of metastatic tumors in lung. We hypothesized that ferumoxytol nanoparticles inhibited cancer metastases by inducing a pro-inflammatory immune response with M1 macrophage polarization. In vitro, Macrophages exposed to ferumoxytol displayed increased mRNA/cytokines associated with pro-inflammatory Th1-type responses. In vivo, intravenous ferumoxytol treatment before intravenous tumour cell challenge prevented development of lung metastasis. Fluorescence-activated cell sorting (FACS) and histopathology studies showed that the observed tumour growth inhibition was accompanied by increased presence of pro-inflammatory M1 macrophages in the tumour tissues. Our results suggest that ferumoxytol could be applied 'off label' to protect the lung from metastatic seeds and potentiate macrophage-modulating cancer immunotherapies.

Materials and Methods: Most experiments were performed with the ultrasmall superparamagnetic iron oxide (USPIO) nanoparticle compound Ferumoxytol (Feraheme®, AMAG Pharmaceuticals Inc., Cambridge, MA, USA). The chemotactic effects were studied using Transwell assays with MMTV-PyMT cancer cells and bone marrow derived macrophages. Real-time PCR, ELISA assays, and ROS measurements were performed in vitro. lung tumors were studied using FACS and followed by histopathology.

Results and Discussion: Through the *in vitro* and *in vivo* experiments, we revealed the following: (1) Low dose ferumoxytol was not cytotoxic. (2) Ferumoxytol increased macrophage migration. (3) Ferumoxytol causes in vivo M1 macrophage polarization. (4) Systemic delivery of ferumoxytol inhibits lung metastases. (5) Pre-treatment with ferumoxytol inhibits development of lung metastases. (6) Ferumoxytol alters macrophage polarization in lung metastasis *in vivo*. (8) α-CSF1 monoclonal antibody treatment inhibits ferumoxytol effects on tumor growth.

Conclusions: We hypothesized that ferumoxytol nanoparticles inhibited cancer growth by inducing a pro-inflammatory immune response with M1 macrophage polarization. The intrinsic impact of nanoparticles on macrophage physiology is not well understood. To our knowledge, no one has investigated intrinsic effects of iron oxide nanoparticles on macrophage physiology in cancer. The results have broad implications for diagnostic and therapeutic nanoparticle applications.

542412 Cell Shape: An Overlooked Factor at the Nanobio Interfaces

Abstract:

Cell Shape: An Overlooked Factor at the Nanobio Interfaces

Morteza Mahmoudi[1], Saeid Zanganeh[2]

[1]Center for Nanomedicine, Department of Anesthesiology, Brigham and Women's Hospital, Harvard Medical School, Boston, MA 02115, United States

[2]Sloan Kettering Institute for Cancer Research, New York, New York 10065, USA

Introduction:

It is now a well-known fact that cells in their native tissue show different morphologies than those grown on a culture plate. In vivo, cells maintain their specific 3D structure, while they almost have a flat body when cultured on the conventional 2D culture plates. Here, we show that the way cells respond to the exact same type of nanoparticles is strongly dependent on their shape. Therefore, the nanoparticles uptake and toxicology data from 2D culture plates cannot exactly predict how the cells would respond in their 3D structures in vivo. In order to better predict the efficacy of the 2D cell culture plates, we propose the use of "pseudo-3D" substrates that could partially mimic the in vivo cell shapes.

Materials and Methods: The polydimethylsiloxane (PDMS) polymer was used as a molding agent. The imprints were prepared from two different kinds of cells; the normal fibroblast cells (HU02), and the cancerous epithelial cells (MCF-7) with spindle and polygonal shapes respectively. To evaluate the GNRs-mediated toxicological responses on the HU02 cells, which were cultured either on the cell-imprinted substrates or the conventional polystyrene culture plate, a wide range of assays were conducted as following: (i) cellular uptake, (ii) cell viability, (iii) reactive oxygen species (ROS) production assay, (iv) comet assay, (v) apoptosis-necrosis detection assay, and (vi) cell cycle assessment. The wells' bed had been covered with the polymeric substrates, including HU02-imprinted ("autologous"), MCF-7-imprinted ("allogeneic"), plain polydimethylsiloxane (PDMS) substrates or without polymeric substrates (control group, conventional polystyrene culture plate).

Results and Discussion: In this study, we clarify that cells cultured on the cell-imprinted substrates presumably receive similar signals from the milieu to that of their natural niche signal and that these cells behave more naturally. Furthermore, this research well explains the fact that the cell toxicity that results from different substrates, is strongly depending on the cells' bed. So, it is safe to say that preparing a topographically optimized niche microenvironment on cell culture substrate, by patterning the natural cell shapes, could be a promising way to simulate the extracellular matrix and to mimic the natural cell architecture, which may further extend our capabilities to investigate and predict the biological effects of nanoparticles such as cell toxicity in a more precise way.

Conclusions: In this study, we draw biomedical communities' attention to another major factor that could diminish the gap between in vitro and in vivo outcomes, the cell shape. We show that the cellular uptake and toxicity of exact same nanoparticles are strongly dependent on the cell shape (i.e., cells which are being cultured on 2D- and cell- imprinted-substrates). Our results suggest that the cell shape effects should be considered in nanobiointerfaces' protocols, enabling them to provide more accurate and precise outcomes.

Research Interests: My research goals are directed toward use of superparamagnetic iron oxide nanoparticles (SPIONs) for various biomedical applications (such as targeted delivery of therapeutic biomolecules to the desired cells, targeted imaging of the cells, protein/gene delivery, and hyperthermia) for earlier detection and better treatment of diseases (such as cancer and neurodegenerative disease).

Teaching Interests: My teaching interests lie at the interface of materials science and engineering, nanotechnology, cell biology, and medicine. I particularly enjoy teaching introductory biomaterials and bioengineering courses to majors and non-majors alike

541902 Nano-Bionics: Polymer and Metal-Organic Thin Films and Particles for Engineering Life

Abstract:

Summary

My research focuses on the interface between biological entities and organic, inorganic, and hybrid materials. Specifically, I am a world expert in layer-by-layer assembly technologies (see my review in *Science*), and am helping to pioneer novel uses for metal-organic materials in biotechnology. I have received superb scientific mentoring from Australia's leading materials scientist, Prof. Frank Caruso, and have learned extensively about industry-focused research during my two-year postdoc at CSIRO.

Academic Background

I have 3 years of postdoctoral research experience at various world class institutions, and I am currently a postdoctoral researcher in Chemical Engineering at the University of Melbourne in Australia under Professor Frank Caruso. I conducted ~2 years of self-driven research and industrial work at CSIRO, with short-term (2-3 month) visits at both University of Tokyo under Associate Professor Hirotaka Ejima and Sungkyunkwan University under Professor Pil J. Yoo. I completed my PhD with Professor Frank Caruso at the University of Melbourne in Chemical and Biomolecular Engineering, and conducted a 3-month research stay the University of Ghent with Professor Bruno de Geest.

Metrics

My research has led to a number of impactful discoveries and resulted in numerous high-impact publications: **12 first author, 4 corresponding author, 51 total peer-reviewed articles, 6 industrial reports, 4 patent applications, 4 invited talks, an H-index of 20, and over 2000 citations**. I have mentored and co-supervised 10 MS students and 6 PhD candidates, and secured a relatively large amount of grant funding for my age (>$500,000 USD excluding scholarships) from the University of Melbourne, the Australian Synchrotron, and the Japanese Society for the Promotion of Science.

Research Interests:

I plan to harness my unique expertise in materials engineering and biotechnology to continue my work in thin films and hybrid materials for biomedicine, and to continue to pioneer the field of nanobionics. I will do this by continuing my work in layer-by-layer assembly and organic-inorganic hybrid materials. I will rely on my diverse and extensive network of collaborators around the globe and I will mentor local and international students and researchers. **Importantly, LbL and metal-organic systems are low-cost and easy to manufacture, and therefore ideal research topics for starting a new lab.**

New Layer-by-Layer (LbL) Assembly techniques: I have a strong background in innovating new LbL assembly methods as this was the focus of my thesis, which led to invitations in *Science* and *Chemical Reviews* to summarize the field. I will investigate new driving forces for colloidal LbL assembly, and also develop assembly and patterning techniques for planar LbL thin films. I will then apply the colloidal materials for anti-counterfeiting, therapeutic delivery, and biosensing, while the planar materials will be used for sensing. research will result in new experimental methodologies and equipment, and also new products and applications. Importantly, LbL assembly is low-cost and can be done with common laboratory equipment, and is therefore perfect for starting up a new lab.

Related work:

- **Richardson J.J.**; Björnmalm, M.; Caruso, F. "Technology Driven Layer-by-Layer Assembly of Nanofilms" *Science* **2015**, *348*, aaa2491.
- **Richardson, J.J.;** et al. "Innovation in Layer-by-Layer Assembly" *Chemical Reviews* **2016**, *116*, 14828.
- **Richardson, J.J.** et al. "Thermally Induced Charge Reversal of Layer-by-Layer Assembled Single-Component Polymer Films" *ACS Applied Materials and Interfaces* **2016**, *8*, 7449.
- **Richardson, J. J.** et al. "Convective polymer assembly for the deposition of nanostructures and polymer thin films on immobilized particles" *Nanoscale* **2014**, *6*, 13416.
- **Richardson, J. J.** et al. "Fluidized Bed Layer-by-Layer Microcapsule Formation" *Langmuir* **2014**, *30*, 10028.
- **Richardson, J. J.** et al. "Immersive Polymer Assembly on Immobilized Particles for Automated Capsule Preparation." *Advanced Materials* **2013**, *25*, 6874.
- **Richardson, J. J.** et al. "Preparation of Nano- and Microcapsules by Electrophoretic Polymer Assembly." *Angewandte Chemie International Edition* **2013**, *52*, 6455.

Organic-inorganic hybrid material synthesis: I have extensive experience in metal-organic systems including metal-phenolic network (MPN) thin films, metal-organic framework (MOF) films and particles, and polymer-stabilized calcium carbonate particles. These materials are all porous and can be used to load large quantities of cargo. Moreover, they are pH responsive and can be used as templates for forming polymeric capsules and films. I will continue to develop new synthesis routes for organic-inorganic hybrid materials with an aim on applying these materials for biotechnology, such as creating biohybrids, forensics, drug delivery, and preserving biomolecules and therapeutics.

Related work:

- Ejima, H.; **Richardson, J. J.** et al. "One-Step Assembly of Coordination Complexes for Versatile Film and Particle Engineering." *Science* **2013**, *341*, 154.
- **Richardson, J.J.*** and coworkers. "Biomimetic synthesis of coordination network materials: Recent advances in MOFs and MPNs." *Applied Materials Today* **2018**, *10*, 93.
- **Richardson, J.J.*** and coworkers. "Biomimetic mineralization of metal-organic frameworks around polysaccharides" *Chemical Communications* **2017,** *53*, 1252.
- Guo, J.=; **Richardson, J.J.=** et al. "Influence of Ionic Strength on the Deposition of Metal–Phenolic Networks." *Langmuir* **2017,** *33*, 10616.
- Guo, J.; ... **Richardson, J.J.** ... "Modular assembly of superstructures from versatile building blocks using polyphenol-based particle functionalization" *Nature Nanotechnology* **2016**, *11*, 1105
- **Richardson, J.J.*** et al "Controlling the Growth of Metal Organic Frameworks Using Different Gravitational " *European Journal of Inorganic Chemistry* **2016**. *27*, 4499.

Drug Delivery: I will continue my work in drug delivery using polymeric and metal-organic drug delivery vehicles. I will primarily focus on novel approaches to drug delivery utilizing my expertise in nanomaterials design. Examples include creating shear responsive drug delivery systems, or vehicles that can cross the blood-brain barrier.

- **Richardson, J.J.=** et al. "Polymer Capsules for Plaque-Targeted In Vivo Delivery" *Advanced Materials* **2016**, *28*, 7703.

- **Richardson, J. J.** et al. "Versatile Loading of Diverse Cargo into Functional Polymer Capsules" *Advanced Science* **2015,** *2*, 1400007.

Nano-bionic Life: Nano-bionic life is an incredibly young field focused on enhancing living organisms with synthetic materials (a sort of complement to synthetic biology and genetic modification). I have helped pioneer this field and published two systems thus far. I plan on pushing the scope of nano-bionics as an independent researcher. I will work on developing injectable technologies (likely subcutaneous) for sequestering toxic metals, gases, or chemical agents from the blood stream, which would be useful for soldiers. And for simpler organisms, I will continue to work on self-assembling hybrid materials in and around the organisms to allow for new metabolic pathways and sensing mechanisms.

Related work:

- **Richardson, J.J.*=** & Liang, K.*= "Nano Biohybrids: In Vivo Synthesis of Metal–Organic Frameworks inside Living Plants." *Small* **2018,** 14 (3).
- Liang, K.; **Richardson, J.J.** et al. "An Enzyme-Coated Metal-Organic Framework Shell for Synthetically Adaptive Cell Survival" *Angewandte Chemie International Edition* **2017,** *29*, 8360.
- **Richardson, J.J.*=** & Liang, K.*= (2018, Jun). Nano-bionic life: A bright future outside the confines of evolution. *Angle Journal*.

Teaching Interests:

I have not taught any university courses, as I have been working in research intensive postdocs, and CSIRO is a research organization without students (non-degree granting). However, I did work as a TA for 3 subjects during my PhD studies (Chemical Process Analysis, Biology for Engineers, and Bioprocess Engineering). I have a unique academic background which has given me a greater breadth of scientific knowledge and communication skills, outside that of a normal chemical engineer. I have a degree in Philosophy, and therefore understand the complexities of formal logic, and can communicate complex thoughts clearly. Therefore, I would be capable of teaching writing classes for engineers. Additionally, my undergraduate career involved classes in all science subjects and extensive research in microbiology and environmental engineering. My postgraduate work encompasses mathematical optimization, business engineering, and chemical and biomolecular engineering. Collectively, this means I have extensive knowledge in logic (symbolic and verbal), mathematics (statistics, linear, and non-linear), chemistry (organic, inorganic, and biochemistry), biology (genetics and microbiology), and writing.

Specific Courses: I would feel comfortable teaching courses dealing with nanotechnology, biotechnology, biomedicine, writing, experimentation (imaging), etc.

Scientific Supervising and Mentoring: I have supervised graduate and undergraduate students on biomedical research projects, and believe that one-on-one mentoring is essential to develop students into world-class leaders in science and engineering. I also think that understanding the implications of science and engineering on society is crucial, which is why I draw from my undergraduate studies in theory of knowledge, the history of science and ethics when supervising and teaching. Importantly, I have lived on 4 continents and have worked with, and supervised, students from over 25 different countries. This has given me a great appreciation for cultural differences between people, and has taught me how to connect and interact with people from all walks of life. I plan to continue supervising postgraduate students and mentoring undergraduate students as I believe this experience is crucial for students to develop into contributing members of society.

522111 Materials Chemistry As Engineering Solutions: Metamaterials, Energy and Water

Abstract:

Research Interests:

Future Research

My research is at the intersection of chemical engineering, chemistry and materials science, relying concepts from nanotechnology, plasmonics, mechanics and mathematics. I aspire to uncover multifarious technological holy grails by fully comprehending their underlying mechanisms and dynamics. My independent research will answer to the question of "how materials chemistry can be engineering solutions for optoelectronics, energy and environmental problems".

Recently, chiroptical meta-materials have gained tremendous attention due to a wide range of advantageous applications from anomalous control of light to sensing of biomolecules. These have mostly been touched by top-down approaches (e.g. lithography), while bottom-up approach has been mostly left as untapped. Hence, to open new doors, I will synthesize optically active inorganic materials, and then assembled them with organic ligands/polymeric backbone for high-performing materials. Eventually, these materials will be used in fabrications of devices.

Efficient, clean energy production and environmental pollutions are constant critical issues of the sustainable future of humanity. I plan to leverage modern synthetic strategies to produce precisely controlled polymeric molecular structures which will be used for selective and efficient transport of various ions/phonons for fuel cells, thermal conduction, and air & water filtration applications. The expertise obtained during these studies can also be applied to water desalination and redox flow batteries.

Post-Doctoral Research

- "Designed Polymers for Energy (Fuel Cell Membranes) and Environment (Hg(II) Removal) Applications"
- Advisor: Prof. Timothy M. Swager, Massachusetts Institute of Technology, Chemistry

Postdoctoral research trained me to become a materials chemist. Through multiple projects, I have demonstrated how rationally designed polymeric systems can contribute remarkably to the development of new engineering tools that tackle environmental problems of our society, especially on energy and water.

First, I demonstrated how free volume promoting moieties and charge delocalized polymer backbone can help enhance the conductivity and stability of anion exchange membranes for fuel cell applications. The addition of unfunctionalized triptycene (free-volume promoting) copolymer into imidazolium poly (ether sulfone) had improved the membrane's conductivity, minimized dimensional changes, and enhanced stability. These enhancements are the result of nanophase separation and internal free volume. Extremely stable pyrozolium type cations are also used as ion conducting moieties.

Another project showed how the use of redox active polyaniline can achieve the electrochemically controlled reversible capture and release of highly toxic Hg(II) in water. Polyaniline naturally adsorbs Hg(II), and a subsequent dose of oxidation potential induces the release. Nanofiber geometry of polyaniline offered faster adsorption kinetics. Additional sulfur chelating groups boosted its uptake capacity.

Graduate Research

- "Self-Organized Nanoparticles for Stretchable Conductors and Chiroptical Materials"
- Advisor: Prof. Nicholas A. Kotov, University of Michigan, Chemical Engineering

My graduate work, "Science and Engineering on Nanocomposites", trained me to become a chemical engineer with interdisciplinary expertise in nanotechnology, materials science, electronics, mechanics and photonics. The projects demonstrated how external stress can re-organized nanoparticles in the composites, for discovering fundamental understandings of mechanisms and dynamics for unique electronic and optical properties.

An excellent stretchable conductors were demonstrated from self-organized gold nanoparticles using a highly stretchable polyurethane matrix. Macroscale free-standing samples were prepared using a layer-by-layer assembly of positively-charged polyurethanes and negatively-charged gold nanoparticles. The dynamic self-organization of polymer-nanoparticle assemblies produced composites with high conductivity and stretchability.

Self-organization of the assemblies also demonstrated chiroptical nanocomposites for applications of photonic metamaterials. The conformal layer-by-layer deposition on pre-twisted substrates offers a distinctive advantage when producing macroscale photonic materials. Chiroptical activity had increased up to ten fold upon stretching samples and was reversibly tunable. S-like chiroptical 3D patterns responsible for optical properties appeared as a consequence of buckling surface instabilities.

Selected Publications

1. **Kim**, T.M. Swager, "Charge Delocalized Anion Exchange Membranes: Extreme Stability and High Conductivity" (Manuscript in Preparation)
2. **Kim**, Z. Lin, I. Jeon, T.V. Voorhis, T.M. Swager, "Polyaniline Nanofiber Electrodes for Capture and Release of Mercury(II)" (To be submitted)
3. **Kim**†, L. Moh†, T.M. Swager, "Anion Exchange Membranes: Enhancement by Addition of Unfunctionalized Triptycene Poly(Ether Sulfone)s" *ACS Applied Materials & Interfaces*, 2017, 9, 42409–42414.
4. **Kim**†, B. Yeom†, O. Arteaga, S.J. Yoo, S.G. Lee, J.G. Kim, N.A. Kotov, "Reconfigurable Chiroptical Nanocomposites with Chirality Transfer from the Macro- to the Nanoscale" *Nature Materials*, 2016, 15, 461-468.
5. **Kim**, J. Zhu, B. Yeom, M.D. Prima, X. Su, J.G. Kim, S.J. Yoo, C. Uher, N.A. Kotov, "Stretchable Nanoparticle Conductors with Self-Organized Conductive Pathways" *Nature*, 2013, 500, 59-63.
6. **Kim**, K. Koh, M.F. Roll, R.M. Laine, A.J. Matzger, "Porous Networks Assembled with Octa Phenyl Silsesquioxanes as Building Blocks", *Macromolecules*, 2010, 43, 6995-7000.

† indicates equal contribution

Teaching Interests:

Teaching and Mentoring Experience

At the University of Michigan, I served as a graduate student instructor for CHE460 Chemical Engineering Laboratory II, a core senior lab class, which requires comprehensive chemical engineering knowledge for undergraduate students (student evaluation: 4.7 / 5.0). My course preferences are thermodynamics, chemical reaction engineering, transport processes, material/energy balances, organic chemistry, physical chemistry, polymer chemistry, and graduate classes for nanoscience and engineering. However, I am capable of teaching other graduate and undergraduate core courses as well. I also look forward to the opportunity to mentor students. During my PhD studies, I had mentored eight students (1 PhD, 2 masters, 4 undergraduates and 1 high school) for off-shoot projects of the electronic and optical nanocomposites. Their mentorship has allowed me to not only gain new technical skills, but also appreciate the positive effect that mentorship can have on a student's professional development.

Teaching Philosophy

My teaching philosophy is underpinned by competence, autonomy, connectedness and diversity. My goal as a teacher, as a mentor, is to inspire innovation, enhance performance and make lab life more enjoyable.

Competence is a prerequisite. I discovered that the best performance is often achieved at an intermediate difficulty task level. For example, for first year graduate students, a good starting project is the synthesis and characterization of nanoparticles. If the student succeeds, then we can move on to more advanced level tasks involving the optimization and assembly for device fabrications.

Autonomy is essential for motivation. Autonomy is the sense that the project emanates from the person and not from an external source. Using punishment or negative feedback will only stifle their enthusiasm. The optimal degree of autonomy is specific to each individual and changes over time. Thus, the advisor and his student need to determine and adjust this together.

Social connectedness. The need for connectedness encompasses the striving to care for others, to feel that others relate to you in mutually supportive ways, and to feel a satisfying involvement with the entire lab. Having an effective communication within the lab is a key to social connectedness. For example, in the Swager lab, we have a mentoring system to pair graduate students and postdocs. Also, we regularly have a dinner together and frequently play tennis or basketball together to build team bonding. These are good examples for lab members to be kept connected.

I am committed to promoting and supporting diversity in my classroom and in my lab. This includes diversity of experience, way of thinking, and ideas in addition to demographic diversity. During my PhD and postdoc programs, I have participated in many outreach activities such as serving as a mentor at alternative high school, volunteers for science day etc. I firmly believe that unparalleled performances can only be achieved with the collaborative efforts from a diverse group of people.

For more information, please visit my website: http://yoonseobkim.com/.

527252 Electricity from Asymmetric Chemical Doping

Abstract:

Research Interests:

I obtained my BS (2014) in Chemical Engineering from California Institute of Technology and am currently working on my PhD with Prof. Michael S. Strano at Massachusetts Institute of Technology. Prior to joining MIT as a graduate fellow, I had studied physical organic chemistry and organometallic chemistry under the mentorship of Prof. John D. Roberts and Prof. Gregory C. Fu, and had spent a summer at Chevron Co. learning zeolite synthesis from Dr. Stacey I. Zones. My research interests include the analysis and control of molecular conformation and engineering of low dimensional materials with applications in chemical transformation and energy generation.

Teaching Interests:

Coming from a family of academics, I aspire to teach materials science and engineering at a collegiate level. My undergraduate experience was in a way very unique, in that I obtained two degrees at two different colleges (BA in Chemistry at Grinnell College and BS in Chemical Engineering at California Institute of Technology). While my experiences at both schools were quite different due to their respective cultures as a liberal arts college and an engineering institute, one consistent theme throughout my academic career was my commitment towards mentoring fellow classmates. Recently, I served as the graduate TA of Transport Phenomena in the Department of Chemical Engineering at MIT. My recitations were very well received. As a result, I was nominated to receive the Goodwin Teaching Medal (awarded annually to one â€œconspicuously effectiveâ€ graduate TA at MIT) and was selected as the best graduate TA of the department. I believe this experience had further honed my teaching skills and prepared me for my future goal as a college professor. If possible, I am planning on TAing more classes during my years at MIT; because for me, giving back what I have learned and seeing it evoke passion in the mentee really is the most satisfying thing there is.

534826 Nano Engineering with X-Ray through Infrared Spectroscopy (NEXIS)

Abstract:

Progress flows from a superior understanding

1st Year Postdoctoral Fellow

Keywords: 2D transition metal oxides (TMOs), bench-scale X-ray absorption fine structure (XAFS), catalysis, collaboration

Research Interests:

2D nanomaterials--especially beyond graphene--are the latest wave in the nanomaterial "gold rush" due to the promise of exciting chemical and electronic properties, high tunability, and amenable geometry over their 0D and 1D counterparts. In particular, manipulation of 2D transition metal oxide (TMO) systems through compositing, doping, and functionalization would be especially useful to fields such as catalysis, photovoltaics, transistors and more where ceramics and semiconductors are already implemented. However, researchers in the 2D TMO field face major challenges: attaining precise structure control and determining how numerous factors affect macroscopic properties. Only through a thorough fundamental scientific understanding can we hope to distinguish truly promising technologies from "fool's gold."

My primary research focus is to engineer new 2D TMO systems by developing new synthesis techniques to precisely control their structure, to study their growth and property relationships using state of the art ex situ and operando XAFS and FTIR spectroscopy, and to boldly use these materials as catalysts and more to solve problems in energy, the environment, and beyond.

Research Experience:

PhD Dissertation: "Tuning the Optical and Catalytic Properties of CuO and Fe_2O_3 Nanosheets for CO_2 Conversion"

Under the supervision of Prof. *Lisa D. Pfefferle* in the Department of Chemical and Environmental Engineering at Yale University I combined synthesis, characterization, and application to better understand and implement 2D TMO systems. I tuned the properties of copper oxide nanosheets and related showed that band gap could be a descriptor for reactivity. I re-pioneered hard-templating in order to synthesize ultra-thin iron oxide nanosheets for CO_2 reduction. I developed new methods for functionalizing copper oxide nanosheets with organic small molecules and showed a drastic increase in the material's hydrophobicity and stability. Throughout my work I utilized a collection of spectroscopic techniques and I collaborated with computationalists to better understand 2D TMOs while improving the modelling of them.

Postdoctoral Research:

Under the supervision of Prof. *Shu Hu* in the Department of Chemical and Environmental Engineering at Yale University I extended my knowledge of synthesis, spectroscopy, and electrochemistry. I used,

- *operando* PM-IRRAS to investigate temperature dependent methane electroxodation
- *in stiu* XAFS to study the local atomic structure of ALD grown TiMnOx for selective H_2O_2 electrodes
- Semiconductor synthesis: CdSe/ZnS quantum dot – porous GaN composites for fabricating μLEDs
- Catalysis: $BaTiO_3$ to study polarization chemistry for high frequency switching catalysis

Future Direction:

As a faculty member I will draw on my experiences from my PhD and post doc to lead a group focused on controlled 2D nanometal oxide synthesis, in situ spectroscopic characterization, and heterogenous catalysis. I will expand the breath and depth of 2D TMO understanding by developing hard templating and functionalization to synthesize new 2D TMOs systems; use in situ X-ray absorption Fine Structure (XAFS) and Fourier-Transform Infrared (FTIR) spectroscopies to understand adsorption, the solid-liquid interface, and map growth and catalytic reaction pathways on material's surface; and by establishing structure-property relationships I will tune these materials for CO_2 reduction and CH_4 oxidation. Of particular interest are synthesizing bimetallic 2D TMO composites, controlling facet exposure, and tuning the number and type of defects present. CO_2 and CH_4 conversion reactions are both important due to the industry relevance of these molecules, the difficulties in their capture and transport, their high stability leading to poor conversion and selectivity in catalytic reactions. The tunability and reactivity of 2D TMOs make them perfect candidates to face the challenge of converting CO_2 and CH_4 into transportable, storable, value added liquid products.

Beyond catalysis, oxide nanomaterials have seeped into nearly every field, from membranes to medicine. My vision for the NEXIS lab is to not only be an engine for nanomaterial engineering and development, but also a hub for collaboration and multidisciplinary research. The NEXIS lab will be home to advanced spectroscopic instruments, such as bench-scale XAFS, which will enable atomic scale characterization only previously achievable using synchrotron radiation. Using existing and future connections, I will work with collaborators in academia and industry, experimentalists and computationalists, to realize a superior understanding of materials and their structure-property relationships in order to enable their implementation.

Teaching Interests:

My goal as an educator is to inspire, challenge, and support students through diverse assignments in order to impart fundamental scientific knowledge while building in them essential skills for a multitude of professions.

My multidisciplinary and collaborative experience and views are not limited to my research. Throughout my career I have TAed for Separations and Purification Processes, Multivariable Calculus, and Fluid Mechanics hosting study session, grading assignments, and guest lecturing on a few occasions. Additionally, I designed my own classes and taught subjects such as philosophy and creative writing as a volunteer teacher in Yale's Sprout and Splash programs. In the lab, I have mentored over 13 undergraduate and high school students who have gone on to become graduate students and lead successful lives. As an educator I am excited to bring my experiences to the classroom balancing tests with other collaborative and interactive assignments in order to foster important skills such as technical writing and public speaking. I feel prepared to teach fundamental chemical engineering classes and look forward to teaching laboratory and collaborative design style classes.

Proposal Writing Experience: Petroleum Research Fund, Samsung: The Global Outreach Program, NSF-CAT, NSF-DMREF, Army Research Office (funded), EPA-STAR

Selected Publications:

1. **Fishman, Z.S.,** Rohr, J.A., He, Y., Batista, V.S., Pfefferle, L.D., Hu, S. "Fine tuning the optical and electronic properties of CuO nanosheets using organic functionalization." *in preparation*

2. He, Y.*, **Fishman, Z.S.***, Yang, K., Ortiz, B., Goldsamt, J., Batista, V.S., Pfefferle L.D., "Hydrophobic CuO Nanosheets Functionalized with Organic Adsorbates" *Am. Chem. Soc.,* 140(5), pp 1824–1833 (2018). ***equal author contributions**

3. Zucker, I., Werber, J.R., **Fishman, Z.S.**, Hashmi, S.M., Gabinet, U.R., Lu, X., Osuji, C.O., Pfefferle, L.D., Elimelech, M., "Loss of Phospholipid Membrane Integrity Induced by Two Dimensional Nanomaterials" *Sci. Technol. Lett.,* 4 (10), pp 404–409 (2017).

4. **Fishman, Z.S.**, He, Y., Yang, K., Rudshteyn, B., Lounsbury, A., Zhu, J., Zimmerman, J.B., Batista, V.S., Pfefferle, L.D., "Hard Templating Ultrathin Polycrystalline Hematite Nanosheets and the Effect of Nanodimension on CO_2 to CO Conversion via the Reverse Water Shift Reaction" *Nanoscale* 9, 12984-12995 (2017).

5. **Fishman, Z.S.**, Rudshteyn B., He, Y., Liu, B., Chaudhuri, S., Askerka, M., Haller, G.L., Batista, V.S., Pfefferle, L.D., "Fundamental Role of Oxygen Stoichiometry in Controlling the Band Gap and Reactivity of Cupric Oxide Nanosheets" *Am. Chem. Soc.*, 138(34), 10978–10985 (2016).

536897 Engineering Nanopores and Nanostructures of Atomically Thin Sheets and Carbon Nanotubes

Abstract:

Research Interests:

Engineering optical properties of nanomaterials allow us to achieve the efficient process of light-material interaction and contribute to improving photochemical reaction and performance of photovoltaics and photodetector. Nanomaterials possess unique properties due to quantum confinement effects, which differs from bulk materials. In particular, quantum well and wire that consists of an atomically thin sheet, for instance, graphene, carbon nanotube, and other related materials, hold great promise for sensing, energy convergence, and other high-tech applications.

My past and current research focused on exploring and engineering physical and chemical properties of the one-dimensional (1D) carbon nanotube and other two-dimensional (2D) atomically thin sheets. My approach includes micro-spectroscopy for the 1D and 2D nanomaterials, which are engineered with photochemical reaction, stacking two different 2D sheets, and fabrication of electrical devices with nano- and micro-lithography. My automated homemade scanning probe spectroscopy has obtained thousands of spectra from individual nanostructures formed in 2D sheets and revealed their collective properties. I aim to translate their unusual characteristics into potential optoelectronic applications.

Postdoctoral Project:

"Characterization of Nanopore Structures in 2D Hexagonal Boron Nitride Using Scanning Probe Photoluminescence Spectroscopy"

"Optical Application of Archimedean Scroll Fiber based on 2D sheets/polymer composites"

PI: Michael S. Strano, Chemical Engineering, MIT

Ph.D. Dissertation:

"Behavior of Photocarrier in Atomically Thin Two-dimensional Semiconducting Materials for Optoelectronics"

PI: Kazunari Matsuda, Institute of Advanced Energy, Kyoto University

Fellowship:

2015-2017 JSPS Research Fellow, Japan Society for the Promotion of Science

Research Experience:

Advanced spectroscopy of 1D and 2D nanomaterials

Carbon nanotube solar cell

Optoelectronics of field effect transistor based on 2D sheets

Synthesis of 1D and 2D nanomaterials

Teaching Interests:

I believe that the role of the teacher in a learning environment is to get students motivated by helping them achieve their learning outcomes. From an engineering viewpoint, my ultimate goal of teaching is to enable students to define essential problems existing in the real world and solve them on their own. For them to achieve it, I am enthusiastic to share my knowledge and experience and to continue growing as a teacher. I will keep attentive to their feedback such as the quiz at the end of lectures and will analyze individual responses in each office hour after the lectures.

I will care classroom environment. For students to get motivated for the lecture, the instructor should help students have a sense of belonging in the classroom. To create an inclusive classroom environment, I believe the active learning can be an essential role. The method promotes students to think on their own, and the discussion in pairs facilitate them to work together so that all students will feel welcome.

Effective teaching facilitates students' learning. My strategy for the effective teaching is to introduce active learning into my teaching. For example, showing a characteristic figure, I will let students in pairs discuss their observation in that figure and ask them to share their idea in front of other students. The active learning is beneficial for students to enhance their understanding and get them refreshed, which keeps them engaged in my teaching.

Future Direction:

As faculty, I will expand my research interests to magnetic properties of nanomaterials in addition to the optoelectronic properties that I have developed so far. I aim at using the nanomaterials as a probe of magnetic response. Since the nanostructures are exposed to environments, these could be sensitive probes. Nuclear magnetic resonance (NMR) and magnetic resonance imaging magnetic (MRI) are typical applications of magnetic properties in materials. I expect that engineering nanostructures of the carbon nanotube and 2D insulator sheets could allow us to manipulate magnetic properties, while the perfect crystals of them do not exhibit such magnetic response. I want to develop how to control the magnetic properties of the nanomaterials with my experience of modification of materials and advanced spectroscopy technique. As magnetic resonance sensor can probe physical parameters such as temperature, strain, and magnetic field as well as the chemical traces, this class of sensor could allow us to reveal unexplored properties. Likely examples are measuring local temperature of the living body in sub-micrometer resolution and visualization of the chemical reaction process, monitoring those physical and chemical parameters. Besides the sensing application, I am also interested in the application of the magnetic properties to quantum communication. Such magnets in the nanostructure could store quantum states for long time. Also the absorption and fluorescence of telecom wavelength of light in carbon nanotube are promising for the direction.

Teaching Experience:

Teaching certificate, MIT Kaufman Teaching Certificate Program

Guest Lecturer of Engineering Nanotechnology in Chemical Engineering at MIT for first-year graduate student

Mentor for one graduate student and one undergraduate student

Selected Publications:

*Son, Y.; ***Kozawa, D.**; Liu, A. T.; Koman, V. B.; Wang, Q. H.; Strano, M. S. A Study of Bilayer Phosphorene Stability under MoS_2 -Passivation. *2D Materials* **2017**, *4* (2), 025091. (*Equally contributed)

Kozawa, D.; Carvalho, A.; Verzhbitskiy, I.; Giustiniano, F.; Miyauchi, Y.; Mouri, S.; Castro Neto, A. H.; Matsuda, K.; Eda, G. Evidence for Fast Interlayer Energy Transfer in $MoSe_2/WS_2$ Heterostructures. *Nano Letters* **2016**, *16* (7), 4087–4093.

Kozawa, D.; Kumar, R.; Carvalho, A.; Kumar Amara, K.; Zhao, W.; Wang, S.; Toh, M.; Ribeiro, R. M.; Castro Neto, A. H.; Matsuda, K.; Eda G. Photocarrier Relaxation Pathway in Two-Dimensional Semiconducting Transition Metal Dichalcogenides. *Nature Communications* **2014**, *5*, 4543.

Kozawa, D.; Miyauchi, Y.; Mouri, S.; Matsuda, K. Exploring the Origin of Blue and Ultraviolet Fluorescence in Graphene Oxide. *The Journal of Physical Chemistry Letters* **2013**, *4* (12), 2035–2040.

Kozawa, D.; Hiraoka, K.; Miyauchi, Y.; Mouri, S.; Matsuda, K. Analysis of the Photovoltaic Properties of Single-Walled Carbon Nanotube/Silicon Heterojunction Solar Cells. *Applied Physics Express* **2012**, *5* (4), 042304.

541819 Tuning Complex Fluids from the Nanoscale

Abstract:

Research Interests:

In mentoring my students and preparing them to be leaders in academia and technology, we will apply fundamental principles from chemical engineering to design devices for various applications ranging from waterborne contaminant sensing and treatment to low-cost diagnostics for use in under-resourced settings. Research in my lab will focus on *tuning nanoscale and microscale properties to adjust and control macroscopic behavior*. Tuning materials and transport properties from the microscale will enable improvement in process control in a variety of industrial settings, particularly those involving complex fluids, such as the pharmaceutical and personal care product industries, and to an increasing degree, the food processing industry as well. Beyond industrial applications, I will leverage my expertise in nano-material characterization and self-assembly to inform challenges in energy, the environment, and biomedical applications.

Research Experience:

My expertise lies in experimental complex fluid transport and dynamics, nanomaterial self-assembly and characterization, and colloidal phenomena, as they apply to a variety of applications. I am an Associate Research Scientist at Yale University, and the founding Director of the Yale Facility for Light Scattering (FLS), where I collaborate with researchers from throughout the University, advising students and postdocs in both experimental design and data interpretation. My scientific expertise has roots in soft matter physics, materials science and chemical engineering. My doctoral research focused on complex fluid flows in microchannels, taking inspiration from physiological fluids. During my postdoctoral research, I approached the study of petroleum engineering through the lens of colloidal aggregation and self-assembly along with complex fluid dynamics. My postdoctoral work alone led to ten first-authored publications, building upon my three prior first-authored publications, all in top journals of fundamental scientific research, including *Soft Matter*, *Journal of Chemical Physics*, *Langmuir*, and *Physics of Fluids*.

Teaching Interests:

Bringing everyday science into the classroom

The best and most memorable science lessons are often taught through demonstrations, hands-on in-class experiments, and through appealing to our everyday experiences. For instance, if we try mixing oil with a dilute acid, we'll have some difficulty: oil and water don't mix. But by adding a polymeric, zwitterionic surfactant to sit at the fluid interfaces and stabilize the system... voila! We have mayonnaise: made from olive oil, lemon juice, and protein from an egg. We've all eaten mayonnaise, know its taste and texture, but do we understand how the microstructure influences our macroscopic experience, how the simultaneous hydrophobic and hydrophilic nature of a protein allows otherwise immiscible fluids to become a homogeneous, smooth mixture? These questions of how microscopic behaviors influence macroscopic

characteristics abound throughout science and engineering, notably in chemical engineering, physical chemistry, and materials science.

More than teaching to any test, I want students in my classroom to feel they are learning information that is relevant, useful, and immediate in their daily lives. While grades of course help assess progress and guide future lesson plans, students should leave the classroom feeling that they have connected to the course material in the present moment: not for the purpose of answering a problem on a problem set, but to understand their physical surroundings better. To that end, abstract concepts, or those describing microscopic phenomena need to be boiled down to physical, *present* concepts in one way or another. In materials science, this means explaining molecular crystal dislocations in terms of macroscopic ball bearing arrays in a demo. In thermodynamics class, we might stretch a rubber band next to our upper lip to demonstrate the equivalence of energy, entropy, and heat.

541832 Nanocomposites Synthesis, Characterization and Its Application in Energy, Environment and Healthcare

Abstract:

Research Interests:

My research in the past several years has focused on Environmental Nanotechnology (biomass to chemicals, biocatalyst, nanocomposites for Energy and Environment) as broad area of research. My primary focus is on the (1) Synthesis of nanocomposite material and its application in Energy (storage device) and Environment (membrane synthesis/water treatment) (2) comprehensive study of the nanocatalyzed bioconversion processes (Green synthesis) (3) photocatalytic treatment. I wish to expand my research in application of the prepared nanocomposites in biomedicine. I have already completed the literature review and in the process of formalizing the project. I am also interested in broader areas like catalysts, composites materials, biosensors, separation technology in general. One of my objective is to convert my basic research innovation to process technology transfer and commercialization.

During my PhD I have proposed a kinetic model for biosorption (*Bioresource. Technol.*, 98, 1781-1787, cited 131 times), continuous column sorption (*Chem. Eng. J.,* 144, 386-390., cited 38 times). Applying the concept extended my research and proposed surface model (*Colloid.Surface. A.,* 329, 95-99, cited 25 times) for biosorption and extended further and studied mass transfer models (*Can. J. Chem. Eng.*, 90, 1011-1016) and review of biosorption (*Can. J. Chem. Eng.*, 89, 889-900) process.

As Master student's dissertation, I have previously developed optimized process condition for the photocatalytic degradation by organic oxidants (PAA, PNBA, MEKP). Our group first time reported this and in recognition one of my Master's students won the Indian National Academy of Engineering-Innovative Student Projects Award – 2010. I considered it as one of my research accomplishment as independent researcher. I have extended this work for industrial wastewater treatment (*Chem. Eng. J.,* 209, 1-6) also and two M. Tech and one PhD degree awarded with a total 11 Journal publications (*Ind. Eng. Chem. Res.*, 49, 3094-3098, *J. Ind. Eng. Chem.*, 18, 249-254., *Sep. Purif. Rev.*, 42, 263-295, *Water. Air. Soil. Poll.*, 224, 1-9, *Water Sci. Technol.,* 67, 440-445, *Water Sci. Technol.*, 67, 2418-2427, *Water Sci. Technol.*69, 1259-1266 etc.) in this area so far.

With my two PhD student's we have biosynthesized (using plants and microorganism resources) metal nanoparticles such as Gold, Silver, Palladium, Selenium, and Platinum (*Ind. Eng. Chem. Res.*, *55*, 9557–9577, cited 32 times; *Ind. Eng. Chem. Res.*, *53*, 13971–13979, cited 42 times; *Ind. Eng. Chem. Res.*, *52*, 18131–18139, cited 48 times; *Ind. Eng. Chem. Res.*, 51, 12993-13328, cited 67 times; *J. Nanopart. Res.*, 15, 1366-76, cited 55 times; *Powder Technol.*, 244, 26-29, cited 47 times) and reported production of valuable chemicals by catalytic conversion. Much scope (little focused so far) is there to explore. So I wish to continue this work further. One PhD student awarded degree on broad area of biocatalysis with 5 Publications (*Appl. Biochem. Biotech.*, 166, 486–520, *Korean. J. Chem. Eng., 31, 1225-1232, Grasas y Aceites,* 63, 202-208, *Chem. Ind. Chem. Eng. Q.*, 20, 127-134 etc.) so far. The review paper published in

Appl. Biochem. Biotech., 166, 486–520 is cited 238 times and I have received e-mail from one of my student doing PhD in Ohio University, USA stating that his supervisor given this paper to them as good study materials and for further use. I considered it as one of my research accomplishment. With my research exposure in separation over the years I understood the global challenges for safe and reliable water provision for the environmental and human well-being. This experience further reinforces my resolve to pursue for funded projects as Investigator in nanocomposite membrane separation. Two M. Tech (*J. Polym. Mater.,* 29, 301-308., *Sep. Sci. Technol.*, 47, 2299-2304) degree is awarded in this area. Two PhD students working in this field and few mentionable publications are *Scientific Reports- Nature,* 8(1) 4976, *J. Ind. Eng. Chem.*, 63, 366-379, *New Journal of Chemistry,* 41,15049-15057. Further work is going on. Two of my recent PhD students are working on applications of nanocomposites in Energy storage device and Biosensors.

Teaching Interests:

My approach to teaching

My teaching philosophy is reasoned thought. I certainly want my students to learn the fundamental content of the course I teach. In the classroom my goal is to encourage thinking in rational ways, so that this can be applied in other, unfamiliar situations.

Another important element of my teaching philosophy is encouraging students to interact to develop logical arguments. A major factor of this is spending my time making suggestions, as students can only learn to present their ideas more effectively.

As I gained experience and confidence as a teacher, I came to regard teaching as my primary professional responsibility. I prefer to teach mathematical subjects where I can easily explain with more than one example. I create a positive atmosphere in the class.

I believe in

"Give a man a fish, and you feed him for a day; show him how to catch fish, and you feed him for a lifetime."

Proposed teaching plan

Teaching courses like to handle : Related subjects in UG & PG

- Advanced Kinetic and Reaction Engineering
- Modelling Chemical, Biochemical and Environmental Engineering
- Physical Chemical Separation Processes.
- Bioseparations and Biochemical Engineering

- Microbial Engineering and Environmental Biotechnology

- Green Engineering
- **New courses related to personal expertise to be taught at PhD level**

- Research Methodology

- **Interdisciplinary courses**

- Environmental nanotechnology
- Nanobiotechnology
- Nanotechnology for water treatment

US University is expected to provide flexibility and abundant opportunities for research to incumbent researcher. Interdisciplinary and collaborative research work is more acceptable today. US University is with world class research facility and thus significant outcome projected.

543830 Functional 2D Material Nanoarchitectures for Sustainable Energy Generation

Abstract:

Functional 2D Material Nanoarchitectures for Sustainable Energy Generation

Sanjay K. Behura

Research Assistant Professor in Chemical Engineering, University of Illinois at Chicago, IL

E-Mail: sbehura1@uic.edu; Website: https://sanjaybehura.weebly.com/

Research Background:

The quantum confinement of charge carriers in two-dimensional (2D) crystals evolve new and, in several cases, superior electronic and optical properties with applicability in energy conversion, nano-optoelectronics, and nano-bio electronics. The challenge is how to controllably nucleate 2D crystals and their quantum heterostructures for engineered properties and advanced device functionality. Owing to the Van Hove singularity in the electronic density of states, the 2D semiconductors exhibit strong light-matter interactions. This enables 2D semiconductors to be surface-coupled with several other material systems including direct or indirect, narrow or wide bandgap inorganic and organic semiconductors to create new energy conversion avenues. For the past 8.5 years of my doctoral and post-doctoral training and research assistant professorship positions, I have built a solid foundation by addressing these multitude of challenging questions.

My graduate research was a result of research exposure from three countries with two scholarships including Commonwealth scholarship for doctoral exchange research at The University of Saskatchewan, Canada and second one is a Department of Science and Technology sponsored travel grant for Ph.D. summer school on 'challenges in 2D materials' at Technical University of Denmark. In the group of Prof. Qiaoqin Yang at The University of Saskatchewan, Canada, I primarily focused on developing strategies for synthesis of vertical graphene films with enhanced electron field emission characteristics. I have developed a novel strategy to produce high-aspect-ratio silicon nanowires directly on SiO_2/Si surfaces *via* oxidation and reduction in hydrogen environment. These nanowires exhibited quantum confinement effect as confirmed by Raman vibrational spectroscopic analysis.

My postdoctoral research with Prof. Vikas Berry at The University of Illinois at Chicago (UIC) involved nanoscale designing and bottom-up synthesis of 2D material heterostructures for low-temperature quantum transport and atomically-precise Moiré fringes. Currently the quantum heterostructures are assembled by transferring the 2D crystals produced on transition metal catalytic surfaces. These polymer-assisted transfer approaches introduce scattering centers for charge carrier transport and mis-orientation of 2D crystals. To address this challenge, I have developed novel surface-chemical-interaction chemistry to nucleate graphene, hexagonal boron nitride (h-BN) and their heterostructures directly on oxide and nitride-based gate-dielectric substrates and investigated low-temperature quantum transport phenomena and Moiré patterns of graphene and h-BN. I have also developed mechanistic understanding of epitaxial graphene formation with ultra-high Raman I_{2D}/I_G ratio on Co (111) surfaces.

In my current role as Research Assistant Professor at UIC, I am developing an 'Emerging Nano-Photovoltaics Research Program on Mixed-Dimensional Junctions'. As Principal Investigator, I have recently received a research grant of **US$ 100,000** and invited to write a commentary article in **Nature Photonics** on this emerging area of nano-energy research program. Being nominated as a graduate faculty, I am advising graduate thesis and undergraduate student projects. In collaboration with Dr. Anirudha Sumant of Center for Nanoscale Materials at Argonne National Laboratory, I am currently investigating an exciting project on the role of nanocrystalline diamond as an electron blocking inter-layer in graphene photovoltaic sensors. This collaboration also resulted a DOE user research proposal at Centre for Nanoscale Materials, Argonne National Laboratory. For the past 8.5 years, I have mentored 15 M.S. and B.S. student projects on 2D materials science and nanoscale photovoltaics which led to several peer-reviewed publications.

Research Interests:

With a solid foundation in the nanoscience of low-dimensional electronic materials and nanoengineering of energy materials and conversion, research projects with high possible throughput are being targeted. My goal is to establish an independent and self-sustaining research program on 'mixed-dimensional complex nanoarchitectures' for (i) surface-junction based emerging nano-photovoltaics, (ii) exciton physics, charge carrier transport, and recombination dynamics for advanced optoelectronics, and (iii) Fermi-level and band-gap engineering *via* chemical, electrical, and biological coupling. Towards achieving my research targets, I aim in developing two major projects in my new lab: (A) New 2D Material Heterostructures and Quantum Electronics: mechanistic understanding of growth chemistry for 1-atom, 2-atom, and 3-atom (soon 4 and 5-atom) thick inorganic 2D materials, their complex heterostructures, fundamental structure-property correlations, and low-temperature quantum transport phenomena and (B) Sustainable Energy Generations: mixed-dimensional emerging heterojunction nano-photovoltaics.

Teaching Interests:

The broad prospective of my research interests are further shaped by my recent teaching experiences. At UIC, I have developed an advanced nanoscience course 'Graphene and Two-Dimensional (2D) Materials (ChE 494)' from Spring 2017 and continue teaching in Fall 2017 and Spring 2018. In this course, I have employed both knowledge-centered (lectures about 2D concepts) and learning-centered (2D laboratory, group collaborations in mini-2D-symposium and science video preparation) techniques to ensure long-term retention of fundamental concepts. Total of 60 graduate and undergraduate students are trained on emerging 2D materials theory, processing and applications. The learning goals are: 2D materials science, synthesis chemistry, nanoscale characterizations, and advanced device functionality. Practical demonstration of concepts is a critical part of active learning and deep understanding, therefore, I included laboratory works: bottom-up synthesis of graphene, its chemical transfer, and Raman spectroscopic characterizations. To inculcate the idea of collaborations among students, I organize mini-symposium and give homework on making innovative science video, where students must interact with fellow students and present as a team. For Spring 2017, the overall teaching rating for this course is **5.00/5.00**, which motivated me to further enrich the course program. For Spring 2018, I am selected to teach 'Science and Technology of Graphene and 2D Materials (Hon 201)' for UIC Honors College. As a new faculty, along with the above course, I will be teaching new courses on: 'Materials for Energy' and 'Engineering Solar Cells at the Nanoscale'. For the past 8.5 years, I have delivered 10 invited lectures and 20 contributed talks at various institutions/conferences in USA, Denmark, Canada, and India. I have presented an invited talk on '2D heterostructure nano-optoelectronics' at 2017 MRS Fall Meeting, Boston in the 'Symposium NM02-Anisotropic Carbon Nanomaterials-Frontiers in Basic and Applied Research'. The invited lectures afforded me the opportunity to further sharpen my teaching skills in front of the broad and diverse audience.

Future Direction:

As an independent investigator, I would like to continue the idea of using experimental techniques and theoretical analysis to understand the structure-property correlations and diverse functionalities of 2D materials and their nanoscale attributes. In this regard, I believe the functional 2D material system I am currently focusing provides a rich experimental toolbox, with a number of fascinating and currently unexplored phenomena to reconnoiter. Potential and immediate future plan includes: (i) mechanistic understanding of charge transport at the interface of biological entities and 2D materials for bio-photovoltaics; (ii) Interfacing of 2D layers with perovskites for novel photojunctions; (iii) origami and kirigami of 2D nanocomposites and device functionality.

The distinctiveness of my methodology is the ability to develop complex 2D heterostructure circuits, characterize their structure-property correlations, and design novel energy junctions *via* nano-bio interfacing. Due to the highly interdisciplinary nature of my work, I envision a strong and active collaboration with fellow faculty in the university to setup specific techniques in my lab.

Selected Publications (6 out of 35 Publications; 280 Citations; *Corresponding Author):

1. **K. Behura***, C. Wang, Y. Wen, and V. Berry*, "Graphene/semiconductor heterojunction sheds light on emerging photovoltaics," **Nature Photonics**, Under 2nd Revision (2018).
2. **K. Behura**, P. Nguyen, R. Debbarma, S. Che, M. R. Seacrist, and V. Berry, "Chemical interaction-guided, metal-free growth of large-area hexagonal boron nitride on silicon-based substrates," *ACS Nano*, Vol. 11, p. 4985-4994 (2017). Highlighted in *Nature India*.
3. **K. Behura**, P. Nguyen, S. Che, R. Debbarma, and V. Berry, "Large-area, transfer-free oxide-assisted synthesis of hexagonal boron nitride films and their heterostructures with MoS_2 and WS_2," *Journal of the American Chemical Society*, Vol. 137, p. 13060-13065 (2015).
4. **K. Behura** and V. Berry, "Interfacial nondegenerate doping of MoS_2 and other two-dimensional semiconductors," *ACS Nano*, Vol. 9, p. 2227-2230 (2015).
5. **K. Behura***, S. Nayak, I. Mukhopadhyay, and O. Jani, "Junction characteristics of chemically-derived graphene/p-Si heterojunction solar cell," *Carbon*, Vol. 67, p. 766-774 (2014).
6. Nguyen#, **S. K. Behura#**, M. R. Seacrist, and V. Berry, "Intergrain diffusion of carbon radical for wafer-scale, direct growth of graphene on silicon-based dielectrics," **ACS Applied Materials and Interfaces**, DOI: 10.1021/acsami.8b07655 (2018). (#Authors with Equal Contributions)

Selected Patents (2 out of 10 invention disclosures):

"Epitaxial growth of defect-free, wafer-scale single-layer graphene on cobalt," V. Berry, **S. K. Behura**, P. Nguyen, and M. R. Seacrist, WO2017058928A1, Priority Date: Oct. 1, 2015, Publication Date: Apr. 6, 2017.

"Direct formation of hexagonal boron nitride on Si-based dielectrics," V. Berry, **S. K. Behura**, P. Nguyen, and M. R. Seacrist, WO2017196559A1, Priority Date: May 12, 2016, Publication Date: Nov. 16, 2017.

542798 Experimental and Numerical Investigations of Particle Flows

Abstract:

Research Interests:

Solid particles are commonplace in a wide range of industries and are used in operations, such as mixing, pneumatic conveying, granulation, heat and mass transfer, fluidized beds, tableting, etc. However, in general, the behavior of particles in these systems is still not well understood. The lack of fundamental understanding can lead to poor system performance, uncertainties in system scale-up, delays to process start-up, ineffective operation design and, in some cases, process failure. For instance, particles have a tendency to segregate by size when in motion which can lead to inconsistent product quality (quality control issues) when particles discharge from hoppers. My research interests include improving the understanding of particle-particle and particle-fluid interactions with the goal of more accurate predictive capabilities, resulting in more efficient process design. Additionally, many industrial operations would benefit from particles with specific properties, such as particle size, cohesion level and/or shape, thus I am interested in improving current methods, as well as developing new techniques, to control particle properties.

Research Experience:

Currently, I am a project leader at Particulate Solid Research Inc. (PSRI), a non-profit research consortium of international companies and national labs. As a project leader, I design and manage the execution of research plans to address various particle-technology related problems, e.g., probe development for measuring flow behavior within fluidized beds. Additionally, I am a principal investigator of a research project that focuses on validating drag force models. Momentum is transferred from the fluidizing gas to the particles via the drag force. Thus, valid drag force models are necessary for accurately predicting fluidized bed behavior. Through a thorough analysis of existing drag models, significant gaps in their validation in the parameter space relevant to industrial fluidized beds was identified. Furthermore, large variations between different drag model predictions exist, which make it difficult to know which drag model is appropriate for a system a priori. The main aim of the work is to assess the accuracy of drag models and develop a microscopic understanding of the differences in various drag model predictions.

As a postdoctoral research assistant in the Chemical and Biological Engineering Department at the University of Colorado, I was the lead experimentalist in Prof. Christine Hrenyaâ€™s research group. I worked on projects related to particle cohesion and heat transfer. The main objective of these various projects was to develop an improved, fundamental understanding of bulk particle behavior by investigating micro-scale (individual) particle properties and interactions. To achieve these aims, I developed experiments and cost-effective characterization techniques to study the effect of individual particle properties on bulk particle properties. For example, in one project, the effects of cohesion between particles in a fluidized bed were measured and understood via the effects of surface roughness on the cohesion level. Additionally, I mentored and supervised undergraduates to carryout high-fidelity measurements that can be used for model validation. During my postdoc, I gained experience writing proposals, collaborating with industry, and mentoring and supervising undergraduate students in research and in applying for grants to pay for their research stipends.

Prior to my postdoc, I received my PhD from the University of Florida Chemical and Biological Engineering Department where I was advised by Prof. Jennifer Sinclair Curtis. During my PhD, I developed and tested physical models for dense phase gas-solid multiphase flow using computational fluid dynamics. Models were verified by simulating the cratering of a particle bed by a subsonic jet of gas. Experimental results from NASA internship and NASA-FAST lunar gravity flight used to further develop and verify the physical model. Experiments performed at the University of Florida to investigate particle property effects on cratering that provide fundamental understanding of particle shape, benchmark data for computational fluid dynamics simulations and improved crater scaling relationships.

Teaching Interests:

Not only does my background in fluid-particle flows make me a strong candidate for teaching standard fluids based chemical engineering courses, but also allows me to incorporate particle technology into the undergraduate curriculum. Training graduate and undergraduate students in this industrially applicable field will help to prepare a relevant workforce. Additionally, my experience in large-scale, experimental systems provides me with the necessary skills to teach students in unit-operations laboratory and process-control courses.

542089 Dynamic Structures in Multiphase Systems: A Pathway Towards Responsive Processes

Abstract:

Profile: Experimental and numerical research in powder technology, heat transfer, fluid mechanics & multiphase flow i.e. spray drying, agglomeration, fluidization, vortex flows. 8 years of experience as engineer, researcher and consultant for global manufacturers of consumer goods, water and energy sectors i.e. P&G, Acciona, ExxonMobil.

Period	*Type*	*Job Title*	*Institution*
2017 / -	Teaching	Teaching Fellow	University College London. UK
2016 / -	Academic Research	Research Associate	University College London. UK
2015	Start up	Consultant	Hexxcell Ltd, Imperial College London. UK
2012/13	Academic Research	Visiting researcher	Imperial College London. UK
2009/12	Corporate R&D	Research Engineer	Procter & Gamble, Technical Centres. UK
2008/09	Production	Engineer/Deputy Plant Director	Acciona Agua. Desalinization. Spain.
2006	Corporate R&D	R&D Manager	Procter & Gamble, Technical Centres. UK
2004	Corporate R&D	Intern	Procter & Gamble, Technical Centres. UK

Period	*Qualification/Field*	*Institution*
2014	EngD - Engineering Doctorate	University of Birmingham. UK.
2008	Master - Environmental Science & Int. Cooperation	School of Industrial Management. Spain
2007	Expert - International Crisis Prevention and management	University of Carlos III, UK.
2005	MEng - Chemical Engineer	University of Salamanca, Spain

Research Interests:

My work looks into complex dynamics in granular media and multiphase flows. I am intrigued by a transition regime between continuum fluid dynamics and solid mechanics where the attractive forces bringing particles/droplets/colloids together and the disruptive stresses pulling them apart are of the same order, resulting in dynamic structures that constantly form and break and change drastically the collective macroscopic behaviour e.g. clustering, dense granular flow, structured fluids. I want to understand how mesoscopic structures appear and react to external stimuli e.g. drag, electric, magnetic. Dynamic structures area common feature of natural processes e.g. nucleation / growth / breakage of particles, drops, snow in a storm, a riverbed, a desert or an avalanche, and industry e.g. clustering / agglomeration / flocculation / deposition / breakage. We can describe some of these phenomena independently, but we do not understand how they interact to form a dynamic system, let along how structure evolves according to various ageing phenomena e.g. drying, sintering, coking. As opposed to scales nano and molecular scales where self-organisation is well-established, in the larger inertial scales typical of the process industry, we simply avoid dynamic effects with a rigid design paradigm that tries to isolate a given process e.g. granulator, grinder. Indeed, neglecting dynamics, we often fail and suffer well-known issues e.g. riser - clustering, heat exchanger - fouling, 3D printer nozzle - clogging. I want to explore a more organic approach to process engineering where we exploit the dynamic features of particulate and colloidal systems to design more efficient and self-regulated processes, materials and devices.

Teaching Interests:

I have experience in various industrial sectors as experimentalist and modeller. I can deliver courses in transport phenomena, powder technology and other core process engineering subjects. Relating back my experience working at multinationals, universities and start-ups to teaching will help me prepare students for the working environment. In addition to invited lectures in the UK and Spain, in 2017/18 I started working as Teaching Fellow at UCL lecturing on fluidization at Masters-level and running the Process Plant Design module for undergraduates i.e. ~150 students. In the future, I would like to build into two areas: a) deliver a comprehensive view of multiphase science from its specific maths to the effects that multiple phases have in transport phenomena (momentum e.g. rheology, mass e.g. emulsions, heat e.g. drying) and the introduction of new rate processes (e.g. agglomeration, breakage) through examples taken from nature and industry, and b) New methods to improve critical thinking. Data interpretation and troubleshooting are essential skills for an engineer. Higher education makes sure that graduates can solve problems and yet, they often struggle to be critical thinkers because they have seldom had the initiative to frame problems themselves. I want to address that by establishing a platform of industrial case studies where they will troubleshoot a real-life challenge in a role-playing exercise. Each group will have a semester to look into a challenge, and unlike a classic "design" exercise, students will face initially incomplete data sets. They will have to think on the given context and ask the relevant staff (me) the right questions to obtain more data, analyse it and interpret their results in a critical manner.

Research Experience:

I have worked in multiphase flow, fluid dynamics, membrane technology and powder processes e.g. agglomeration, fluidization, spray drying. My doctorate dealt with droplet/particle agglomeration during spray drying. I developed new methods base in sonic anemometry to study turbulent air flows [1] and characterised the flow employed in P&G drying towers [2]. I identified distinct recirculation regimes associated to the flow over rough walls and drew new scale up rules accordingly [3]. My research led to fundamental changes in the traditional view of particle formation in spray drying. I proved that most of the product in swirl dryers forms through deposition and breakage of material in a dynamic structure of deposits [4], and based on this insight, redefined the study of agglomeration sources [5], proposed new methods to quantify how sprays interact [6] and optimized the operation of P&G dryers [7]. On these grounds, I won funds to move to Imperial College London and work in a numerical frame that I now follow up in collaboration with University of Salamanca. In my postdocs, I expanded my numerical skills first working in the optimization of another fouling phenomenon in the oil industry and later joining University College London where I work in granular physics. Here, we are using oscillating flows to create new designs for gas-solid processes e.g. foods, household, pharma, energy. By using a periodic perturbation to the inlet gas flow of a fluidised bed, we can suppress the flow instability that leads to chaotic bubbling and create very stable and reproducible bubble patterns in reactors or dryers. In the last two years, I have conducted/supervised experiments in pilot scale units using ultrahigh speed cameras, developed image analysis and pattern recognition codes and numerical models e.g. CFD-DEM, TFM. Several publications are being prepared: 1) identification of the regime where one can "dynamically structure" a bubbling fluidised bed, 2) methods to quantify the level or order in a bubbling system, 3) analysis of the role of friction in the self-organisation of gas bubbles in granular media and 4) quantification of mixing and heat transfer efficiency in dynamically structured beds.

Research Plans:

The mesoscale is one crucial challenge in multiphase science. It is responsible of the limits between dilute, dense and quasi-static regimes in granular flow, the rheology of structured fluids or the transport in gas-solid fluidized beds to mention a few critical questions. When a set of particles consolidates, multi-particle sustained contacts become frequent and shape, long-range effects and anisotropy more important. It is hard to replicate these effects in non-fully resolved models and it is very complex to include any structure in kinetic frames because classic formulations do not deal with correlated velocities or inhomogeneous contact distributions. This is complicated further in practical scenarios as particles evolve e.g. sinter, dry, cure, cake. We use detailed numerical tools e.g. Direct Numerical Simulation DNS, Lattice- Boltzmann LB, Large Eddy Simulations LES, Discrete Element Method DEM, to look into these phenomena at a microscale and then make a rather uncontrolled leap onto multi-scale approaches and coarse grain models in which we rarely extract constitutive equations or rate kernels for application in macroscopic models e.g. Population Balance Equation, let alone 1D formulations for optimization engines e.g. GPROMS. I want to focus on this gap: design experiments and use microscale models to derive new closures, constitutive equations and kernels that will connect physics naturally occurring at a mesoscale with multi-scale models acrtoss different disciplines. My programme will cut across the following themes:

1. **Dynamic Processes**: Intermittent structures at the bulk of multiphase flows, how they affect transport and how external fields may be used to control them.
2. **Dynamic Fouling:** Dynamic structures at the boundaries of a multiphase flow; deposition, consolidation and resuspension of multilayers for application at consumer goods, energy, medicine.
3. **Smart Media:** Transition of clustered structures into flow e.g. multifunctional inks, self-healing materials.
4. **Industry 4.0:** Hybrid optimization approaches where classic multi-scale modelling frameworks are used in combination with a machine learning frame for the estimation of complex physical magnitudes.

Theme I will initially follow up my current collaborations in the energy sector. My first initial bid as PI will focus in **Theme II**, launching studying particulate multilayers as combination of deposition, consolidation/sintering and removal/disengagement mechanisms. I will do so with support from previous partners in consumer goods and energy, and expand it to foods, pharma and healthcare e.g. biomass, heat exchangers, caking, gas-solid reactors, biological fluids. My goals include transferable outputs to stochastic e.g. collision rates, aggregation probability, or rate-based models e.g. diffusion coefficients, agglomeration/breakage kernels. My connexion with industry and SMEs will develop **Theme IV** and after having acquired expertise in the domain of processes, **Theme III** will pose exciting questions on how we can develop new products.

526153 Effects of Complex Particle Interactions on Fluid-Particle Flows

Abstract:

Effects of Complex Particle Interactions on Fluid-Particle Flows J. Kolehmainen[1]

[1]*Department of Chemical and Biological Engineering, Princeton University, Princeton, New Jersey 08540, USA*

Research Interests: My research explores granular electrostatics resulting from contact charging and its effects on flow; and coarse modeling of momentum and scalar transport in gas-particle flows where meso-scale structures arise due to complex fluid-particle and particle-particle interactions, which include van der Waals force, liquid bridge forces and electrostatic interactions. I pursue an integrated approach combining theoretical tools, numerical simulations, and experimental methods. I am also interested in using machine learning techniques to formulate constitutive models for coarse transport equations.

In the future, I intend to continue my research on granular electrostatics and also broaden the scope of my research to liquid-gas flows in trickle beds and liquid-gas flow in porous media. I plan to study these flow problems through lattice Boltzmann simulations, and then formulate constitutive relations for continuum models using machine learning techniques.

Prior Research: Good understanding of gas-solid flows is critical for a wide range of industrial applications ranging from fine laser machining [1] to fluidized beds [2]. For many of these applications, the flows are affected not only by fluid-particle interactions, but also by phase transitions and inter-particle forces. In this poster, I will outline how I have used experimental, theoretical and numerical tools to probe these effects in various systems.

Particle contact charging is a common problem in polyethylene reactors [3], but some applications, such as triboelectric generators [4] and triboelectric separation of fly-ash [5] or municipal waste [6], are based on it. While the underlying physical principals of electrostatic interactions are well understood, the contact charging remains an unsolved problem [7]. I have investigated these effects in fluidized beds and vibrated beds via experimental techniques [8,9] and numerical modeling [10-13] to uncover the underlying phenomena.

Spatter deposition is a major problem in solar panel production where individual elements are separated by a laser. The deposit are usually removed by an assist gas jet in conjuction with the laser head, but optimization of process is challenging due to very small timescale and phase transitions (plasma-gas-solid). I developed 3D particle-tracking-velocimetry (PTV) to capture the spatter movement in 3D space to guide the engineering work [1,14,15].

Teaching Interests: I enjoy greatly helping students learn new concepts and seeing them flourish in their studies. I have a fairly wide educational background ranging from pure mathematics to chemistry, which I believe help in showing students how the topic at hand fits in the larger picture. I have teaching experience in fluid mechanics, heat and mass transfer, thermodynamics, and mathematics classes, all of which have been profoundly enjoyable.

[1] Okamoto, Yasuhiro, et al. "Velocity and angle of spatter in fine laser processing." *Physics Procedia* 39 (2012): 792-799.

[2] Sundaresan, Sankaran, Ali Ozel, and Jari Kolehmainen. "Toward Constitutive Models for Momentum, Species, and Energy Transport in Gasâ€"Particle Flows." *Annual review of chemical and biomolecular engineering* 0 (2018).

[3] Hendrickson, Gregory. "Electrostatics and gas phase fluidized bed polymerization reactor wall sheeting." *Chemical Engineering Science* 61.4 (2006): 1041-1064.

[4] Wang, Zhong Lin. "Catch wave power in floating nets." *Nature* 542.7640 (2017): 159.

[5] Dwari, R. K., et al. "Studies on the effect of electrode plate position and feed temperature on the tribo-electrostatic separation of high ash Indian coking coal." *Advanced Powder Technology* 26.1 (2015): 31-41.

[6] Wu, Guiqing, Jia Li, and Zhenming Xu. "Triboelectrostatic separation for granular plastic waste recycling: A review." *Waste Management* 33.3 (2013): 585-597.

[7] Lacks, Daniel J., and R. Mohan Sankaran. "Contact electrification of insulating materials." *Journal of Physics D: Applied Physics* 44.45 (2011): 453001.

[8] Kolehmainen, Jari, et al. "Effect of humidity on triboelectric charging in a vertically vibrated granular bed: Experiments and modeling." *Chemical Engineering Science* 173 (2017): 363-373.

[9] Sippola, Petteri, et al. "Experimental and Numerical Study of Wall Layer Development in a Tribocharged Fluidized Bed." *Journal of Fluid Mechanics* (2018): Under Review.

[10] Kolehmainen, Jari, et al. "A hybrid approach to computing electrostatic forces in fluidized beds of charged particles." *AIChE Journal* 62.7 (2016): 2282-2295.

[11] Kolehmainen, Jari, et al. "Triboelectric charging of monodisperse particles in fluidized beds." *AIChE Journal* 63.6 (2017): 1872-1891.

[12] Kolehmainen, Jari, et al. "Eulerian Modeling of Gas-Solid Flows with Triboelectric Charging." *Journal of Fluid Mechanics* (2018): Under Review.

[13] Kolehmainen, Jari, et al. "Effects of Polarization on Particle-Laden Flows." *Physical Review Letters* (2018): Under Review.

[14] Kolehmainen, Jari Tapani, et al. "Measurement of the spatter velocity in fine laser cutting." *Proceedings of JSPE Semestrial Meeting 2012 JSPE Spring Conference*. The Japan Society for Precision Engineering, 2012.

[15] Viitanen, Timo, et al. "Spatter tracking in laser machining." *International Symposium on Visual Computing*. Springer, Berlin, Heidelberg, 2012.

537475 Two-Component Polymeric Systems That Provide High Performance, Easy Operation, Environmental Friendliness, and Health Benefits

Abstract:

Coatings, adhesives, sealants and elastomers (CASE) are used in an easy operation processes and exhibit high performance. Thus, they are readily used in applications including, military, aerospace, electronics, automotive, marine equipment, energy, and infrastructure. Two-component (2K) polymeric liquid or semi-liquid resins are the main choice of CASE due to their cure-on-demand nature that provides high performance. Common 2K systems include, epoxy, thiol-ene, polyurethane, polyurea, polyester, and so forth. However, there is increasing interest in overcoming the environmental and health concerns associated with CASE. Here, I will present my development sustainable resins and 2K polymers. The epoxy resins are derived from the plant-derived phenolic acid are an ideal replacement of bisphenol A, which are a concern to the public and the industry due to their health effects and nonrenewable petroleum source. Also, a chemical-structure-properties study on sustainable thiol-ene elastomer based on various phenolic acids will be discussed for a methodology to easily prescreen resins for high-performance 2K polymers merely based on their chemical structure before a comprehensive design of experiment. Additionally, I will showcase polymer particles fabricated through colloidal engineering. A novel particle containing reactive resin was developed in a scalable suspension polymerization reactor. This technique enabled high-performance 2K coatings at room temperature without releasing the toxic volatile organic compounds (VOCs) that are emitted from conventional solvent-based coatings, thus, reducing environmental and human health risks.

Research Interests:

Additive manufacture and 3D printing materials

High-performance and sustainable polymer materials

Particles Engineering Through Colloidal Processing

Teaching Interests:

Chemical Engineering and Polymer Materials

542531 Modeling Transport and Rheology in Polymers and Particle-Polymer Mixtures to Enable the Rational Design of Novel Soft Materials

Abstract:

Research Interests:

Polymer solutions and melts are complex fluids ubiquitous in both nature (e.g. DNA, polypeptides) and technological applications (e.g. in personal care products, food processing). These materials are comprised of large macromolecules whose building blocks (molecular repeating units or monomers) are many orders of magnitude smaller than the macromolecule size. This breadth of length scales enables the design of polymers with a multitude of microstructures (physical architecture). These microstructures can give rise to vastly different transport and mechanical properties in materials that have a similar chemical composition. Further tailoring of material and flow properties of polymer solutions and melts can be achieved through suspension of hard and/or soft particles. By perturbing the polymer flow field and forming a microstructure of their own, the suspended particles create materials with enhanced viscoelastic properties. Developing theoretical models that connect the microstructure to material and flow properties in polymers solutions and melts as well as particle-polymer mixtures remains an important problem at the interface between polymer physics, statistical physics and fluid mechanics. Development of such models will enable "designer" materials for next generation technological applications and a fundamental understanding of the different physical processes (e.g. the interplay between entropic and elastic forces) that give rise to their viscoelastic properties.

My research interests focus on developing predictive models of transport and rheology in polymers and particle-polymer mixtures. Utilizing theoretical techniques such as scaling theories, dynamic simulations, micro-mechanical models, and constitutive models, my goal is to develop predictive theory connecting microscopic structure to properties near and far from equilibrium. These models aim to reveal the underlying physical mechanisms driving such behaviors and serve as a basis to design tailored materials for next generation technological applications.

As a faculty member I will begin by focusing on three research thrusts:

1) The first is the development of charged polymer elastomers with enhanced conductive and mechanical properties. Charged polymer elastomers have potential applications in energy storage and as soft actuators. However, current materials have prohibitively low conductivities, hindering their use in practical applications. My goal is to design new charged polymer elastomers with improved conductivities and predict their properties via the use of dynamic simulations and scaling theories.

2) The second item on my research agenda is to develop predictive models for the non-linear rheology of complex coacervates. Complex coacervates are formed from mixtures of oppositely charged polyelectrolytes that phase separate into a polymer rich coacervate and a polymer-depleted supernatant. Coacervates are

utilized in the food and personal care industry, and have shown promise in drug and gene delivery applications. There is burgeoning interest in understanding the rheological behavior of these materials far from equilibrium, as it is the most relevant regime for many practical applications. Understanding the physical mechanisms governing their non-equilbrium behavior also allows for rational design of their viscoelastic properties. My goal is to utilize dynamic simulations to inform the development of analytical models predicting the non-linear rheology of complex coacervates.

3) My third project aims to develop predictive models for the rheological behavior of particle laden polymer flows. Despite their importance in many industrial processes, the non-linear rheological behavior of particles suspended in non-newtonian solvents, such as polymer solutions and melts, remains poorly understood. Through a combination of constitutive and micro-mechanical models, my goal is to connect solvent viscoelasticity, particle microstructure, and hydrodynamic interactions to describe the rheological properties far from equilibrium. This will enable improved industrial processing, and provide rigorous interpretations of non-linear rheological characterization techniques such as large amplitude oscillatory shear.

Research Background:

My current research, carried out as a postdoctoral associate at Duke University under the supervision of Professor Michael Rubinstein, focuses on the linear viscoelasticity of complex coacervates. I developed a scaling theory for the dynamic behavior of asymmetric liquid-like coacervates formed from oppositely charged polyelectrolyte solutions. Depending on the degree of polymerization, the asymmetric liquid coacervate can form either an interpenetrating double-semidilute structure, wherein both polyanion and polycation are found above their overlap concentration, or a dilute-semidilute structure, where only one of the polyelectrolytes is found above their overlap concentration. I developed a scaling theory for polymer dynamics, providing predictions for the relaxation modulus and steady state shear viscosity of the coacervate, and predicting the diffusivity of the polyelectrolyte chains. The scaling theory highlights the different dynamical regimes of the system, and how the dynamic properties can be tuned from experimentally controllable parameters such as the degree of polymerization, the number density of charges of the polyanion and polycation, and the strength of electrostatic interactions. To complement the scaling theory we conduct molecular dynamics simulations to determine the structure and dynamical properties of the coacervate and confirm the different dynamical regimes.

For my PhD thesis, carried out at Cornell University under the supervision of Professor Roseanna Zia, I studied the diffusion in and rheology of hydrodynamically interacting colloids confined by a spherical cavity via dynamic simulation, as a model of intracellular and other confined biophysical transport. The modeling of transport and rheology in such confined inhomogeneous soft materials requires an accurate description of the microscopic forces driving particle motion, such as entropic and hydrodynamic forces, and of particle interactions with nearby boundaries. Previous models of such micro-confined transport behavior had been limited primarily to a single particle inside a spherical cavity. Although attempts had been made to extend such models to more than one confined particle, none had yet successfully accounted for the effects of hydrodynamics, owing to the difficulties of modeling many-body long-ranged interactions. To accurately model spherically confined suspensions, we derived new far-field mobility functions and, together with the appropriate near-field resistance functions, implemented them in a Stokesian-dynamics like approach. The method fully accounts for all many-body far-field interactions and near-field interactions both between the particles themselves and between particles and the enclosing cavity. Utilizing our newly developed method, we studied short- and long- time self-diffusion at equilibrium, with a focus on the dependence of the former on particle positions relative to the cavity, and of both on volume fraction and size ratio. We found the cavity exerts qualitative changes in transport behavior, such as a position dependent and anisotropic short-time self-diffusivity and anisotropic long-time transport behavior. Such qualitative changes suggest that careful interpretation of experimental measurements in 3D confined suspensions requires accounting for such confinement induced behaviors. To elucidate the effects of confinement on inter-particle hydrodynamic interactions, we utilized our method to determine the concentrated mobility of particles in the spherically confined domain. We found that confinement induces qualitative changes in the functional dependence of particle entrainment with inter-particle separation. We developed a scaling theory to predict the effects of confinement on particle entrainment and utilized the theory to develop a more accurate framework for two-point microrheology measurements near confining boundaries.

Selected Publications

1. "Simulation of hydrodynamically interacting particles confined by a spherical cavity", **Christian Aponte-Rivera**, Roseanna N. Zia. Physical Review Fluids 2016. 1(2):023301.
2. "Equilibrium structure and diffusion in concentrated, hydrodynamically interacting suspensions confined by a spherical cavity", **Christian Aponte-Rivera**, Yu Su, Roseanna N. Zia. Journal of Fluid Mechanics (2018). 836:413–50

Awards and honors:

- Bouchet Graduate Honor Society, 2018
- National Science Foundation Graduate Research Fellowship, 2013
- Colman Fellow, 2012

Teaching Interests:

My teaching interests focus on engineering core courses (e.g. transport phenomena and fluid mechanics, mathematical methods) as well as interdisciplinary courses with a focus on soft matter physics (including topics such as polymer physics, micro-hydrodynamics, and colloidal suspension rheology).

542160 Understanding the Remarkable Physical Chemistry of Novel Polymer Materials: How Does Intricate Chemical Functionality Enhance Material Properties?

Abstract:

Research Interests:

Future research directions

Advances in polymer synthesis techniques have enabled the creation of soft materials with features once thought to be impossible: precise structures and dynamic responses across multiple length scales, mechanical robustness that does not inhibit processability, and well-defined transport properties that can be tuned via external stimuli. These next-generation polymers show great promise to solve current and future societal needs affecting the environment, energy, health, and food accessibility. Making these novel materials useful, however, requires deep understanding of their fundamental physics.

My lab will investigate how the enhanced functionality of technologically important polymer materials alters their thermodynamics, dynamics, and structure-property relationships. To study these complex systems, we will routinely – but not exclusively – use cryogenic and analytical electron microscopy, scattering (soft and hard X-rays, neutrons, light, and electrons), rheology (linear and non-linear), and theoretical modeling. Initially, we will focus on three project areas:

1) Polymerization induced encapsulation

2) Vitrimer gels - elucidation of nanostructure and dynamics

3) Interfacial dielectric properties of polymer nanocomposite capacitors

Graduate research

"Characterization of hydroxypropyl methylcellulose acetate succinate solid dispersions in the solid-state and during dissolution"

Department of Chemical Engineering and Materials Science, University of Minnesota-Twin Cities Advisors: Timothy P. Lodge and Marc A. Hillmyer

During my PhD thesis, I developed novel strategies for characterizing the structure-property relationships of solid dispersions, which are blends of a pharmaceutical drug and a polymer. First, I pioneered the use transmission electron microscopy techniques to map the spatial distribution of the drug and polymer in the bulk [1-2]. Second, I discovered – using novel X-ray scattering and electron microscopy assays – that the drug and polymer form metastable nanostructures during dissolution in aqueous solvent [3]. My work conclusively showed that nanoscale phenomena directly affect the efficacy of solid dispersions, and the characterization strategies I designed will guide the rational design of future solid dispersion formulations [4].

Postdoctoral research

"Understanding the fundamental structure and dynamics of next-generation vitrimer materials" Matière Molle et Chimie, ESPCI Paris

Advisors: Ludwik Leibler and Michel Cloître

For my postdoctoral work, I investigated the fundamental physics of vitrimers, a novel class of polymers that are crosslinked like thermosets, but flow when heated like thermoplastics. While vitrimers are robust and versatile, poor understanding of their microstructure hinders their transfer from the laboratory to commercialization. Using a suite of experimental tools and theoretical modeling, I discovered that polyethylene vitrimers exhibit unique self-assembly behavior, which has significant repercussions on processing and material properties [5]. Currently, I am developing experimental and theoretical tools to understand the linear viscoelastic properties of model vitrimer materials.

Teaching Interests:

Teaching experience

As an undergraduate at UT Austin, I served as a tutor/grader for several courses, including Material and Energy Balances, Transport Phenomena, and Unit Operations. As a graduate student at the University of Minnesota, I was a TA for courses in Reaction Engineering/Kinetics and Plant Design. I also served as a course instructor for Mass Transfer and Separations. In this role, I taught semi-weekly recitation sessions for a class of 20 students.

In addition to formal courses, I also devoted significant time to scientific outreach. During my Ph.D. I founded Science For All, a nonprofit program dedicated to teaching science to the urban Twin Cities community. For this program, I designed scientific lessons and experiments for K-12 students, managed relationships with partner schools, and raised funding for supplies and transportation. Since its founding in 2014, SFA has grown to include 30 mentors and works with almost 100 K-12 students. Furthermore, from 2012-2014 I served as the head coordinator for Polymer Day, a day-long outreach event focused on teaching high school students about polymer science. I organized the event schedule, recruited mentors, and designed experiments that highlighted concepts from polymer chemistry and physics. In 2016 I co-authored a journal article describing the program and its success at teaching high school students about polymer science [6].

Future teaching goals

Because of my varied teaching experiences during my academic career, I can teach all core undergraduate chemical engineering courses. I am also interested to teach elective/graduate courses in areas related to soft materials, such as polymer chemistry and physics, colloidal science, transmission electron microscopy, scattering techniques, rheology, nanoscience, etc.

Furthermore, I plan to promote diversity in chemical engineering by starting a mentorship program for underrepresented minorities in the sciences. The program will include two branches. One branch will focus on in-class lessons and experiments for K-12 students (similar to Science For All). The other branch will aim at placing high school students in academic and industrial internships. This program will not only strengthen the relationship between the public and scientific communities, but also provide opportunities for graduate students and postdocs to act as mentors.

Selected successful proposals:

European Synchrotron Radiation Facility: ID02 Beamline (2018)

Main proposer

Awarded 72 hrs of ultra small-angle X-ray scattering time to study vitrimer nanostructure

Advanced Photon Source at Argonne National Laboratory: DND-CAT Beamline (2018)

Main proposer

Awarded 72 hrs of simultaneous small- and wide-angle X-ray scattering time to study vitrimer nanostructures

SOLEIL Synchrotron: SWING Beamline (2018)

Main proposer

Awarded 72 hrs of small-angle X-ray scattering time to study vitrimer nanostructures

Campus France PRESTIGE Co-financing Grant Award (2017)

Sole proposer

Awarded 41000 € to conduct research on the fundamental physics of vitrimer materials

National Science Foundation Graduate Research Fellowship Program Award (2011)

Sole proposer

Awarded $100,000 to conduct graduate research in chemical engineering

Selected publications:

1) **Ricarte, R.G.**; Lodge, T.P.; Hillmyer, M.A. "Detection of pharmaceutical drug crystallites in solid dispersions by transmission electron microscopy." *Molecular Pharmaceutics* **2015,** *12*, 983–990.

2) **Ricarte, R.G.**; Lodge, T.P.; Hillmyer, M.A. "Elucidation of the spatial distribution of small molecules in amorphous polymer matrices by electron energy-loss spectroscopy." *Langmuir* **2016**, *32*, 7411–7419.

3) **Ricarte, R.G.**; Li, Z.; Johnson, L.M.; Ting, J.M.; Reineke, T.M.; Bates, F.S.; Hillmyer, M.A.; Lodge, T.P. “Direct Observation of Nanostructures During Aqueous Dissolution of Polymer/Drug Particles.” *Macromolecules*, **2017**, *50*, 3143–3152.

4) **Ricarte, R.G.**; Li, Z.; Johnson, L.M.; Van Zee, N.; Porter, Schmitt; R.; W.W.; Hillmyer, M.A.; Lodge, T.P. “Recent advances in understanding the nanoscale phenomena of solid dispersions.” *In preparation*.

5) **Ricarte, R.G.**; Tournilhac, F.; Leibler, L. “Semi-crystalline polyethylene vitrimers: macro- and micro-phase separation create hierarchical structure.” *In preparation*.

6) Ting, J.M.; **Ricarte, R.G.**;* Schneiderman, D.K.;* Saba, S.A.; Jiang, Y.; Hillmyer, M.A.; Bates, F.S.; Reineke, T.M.; Macosko, C.M.; Lodge, T.P. “Polymer Day: Outreach Experiments for High School Students.” *Journal of Chemical Education*, **2017**, *94*, 1629–1638. (*Equal authorship)

542187 Understanding and Controlling Self-Assembly in Polymer and Colloidal Systems through Simulation, Theory, and Experiment

Abstract:

Research Interests:

Functional polymer materials play a central role in a vast array of fields, ranging from polymer nanocomposites for aerospace materials, to electron and ion-conducting polymers for electronics and energy, to biocompatible polymers for drug delivery. To satisfy performance metrics in these myriad applications, diverse new polymer chemistries, architectures, assembly paradigms, and processing schemes are constantly being developed. However, controlling the complex interplay between equilibrium driving forces and kinetic limitations in polymer and colloidal self-assembly, though crucial to appropriately harnessing these new techniques, remains a challenge. My independent research program will address this gap by predicting the equilibrium structure of new functional polymer and colloidal systems and uncovering how to use processing techniques, additives, or chemical modifications to shift the equilibrium state and/or manipulate assembly kinetics to trap useful nanoscale and microscale structures. My group will employ a unique combination of molecular simulation and theory techniques supported by key experimental characterization. This combined simulation/theory/experiment approach will allow me to efficiently interrogate the molecular-level phenomena behind soft materials assembly, enabling the design of optimized processes and strategies to navigate the structure/function pipeline of materials development.

In my Ph.D. research with Prof. Arthi Jayaraman, I used a combination of molecular simulation, theory, and scattering techniques to study a variety of synthetic and bioinspired soft materials systems—this experience forms the basis of my future research program. From a computational methodology standpoint, I leveraged statistical thermodynamics to develop new Gibbs ensemble simulation methods (6) to enable efficient study of the phase equilibria of polymer solutions, and I also contributed an open-source Python implementation of the Polymer Reference Interaction Site Model (PRISM) theory (1,9). Using these techniques, I investigated the structure and thermodynamics of polymer solutions of varying architecture (linear, cyclic, star) (2,11) and connected *intra*molecular structure to *inter*molecular assembly in films and dense solutions (2). In another effort, I developed simulation methods to study the colloidal self-assembly of bioinspired photonic materials in evaporative thin-film and reverse emulsion assembly processes (10). Using these techniques, I uncovered key physical insights that helped explain the structure of the photonic assemblies and clarified the differences between the film and emulsion techniques. My simulation and theory expertise is complemented by my experience performing experiments in Prof. Thomas Epps's lab, where I used neutron and x-ray scattering to study lithium salt-doped block polymer materials for battery electrolytes (3,4), rationalizing the self-assembly of the salt-doped block polymers through comparison to molecular simulations and strong segregation theory. In each project I pursued a holistic understanding of the phenomena that control the structure and thermodynamics of polymer and colloidal materials by combining simulation, theory, and experiment. Furthermore, I successfully engaged in a diverse set of working styles, ranging from simulation-only projects to joint efforts with 1-2 experimental collaborators, to large multi-institution collaborations that connected multiple materials synthesis, characterization, and computational groups. Prior to graduate school I worked for three years in industrial R&D in semiconductor manufacturing, in which I received recognition for my creative research contributions—from this industry experience I learned how to design and pursue industrially-relevant and applications-oriented research. As a faculty member I will leverage this diversity of methodological and topical expertise from my academic and professional background to enrich and strengthen my independent research program.

My future research will focus on three key areas: 1) macromolecular design and architecture, 2) dynamic non-covalent bonding and supramolecular assembly, and 3) 'alchemical' transformations in polymer and colloidal systems. I will pursue these topics with an emphasis on applications in optical and electronic materials, nanotemplating and patterning, and drug delivery, areas where precisely tunable self-assembly, dynamic responses to the environment, and switchable structure and functionality are crucial. Manipulating polymer architecture is a longstanding technique to tune phenomena ranging from self-assembly to macroscopic material properties, and new advances in polymer synthesis have greatly expanded the available library of architectures to attain new states, structures, and functionalities. Supramolecular self-assembly is a rich area for new materials development, as the dynamic nature of non-covalent bonding provides valuable characteristics like fast assembly, self-healing capabilities, and 'smart' responsiveness to environmental and/or actuated triggers. Furthermore, modulating soft materials assembly through 'alchemical' transformations in polymer chemistry and/or architecture is an emerging field that can produce unexplored non-equilibrium or metastable structures, and unlocks new paradigms in switchability of structure and function. In my research on each of these topics I will synergistically apply molecular simulation and PRISM theory to quickly screen parameter spaces and provide mechanistic insights, while validating model development and key conclusions through comparison to scattering experiments. In parallel, I will also develop and leverage non-equilibrium simulation techniques to guide the design of polymer processing schemes. This combined computation/theory/scattering research framework will help direct synthetic efforts and enable improved process and materials design, furthering the development of new functional polymer materials solutions for photonics, medicine, nanomaterials technology, and energy.

Teaching Interests:

I consider effective teaching to be of equal importance to innovative research, as I believe that impactful faculty-student interactions can have wide-reaching effects in fostering an informed and interested cohort of future scientists, engineers, and policymakers. As a faculty member I am excited to act on my commitment to teaching by creatively connecting students' classroom work to the world around them via tangible and relatable examples, assignments, and projects. In my teaching I plan to draw from my hybrid academic and professional background to foster interest in academic research while maintaining an eye on the industrial relevance of the subject matter. When explaining key concepts, I will draw from the diversity of career paths and research interests within chemical engineering to help foster curiosity and engagement in my students.

My formal teaching background centers around experiences as a teaching assistant for courses in thermodynamics and polymer science and engineering; in both courses I delivered lectures, hosted office hours and review sessions, and fielded individual questions from students. I have also tutored an undergraduate student in thermodynamics, fluid mechanics, heat and mass transfer, and chemical kinetics. Though my varied research, teaching, and professional background has prepared me to teach several core chemical engineering subjects, I am particularly interested in thermodynamics at the undergraduate and graduate level. Through my teaching assistantship experience I gained a keen appreciation for the challenges and sticking points for students when approaching key concepts in thermodynamics—for example, choosing from the myriad available activity coefficient models in phase equilibria calculations. In my thermodynamics curriculum I hope to clarify the process of problem-solving, to help students navigate available methodological choices and develop an appropriate problem-solving workflow. In addition, I am interested in developing courses in polymer physics, statistical thermodynamics, and molecular simulation and theory with a polymer and/or soft materials focus to help train the next generation of soft materials scientists.

Publications (7 current publications, 4 additional in preparation or review):

1. T.B. Martin, **T.E. Gartner, III**, R.L. Jones, C.R. Snyder, A. Jayaraman, "*py*PRISM: a computational tool for liquid-state theory calculations of macromolecular materials," *Macromolecules,* **2018,** *51*(8), 2906–2922
2. **T.E. Gartner, III**, A. Jayaraman, "Macromolecular 'size' and 'hardness' drives structure in solvent-swollen blends of linear, cyclic, and star polymers," *Soft Matter,* **2018,** *14*, 411-423
3. **T.E. Gartner, III**,* M.A. Morris,* C.K. Shelton,* J.A. Dura, T.H. Epps, III, "Quantifying lithium salt and polymer density distributions in nanostructured ion-conducting block polymers," *Macromolecules,* **2018**, *51* (5), 1917-1926; *Equal contributions
4. **T.E. Gartner, III**, T. Kubo, Y. Seo, M. Tansky, L.M. Hall, B.S. Sumerlin, T.H. Epps, III, "Domain spacing and composition profile behavior in salt-doped cyclic vs linear block polymer thin films: a joint experimental and simulation study," *Macromolecules,* **2017**, *50* (18), 7169-7176
5. M.A. Morris,* **T.E. Gartner, III**,* T.H. Epps, III, "Tuning block polymer structure, properties, and processability for the design of efficient nanostructured materials systems,"** *Macromol. Chem. Phys.,* **2017**, *218* (5), 1600513; *Equal contributions; **Cover article
6. **T.E. Gartner, III**, T.H. Epps, III, A. Jayaraman, "Leveraging Gibbs ensemble molecular dynamics and hybrid Monte Carlo/molecular dynamics for efficient study of phase equilibria," *J. Chem. Theory Comput.,* **2016**, *12* (11), 5501-5510
7. H. Wu, H. Pan, M.A. Green, D. Dietderich, **T.E. Gartner, III**, H.C. Higley, M. Mentink, D.G. Tam, F.Y. Xu, F. Trillaud, X.K. Liu, L. Wang, and S.X. Zheng, "The resistance and strength of soft solder splices between conductors in MICE coils," *IEEE Trans. Applied Superconductivity,* **2011**, *21* (3), 1738-1741

In Preparation or Review:

8. H. Kuang, **T.E. Gartner, III**, A. Jayaraman, E. Kokkoli, "Controlling the Length and Diameter of DNA Nanotubes Formed by ssDNA-Amphiphiles," *Submitted to ACS Appl. Nano Mater. (in revision)*
9. T.B. Martin, **T.E. Gartner, III**, R.L. Jones, C.R. Snyder, A. Jayaraman, "Design and Implementation of pyPRISM: A Polymer Liquid-State Theory Framework," *Submitted to Proceedings of the 17th Python in Science Conference. (in review)*
10. M. Xiao,* **T.E. Gartner, III**,* Z. Hu,* X.Yang, W. Li, A. Jayaraman, N.C. Gianneschi, M.D. Shawkey, A. Dhinojwala, "Surface segregation of binary particles in photonic supraballs and films," *Equal contributions *(in preparation)*
11. **T.E. Gartner, III**, M.J.A. Hore, A. Jayaraman, "Solvent quality effects on the thermodynamics and scaling of linear and cyclic polymer solutions," *(in preparation)*

542215 Designing New Functional Soft Materials with Molecular Simulations

Abstract:

Research Interests:

Soft condensed matter plays an increasingly important role in physics, chemistry, biology, and all engineering disciplines. Examples include polymers in the bulk and in thin films, as well as colloidal suspensions. Due to the vast range of possible applications, it is essential to improve understanding and gain insights into the complex interplay of processing, structure and properties of soft matter. Molecular simulations are a valuable tool to make significant advances by filling in the missing links between molecular structure and the resulting properties of soft matter.

My doctoral research (with Prof. Kurt Binder, University of Mainz) was focused on homogeneous crystallization of colloids, which provide a perfect model system for research into soft matter. I developed a novel method to determine nucleation barriers in simulations without the need to calculate the anisotropic interfacial tension or to identify the phases on a microscopic level.4 While I used a colloidal model to validate the method, it is applicable to non spherical and anisotropic crystals and can be, for example, extended to model ice and water. As a postdoctoral fellow (with Prof. Athanassios Panagiotopoulos, Princeton University) I applied my background in simulations to polymer systems, where I am especially interested in non-equilibrium phenomena like evaporation or flow of polymer solutions.[1,2] We showed the previously under-appreciated role of solvent hydrodyamics on the stratification of drying films, which is important for coatings, polymer thin film formation, and paints.

Building upon my previous expertise I will aim my research efforts at soft materials like colloidal dispersions, polymers solutions, polymer networks/gels, and polymer nanocomposites. Initially, I will focus on the following topics:
(1) The influence of hydrodynamics on the evaporation process of mixtures. Because this influence is not fully understood, developing a predictive model will accelerate the material design process for multicomponent coatings and thin films.
(2) The degree of crystallization in organic, molecular glasses can be tuned using polymer additives. Organic, molecular glasses are used in electronics, pharmaceutics, and biodegradable coatings. For many of these applications, it is desirable to suppress crystallization. Coarse grained molecular simulations provide a efficient way to systematically explore different chemistries and compositions and to guide experimental design to enhance or suppress the structure of the organic molecular glass.
(3) Star or brush shaped polymers are promising building blocks for flexible, multicomponent organic transistors, they form networks/gels which can be used for filtering and slow drug release. Understanding and controlling the structure, elasticity and flow behavior of these soft materials from the molecular level is going to open up new possibilities to design new functional soft materials.

I will build a vibrant research group operating at the intersection of engineering, physics, and chemistry, which will enable us to address both fundamental and applied questions. This why I believe a chemical engineering department would provide the perfect environment for my group. My goal is to translate the insights gained by advanced simulation and modeling techniques to design strategies and practical knowledge. In collaboration with experimental groups, this research will help design new functional soft materials to meet the emerging needs in energy, environmental science, engineering, and medicine. **Teaching Interests:**

Because my passion for research is strongly intertwined with my passion for teaching, I am looking forward to having the opportunity to teach at the undergraduate and graduate levels. Learning happens best if the students know why the topic matters and are invested in understanding it. That is why I aim to communicate not only topics and principles when teaching, but also why these topics and principles matter. I strive to create a supportive, inclusive learning environment for every student, encourage them to think critically, and take active part in the learning process.

As a graduate student, I worked as a teaching assistant, giving me the opportunity to lead weekly review and homework sessions as well as designing and grading the exam and homework problems. During both my time as a grad student and postdoctoral fellow, I had the pleasure to mentor the research of three undergraduate students and two graduate students, some resulting in co-authored publications. Since I am determined to increase my experience and become an outstanding teacher, I volunteer with the Prison Teaching Initiative at the federal prison in Ft Dix. I have taught Environmental Science, Beginning Algebra, and Biological Science Concepts, which are accredited through Mercer County Community College. At Princeton University, I have had the opportunity to teach a part of the "Numerical Algorithms for Scientific Computing" graduate course.

My personal education and scientific background give me confidence in teaching core undergraduate chemical engineering courses as well as courses such as transport phenomena and thermodynamics. In addition, I am interested in teaching a range of advanced undergraduate and graduate-level courses in statistical mechanics, simulation methods, and scientific computing. I would like to include developing elective courses focusing on soft matter, polymers and simulation methods as an essential part of my responsibilities as a faculty member. **Selected Publications:** (19 total, 9 first author, 168 citations)

1. A Statt, MP Howard, AZ Panagiotopoulos, "Influence of hydrodynamic interactions on stratification in drying mixtures" arXiv:1804.00066

2. A Statt, MP Howard, AZ Panagiotopoulos, "Solvent quality influences surface structure of glassy polymer thin films after evaporation", J. Chem. Phys. 147 (18), 184901

3. A Statt, R Pinchaipat, F Turci, R Evans, CP Royall, "Direct observation in 3d of structural crossover in binary hard sphere mixtures", J. Chem. Phys. 144 (14), 144506

4. A Statt, P Virnau, K Binder, "Finite-size effects on liquid-solid phase coexistence and the estimation of crystal nucleation barriers" Phys. Rev. Lett. 114 (2), 026101

542251 Nanostructural Engineering Towards on-Demand Manipulation of Polymers and Their Derivatives Functionality

Abstract:

Research Interests: Polymers self-assembly has emerged as leading patterning technique due to its ability to generate various structures with spontaneous nanoscale order, but the implementation of these organized features into energy-related applications is still largely unexplored. My research program will focus at bridging the polymer chemistry and assembly physics with functional inorganic materials synthesis as well as employing advanced techniques for understanding materials structure-property relationships through in-situ characterizations. We aim to address many grand challenges in sustainable chemistry such as the lack of efficient routes for low-dimensional materials synthesis, transfer and doping at large-scale for their practical applications and elucidating the mechanism of materials performance decaying upon external stimuli. With integrating fundamental mechanistic studies with development of scalable nanomanufacturing platforms, we will provide new engineering solutions for advanced materials design, syntheses and characterizations with precise nanoscale control.

Postdoctoral Project:

1) Understanding polymer assembly dynamics by in-situ visualization its structural evolution using super-resolution microscopy; 2) Simultaneous in-film polymer chemistry with self-assembly physics for hierarchical nanopatterns. Department of Chemical and Biological Engineering, Northwestern University (Advisor: Professor Muzhou Wang).

PhD Dissertation:

A generalized method for aligning functional block copolymers and blends thin film and large-scale roll-to-roll fabrication of ordered porous materials for energy storage applications. Department of Polymer Engineering, University of Akron (Advisors: Professor Bryan D. Vogt and Professor. Kevin A. Cavicchi)

Research Experience: My research focuses on the development of new methodologies of engineering nanostructured polymeric materials towards improved material properties, especially for energy storage applications, from understanding their fundamental assembly dynamics using advanced in-situ characterization techniques including super-resolution microscopy and scattering measurements.

During my PhD research, **I demonstrated a generalizable method (solvent vapor annealing with shear, SVA-S) for aligning block copolymers (BCPs) and their blend thin films that enables functional nanopatterns upon proper selection BCPs and functional precursors**. Using this approach, we can generate arbitrary line based nanofeatures such as rhombic arrays and porous carbon films with anisotropic electrical conductivity. Additionally**, I developed a large-scale fabrication (> kg) methodology for a wide variety of functional mesoporous materials through a continuous roll-to-roll processing**, which is compatible with the traditional evaporation-induced self-assembly process.

The ability to generate large quantities of functional materials including highly doped mesoporous carbons and ordered porous carbon/nanoparticle composites using low-cost raw materials has enabled high-performance devices for sustainable energy storage.

In my postdoctoral research, **I demonstrated a photo-controlled *in-film* polymerization process for synthesizing homopolymers in self-assembled BCP films, which successfully integrates the polymer chemistry with self-assembly physics.** The region and kinetics of in-film polymerization can be modulated by the location and power of irradiation using photomasks for obtaining distinct morphologies on one substrate, providing a new platform for hierarchically manipulating nanopatterns from the self-assembled BCP thin film. Additionally, **I developed a platform multifunctional photoswitchable fluorophores for its incorporation into a super-resolution microscopy experiment, which greatly streamlines labeling strategies into a single step,** where a small quantity is mixed with standard monomers and copolymerized to yield the functionalized material with no further purification. This allows the in-situ nanoscale imaging of polymer self-assembly upon external stimuli such as solvent-swollen for revealing assembly kinetics and defects annihilation. This work holds the potential as an imaging platform for promoting widespread adoption of super-resolution microscopy in the polymer community.

Future Research Directions: As an independent investigator, I will work at the interface of self-assembled polymers, low-dimensional functional materials and their applications for renewable energy storage. Initial investigations in my laboratory will compose of three different projects/aims (1) scalable synthesis of 2D materials through polymer templating; (2) in-operando characterizing polymer nanostructure evolution under external stimuli, and (3) on-demand, on-surface polymer synthesis and pattern formation with dynamic linkages. These proposed works will require a multidisciplinary approach, combining my experience in nanostructure engineering, large-scale roll-to-roll processing, electrochemical energy storage and advanced characterization with fluorophore-based techniques.

Mentoring Experience: During my Ph.D. and postdoc studies, I have mentored 4 graduate students, 1 undergraduate student, 1 REU student, 1 high school teacher and 1 high school student on many different projects including materials synthesis, novel rapid microwave processing and mechanical characterizations of elastomeric materials. The experience and relationship with these mentee from diverse research/education backgrounds not only builds up my communication and teaching skills with students at different levels, but also helps me understand how to design a research project that fits with different backgrounds. These lead to the fact that 7 out of 8 mentees have co-authored publications with me after finishing up their projects.

Teaching Interests: In engineering education, I believe the goal as a teacher is to not only provide fundamental knowledges/equations, but more importantly is to educate them with strategies for problem-solving that can be extended to the future. Although the subjects and teaching styles can be diverse, the key attributes in my teaching philosophies include caring about students, teaching with enthusiasm and energy and keeping an open mind with alternative methods, solutions and answers. My education background in polymer engineering enables me to teach any core courses in chemical engineering related to thermodynamics, reaction kinetics and advanced characterizing techniques, Besides, my postdoc experience in Chemical Engineering at Northwestern makes me interested in teaching any existing class in the department, as well as design new classes that provides connections between polymers and energy storages.

Grant Writing Experiences: 1) ACS Petroleum Research Fund **(Awarded for $110,000)**; 2) National Science Foundation; 3) Academic Year Undergraduate Research Grants, Northwestern University **(Awarded)**

Selected Awards and Honors: 1) Selected as PMSE Future Faculty Scholars, ACS, 2018; 2) Excellence in Graduate Polymer Research, ACS, 2017; 3) Polymer Physics Best Poster Prize, APS, 2016; 4) Finalist, Frank J. Padden Jr. Award for Excellence in Polymer Physics Graduate Research, APS, 2015

Selective Publication Lists: (29 published, 2 under revision, 2 in preparation)

Z. Qiang, S. Akolawala and M. Wang "Simultaneous in-film polymer synthesis and self-assembly for hierarchical nanopatterns" *ACS Macro Letters*, 2018, 7, 566-571

Z. Qiang, Y. Xia, X. Xia and BD. Vogt "Generalized synthesis of a family of highly heteroatom-doped ordered mesoporous carbons" *Chemistry of Materials*, 2017, 29 (23), 10178-10186

Z. Qiang, Y. Chen, Y. Xia, W. Liang, Y. Zhu and BD. Vogt "Ultra-long cycle life, low-cost room temperature sodium-sulfur batteries enabled by highly doped (N, S) nanoporous carbons" *Nano Energy,* 2017, 32, 59-66

Z. Qiang, X, Liu, F. Zou, KA. Cavicchi, Y. Zhu and BD. Vogt "Bimodal porous carbon-silica nanocomposites for low-cost, high rate, long life-time Li-ion batteries" *Journal of Physical Chemistry C*, 2017, 121 (31), 16702-16709

Z. Qiang, Y. Chen, B. Gurkan, Y. Guo, M. Cakmak, KA. Cavicchi, Y. Zhu and BD. Vogt "Cooperatively assembled, nitrogen-doped, ordered mesoporous carbon/iron oxide nanocomposites for low-cost, long cycle life sodium ion batteries" *Carbon,* 2017, 116, 286-293

Z. Qiang, L. Zhang, GE. Stein, KA. Cavicchi and BD. Vogt "Unidirectional alignment of block copolymer films induced by expansion of a permeable elastomer during solvent vapor annealing" *Macromolecules*, 2014, 47(3), 1109-1116

Z. Qiang, M. Wadley, KA. Cavicchi and BD. Vogt "Facile non-lithographic route to sub-10 nm, highly aligned silica nanopatterns using unidirectionally aligned polystyrene-*block*-polydimethylsiloxane films" *Journal of Polymer Science Part B: Polymer Physics* 2015, 53(15), 1058-1064

Z. Qiang, Y. Guo, H. Liu, SZD. Cheng, M. Cakmak, KA. Cavicchi and BD. Vogt "Large scale roll-to-roll fabrication of mesoporous materials" *ACS Applied Materials & Interfaces* 2015, 7(7), 4306-4310

Z. Qiang, Y. Zhang, Y. Wang, S. Bhaway, KA. Cavicchi and BD. Vogt "Highly aligned, large pore ordered mesoporous carbon film by solvent vapor annealing" *Carbon*, 2015, 82, 51-59

Z. Qiang, Y. Zhang, JA. Groff, KA. Cavicchi and BD. Vogt "A generalized method for alignment of block copolymer films: solvent vapor annealing with soft shear" *Soft Matter*, 2014 10(32), 6068-6076

S. Bhaway, **Z. Qiang**, Y. Xia, X. Xia, B. Lee, KG. Yager, L. Zhang, K. Kisslinger, Y. Chen, K, Liu, Y. Zhu and BD. Vogt "Operando grazing incidence small angle X-ray scattering/X-ray diffraction of modeled ordered mesoporous lithium-ion batteries anodes" *ACS Nano,* 2017, 11(2), 1443-1454

P. Cao, B. Li, T. Hong, J. Townsend, **Z. Qiang**, K. Xing, K.D. Vogiatzis, Y. Wang, J. Mays, A.P. Sokolov and T. Saito "Super-stretchable, self-healing polymeric elastomers with tunable properties" *Advanced Functional Materials,* 2018, 1800741

K. Staggs, **Z. Qiang**, K. Madathil, C. Gregson, Y. Xia, BD. Vogt and D. Nielsen "High efficiency and facile butanol recovery with magnetically responsive micro/mesoporous carbon adsorbents" *ACS Sustainable Chemistry and Engineering* 2017, 5(1), 885-894

G. Deng, Y. Zhang, C. Ye, **Z. Qiang**, GE. Stein, KA. Cavicchi and BD. Vogt "Bicontinuous mesoporous carbon thin film via an order-order transition" *Chemical Communication*, 2014, 50,12684-12687

Book Chapters

BD. Vogt, **Z. Qiang**, J. Xue, G. Deng, A. Karim and KA. Cavicchi "Structural control in block copolymer templated nanoporous carbon films" in Polymer Precursor Derived Carbon, 2014, *ACS Symposium Series,* Vol. 1173, Chapter 3, 35-60.

542377 From Chemical Bond Forces and Breakage to Macroscopic Fracture of Soft Materials

Abstract:

Research Interests:

Soft materials are irreplaceable in engineering applications where large reversible deformations are needed, and in life sciences to replace a variety of living tissues. While mechanical strength may not be essential for all applications, excessive brittleness is a strong limitation. Yet predicting if a soft material will be tough or brittle from its molecular composition relies on empirical concepts due to the lack of proper tools to detect the damage prior failure. Taking advantage of the recent advances in materials science and mechanochemistry, we propose a groundbreaking method to investigate the mechanisms of fracture of tough soft materials. By using mechanoluminescent molecules incorporated in model materials, we will gain an unprecedented molecular understanding of where and when bonds break as the material fails and the crack propagates, and will then be able to establish a direct relation between the architecture of soft polymer networks and fracture energy, leading to a new molecular and multi-scale vision of macroscopic fracture of soft materials. Such advances will be invaluable to design soft but tough materials to replace living tissues and make lightweight tough and flexible parts for energy efficient transport.

Teaching Interests:

Topics that I am particularly interested in teaching include core chemical engineering courses such as heat and mass transport, thermodynamics, and chemical reaction engineering; as well as more specialized courses in my areas of expertise related to polymer physics, polymer chemistry, and mechanics of soft materials.

541883 Gradient Double Network Gels for Medical Implants

Abstract:

Research Interests:

Gradient materials exhibit a continuous variation in any of their physical or chemical properties. For example, tendons, fibrous tissues found in our bodies, illustrate naturally occurring gradient materials; they exhibit a continuous change in modulus, which helps to achieve efficient connectivity between soft muscle tissues and hard bones. Mimicking such gradients using synthetic hydrogels with tunable stiffness will have a tremendous impact on the development of medical implants. The research program that I envision to develop in my research group will explore new methods for generating tough gradient hydrogels. The strategy will be to synthesize double network hydrogels (DN) and elastomers by incorporating degradable network strands, wherein the first network can be degraded under acidic conditions while maintaining the second network intact. Such a system will also allow for complete characterization of DN strands after their formation by selective degradation of either one of the two networks to examine the network properties of the other gel. After gaining fundamental insights into the structure of DN, the project will focus on the development of gradient DN (gDN), demonstrating a continuous variation in network density and modulus. A key advantage of gDN is that a single specimen can be used to investigate the effect of varying a given parameter systematically (e.g., elasticity or molecular concentration). I am also interested in exploring the non-thrombogenic coatings, and the detection method for thrombi formation in blood contacting devices, e.g., left ventricular assist devices (VADs). VADs are used as life support for many patients with heart failure or those waiting for the heart transplantation, which can take months to several years. This project will develop a new technique to target platelets using biocompatible microbubbles (MBs) as ultrasound contrast agents. Such a noninvasive and low-cost molecular imaging technique will have a tremendous impact on blood contacting device industry, where monitoring thrombus formation (in-vivo) is of utmost importance. The interdisciplinary nature of my research interest would draw a potential funding from NIH (NIEHS, NIBIB), NSF (BMAT, CMMT, EBMS, POLY, MRSEC), ONR and many industries.

Selected Publications:

- C. K. Pandiyarajan, Jan Genzer, "One-Pot Synthesis of Surface-Anchored Network Coatings", article in preparation 2018.
- Joachim Jelken, C. K. Pandiyarajan, Jan Genzer, Nino Lomadze, Svetlana Santer, "Polymer nano-membranes with reversible tunable pore size", ACS Applied Materials 2018, Submitted.
- C. K. Pandiyarajan, Jan Genzer, "Effect of network density in surface-anchored poly (N-isopropyl acrylamide) hydrogels on adsorption of fibrinogen", Langmuir 2017, 33, 1974-1983 (Link).
- C. K. Pandiyarajan, Oswald Prucker, Jürgen Rühe, "Humidity Driven Swelling of the Surface-Attached Poly (N-alkylacrylamide) Networks", Macromolecules 2016, 49 (21), 8254-8264 (Link).
- C. K. Pandiyarajan, Michael Rubinstein, Jan Genzer, "Surface-Anchored Poly (N-isopropyl acrylamide) Orthogonal Gradient Networks", Macromolecules 2016 49 (14), 5076-5083 (Link).
- C. K. Pandiyarajan, "Biofunctionalized Hydrogels for Tissue Engineering", Advanced Science News: Regenerative Medicines, 2016 (Link)
- Ke Li, C. K. Pandiyarajan, Oswald Prucker, Jürgen Rühe, "On the lubrication mechanism of surfaces covered with Surface-attached hydrogels", Macromolecular Chemistry and Physics 2016, 217, 526-536 (Link).
- C. K. Pandiyarajan, Oswald Prucker, Barbara Zieger, Jürgen Rühe, "Influence of the Molecular Structure of Surface-Attached poly (N-alkylacrylamide) Coatings on Blood Platelet Adhesion", Macromolecular Bioscience, 2013, 13, 873-884 (Link).
- C. K. Pandiyarajan, Oswald Prucker, Barbara Zieger, Jürgen Rühe, "Influence of the Molecular Structure of Surface-Attached poly (N-alkylacrylamide) Coatings on Blood Platelet Adhesion", Macromolecular Bioscience, 2013, 13, 7, 817 [Front cover page, Link].
- N. Somanathan, C. K. Pandiyarajan, W.A. Goedel, W. C. Chen, "Physico-mechanical Studies on the Langmuir-Blodgett Films of Polythiophene Containing Mesogenic Side Chain", Journal of Polymer Science B: Polymer Physics, 2009, 47, 2, 173 (Link).

Books:

- C. K. Pandiyarajan, "The interaction of Blood Proteins and Platelets on Surface-Attached poly (N-alkyl acrylamide) Networks", Der Andere Verlag (CPI Band 23) 2014, ISBN:978-3-86247-408-0

Teaching Interests:

I am qualified to teach any undergraduate major courses on Chemical Engineering and chemistry:

- Chemical Engineering Thermodynamics
- Chemical Reaction Engineering and/or Reaction Kinetics
- Macromolecular Science and Engineering
- Material and Energy Balances and/or Materials Science
- Computer-Aided Chemical Engineering
- Principles of Chemical Engineering
- Biomaterials
- Polymer science and Technology
- Chemistry for Engineers (My course design for undergraduates)
- Organic Chemistry and Bio-organic chemistry
- In-organic Chemistry, Physical Chemistry and Analytical Chemistry

For graduate students, considering my previous experiences, I can teach (or design) the following courses:

- Polymer Chemistry & Physics
- Polymer at Interfaces
- Proteins at Interfaces
- Biomaterials
- Spectroscopic methods for Molecular Characterization
- Surface-Analytical Techniques for Soft Matters

541892 Structure and Design of Soft Materials for Stretchable Electronics

Abstract:

Research Interests:

Soft materials, such as polymers, have shown considerable potential in a wide range of electronic applications, including light-weight, flexible organic field-effect transistors (OFETs), organic photovoltaics (OPVs), and organic light-emitting diodes (OLEDs). Although current electronics using rigid silicon-based components provide outstanding performance, there is growing demand for the next generation of organic electronics that is deformable and stretchable. Polymers are promising materials enabling both high electrical performance and intrinsic stretchability. Thus, the primary aim of my research is to design polymeric materials with mechanical and conductive properties and develop their structures for stretchable electronics (e.g., artificial muscles, electronic skins). As a basic unit in electronics, OFETs have attracted a significant amount of academic and industrial research attention over the past three decades. To realize the high performance of stretchable OFETs, the polymer materials can be served as two parts (i.e., channel, dielectric layers). First of all, semiconducting polymers as channel materials (i.e., conjugated polymers) have experienced huge developments in the ï¬elds of OFETs. As an example of the developments, the alkyl side chains of these conjugated polymers are no longer considered as solubilizing groups, but as key players that can have a profound impact on the molecular packing density and electronic performance. Furthermore, the stretchability of the conjugated polymers depends on the length and size of the side-chains, and thus the design of conjugated polymers play an important role in the development of stretchable electronics. The polymer materials can also contribute to stretchable and mechanically robust dielectrics. Recently, the use of polymer electrolytes as dielectrics has attracted growing attention due to a low voltage operation and electrochemical properties. In particular, the polymer electrolytes are electronically insulating, but ionically conductive so that they are applicable as stimuli-driven materials. Therefore, the polymer electrolytes are significantly attractive for more diverse stretchable electronics (e.g., sensors, actuators) beyond organic transistors.

As described, polymers have significant potential to serve the next generation of stretchable organic transistors with improved performance. However, polymer-based stretchable electronics still remain challenging to be commercialized, because they have a rigid polymer backbone that potentially limits the elastic stretchability, and are highly susceptible to damage from environmental conditions. In order to establish the environmentally stable and mechanically stretchable polymer electronics, my career objectives will focus on three specific research efforts: (1) designing the environmentally stable conjugated polymers and evaluating the impact of molecular architectures by X-ray characterization, (2) developing the polymer electrolytes as dielectrics and elucidating the relationship between molecular structure and ionic properties, and (3) investigating a polymer composite material consisting of two or more components and illustrating the interfacial structure for stretchable organic transistor applications. Because extrinsically stretchable polymer composites will present another great opportunity that cannot be fulfilled by a single material as well as the rational design of intrinsically stretchable polymers. Beyond the macroscopic investigation by microscopes (e.g., scanning electron microscopy (SEM), atomic force microscopy (AFM)), the nanoscopic characterization by X-ray scattering will be employed for the information of these polymers such as molecular packing density, degree of crystallinity and orientation of domains. In addition to my three primary areas, my future research interest involves the development of sensors, actuators, and skin electronics system with various functionality.

Research Experience:

My research at both Seoul National University (Mentor: Professor Kookheon Char) and Purdue University (Mentor: Professor Bryan W. Boudouris) has involved the design of soft materials and their integration into next-generation flexible electronic devices. In brief, I have focused on: (1) determining the impact of nanostructure in phase-separated organic memory diodes in order to advance flexible logic circuits; (2) utilizing interfacial engineering with radical polymers in organic field-effect transistors for enhanced device performance; (3) and elucidating the charge transport mechanism of radical polymers in order to generate ambipolar, transparent thin film transistors. As a graduate student, I have a significant amount of experience with respect to the design, fabrication, and optimization of organic electronic devices. Moreover, I have been able to utilize these devices to probe the fundamental physics of charge transport in a unique and emerging conducting polymer system (i.e., open-shell macromolecules).

I intentionally switched from more synthetic and device-focused work during my graduate work to a research position that is centered on polymer physics during my postdoctoral career at University of Delaware (Mentor: Professor Thomas H. Epps III). Currently, my work mainly involves the design, synthesis, and nanoscale characterization of macromolecules for integrated electronic circuit configurations. Particularly, I am focusing on the behavior of fluorinated block polymers and the control of nanoscale morphologies using a straightforward solvent vapor annealing (SVA) process

Thus, my work to date has focused on bridging nanoscale scientific phenomena to device engineering at the macroscale, and I anticipate utilizing this previous experience to launch my own research group in order to impact the realms of flexible and stretchable electronics in the near future.

Teaching Interests:

Aside from my research career, I also have considerable teaching experience and have previously served as a teaching assistant for undergraduate courses in Engineering Mathematics, Transport Phenomena, and Design of Staged Separation Processes. This previous experience has allowed me to instruct some of the core, fundamental courses in chemical engineering at my previous institutions of Seoul National University and Purdue University. I was able to gain a deep insight into how to instruct undergraduate students in the fundamentals of chemical engineering from some of the most respected instructors in the field (e.g., Professor Phillip Wankat at Purdue University). Specifically, I delivered lectures for students, held regular office hours, and developed course materials. Thus, I can contribute to delivering any courses in the core chemical engineering curricula and more specific electives, such as polymer chemistry and organic electronics. Additionally, I participated in the mentorship for undergraduate students as a part of their undergraduate thesis at Seoul National and Purdue. Based on my previous experiences, I am more than ready to lead the new generations of students to the high level of learning and training.

541955 Transport and Structure in Polymer Membranes for Energy-Efficient Separations

Abstract:

Research Interests:

With the worldâ€™s population growing rapidly, the need for clean water is greater than ever. Among water purification methods, membrane-based separations draw attention because of their energy efficiency (they do not require an energy-intensive phase-change step, as in distillation), small footprint, and ability to eliminate trace contaminants. However, only a few membrane materials and formation methods are used in industry today, and current membranes often have inadequate efficiency (i.e., selectivity), productivity (i.e., permeability), and chemical stability to be used with water from all sources. To overcome these challenges, new polymer membrane materials and formation methods are needed that will allow us to precisely tune pore size and pore size distributions and to control chemical stability for better separation performance.

Transport of small molecules and ions through polymer membranes is predominantly controlled by the polymer nanostructure. Therefore, morphological control of the membrane structure by tuning chain architecture, synthetic routes, and post-synthesis treatment is a critical aspect of membrane system design. My graduate research focused on the water and salt transport properties (permeation, diffusion and sorption) of ion-containing polymers (sulfonated aromatic hydrocarbons) prepared by a new membrane formation method, i.e., solvent-free melt processing. My postdoctoral research centers on extending my understanding of a polymerâ€™s physical structure by designing and synthesizing highly-structured polymeric systems and performing extended transport property studies, including sorption behavior of biomolecules and ionic conductivity for applications in bioengineering and electrochemical systems. These porous conducting polymers can be used as polymer electrolytes in electrochemical systems and as blood filtering devices for biomedical applications.

My research is directed at understanding the fundamental structure-property relationships of water and ion transport in polymers and composite materials. These interests span disciplines from chemical engineering and chemistry to materials science. The advancement of the technologies mentioned above is highly dependent on the development of polymer membranes that selectively permeate only desired components, while maintaining their chemical stability. My goal is to develop correlations between design, synthesis, structure, and transport properties of polymers for targeting applications in energy-efficient separations, batteries and biomedical devices. Synthetic work will focus on such highly-structured polymers as ion-containing polymers, block copolymers, and random copolymers, and polymer characterization will focus on water purification, electrochemical systems, and biomedical devices.

Teaching Interests:

I find teaching and mentoring the next generation of scientists to be the most satisfying aspect of working in academia. Both in classroom instruction and in a research laboratory, my goal is to help students identify important problems and to arrive at intelligent solutions by using critical thinking skills.

My education and experiences have prepared me well to teach both undergraduate and graduate courses in the Chemical Engineering department. Additionally, as a graduate student and a postdoctoral researcher, I have supervised the research projects of three undergraduate students and several high school students. Seeing these students succeed has been amazingly rewarding to me, and I look forward to continuing this mentorship with my future graduate students and post-docs.

541957 Synthetic Polymeric Materials for Energy Storage and Gas Separation

Abstract:

Research Interests:

Synthesis of polymers with unconventional architectures and unique properties towards both fundamental studies and various applications including energy storage, gas-separation, and flexible electronics.

Research areas includes:

Mechanical interlocked block copolymers: synthesis the "mechanically interlocked" block copolymers, and study this "topology" effect on their physical properties for fundamental studies and potential applications.

Single-ion conducting polymer electrolyte: develop a series of synthetic SLiPE with low interfacial impedance, enhanced mechanical property and high ionic conductivity for a safe lithium-ion battery.

High-performance polymer binder for high capacity anode materials: develop the polymeric materials that have both adhesion capacity with active materials (Si for Li-ion battery, P for sodium-ion battery) and mechanically robustness.

Intrinsic self-healing polymer materials: develop intrinsic self-healing polymer elastomer with good stretchability, high mechanical strength, faster self-healing kinetics and recyclability.

Teaching Interests:

Based on my previous training in Department of Chemistry and Department of Macromolecular Science and Engineering, I am able to teach core courses such as: Polymer Chemistry, Polymer Engineering/Physics Environmental Chemistry.

Based on my previous and current research work, I find teaching Polymers for Energy storage, Nanomedicine development and Oil-gas Industry will also be very interesting: Polymer for Energy Storage, Application of Polymers in Oil-gas industry, Polymer for Drug Delivery, Stimulus-responsive Polymers

Moreover, I should be flexible for the course work topics, and I am willing to teach other related topics during any semesters.

Poster Abstract:

Synthetic Polymeric Materials for Energy Storage and Gas Separation

Peng-Fei Cao,[a] *Alexei Sokolov*[a,b] *and Tomonori Saito*[a]

[a] Chemical Sciences Division, Oak Ridge National Laboratory,

[b] Department of Chemistry, University of Tennessee

Rechargeable lithium ion battery (LIB) has become one of the most successful energy storage devices. Unfortunately, LIBs with common electrolyte still have various problems, such as flammability, toxicity and non-uniform lithium deposition that eventually cause the failure of the cell. Single-ion conducting polymer electrolytes (SCPEs) are well recognized for increased energy efficiency and prolonged cell lifetime due to their capability to mitigate electrode polarization and lower electrolyte loss. Herein, we propose a versatile approach on the fabrication of a molecular-level stretchable functional material *via* rational polymer design (Figure A): covalent attachment of the functional molecules as the side groups (but main component) of intrinsically stretchable polymer network will result the polymeric materials with defined functionality and well-tunable mechanically property. Despite the numerous reports on the solid- or gel- form SCPEs for lithium-ion(metal) battery applications, fabrication of SCPE based materials with mechanically stretchable has never been achieved. The achievement will definitely benefit the future stretchable batteries/electronics considering the advantageous electrolyte performance of the SCPE based systems. Moreover, the molecular-level design of intrinsically stretchable polymer is also suitable for other functional polymer materials.

Self-healable polymer elastomers with tunable mechanical properties are especially attractive for a variety of applications. Herein, we report a series of urea functionalized poly(dimethyl siloxane)-based elastomers (U-PDMS-Es) with extremely high stretchability, self-healing mechanical properties and recoverable gas-separation performance (Figure B). Tailoring the molecular weights of PDMS or weight ratio of elastic cross-linker offers tunable mechanical properties of the obtained U-PDMS-Es, such as ultimate elongation (from 984 to 5,600%), Young's modulus, ultimate tensile strength, toughness and elastic recovery. The U-PDMS-Es can serve as excellent acoustic and vibration damping materials over a broad range of temperature (over 100 °C). The strain dependent elastic recovery behavior of U-PDMS-Es was also studied. After mechanical damage, the U-PDMS-Es can be healed in 120 mins at ambient temperature or in 20 mins at 40 °C with completely restored mechanical performance. The U-PDMS-Es were also demonstrated to exhibit recoverable gas-separation functionality with retained permeability/selectivity after being damaged.

541960 Functional Designer Polymers for Integrating Advanced Synthetic and Biological Materials

Abstract:

Research Interests: Polymers have entered into virtually every aspect of modern society, from familiar commodity items such as disposable plastic bags to high-value technological applications in electronics, energy, and medicine. As the vibrant field of polymer science progresses to new frontiers, the incorporation of more advanced macromolecules into emerging health applications presents several longstanding challenges. Two of the most prominent issues include (1) how to rationally compartmentalize sensitive biomolecules into polymeric domains and (2) how to predictively translate such strategies to harsh physiological environments, which are often complex, dynamic, and responsive.

My long-term goal is to apply my expertise in *polymer science*, *molecular engineering*, and *precision drug formulation* to develop multifunctional materials with prescribed chemical attributes to address these unfulfilled needs in the biomedical landscape. To achieve this, I will employ synthetic tools and ingredients to construct customized materials and elucidate the contributions of structure to function across hierarchical length and time scales. As an independent investigator, my research will focus on designer storage and delivery, a convergence science approach to develop nanocarriers with controlled functionality and programmable performance. This idea combines multimonomer selection, architecture assembly, and structure-property evaluation from a formulation standpoint to enhance the translation of pipeline therapeutics into accessible medicines for treating chronic, infectious, and inheritable disease.

The theme of my future research builds on my interdisciplinary training in polymer chemistry, engineering, and both physical/life sciences to formulate ideal assemblies around drugs, DNA, peptides, and proteins. These well-defined, responsive systems can enable direct investigations into resultant spatiotemporal behavior in solid- and solution-states. Such motifs are also broadly applicable to enabling analogous encapsulation/release strategies in energy management, agricultural/food production, and personal care. Moving forward, I anticipate that my laboratory infrastructure can be readily adapted to host such avenues.

For initial investigations, I will leverage my broad skillset into three main aims:

(1) Accelerating Soft Materials Discovery, Development, and Deployment

(2) Surmounting Biological Barriers to New Therapeutics and Genome Editing

(3) Orchestrating Multi-Responsivity in Biocompatible, Artificial-Living Soft Matter

PhD Dissertation: *Tunable Polymers as Specialized Excipients for Oral Drug Delivery*, Department of Chemical Engineering and Materials Science, University of Minnesota (Advisors: Professor Frank S. Bates and Professor Theresa M. Reineke)

Postdoctoral Project: *Static and Dynamic Properties of Polyelectrolyte Complexes*, Institute for Molecular Engineering, University of Chicago (Advisor: Professor Matthew V. Tirrell)

Funding, Successful Proposals, Achievements: As potential funding sources, the scientific challenges in these projects align with the major initiatives of national funding agencies, including NSF (NSF-DMR, NSF-DEMS), NIH (NIGMS, NIH Director's New Innovator Award, NIBIB), DOE (DOE-MSE), DOD (CDMRP, ARO), and private foundations (Camille and Henry Dreyfus Foundation, Baxter International Foundation, HHMI, Pew Charitable Trusts). My previous record of successful proposals and select distinctions are chronologically summarized:

Proposals: Oakridge National Laboratory Center for Nanophase Materials Science Research Proposal "SANS based molecular exchange kinetics in polyelectrolyte complex micelles" (2017); Argonne National Laboratory APS User Proposal "Microstructural Characterization of Polyelectrolyte Complexes and Polyelectrolyte Complex Based Micelles & Hydrogels" (2016).

Distinctions: PMSE Future Faculty Scholar (2018); NIST-CHiMaD Postdoctoral Fellowship (2016); 1st AIChE Pharmaceutical Discovery, Development and Manufacturing Student Award (2015); Doctoral Dissertation Fellowship (2015); NSF Graduate Research Fellowship (2011).

Research Background: My research has relied on not only my technical training in chemical engineering, but also overarching aspects of chemistry, biomedical engineering, and materials science. I have led multidisciplinary projects supported by both industry and national laboratories, which I believe positions me to initiate new basic and applied research partnerships. Because of this, my perspective on translating advanced polymeric materials into end-use technologies also can facilitate ideal collaborations with existing research endeavors in Centers.

My graduate research involved the (1) controlled synthesis of polymer systems that were judiciously tailored to drugs, (2) preparation of amorphous solid dispersions, and (3) *in vitro / in vivo* dissolution experiments to assess the oral solubility advantage. This enabled the advancement of multicomponent copolymerization[1] and practical guidelines for solid dispersion formulation.[2-5, 8] Macromolecular systems were engineered from the molecular level up to impart favorable non-covalent interactions for stabilizing otherwise-intractable anti-cholesterol, anti-seizure, and anti-cancer drugs. In one example, I led high-throughput synthesis and screening campaigns and identified a promising system that ultimately exhibited excellent correlation to animal studies in rat models,[4] which is now under scale up and production by Dow Chemical.

My postdoctoral work aims to accelerate the pace of next-generation materials discovery into industrially-relevant products. Polyelectrolyte complexes, mixtures of oppositely-charged polymers in water, have highly tunable materials properties that are appealing for numerous bioapplications. However, harnessing these materials for clinical applications remains an enduring challenge. To this end, I am using carefully-selected building blocks to explore chemical, ionic, and architectural effects in well-defined polymer libraries. In this rich state space, monomer constituents are being systematically combined in micelle/hydrogel platforms,[10] with time-resolved scattering, spectroscopy, and fluorescent imaging investigations underway.

Teaching Interests: My interests include courses related to polymers (e.g., Polymer Chemistry, Polymer Physics), including topical courses I have taken like Rheology (under Prof. Chris Macosko), Scattering (under Prof. Timothy Lodge), Colloids/Interfaces (under Prof. Chris Macosko), or Biomaterials (under Prof. Chun Wang). I am also confident in teaching core courses in chemical engineering, such as Transport Phenomena or Thermodynamics.

My teaching philosophy focuses on mentoring students to become independent leaders, equipped with critical thinking and creative problem-solving skills for successful careers. In this manner, I strive to be cognizant of different ways to guide their overall education while supporting each student's individuality and valuing their diversity. At the University of Minnesota, I was a Teaching Assistant in Thermodynamics and Polymer Chemistry. Each course involved grading for ~120 undergraduate students. I also held review sessions and met with students to review unclear topics. I was awarded a faculty-nominated Chemical Engineering Outstanding TA Award and a student-nominated Council of Graduate Students Award for

Graduate Teaching. I was also a Recitation Teaching Assistant in Transport Phenomena. Here, I led supplementary lectures in weekly recitation classes to ~40 students. Seeing mentored students succeed gives me immense gratification and strengthens my commitment to higher education.

Furthermore, I led a collaborative effort in an open-access Journal of Chemical Education article,[6] outlining the motivation, methodology, and measured outcomes of *Polymer Day* to spark outreach ideas for not only other universities, but also for high school teachers. This positive experience taught me to tailor my science and teaching directions toward broader impacts, which I believe will help me develop junior faculty application packages like the NSF CAREER award.

Outreach and Diversity: Public outreach and commitment to diversity will be hallmarks of my research group, with a central goal of inspiring students and their families from diverse backgrounds to proactively engage with science and engineering. This is particularly important to me because I believe that the key to solving pressing global challenges ahead is investment in inclusion of an increasingly diverse workforce, especially for under-represented groups such as women, people of color, LGBTQ+, or persons with disabilities. In my career, I want to be known for training the next generation of academic and industrial leaders to develop curiosity-driven scientific solutions for grand challenges in biotechnology, human health, and welfare.

Select Publications:

1. Ting, J. M.; Navale, T. S.; Bates, F. S.; Reineke, T. M. "Precise Compositional Control and Systematic Preparation of Multimonomeric Statistical Copolymers" *ACS Macro Lett.* **2013**, *2*, 770–774.

2. Ting, J. M.; Navale, T. S.; Bates, F. S.; Reineke, T. M. "Design of Tunable Multicomponent Polymers as Modular Vehicles to Deliver Highly Lipophilic Drugs" *Macromolecules* **2014**, *47*, 6554–6565.

3. Ting, J. M.; Navale, T. S.; Jones, S. D.; Bates, F. S.; Reineke, T. M. "Deconstructing HPMCAS: Excipient Design to Tailor Polymer–drug Interactions for Oral Drug Delivery" *ACS Biomat. Sci. Eng.* **2015**, *1*, 978–990. (**ACS Editors' Choice: Open-Access**)

4. Ting, J. M.; Tale, S.; Purchel, A. A.; Jones, S. D.; Widanapathirana, L.; Tolstyka, Z. P.; Li, G.; Guillaudeu, S. J.; Bates, F. S.; Reineke, T. M. "High–throughput Excipient Discovery Enables Oral Delivery of Poorly Soluble Pharmaceuticals" *ACS Cent. Sci.* **2016**, *2*, 748–755. (**ACS Author Choice: Open-Access**)

5. Ricarte, R. G.; Li, Z.; Johnson, L.; Ting, J. M.; Reineke, T. M.; Bates, F. S.; Hillmyer, M. A.; Lodge, T. P. "Direct Observation of Nanostructures during Aqueous Dissolution of Polymer/Drug Particles" *Macromolecules* **2017**, *50*, 3143–3152.

6. Ting, J. M.; Ricarte, R. G.; Schneiderman, D. K.; Jiang, Y.; Saba, S. A.; Hillmyer, M. A.; Bates, F. S.; Reineke, T. M.; Macosko, C. W.; Lodge, T. P. "Polymer Day: Outreach Experiments for High School Students" *J. Chem. Educ.* **2017**, *94*, 1629–1638. (**ACS Author Choice: Open-Access**)

7. Acar, H.; Ting, J. M.; Srivastava, S.; LaBelle, J. L.; Tirrell, M. V. "Molecular Engineering Solutions for Therapeutic Peptide Delivery" *Chem. Soc. Rev.* **2017**, *46*, 6553–6569.

8. Ting, J. M.; Porter III, W. W.; Mecca, J. M.; Bates, F. S.; Reineke, T. M. "Advances in Polymer Design for Enhancing Oral Drug Solubility and Delivery" *Bioconjugate Chem.* **2018**, *29*, 939–952.

9. Schneiderman, D. K.; Ting, J. M.; Purchel, A. A.; Miranda, R. Jr.; Tirrell, M. V.; Reineke, T. M.; Rowan, S. J. "RAFT Polymerization in Complex Solvents: from Whisky to Fermentation Broth" *ACS Macro Lett.* **2018**, *7*, 406–411.

10. Ting, J. M.; Wu, H.; Herzog-Arbeitman, A.; Srivastava, S.; Tirrell, M. V. "Synthesis and Assembly of Designer Styrenic Diblock Polyelectrolytes" *ACS Macro Lett.* **2018**, *7*, 726–733.

541700 Multiscale Structure and Dynamics of Polymers and Biological Soft Matter

Abstract:

Research Interests:

A grand challenge in polymer engineering lies in integrating the design, characterization, analysis, and prediction of polymer structure and dynamics from nanometers to meters. My research addresses this need by investigating the molecular-scale mechanisms that underlie macroscopic properties of soft materials. Polymers exhibit disparate behaviors across a broad range of length and time scales, such that molecular properties intertwine with microstructure and non-equilibrium dynamics to dictate a material's macroscopic response and applicability. My work probes materials across time and length scales by integrating bioinspired polymer engineering with molecular theory, coarse-grained simulation, and multiscale experimental characterization, including single molecule fluorescence microscopy, light and neutron scattering, and opto-mechanical testing. This approach enables strategic iterations through the materials design loop to develop responsive and functional soft materials. Initial research areas in my proposed research program include programmable bio-coordinated polymer hydrogels, responsive biopolymer brushes, and muscle-mimetic recombinant protein materials.

Research Experience:

Nucleoporin-based materials for selective biomolecular separations, *Department of Chemical Engineering, Massachusetts Institute of Technology* (advised by Bradley D. Olsen)

My postdoctoral research has focused on engineering materials with selective biomolecular transport properties. Controlling biomolecular transport is central to numerous technologies including bio-sensing, bio-separations, and tissue engineering. Natural systems such as nuclear pores routinely regulate molecular transport with remarkable specificity (>99.9% of proteins rejected) and speed (1,000 proteins per second per pore), which inspired the development of a molecular transport theory to identify design criteria for synthetic polymers with selective biomolecular permeability. A quantitative transport model predicted two key principles for specific biomolecular transport by a polymer network: (1) entropic repulsion of non-interacting molecules and (2) affinity-mediated permeation of interacting molecules through a walking mechanism. The model guided the design and synthesis of artificial nucleopore-inspired polymer hydrogels that replicate the selective transport function of nuclear pore proteins. Biophysical characterization of nucleopore-inspired hydrogels using small-angle neutron scattering, protein permeability assays, and fluorescence recovery after photobleaching (FRAP) confirmed the importance of entropic size exclusion, moderate binding affinity, and bound-state diffusive mechanisms in selective hydrogel permeability and transport, in agreement with the model predictions. This work presents a new paradigm for selective transport that critically enables the design of polymer hydrogels to control the transport of multi-receptor biomolecules including therapeutic proteins, immunoglobulins, and broad classes of biotoxins.

Single molecule studies of branched polymer dynamics, *Department of Chemical and Biomolecular Engineering, University of Illinois at Urbana-Champaign* (advised by Charles M. Schroeder)

My dissertation research extended the field of single polymer dynamics to include DNA comb polymers, effectively bridging a knowledge gap between molecular-scale structure and behavior of topologically complex polymers. The vast majority of single polymer studies focus on linear DNA molecules, but branched polymers generally exhibit more complex flow behavior compared to their linear counterparts. Comb-shaped polymers consist of a main polymer backbone with grafted side chains (branches), and this molecular topology enables chemical versatility for drug delivery, biodegradable sutures, and antifouling coatings. My Ph.D. work included the design, synthesis, characterization, single molecule visualization, and coarse-grained simulation of DNA comb polymers. Microfluidic devices were used to manipulate comb-shaped DNA molecules in extensional flow fields, which are common in polymer processing, while directly visualizing the conformational dynamics of single polymers during flow-induced stretching and relaxation. Single molecule experiments revealed the impact of local constraints (branches) on global behavior (conformations, longest relaxation time), in agreement with Brownian dynamics simulations of DNA comb polymers. This research provided fundamental understanding of branched polymers in flow, as well as an unexpected comb polymer stretching mechanism that suggests a major role of dynamic heterogeneity in the known macroscopic features of topologically complex polymers, including enhanced strain hardening and stress overshoots in the startup of shear flow.

Successful Proposals: Oak Ridge National Laboratory High Flux Isotope Reactor, Bio-SANS CG-3 (September 2017, May 2018); NIST Center for Neutron Research, NG-7 SANS (March 2018)

Selected Publications:

- **D. J. Mai**, Y. J. Yang, A. C. Huske, S. Li, B. S. Souza, B. D. Olsen, *in prep.*, (2018).
- R. K. Avery, A. Velmahos, **D. J. Mai**, B. D. Olsen, *in prep.*, (2018).
- C. E. R. Edwards,* **D. J. Mai**,* S. Tang, B. D. Olsen, *in prep.*, (2018). **equal contributions*
- **D. J. Mai**,* Y. J. Yang,* T. J. Dursch, B. D. Olsen. "Nucleopore-inspired polymer hydrogels for selective biomolecular transport", *submitted*, (2018).
- **D. J. Mai**,* A. Saadat,* B. Khomami, C. M. Schroeder. "Stretching dynamics of single comb polymers in extensional flow", *Macromolecules*, 51(4), 1507-1517, (2018).
- **D. J. Mai**, C. M. Schroeder. "Single polymer dynamics of topologically complex DNA". *Current Opinion in Colloid and Interface Science*, 26, 28-40, (2016).
- **D. J. Mai**, A. B. Marciel, C. E. Sing, C. M. Schroeder. "Topology-controlled relaxation dynamics of single branched polymers". *ACS Macro Letters*, 4, 446-452, (2015).
- A. B. Marciel, **D. J. Mai**, C. M. Schroeder. "Template-directed synthesis of structurally defined branched polymer architectures". *Macromolecules*, 48(5), 1296-1303, (2015).
- **D. J. Mai**, C. A. Brockman, C. M. Schroeder. "Microfluidic systems for single DNA dynamics". *Soft Matter*, 8, 10560-10572, (2012).

Selected Awards:

- Arnold O. Beckman Postdoctoral Fellows Award (2017 – present)
- ACS Polymeric Materials Science and Engineering Future Faculty Scholar (2018)
- 1st Place Poster Prize, APS Division of Polymer Physics (2018)
- National Science Foundation Graduate Research Fellowship (2013 – 2016)
- Illinois Distinguished Fellowship (2011 – 2016)
- Mavis Future Faculty Fellowship (2013 – 2015)
- Excellence in Graduate Research Travel Award, APS FGSA (2016)
- AIChE Women's Initiative's Committee (WIC) Travel Award (2015)
- Lam Research Outstanding Graduate Student Award (2015)

Teaching Interests:

As a formally trained chemical engineer, I am excited and qualified to teach core coursework in thermodynamics, fluid mechanics, transport phenomena, separations, chemical kinetics and reactor design, and process dynamics and control. I also look forward to leveraging my research experience and interests to develop and teach electives in polymer science and engineering, soft matter physics, biomolecular engineering, non-Newtonian fluid dynamics, and/or applied mathematics for chemical engineers. My teaching extends beyond the classroom into the laboratory, such that students and postdoctoral researchers in my research group will engage in an interdisciplinary approach toward polymer science and engineering by drawing from expertise in chemical engineering, materials science, biology, chemistry, and physics.

Teaching Experience:

- Structure of Soft Matter (MIT ChemE), Guest Lecturer (2016)
- Applied Mathematics in Chemical Engineering (Illinois ChBE), Teaching Assistant and Guest Lecturer (2012, 2015)
- Polymer Science and Engineering (Illinois ChBE), Guest Lecturer (2013)
- Thermodynamics (Illinois ChBE), Teaching Assistant (2013)
- Chemical Engineering Thermodynamics (Michigan ChE), Undergraduate Instructor (2010)

Service:

I am passionate about recruiting the world's top talent to pursue careers in engineering and science in order to solve society's most pressing needs. To this end, I am committed to broadening access to mentorship for talented individuals from all walks of life. At the University of Illinois, I co-founded a campus-wide program to provide support, resources, and advocacy for prospective STEM-field faculty candidates. This program cultivated a community of scholars through a series of faculty-led seminars, panels, and facilitated group discussions. At least seven alumnae have earned tenure-track positions since the program began in 2014, and the program continues to operate under new student leadership. I have also shared my experiences with the next generation of engineers by participating in panels for the AIChE Women's Initiative's Committee (2015 Annual Meeting in Salt Lake City) and the Multi-Cultural Engineering Recruitment for Graduate Education (MERGE) Program at Illinois. At MIT, I supervised under-represented students at the high school, community college, and undergraduate levels, with aims to provide mentorship and to support the students' continued engagement with the scientific method in their careers. Finally, in my pursuit to recognize the contributions of diverse researchers in polymer physics, I accepted a role as co-chair of the 2018 Polymer Physics Gordon Research Seminar. Overall, I deeply value the richness of scientific communities, and I hope to contribute to their sustained growth and diversity.

541856 Designing Polymers As Molecular Recognition Agents for Diagnostic Biosensing and Imaging

Abstract:

Research Interests:

In most diagnostic tests, whether biosensor or imaging based, biomolecules are used for molecular recognition of biomarkers. While antibodies and enzymes exhibit unparalleled specificity, they are expensive and thermally unstable, requiring cold-chain methods for transportation and storage. As the healthcare field move towards point-of-care testing and personalized diagnostics, there is a need to make diagnostic tests more affordable and environmentally robust. To this end, I am interested in engineering polymers capable of molecular recognition. During my PhD, I designed and tested different strategies for synthesizing hydrogel particles capable of protein biomarker detection ("*Investigation of Micro- and Nanoscale Hydrogels as Protein Receptors for Use in Diagnostic Biosensors*", advisor Nicholas Peppas). I then chose to pursue a postdoctoral position under the advisement of Christopher Bowman to learn how to synthesize tandem repeat "click" nucleic acids for use in disease diagnosis and medical imaging.

As a faculty candidate, I aim to apply my knowledge and previous research experience in polymer chemistry, biophysical methods, and nanotechnology to continue advancing the field of polymer-based diagnostics. The multidisciplinary nature of this type of research aligns with my interests across the fields of chemistry, materials science, nanotechnology, and statistics. Future directions that excite me include, but are not limited to, the following:

- **Differential sensing using cross-reactive polymers**. I began investigating this relatively unexplored field during my PhD. By applying multivariate statistical methods, signal patterns generated from the cross-reactive polymers could be linked to disease-related changes in biomarker levels in human tears. By further tuning polymer chemistry and architecture, I expect that a wide-range of cross-reactive receptors for different biomarker panels could be designed.
- **Dynamic covalent chemistry for template-directed polymer synthesis**. Dynamic covalent chemistry is emerging as a powerful tool for manipulating material properties. In parallel to the first direction (i.e., differential sensing), I am excited to explore dynamic covalent chemistry approaches to improve selectivity of artificial molecular recognition agents.
- **Applications of click nucleic acids in diagnostic imaging**. The click nucleic acids (CNAs) being developed in the Bowman Laboratory are an emerging class of synthetic nucleic acids. Unlike DNA and peptide nucleic acids, tandem repeat sequences of CNAs can be efficiently and affordably synthesized. As such, CNAs will have great utility in imaging and detection of disease-associated, polymorphic tandem repeat sequences.

Teaching Interests:

With a B.S. in Materials Science and Engineering and a Ph.D. in Biomedical Engineering, I am well-suited to teach courses covering a range of chemical and biological engineering topics. I am particularly interested in teaching courses covering topics such as biophysical methods, biomaterials and biotechnology, polymer science, nanotechnology, and statistical analysis for engineers, as these are the topics I am most intimately familiar with given my research experiences. In addition to leading lectures, I would be interested in leading laboratory courses, including designing new experiments to help students better grasp complex topics.

Regardless of the topic, my approach to teaching will be centered around designing problem sets that, while challenging, will ultimately lead to a better understanding of the topics covered in class. While at the time I may have said otherwise, working tirelessly through problem sets with my peers were some of my favorite memories from my undergraduate education. I learned so much through these problem sets, not only about the concepts covered in class, but also about the importance of teamwork and learning new strategies for problem solving.

Research Experience:

Based on my undergraduate research experiences, I entered the Biomedical Engineering graduate program at University of Texas at Austin with a strong interest in polymer chemistry and nanomaterials. The objective of my dissertation research in the Peppas Laboratory was to develop a low-cost, non-invasive screening test for Sjögren's syndrome (SS), an autoimmune disease that primarily affects the exocrine system. My hypothesis was that, by using polymers as molecular recognition agents and nanomaterials for signal transduction, it would be possible to develop an inexpensive biosensor for detecting concentration changes of SS protein biomarkers. During the first three years of my PhD, the synthetic receptors I focused on were molecularly imprinted polymers (MIPs). MIPs are made by incubating a template molecule, such as a protein biomarker, with monomers that form favorable non-covalent interactions with the template. I worked on developing MIPs for lysozyme, a biomarker of SS, with an emphasis on characterizing the selectivity of the MIPs for lysozyme. I found that MIPs had higher adsorption capacity for proteins than non-imprinted polymers (NIPs), but the imprinting process did not improve selectivity for lysozyme. Additionally, the data showed that the MIPs and NIPs were cross-reactive (i.e., semi-selective) for proteins that had similar charges as lysozyme.

Given the added time and cost associated with synthesizing MIPs without improvement of selectivity, I transitioned my research to embrace the cross-reactivity of charge-containing hydrogels. Specifically, I shifted from the idea of developing a biosensor using highly selective molecular recognition agents to the idea of using differential sensor arrays. In this new approach, addition of an analyte solution or mixture to a set of cross-reactive receptors produces a signal pattern that corresponds to the binding affinity of different receptors for different analytes. The cross-reactive receptors I developed were poly(N-isopropylacrylamide) (poly(NIPAM)) nanogels with different ionic functional groups. These studies demonstrated that by using turbidity changes as input for multivariate analysis, it was possible to differentiate eleven proteins and detect clinically relevant changes in lysozyme concentration with as few as two nanogels. The last major thrust of my dissertation research was to synthesize these poly(NIPAM) nanogels on the surface of gold nanomaterials for use in a localized surface plasmon resonance (LSPR)-based biosensor. Varying the concentration of two SS biomarkers, lysozyme and lactoferrin, in human tears led to concentration-dependent shifts in the LSPR wavelength of poly(NIPAM-co-MAA) coated AuNSs.

After completing my PhD, I still had a strong interest in biosensing using synthetic polymers for molecular recognition. To this end, I joined the laboratory of Dr. Christopher Bowman at University of Colorado Boulder, where I am working on characterizing and applying "click" nucleic acids (CNAs) for biosensing applications. In addition to characterizing CNA-DNA binding, I am working towards applying the CNAs as diagnostic imaging probes.

Publications:

1. H.R. Culver, I. Sharma, M.E. Wechsler, E.V. Anslyn, N.A. Peppas, "Charged poly(N-isopropylacrylamide) nanogels for use as differential protein receptors in a turbidimetric sensor array," *Analyst* **2017**, *142*, 3183-3193.
2. H.R Culver, N.A. Peppas, "Protein-imprinted polymers: The shape of things to come?" *Chem. Mater.* **2017**, *29*, 5753-5761.

3. H.R. Culver, J.R. Clegg, N.A. Peppas, "Analyte-responsive hydrogels: Intelligent materials for biosensing and drug delivery," *Acc. Chem. Res.* **2017**, *50 (2)*, 170-178.
4. H.R. Culver, S.D. Steichen, N.A. Peppas, "A closer look at the impact of molecular imprinting on adsorption capacity and selectivity for protein templates," *Biomacromolecules* **2016**, *17 (12)*, 4045-4053.
5. H.R. Culver, S.D. Steichen, M. Herrera-Alonso, N.A. Peppas, "A versatile route to colloidal stability and surface functionalization of hydrophobic nanomaterials," *Langmuir* **2016**, *32*, 5629-5636.
6. J.L. Santos, Y. Li, H.R. Culver, M.S. Yu, M. Herrera-Alonso, "Targeting denatured collagen through collagen mimetic peptide decorated conducting polymer nanoparticles," *Chem. Commun.* **2014**, *50,* 15045-15048.
7. H.R. Culver, A.M. Daily, A. Khademhosseini, N.A. Peppas, "Intelligent recognitive systems in nanomedicine," *Curr. Opin. Chem. Eng.* **2014**, *4*, 105–113.

Proposal writing experience:

Fellowships:

- NSF Graduate Research Fellowship (**funded**)
- NSF Graduate Opportunities Worldwide Fellowship (**funded**)
- USAID Development Research and Innovation Fellowship (**funded**)
- F32 NS106791 (not discussed)
- F32 GM130082 (scored, not funded – currently in revision)

Grants (Assisted PIs):

- NIH R01 EB019472 (scored, not funded)
- NIH R01 EB022025 (original scored, not funded; **resubmission funded**)
- NIH R21 NS107984 (scored, not funded)
- DARPA proposal in response to Broad Agency Announcement #HR001118S0034 (under review)

Awards:

- Donald D. Harrington Dissertation Fellowship
- Philanthropic Educational Organization Scholar Award
- Thrust 2000 – O.L. Chenoweth Endowed Graduate in Engineering Fellowship
- BMES Innovation and Career Development Award
- The Materials Science and Engineering Achievement Award, Johns Hopkins University

542192 Process Design and Optimization Leveraging Multiscale Modeling and Machine Learning

Abstract:

Research Interests:

Computation and modeling have played an important role in chemical engineering on multiple scales. For example, microkinetic modeling provides ways of exploring complex reaction systems for which experimental measurements are challenging; Optimization algorithms can be used to guide efficient design of experiments and improvement of processes. More recently, the advance of data science and machine learning has also demonstrated their strength in elucidating complicated patterns from molecular structures and reaction systems. While they have been powerful individually, the synergies between these multi-scale modeling techniques would provide much more benefits. I am excited about combining these modeling techniques, both theoretical and data-driven, to study chemical reaction and process systems, in order to better elucidate the underlying implications of experimental results and thus enabling more effective experimental efforts and speed up process development and optimization.

Research Experiences:

My PhD dissertation at Northwestern University focuses on simulation and optimization of radical polymerization processes, combining mechanistic kinetic modeling and optimization algorithms. I developed schemes to accelerate kinetic Monte Carlo (KMC) method, which is used to simulate the explicit sequence of polymer chains during polymerization process. These methods significantly reduced the computational cost of KMC simulations so that they were successfully incorporated into derivative-free optimization algorithms to design and optimize polymer sequence distributions.

During my post-doc at MIT, I worked on a large collaborative project on synthesis route design and evaluation for organic molecules. Specifically, I developed a hierarchical neural network model trained on large reaction databases to recommend conditions (catalyst, reagent and solvent) for given organic molecules with a high accuracy. Embeddings extracted from the model were used to quantify the similarity of solvents/reagents/catalysts and compatibility of reactions, which aids pathway-level evaluation of synthetic routes. I also quantified model uncertainty to indicate whether or not a reaction is out of scope, which can inform reaction evaluation and retrosynthesis.

Teaching Interests:

As a chemical engineer by training, I am capable of and passionate about teaching core chemical engineering courses in general. I am especially enthusiastic about and good at developing interactive class sessions that provokes creative and critical thinking. During the 4th year of my PhD, I participated in the Teaching Apprenticeship Program at Northwestern University and taught half of the graduate class, *Chemical Kinetics and Reactor Design*, with Prof. Justin Notestein. Some lectures on stochastic kinetic modeling were very popular and I was invited to give the lectures again the next year. I have also been the teaching assistant for multiple classes including *Phase Equilibrium and Staged Separations, Analysis of Chemical Process Systems,* and *Chemical Engineering Design Projects*.

Publications:

- **Gao, H.**, Struble, T. J., Coley, C. W., Green, W. H., Jensen, K. F. (2018). Using Machine Learning to predict suitable conditions for organic reactions. Submitted for publication in *ACS Central Science*.
- **Gao, H.**, Konstantinov, I. A., Arturo, S. G., Broadbelt, L. J.(2018). A Simulation-based Derivative-free Optimization Framework Using the Kinetic Monte Carlo Method for Controlling Polymer Molecular Weight and Sequence Distribution Synthesized via Free Radical Polymerization. *Computers & Chemical Engineering,* 108, 268-275.
- **Gao, H.**, Konstantinov, I. A., Arturo, S. G., Broadbelt, L. J. (2017). On the Modeling of Number and Weight Average Molecular Weight of Polymers. *Chemical Engineering Journal,* 327, 906-913
- **Gao, H.**, Konstantinov, I. A., Arturo, S. G., Broadbelt, L. J. (2016). Acceleration of Kinetic Monte Carlo Simulations for Free Radical Copolymerization: a Hybrid Approach with Scaling of Kinetic Parameters. *AIChE Journal,* 63(9), 4013-4021.
- **Gao, H.**, Oakley, L. H., Konstantinov, I. A., Arturo, S. G., Broadbelt, L. J. (2015). Acceleration of Kinetic Monte Carlo Method for the Simulations of Free Radical Copolymerization through Scaling. *Industrial & Engineering Chemistry Research*, vol. 54(48), 11975-11985. Selected as the ACS Editorâ€™s Choice for October 26th, 2015
- Regatte, V. R., **Gao, H.**, Konstantinov, I. A., Arturo, S. G. and Broadbelt, L. J. (2014). Design of Copolymers Based on Sequence Distribution for a Targeted Molecular Weight and Conversion. *Theory Simul.*, vol. 23 (9), 564-574
- Leperi, K., **Gao, H.**, Snurr, R.Q., & You, F. (2014). Modeling and Optimization of a Two-stage Mof-based Pressure/vacuum Swing Adsorption Process Coupled with Material Selection. Proceedings of the 17th Conference on Process Integration, Modelling and Optimization for Energy Saving and Pollution Reduction (PRES). *Chemical Engineering Transactions*, 39, 277-282
- Zhang, G.; Zhang, L.; **Gao, H.**; Konstantinov, I. A.; Arturo, S. G.; Yu, D.; Torkelson, J. M.; Broadbelt, L. J. (2016). A Combined Computational and Experimental Study of Copolymerization Propagation Kinetics for 1-Ethylcyclopentyl methacrylate and Methyl methacrylate. *Macromol. Theory Simul.*, vol. 25(3), 263-273

542457 Building a New Computational Toolbox for Bioengineering and Advanced Manufacturing

Abstract:

Research Interests:

Experience

2017-Pres., Johns Hopkins University and Princeton University (Postdoctoral Fellow/Visiting Researcher)

2016-2017, Delaware Biotechnology Institute, Postdoctoral Researcher

2012-2016, University of Delaware, Graduate Research Assistant

Keywords: complex and nonlinear systems, process control, data mining and machine learning, computational systems biology

Application areas: bioengineering and biomanufacturing, computer-interfaced biosystems, photovoltaics and semiconductor manufacturing

Process systems engineering has enabled routine high performance and high reliability in modern chemical manufacturing. Partly because of this success, the interests of chemical engineers have expanded to more fields than before, including biological systems, advanced materials, and renewable energy production and storage. Powerful new experimental tools have been developed for the design and analysis of such systems, including synchrotron methods, scattering and diffraction techniques, metabolic flux analysis, and a large suite of "-omics" techniques. Terabytes of data are produced, and analysis is not a human-scale problem any longer.

These experimental advances, coupled with new computational tools enabled by faster computer hardware and improved algorithms, motivate investigating new biological systems and advanced materials. Therefore, I will pursue two overall research thrusts:

- Computer-interfaced Biosystems
- Machine Learning Driven Material Design

For the first thrust, I plan to collaborate with expert synthetic biologists and metabolic engineers to develop new ways for integrating biological control systems with a computer in the loop. I will use the intrinsic ability of organisms and their native or engineered signaling pathways to sense and respond to stimuli in tandem with a human-designed hardware-driven control system. I expect that this approach will be useful for metabolic engineers seeking improved product yield, and for medical scientists seeking to understand and repair the control systems built-in to humans. For the second thrust, I plan to build models of important physical phenomena—such as thin film deposition kinetics or electrical performance of semiconductor devices—directly from data. These models will be valuable for understanding and designing new materials for many applications including energy storage and production. Here, I will review my current postdoctoral project, where I am developing tools for biomanufacturing using optogenetics, and my PhD project, where I designed a process for producing thin film solar cells.

1: Postdoctoral Project: Modeling and control of bioprocesses using optogenetics

Advised by Yannis G. Kevrekidis at Johns Hopkins University and José Avalos at Princeton University

Recent advances in synthetic biology enable a new biomanufacturing platform where high-resolution process inputs and outputs can be used for online optimization and control. In particular, biosensors and optogenetics allow for measuring and manipulating the state of a bioreactor with more precision than ever before. We are developing a "photobioreaction" process, where the reactor state is observed using biosensors and actuated with LEDs to manipulate optogenetic circuits. Abundant state information from biosensors can be used for data-assisted modeling and data-driven control system design. We are combining mechanistic models and new methods for dynamic system identification with tools from control systems theory to optimize photobioreactor performance. The effectiveness of these methods will be demonstrated through production of the advanced biofuel isobutanol. We are developing tools can be used for highly complex, multiple input, multiple output biological circuits that will be used for producing high-value products, including fuels, chemicals, and pharmaceuticals.

2. PhD Dissertation: Rapid thermal processing for production of chalcopyrite thin films for solar cells: Design, analysis, and experimental implementation

Advised by Babatunde A. Ogunnaike and Robert W. Birkmire at the University of Delaware

Due to their potential for low cost and high efficiency, thin film solar cells are a strong candidate for utility-scale electricity production. One attractive manufacturing route for thin film solar cells is rapid thermal processing (RTP). For my PhD research, I designed an RTP reactor for producing thin film solar cells and deployed a novel control system using an advanced model-based controller and a nonlinear observer. With the reactor in place, I developed a novel stochastic model to describe thin film deposition in the reacting system. I derived expressions to relate simulation parameters to physical properties, showed that the model could predict film composition—including composition gradients that affect solar cell performance—as observed in the lab, and used the stochastic nature of the model to characterize lateral heterogeneity in the film. I also showed that the model predictions correspond with experimental data. Finally, I applied statistical design of experiments to probe the effects of process variables on the material properties of the films and optimize the efficiency of solar cells.

Teaching Interests:

Experience

2016, University of Delaware, Assistant Instructor
Courses: Engineering Statistics (graduate), Chemical Engineering Laboratory (undergraduate)

2012, University of Delaware, Graduate Teaching Assistant
Course: Process Dynamics and Control (undergraduate)

I am dedicated to training future engineers and take seriously the responsibility of effective teaching. To ensure that I would become an effective educator, I sought out opportunities to build teaching experience throughout my academic career. I have experience as the instructor of computer labs (as a TA) and experimental labs (as an instructor), as well as lecturing experience at the graduate level. Beyond classroom experience, I helped develop an undergraduate laboratory project where students produced biodiesel from soybean oil using bench-scale and pilot-scale reactors, and presented this work at AIChE

Engineering lies in a unique position among most university departments. It demands technical proficiency in science and mathematics, yet also requires innovative problem-solving where core concepts are applied to new situations. I therefore plan to structure all my courses to include experiential learning, which is effective at helping students understand how fundamental theory can be applied. In experiential education, students first complete a realistic task (such as a laboratory experiment or a computer simulation) before being taught the underlying theory. Next, they are asked to reflect on the results, and, with guidance from the instructor, they can abstract and generalize the concepts. Finally, students are asked to apply the new concepts to new situations. I plan to use my experience in computer programming to build realistic computer simulations so that I can apply experiential education even in courses without a laboratory component.

Every class is composed of students with diverse experiences, learning styles, and academic preparation. To therefore ensure effective learning, I will develop and apply adaptive teaching methods, such that each class is structured flexibly and can accommodate all students. I will come to class prepared with different tools, including example problems, analogies, and visual demonstrations. In the classroom, instead of strictly adhering to a fixed syllabus, I will ensure that every learner in class has grasped the concepts before moving ahead. To achieve this, I will use example problems as a tool for class discussions, and make sure that everyone is participating.

I can confidently teach the entire chemical engineering undergraduate and graduate core curriculum: from introductory engineering through senior design and graduate-level kinetics, thermodynamics, and transport phenomena. Given my research background and teaching experience, I would choose to teach courses in process dynamics and control, engineering statistics, or other courses with a significant mathematical modeling component. For these courses, my research experience will allow me to draw examples from contemporary problems in semiconductor manufacturing, bioreactor control, and more, which will make the class more relevant and more interesting to students. Even for courses outside the domain of my research, however, my academic preparation has given me an appreciation for the importance of each course in the curriculum and I will include contemporary topics and current applications in the classroom.

Publications

2018, (*under review*), "Overcoming carrier concentration limits in polycrystalline CdTe thin films with *in situ* doping." B. E. McCandless, W. A. Buchanan, C. P. Thompson, G. Sriramagiri, **R. J. Lovelett**, J. Duenow, D. Albin, S. Jensen, E. Colgrove, J. Moseley, H. Moutinho, S. Harvey, M. Al-Jassim, W. K. Metzger.

2017, AIChE Journal, "Hierarchical monitoring of industrial processes for fault detection, fault grade evaluation and fault diagnosis," L. Luo, **R. J. Lovelett**, B. A. Ogunnaike. www.dx.doi.org/10.1002/aic.15662

2016, Journal of Process Control, "Design and experimental implementation of an effective temperature control system for thin film $Cu(InGa)Se_2$ production via rapid thermal processing," **R. J. Lovelett**, G. M. Hanket, W. N. Shafarman, R. W. Birkmire, B. A. Ogunnaike. www.dx.doi.org/10.1016/j.jprocont.2016.07.005

2016, AIP Advances, "A stochastic model of solid state thin film deposition: Application to chalcopyrite growth," **R. J. Lovelett**, X. Pang, T. M. Roberts, W. N. Shafarman, R. W. Birkmire, B. A. Ogunnaike. www.dx.doi.org/10.1063/1.4948404

2016, University of Delaware Ph.D. Dissertation "Rapid thermal processing for production of chalcopyrite thin films for solar cells: Design, analysis, and experimental implementation," **R. J. Lovelett**, http://udspace.udel.edu/handle/19716/21450

2016, IEEE Photovoltaic Specialists Conference, "Growth of $Cu(In,Ga)(S,Se)_2$ films: Unravelling the mysteries by in-situ x-ray imaging," B. West, M. Stuckelberger, L. Chen, **R. J. Lovelett**, B. Lai, J. Maser, W. Shafarman, M. Bertoni. www.dx.doi.org/10.1109/PVSC.2016.7749650

2015, IEEE Photovoltaic Specialists Conference, "A stochastic model for $Cu(InGa)(SeS)_2$ absorber growth during selenization/sulfization," **R. J. Lovelett**, W. N. Shafarman, R. W. Birkmire, B. A. Ogunnaike. www.dx.doi.org/10.1109/PVSC.2015.7356226

542490 Process System Engineering (PSE): *Continuous Pharmaceutical and Bio-Pharmaceutical Manufacturing*

Abstract:

My main research focus is process system engineering (PSE) and current application domain is continuous pharmaceutical and bio-pharmaceutical manufacturing. However, the PSE methods and tools that I developed are generic and can be applied to any manufacturing industries. Currently, pharmaceutical and bio-pharmaceutical industries are going through paradigm shift from conventional batch to continuous manufacturing process in which PSE can play a very significant role. Furthermore, there is a need of advanced modular continuous manufacturing platform technology for fast and robust development of pharmaceutical and bio-pharmaceutical products. Master formulations, process models, standardized sensing and control architecture, superior automation and novel methods and tools need to be developed to enable platform technology. This area of research and education are highly encouraged and supported by pharmaceutical and biopharmaceutical companies, academic institutions, regulatory authority (FDA) and national science foundation (NSF).

In this work, a scientific foundation for the optimal design, optimization and control of the modular continuous process involving structured organic particulate systems has been developed. The mathematical models for continuous manufacturing of active pharmaceutical ingredients and final product tablet, has been developed for virtual experimentation and applied for plant design, optimization and control. The developed continuous tablet manufacturing plant and process model has been adapted by several pharmaceutical companies. A flexible monitoring and control system incorporating different PAT techniques and tools and control strategies for continuous pharmaceutical manufacturing process has been designed and implemented to Rutgers pilot-plant. Different control strategies such as PID, PID coupled with dead time compensator, and MPC has been compared and evaluated practically for pharmaceutical process control. A new hybrid MPC-PID control strategy has been developed and implemented into the pharmaceutical tablet manufacturing plant and its performance has been found better than the other control strategies. A novel RTD based control system has been developed and implemented into the pilot-plant. Novel sensing methods have been developed for real time measurements of powder density and powder level. Feedforward control system has been integrated with feedback control system to take proactive mitigation actions on raw materials and process variability. A moving horizon based Dynamic Real Time Optimization (MH-DRTO) method has been also developed to provides the optimal operational set points for the control system. The developed control architecture has been implemented into pilot-plant.

Biosketch: I am working as an Assistant Research Professor at C-SOPS, Department of Chemical and Biochemical Engineering, Rutgers University, NJ, USA. I am the recipient of prestigious EFCE Excellence Award given in Recognition of an Outstanding PhD Thesis, from European Federation of Chemical Engineering. I received several grants as a PI/Co-PI and have well stablished research group. I have published more than 60 scientific research papers, written 12 book chapters, presented at over 95 international conferences and edited one pharmaceutical book published by Elsevier. I am actively serving as a conference session chair, Journal reviewer and editor.

Research Interests: My main research area is process system engineering with focus on continuous pharmaceutical and biopharmaceutical manufacturing. I have contributed significantly in the field of process monitoring and control, process modelling and simulation, optimization, PAT, QbD, new methodology and software tools development.

Teaching Interests: My teaching experience that I acquired serving as a teaching assistant, lecturer, and research assistant professor makes me confidant to develop and teach any courses of chemical and biochemical engineering. The industrial invited crash courses that I have given provide me opportunity to extend my teaching skills to cover more applied courses and to interact with highly experienced scientist and operators from manufacturing companies. I demonstrated my skills to instruct hands-on exercise through several workshops that I conducted. Several training courses that I conducted for industrial and academic participants make me instrumented with additional capability of building highly skilled workforce. I developed two online course modules for FDA where I learn how to develop and give online courses. I have dual ability to teach theoretical as well as practical courses. I am open to teach any courses that your university will assign me. If needed then I can develop new courses based on my research outcomes.

Grants: Extensive experience in funding proposals writing.

(1). 2018 Glaxo Smith Kline (GSK) (granted)

Development of Computational Models for Continuous Tablet Manufacturing

Role: PI. Source: GSK Company. Amount: **$539,991 (1st year);** Total: **$2.5 Million**.

(2). 2018 Siemens (granted)

Cyber-physical attack in the Pharmaceutical Manufacturing Process

Role: PI. Source: Siemens Company. Amount: $78,750.

(3). 2018 Diversitech Corporation (granted)

Fluid bed drying trials

Role: PI. Source: Diversitech Company. Amount: **$1600**.

(4). 2018 Rutgers Global International Travel Grant (granted)

Role: PI. Amount: $750.

(5). 2018 CNH (granted)

Wear prediction and validation with fine particle interaction

Role: Co-PI. Source: CNH Company. Amount: **$118608.**

(6). 2017 Glaxo Smith Kline (GSK) (granted)

Development of tablet press feed frame RTD model and BU to CU relation

Role: Co-PI. Source: GSK Company. Amount: **$150000**.

(7). 2017 Grant to stablish international collaboration (granted)

Grants to Support the Initiation of International Collaboration.

Role: Co-PI. Source: Hamburg University of Technology. Amount: **â‚¬1610.**

(8). 2016 Integra Continuous Manufacturing Systems (granted)

Development of sensing and control strategy for direct compaction continuous pharmaceutical tablet manufacturing process. **Role: PI**. Source: Acumen Company. Amount: **$5114**.

(9). 2016 Research Council Grant Award (granted)

Development and validation of advanced model predictive feedforward/feedback control strategy for pharmaceutical process. **Role: PI**. Source: Rutgers. Amount: **$3800**.

(10). 2018 Food & Drug Administration (FDA) (submitted)

Advanced Bio-manufacturing

Role: Co-PI. Source: FDA. Amount: **$1.8 Million**

(11). 2018 Food & Drug Administration (FDA) (submitted)

Industry 4.0 Implementation in Continuous Pharmaceutical Manufacturing

Role: Co-PI. Source: FDA. Amount: **$3.96** Million

Publications: Scientific articles (59); Book (1); Book chapters (12); International conference presentations (95); Citations>984, h-index>17, i10-index >26.

Activities in AIChE 2018 conference

Session chair/co-chair

1. **Chair**: Predictive Control and Optimization II
2. **Chair**: Big Data in Chemical and Pharmaceutical Processes
3. **Co-chair:** Integrated Product and Process Design.

Oral presentations

1. **Singh, R.,** Muzzio, F. J. **(2018)**. RTD Based Control System for Continuous Pharmaceutical Manufacturing Process. Oral presentation at AIChE, Pittsburgh, USA, 28 October - 2 November, 2018.
2. **Singh, R. (2018)**. Implementation of Advanced Process Control System into Continuous Pharmaceutical Manufacturing Pilot-Plant. Oral presentation at AIChE, Pittsburgh, USA, 28 October - 2 November, 2018.

Contact:

Email: *ravendra.singh@rutgers.edu*, ***Phone:*** *(001)848-445-4944;* ***Fax***: *(001)732-445-2581*

Websites: *http://cbe.rutgers.edu/faculty; http://ravendrasingh.wixsite.com/ppse*

542492 Novel Strategies for Real-Time Stochastic Optimization, Quantification of Model Uncertainty and Estimation of the Physical Properties of Biologics

Abstract:

Research Interests:

In the last decade, there has been a rapid growth in the application of optimization techniques to the solution of problems of industrial importance. In particular, the process and related industries have recently expanded the systematic use of optimization in supply-chain management, operational planning and scheduling, and optimal process control. Even pharmaceutical companies, which have been conservative in model-based decision support innovations, have turned to optimization for product design and drug delivery purposes. This global trend, primarily motivated by increasing competition, calls for further developments in optimization and alike research fields, e.g. dynamic and mixed-integer optimization, model uncertainty quantification, optimization under uncertainty and risk assessment.

In view of these considerations, my passion for applied mathematics and statistics, and my rigorous chemical engineering education, my research expertise and current research interests span the following topics:

- Supply-chain and enterprise-wide optimization applied to continuous processes;
- Operational planning and scheduling of batch processes;
- Deterministic and stochastic online optimization of batch processes;
- Quantification of parametric uncertainty in nonlinear models;
- Identification of optimal uncertainty sets in stochastic dynamic optimization problems;
- Prediction of the physical properties of formulations of biologics;

Specific examples of my research achievements include a framework for simultaneous optimization of the supply-chain and production systems of industrial gas producers [1], two new strategies for deterministic and stochastic dynamic optimization of single batch units ([2,3] and [4,5]), and a novel approach to the integration of scheduling, dynamic optimization and control in multi-unit batch processes [6]. The algorithms, developed in [2,3] and [6], assume that the process model is accurate while the approach, proposed in [4,5], explicitly accounts for model uncertainty, through a novel dynamic scenario selection procedure. These three strategies are very flexible, and will accommodate virtually any batch processing system (reactors and bio-reactors, crystallizers, freeze-drying units, home appliances, ...) and performance metric (economic functions, indicators of process environmental impact, ...). Thus, they constitute a general, robust optimization platform for batch processes.

Additional research work, completed in the last year, comprises an innovative method for rapid estimation of the probability distribution function (PDF) of the uncertain parameters of ODE/DAE models [7,8] and a strategy for dynamic selection of the best uncertainty set in stochastic dynamic optimization problems [9]. The first method relies on uncertainty back-propagation and projection techniques, and offers a very good trade-off between accuracy and computational cost. The second is based on a combination of fast PDF estimation, multi-point sensitivity analysis and ranking strategies, and allows application of stochastic dynamic optimization to systems with stochastic and time-varying features (reactors/crystallizers subject to fouling, food processing units, ...). These two approaches seek to exploit the rapid growth in the availability of process data, which makes it possible to model the underlying uncertainties that arise in real-life process applications. Therefore, they constitute the backbone of a structured approach to the analysis and utilization of big data sets.

My current research efforts are devoted to developing a general strategy for prediction of the thermodynamic and transport properties of mixtures of monoclonal antibodies (mABs), excipients and water. This approach, based on the SAFT equation of state, is an essential component of a more complex framework, aimed at identifying the most promising mAB formulations without expensive and time-consuming experimental analyses.

My research plan for the near future includes several activities, which can be summarized as follows:

- Application of stochastic optimization to drug delivery problems (administration of mABs, chelation therapy, pediatric drug administration, ...);
- Application of stochastic optimization to critical food processing operations (freeze-drying, addition of preservatives, pasteurization, ...);
- Application of stochastic optimization to the optimal design of mAB formulations;
- Development of stochastic dynamic optimization strategies able to account for both parametric and structural model uncertainty;
- Development of frameworks for multi-objective, robust, integrated scheduling, dynamic optimization and control of batch processes;
- Development of robust soft sensors for real-time risk assessment in pharmaceutical systems.

All of these activities will require use and extension of the frameworks, described previously, as well as development of completely new ones. This tentative research plan will of course be modified based on funding availability and potential collaborations with other faculty members.

Teaching Interests:

My research work and graduate studies have provided me with deep knowledge of control theory, numerical methods, dynamic and steady-state modelling, coding/parallel computing (especially in C++), and statistics. Therefore, I would be interested in teaching any course broadly related to these subjects, e.g. applied statistics, process control, numerical methods, dynamic and steady-state modelling, and process optimization. I would also like to introduce and teach two advanced courses, whose topics lie at the boundaries of Chemical Engineering, Statistics and Computer Science:

- Applied statistics and big data analytics for chemical engineers;
- Applications of parallel computing to chemical engineering problems.

Since optimization strategies have become an essential tool for the solution of many problems of industrial relevance in the last few years, I strongly believe these interdisciplinary courses would provide new chemical engineers with useful and valuable knowledge.

References:

1. Rossi F, Manenti F, Reklaitis G. A general modular framework for the integrated optimal management of an industrial gases supply-chain and its production systems. *Computers and Chemical Engineering*. 2015;82:84-104.
2. Rossi F, Manenti F, Buzzi-Ferraris G. A novel all-in-one real-time optimization and optimal control method for batch systems: Algorithm description, implementation issues, and comparison with the existing methodologies. *Industrial and Engineering Chemistry Research*. 2014;53:15639-15655.

3. Rossi F, Copelli S, Colombo A, Pirola C, Manenti F. Online model-based optimization and control for the combined optimal operation and runaway prediction and prevention in (fed-) batch systems. *Chemical Engineering Science*. 2015;138:760-771.
4. Rossi F, Reklaitis G, Manenti F, Buzzi-Ferraris G. Multi-scenario robust online optimization and control of fed-batch systems via dynamic model-based scenario selection. *AIChE Journal*. 2016;62:3264-3284.
5. Rossi F, Manenti F, Pirola C, Mujtaba I. A robust sustainable optimization & control strategy (RSOCS) for (fed-)batch processes towards the low-cost reduction of utilities consumption. *Journal of Cleaner Production*. 2016;111:181-192.
6. Rossi F, Casas-Orozco D, Reklaitis G, Manenti F, Buzzi-Ferraris G. A computational framework for integrating campaign scheduling, dynamic optimization and optimal control in multi-unit batch processes. *Computers and Chemical Engineering*. 2017;107:184-220.
7. Rossi F, Mockus L, Manenti F, Reklaitis G. Assessment of accuracy and computational efficiency of different strategies for estimation of probability distributions applied to ODE/DAE systems. *Computer Aided Chemical Engineering. Accepted.*
8. Rossi F, Manenti F, Buzzi-Ferraris G, Reklaitis G. A novel strategy for rapid estimation of the probability distribution of the uncertain parameters of ODE/DAE systems. *In preparation. Estimated submission at the end of 2018.*
9. Rossi F, Manenti F, Buzzi-Ferraris G, Reklaitis G. Stochastic NMPC/DRTO of batch operations: batch-to-batch dynamic identification of the optimal description of model uncertainty. *Computers and Chemical Engineering. Under review.*

542161 Optimization-Based Control of Complex Process Networks in Smart Manufacturing: The Appearance of Cyber-Physical Systems, Cloud Computing, and Big Data Analytics

Abstract:

Research Interests:

Large-scale process networks are becoming the rule rather than the exception in energy and chemicals production industries due to the increasing demand for integrated and intensified process operation, which provides sustainable productivity improvement benefits. Examples of such systems include chemical processes, power plants, electrical grid, water distribution networks and many more. Process integration and intensification, achieved through energy/material recycling and heat/mass transfer optimization, has been extensively targeted as a means to reduce raw material consumption and lower utility requirements, and thus promote sustainable designs. However, these benefits are accompanied by major challenges in automated operation and control, specifically in smart process manufacturing which requires high levels of control and adaptivity.

In the architecture of the process control systems, different components need to communicate with each other (e.g., the sensors transmit their measurements to the controllers, the controllers use this information to compute the control input, and the control input is then sent to the actuators so that it can be physically implemented). For this communication to take place, a shared communication network is usually deployed across the plant to be controlled. Although wired networks have been traditionally used for this purpose, an increasing number of industrial control systems now use wireless networks since they are easier to deploy and to maintain. In addition, these networks are sometimes connected to the corporate intranet, and in some cases even to the Internet, to utilize off-site and cloud computational resources for system identification (by analyzing a significant amount of process data) and computing control inputs (by solving real-time nonlinear constrained dynamic optimization problems). Consequently, the modern control systems are becoming more open to the cyber-world, and as such, are more vulnerable to attacks that can cause faults and failures. Real-world attacks on control systems have in fact occurred in the past decade and have in some cases caused significant damage to the targeted physical processes, indicate the apparent need for strategies and mechanisms to deal with new security challenges related to the size of the large-scale process networks.

Regarding my future research efforts, I intend to extend my computational work to modeling, identification, and optimization-based estimation and control of large-scale process networks in sustainable and smart manufacturing. Specifically, the intellectual objective of my future research is to develop enabling technologies designed to enhance control and real-time decision making for smart chemical and energy systems in the 4th generation industrial revolution (Industry 4.0). I aim to bring theoretical tools from cyber-physical systems, internet of things, and cloud computing to solve the communication and real-time computational challenges in networked control of large-scale chemical and energy systems. I plan to use the traditional tools in the big data analytics to address model uncertainty issues. The proposed research will introduce systematic computationally efficient process identification methods to estimate system parameters required by the model predictive control to compute the optimal control action subject to a continuous decrease in the identification error. Note that transferring such large amount of data over the shared communication network must be managed by smart communication protocols and algorithms. Also, my proposed research will cover theoretical and algorithmic aspects of estimation and control of the process networks subject to adversarial attacks and guarantee the maximum resilience of the system through secure distributed control.

Teaching Interests:

I view teaching as an integral part of my overall academic activity. Given my chemical engineering background and my experience as a lecturer, lab supervisor, and teaching assistant in several undergraduate courses, laboratories, and workshops, I feel confident that I can effectively teach every course in the undergraduate chemical engineering curriculum, and I will be happy to do so. I believe that problem-solving is the crucial skill that engineering students should acquire in any class. In my future teaching career, I plan to incorporate active problem-solving in the undergraduate classroom through group discussions and conversation which will engage more students and improve educational outcomes. Given my expertise in the areas of process dynamics, control, design and applied mathematics, my immediate teaching plans concern the improvement and development of computer-aided courses pertaining to process dynamics and control, process design, and analytical/numerical methods for solving chemical engineering problems.

I look forward to strengthening the mathematical and computational content of core chemical engineering courses such as transport phenomena, reaction engineering, and thermodynamics. This strategy will provide the students with required skills for precise modeling of the velocity/pressure fields, heat/mass transfer, equilibrium conditions, and reaction mechanisms of complex systems in real-world applications. It will also supply the necessary tools for the students to develop computationally efficient programs to solve the resulting nonlinear algebraic, ordinary, and partial differential equations. Besides the direct impacts on improving the problem-solving ability in the undergraduate core courses, such an approach can inspire students to employ the power of modern numerical mathematics to provide solutions to cutting-edge research problems across science and engineering. Also, I intend to develop my own graduate-level courses focusing on control and optimization of large-scale process networks, sustainable approach to process control, secure estimation and control of the processes in smart manufacturing, and advanced tools for process data analytics.

542344 Development and Assessment of New Sustainable Processes for the Production of Bio-Products

Abstract:

Research Interests:

Research Experience:

My academic research is a blend of fields of process engineering, economics, and sustainability. My formal training was in process simulation, techno-economic analysis, economic risk assessment, and life-cycle analysis. I have worked on quantifying financial risk of investments for the development of platform technologies to produce biobased products, introducing a new methodology for the economic and environmental assessment of early stage renewable chemical technologies, and determining generalities to evaluate and guide the development of biocatalytic technologies as well as chemical catalytic pathways for the transformation of platform chemicals. As a result of these projects, I have acquired the knowledge in the areas of chemical process development, cost and benefit analysis, biofuel and biorenewable chemical production, and sustainability assessment.

Postdoctoral Projects: Economic and modeling analysis of product-and energy driven biorefineries, Assessment of the viability of platform chemical concept, Effect of biomass densification technologies on the economics of biofuel production processes, and Finding new applications of bio-char

Under supervision of M. Clayton Wheeler, William DeSisto, and Hemant Pendse, Chemical and Biological Engineering, Forest Bioproducts Research Institute, University of Maine at Orono.

PhD Dissertation: "*Techniques for Evaluating and Guiding Development of Renewable Chemical Technologies for Sustainable Products*"

Under supervision of Robert Anex, Biological Systems Engineering, University of Wisconsin at Madison

Research interests and Future direction:

1. Technology assessment

As faculty, I will continue working on process modeling, economic analysis, and sustainability assessment of new renewable technologies. Additionally, I will focus on incorporating social costs, policy interventions, and policy uncertainty into the techno-economic modeling for the evaluation of renewable energy and chemical technologies. I will quantify social costs of making renewable gasoline and diesel via fast pyrolysis type technologies from a wide range of feedstocks. Such quantification can facilitate comparison of renewable fuel production against the fossil fuel production in terms of total costs that include both process and social costs. Determining the difference between the total costs of making renewable fuels and fossil fuels will help making future policies that encourage the production of renewable fuels by penalizing the production of fossil fuels, if the total costs of making fossil fuels are found to be higher than that of producing renewable fuels. However, penalizing fossil fuels might increase the food prices that can lead to food insecurity. Thus, impact on food prices need to be considered along with the social costs while making policies to penalize the production of fossil fuels. I plan to get external funding for this work from USEPA and NSF- Faculty career development agencies.

2. Development of separation and purification processes

The lack of processes to purify low-value added biorenewable products from clarified fermentation broths with a larger number of low concentrated impurities such as amino acids as well as the use of energy intensive distillation or solvent extraction techniques to extract low volatile biorenewable products from dilute solutions have been limiting the commercialization of new biorenewable technologies. There is a research need for developing less energy intensive separation processes to overcome such limitations. The new separation processes must achieve desired separation efficiency as well as a high throughput. As faculty I will work on developing new less energy intensive separation techniques by focusing on the molecular properties of interested product and impurities. In near term, my focus will be on modifying bio-char for use as an adsorbent for removing minor impurities such as amino acids from a clarified fermentation broth. Additionally, I will work on finding inducers to lower pH of the clarified fermentation broths for the easy separation of organic acids. After obtaining preliminary results from laboratory experiments for bio-char modification and altering pH with inducers, I plan to get external funding for the further technology development from NIST Advanced Manufacturing Technology Consortia (AMTech) program.

3. Process research and Valorizing waste streams

The use of sulfuric acid as an acid catalyst in the production of biorenewable products from renewable feedstock creates a problem of recovering it for reuse. Neutralizing sulfuric acid with calcium salts instead of recovering it may not be a viable option for the large-scale production of biorenewable products, as the storage, handling, and transportation of tones of sulfuric acid can be hazardous. Thus, future research is necessary to provide acid environment for the production biorenewable products without the addition of an acid catalyst.

The use of an inexpensive feedstock like food waste to make bioproducts can reduce the economic barriers to commercialize the new biofuel production technologies. In this study, the potential of converting food waste to bioproudcts via combined acid hydrolysis and dehydration will be investigated. The option of replacing sulfuric acid catalyst with easily recyclable organic as well as aromatic organic acids will be studied. The results of sulfuric acid replacement will be compared against the technical and economic performance of the base case involving sulfuric acid as a catalyst. Finally, the detailed techno-economic analysis (TEA) and life cycle assessment (LCA) will be performed to assess the economic performance and environmental viability of upgrading food fraction

4. Water Food Energy Nexus

The agriculture in Arizona has begun shifting towards the production of the water-intensive alfalfa crop. This shift to alfalfa production has been associated with the growth of local dairy. The water shortage in Arizona has raised concerns with the production of a highly water intensive alfalfa crop. Replacing alfalfa as an animal feed with the algae crop can alleviate stress on water resources of Arizona. However, the production of algae must be economically competitive and environmentally sustainable to use as an animal feed. Additionally, dairy owners must be made aware of the benefits of using algae as an animal feed over alfalfa. The life cycle analysis and techno-economic studies will be performed to assess the environmental and economic sustainability of synergistic integration of algae production with wastewater treatment plants in Arizona. I plan to write proposals for NSF and USDA water food energy nexus grant. Additionally, there are grant opportunities from international agencies such as UN for the food energy water nexus research.

Refereed Journal Publications

1. **S. Gunukula**, P. Keeling, R. Anex, Risk advantages of platform technologies for biorenewable chemical production, Chem. *Eng. Research Design*. 2016, 107, pp. 24-33.
2. **S. Gunukula**, R. Anex, Evaluating and guiding the development of sustainable biorenewable chemical technologies, *Biochem. Eng. J.* 2017, 119, pp.74-83
3. J.L. Carrasco, **S.Gunukula**, A.A. Boateng, C.A. Mullen, W.J. DeSisto, M.C. Wheeler, Pyrolysis of forest residues: An approach to techno-economics for bio-fuel production, *Fuel*. 2017, 193, pp.477-484.
4. **S. Gunukula**, R. Anex Techno-Economic Analysis of Multiple Bio-based Routes to Adipic Acid, *Biofuels Bioprod. Bioref*. 2017, 11 (5), pp. 897-907
5. **S. Gunukula**, T. Runge, R. Anex, Assessment of biocatalytic production parameters to determine economic and environmental viability, *ACS Sustainable Chem. Eng*. 2017, 5 (9), pp. 8119-8126
6. H. AlMohamadi, **S. Gunukula**, W.J. DeSisto, M.C. Wheeler, Formate assisted pyrolysis: An economic analysis, *Biofuels Bioprod. Bioref.* 2018, 12 (1), pp. 45-55.
7. **S. Gunukula**, W.J. DeSisto, H. Pendse, M.C. Wheeler, Techno-Economic Analysis of Thermal Deoxygenation Based Biorefineries for the Coproduction of Fuels and Chemicals, Applied Energy, 2018, 214, pp. 16-23.
8. **S. Gunukula**, W.J. DeSisto, H. Pendse, M.C. Wheeler, Heuristics to Guide the Development of Sustainable, Biomass-Derived Platform Chemical Derivatives, ACS Sustainable Chem. Eng. 2018, 6 (4), pp. 5533-5539.
9. Dalvand, J. Rubin, **S. Gunukula**, M.C. Wheeler, G. Hunt, Economics of Biofuels: Market Potential of Furfural and its Derivatives, Biomass and Bioenergy. 2018, 115, pp. 56-63.

Invited Talks

1. Gunukula, S., Tools to Evaluate and Guide the Sustainable Production of Renewable Fuels and Chemicals for RAS. US Army Research Lab, 2018 (January 5th)
2. Gunukula, S., Evaluating and Guiding the Development of Biorefinery Technologies, Distinguished Lecture Series, University of Maine, 2018 (February 16th)
3. Gunukula, S., Energy and Sustainability, Arizona State University, 2018 (April 5th)

Awards and Grants

1. Determining the different reaction engineering factors that affect the economic feasibility of the production of alpha olefins from carboxylic acids through techno-economic analysis. CBiRC Student-led Research Grant, Iowa State University, 2012.
2. US Army Research Lab Fellowship to work on the development of in-fuel generation systems, 2018
3. Valorizing Food Waste to Energy Dense Liquid Fuel and Bioproducts, Topic 4 of DE FOA 0001916, Department of Energy, 2018 (Pending)

Teaching Interests:

I worked as a teaching assistant during my PhD study. I have given a couple of lectures and helped in building the course material for Quantitative Techniques for Biological Systems and Bioprocessing Unit Operations classes. As faculty I would like to help students to develop their creative abilities, critical thinking and communication skills, and understanding ethics and rationality within sciences. I will try to propel students into new fields of learning and promotes interdisciplinary understanding. I mentored a couple of undergraduate students and currently co-supervising a couple of graduate students.

542087 Optimal Design of Petroleum Refinery Configuration Using a Model Based Mixed-Integer Programming Approach with Practical Approximation

Abstract:

Research Interests: Petroleum Refining

Teaching Interests: Petroleum Refining, Mass Transfer Unit Operations

We present a model-based optimization approach to determine the configuration of a petroleum refinery for grassroots (new) or existing site that considers a large number of commercial technologies particularly for heavy oil processing of crude oil residue from an atmospheric distillation unit. First, we develop a superstructure representation for the refinery configuration to encompass all possible topology alternatives comprising 96 technologies and their interconnectivities. The superstructure is postulated by decomposing it to incorporate representative heavy oil processing scheme alternatives that center on the technologies for atmospheric residual hydrodesulfurization (ARDS), vacuum residual hydrodesulfurization (VRDS), and residual fluid catalytic cracking (RFCC). We formulate a mixed-integer linear program (MILP) based on the superstructure by devising logic propositions on design and structural specifications that represent these processing options to aid convergence to an optimal refinery configuration. A numerical example is illustrated to implement the proposed technique in which an equivalent of more than two million refinery plot plans is evaluated. To assess the applicability and value of the approach, we validate the results against the literature as well as compare with existing real-world refinery configurations. A main contribution of this work is to demonstrate how a mixed-integer programming approach can be applied to a large-scale petroleum refinery design problem with suitable approximations informed by practical considerations to obtain results with reasonable computational load.

543264 Active Process Control in Pharmaceutical Continuous Manufacturing â€" the Quality By Control (QbC) Paradigm

Abstract:

Research Interests:

Pharmaceutical manufacturing is traditionally operated in batch mode by processing a certain amount of raw material, defined as a batch or lot, through various steps of unit operations, such as reaction, crystallization, filtration, drying, blending, tableting, etc., to obtain final drug products, for example., oral solid dosages. Quality attributes of in-process material or final product are tested batch wise at the end of each processing step, or known as the *Quality by Testing* (QbT) [1]. One example is in the blending step, where powders of active ingredients and excipients are mixed together in a rotating Y- or V-shaped vessel before the mixing uniformity end-point is detected at the batch end [2]. Although various process monitoring and control strategies are proposed for batch manufacturing in pharmaceutical or other manufacturing industries to achieve the batch-end product quality, for example, the multiway partial least squares for batch process monitoring [3, 4], a shrinking horizon nonlinear model predictive control (NMPC) for batch-end product quality control [5], etc., it should be noted that many of them are relied on the known priori optimal or nominal process operation trajectories (or recipes). It is the optimizing and tracking of these operational trajectories during the batch that lead to the batch-end product quality. Rarely, rework or recycle of the material at the batch end, particularly at the secondary manufacturing process, is allowed in the highly regulated pharmaceutical manufacturing industry. Hence, the loss of a whole batch is possible after the end-testing and remedial control actions can only start from the next batch. In other words, there is a large time delay of the entire batch time in controlling the batch-end product qualities by mere end-testing, which is usually known as the batch-to-batch control strategy [6]. Over the years, the US Food and Drug Administration (FDA) recognized that increased testing does not necessarily improve product quality and quality must be built into the product [7].

Quality by Design (QbD) is a concept first developed by the quality pioneer Dr. Joseph M. Juran, who suggests that quality should be designed into a product, and that most quality crises and problems relate to the way in which a product was designed in the first place [8]. Over the years, pharmaceutical QbD has evolved with the issuance of ICH Q8 (R2) (Pharmaceutical Development), ICH Q9 (Quality Risk Management), ICH Q10 (Pharmaceutical Quality System), and ICH Q11 (Development and Manufacture of Drug Substance). These documents provide high level directions with respect to the scope and definitions of QbD as it applies to the pharmaceutical industry [7]. QbD elements include the following: (1) a quality target product profile (QTPP) that identifies the critical quality attributes (CQAs) of the drug product; (2) product design and understanding including identification of critical material attributes (CMAs); (3) process design and understanding including identification of critical process parameters (CPPs), linking CMAs and CPPs to CQAs; (4) a control strategy that includes specifications for the drug substance(s), excipient(s), and drug products as well as controls for each step of the manufacturing process; and (5) process capability and continual improvement. QbD tools and studies include priori knowledge, risk assessment, mechanistic models, design of experiments (DoE) and data analysis, and process analytical technology (PAT).

Though the QbD principles are equally implemented in the conventional pharmaceutical batch manufacturing [9], at the meantime of the QbD issuance, there has been significant advancements in science and engineering to support the implementation of pharmaceutical continuous manufacturing over the past decade [10]. The conventional batch manufacturing faces many challenges in scaling-up, manufacturing cost, product quality variance, etc.; while the continuous manufacturing, defined as processing of raw materials without interruption and with continuity of production over a sustained period of time, offers many benefits in minimization of uncertainty in scaling-up, reduced space and capital requirement, and improved quality with consistent operations, etc. [11]. A great deal of potentials of improved agility, flexibility, and robustness in the continuous manufacture of pharmaceuticals have also been demonstrated. These investments along with the adoption of the QbD paradigm for pharmaceutical development and the advancement of PAT for designing, analyzing, and controlling manufacturing have progressed the scientific and regulatory readiness for continuous manufacturing [12]. For example, pharmaceutical continuous manufacturing has already been identified as an emerging technology by FDA and it is now staging from conceptual designs to pilot or production processes. Recently, a few continuous manufacturing facilities and drug have also been approved by FDA, e.g., the Orkambi (lumacaftor/ivacaftor) from Vertex in 2015 and the darunavir from Janssen in 2016 [13].

Beside the flexibility and cost-saving features with pharmaceutical continuous manufacturing, a unique advantage of continuous manufacturing compared to the batch manufacturing when both under the QbD guidance is that those identified CQAs and CPPs can be continuously monitored and controlled in real time, which therefore paves the way for a superior real-time product quality assurance, viz., the real-time release (RTR) strategy. For example, in batch blending, CQA variable of final mixing uniformity can only be available at the end of each batch blending step, even though this CQA variable is also continuously measured during the batch. Whereas this final CQA can be continuously measured at the outlet of a continuous blender and sent back to the control system as feedback signals to support a closed-loop control of this CQA variable by adjusting CPP variables, e.g., the blender rotation speed. In such, it is acknowledged that continuous manufacturing, integrated with online/inline-PAT tools and efficient control systems, can accelerate the comprehensive implementation of the QbD principles for the next generation of pharmaceutical products [14, 7]. It is also worth mentioning that active process control strategies are not new in other continuous manufacturing processes, e.g., in bulk chemicals and petrochemical industries. However, despite the feasibility of implementing active process control, which has been demonstrated for a number of pharmaceutical continuous manufacturing facilities, e.g., continuous crystallization, continuous direct compaction, etc., there remains hesitation in widespread adoption of active process control strategies due to existing investments in established and mature batch technologies [15]. Besides, most of the pharmaceutical continuous unit operations are still operated locally with very limited flexibility to implement the much more advanced active process control system.

Hence, owing to the substantial change in the control strategy for QbD implementation in pharmaceutical batch and continuous manufacturing on the one hand, and the rare application of active process control system in the pharmaceutical industry on the other hand, a *Quality by Control* (QbC) paradigm has since been evolving and nurturing the implementation of efficient advanced active process control of the CPPs and CQAs in continuous manufacturing. It has also been shown recently that the QbC is indispensable to the QbD implementation in pharmaceutical continuous manufacturing, and that it ensures more robustness and efficiency than the techniques used in conventional batch manufacturing, providing the key ingredient towards the comprehensive implementation of QbD [16].

To this end, my three tentative research plans in pharmaceutical continuous manufacturing and crystallization are listed as follows.

1. **On-site crystallization process monitoring, modeling, and control**

Crystallization is an important unit operation for separation and purification of pharmaceutical and fine chemicals. With the advent of in situ real-time process analytical technologies in the past decades, e.g., Raman spectroscopy, Focused Beam Reflectance Measurement (FBRM), or the Attenuated Total Reflection Fourier-transform infrared spectroscopy (FTIR), process knowledge on the crystallization kinetics of nucleation and crystal growth has been advanced unprecedentedly by using mathematical modelling tools of population balance equations, mass balance equations, etc. Specifically, the software package of gCRYSTAL from Process Systems Enterprise Ltd. (PSE), has initiated the development of general and convenient mathematical tools for the crystallization community, compared to the most of in-house developed codes or toolkits in the academic sector.

It should be noted that there is an obvious gap usually observed between the experimental and modelling works in crystallization, viz., the pale description and explanation of experimental observations without insights in crystallization kinetics and dynamics, or the plausible accuracy and reliability of theoretical predictions without strength of proofs in crystallization experiments. One of the examples is in the use of FBRM which measures the chord length distribution (not exactly the crystal size distribution), experimental authors believe the gradual increase in total counts represents the occurrence of secondary nucleation, while the modelling authors may believe the growth of crystal size and morphology changes accountable. One of the root causes for the gap is that mathematical tools are not readily available on site of the experimental set-ups. For example, the gCRYSTAL modelling package is not yet efficiently integrated with those in situ PAT sensors, and those house-made toolkits, e.g., the CryPIN from Loughborough University or CryMOCO from Purdue University, are lack of powerful mathematical solutions.

To equip researchers with both experimental and mathematical expertise, it is proposed here an integrated computational platform for crystallization communication, monitoring, modelling, and control, i.e., communicating with the experimental instruments, statistical monitoring of the measurements, mathematical modelling of crystallization process, and implementing advanced model-based control strategies to the process. The platform would stand out with many innovative features such as using industrial standard OPC servers for instrument communication, on-site modelling, moving-window kinetic parameters and state variables estimation of crystallization process, robust product quality control with nonlinear model-predictive control strategies. The proposed platform would surely support the practise of FDA Quality by Design (QbD) guidance which emphasizes the process knowledge in product quality control.

Dr. Su has solid research background in crystallization process modelling and control with working experience in UK EPSRC Continuous Manufacturing and Crystallization (CMAC) center and under Prof. Zoltan Nagy at Purdue University, who is a distinguished contributor to the crystallization process control. Dr. Su is also a fluent user and developer in gCRYSTAL software, whose work involving gCRYSTAL was awarded the PSE Model-Based Innovation Award in 2017. Dr. Su is also responsible for the plant-wide control strategy development in a research pilot plant for pharmaceutical continuous manufacturing at Purdue University and has the problem solving-skills in Emerson DeltaV OPC server for control strategy development. The proposed platform would combine his unique professional knowledge and supports from industrial and academic partners to promote the development and implement of process system engineering tools in crystallization community, narrowing the gap between experimental and modelling practise, and advancing the research in crystallization towards a sound scientific understanding.

2. **Process-oriented high reliable Process analytical technology (PAT) sensors for pharmaceutical continuous manufacturing**

Decade after the US Food and Drug Administration (FDA)â€™s Process Analytical Technology initiative, its implementation and advantages of monitoring and controlling of critical quality attributes (CQAs) in pharmaceutical manufacturing processes are world-wide recognized in industry and academy. For example, the Near Infrared (NIR) spectroscopy has been extensively reported in monitoring several CQA variables, e.g., active pharmaceutical ingredient (API) mass fraction in formulated powder, ribbon density after roller compaction, or the water content in wet granulation. Worth mentioning is that the pharmaceutical industry is now undergoing a mindset change from batch to continuous manufacturing, which depends more and more on the high reliability and availability of in situ PAT sensors to maintain product quality in real time.

However, the acute operating conditions of phase separation, e.g., crystallization, or the involvement of fine cohesive powders, e.g., API, has led the fouling one of the challenging issues in many PAT sensor applications. Furthermore, besides the chemometric model calibration for most of the PAT sensors based on spectroscopy, common raw spectra data pretreatment methods, e.g., Savitzky-Golay filter, Extended Multiplicative Scatter Correction, are inefficient in handling environmental uncertainties, raw material property variations, or process disturbances, e.g., humidity, powder flowrate, particle size distribution, etc. For example, the NIR spectroscopy collects the complex diffusions that are sensitive to both chemical compositions and physical properties, making a reliable NIR sensor especially challenging. Specifically, though the NIR sensing has been acknowledged as a convenient spectroscopy method without the need of sample pretreatment, except the various in-house designs of NIR boxes or flow chutes in academic research centers, there is a limited number of reports in open literature in terms of reliable application of NIR sensors in the pharmaceutical industry.

Hence, a complete process engineering solution to the secondary development of most commercially available spectroscopic probes are indeed necessary to produce reliable PAT sensors. It is proposed here that systematic design and risk analysis of PAT sensors, including sensor selection, sampling automation, sensor placement, chemometric calibration, model maintenance, sensor redundancy and network design, data reconciliation, are critically important for a high reliable PAT sensor to keep a good track of CQAs in pharmaceutical manufacturing. Unlike the previous works that are interested in the demonstration of sensing mechanisms, the proposed systematic design and risk analysis are process oriented and are focused more on the practical engineering solutions for reliable PAT sensor design, and should have a great impact on the industrial applications of PAT sensors. The proposed work also has the potentials for several patent applications.

Dr. Su has been financially supported by an US FDA funded project (grant number DHHS-FDA U01FD005535-01) on real-time release in continuous solid dose manufacturing: systematic characterization of material properties, and optimal design of sensing and control methods. As a main contributor to the risk-based monitoring and control strategies in this project, Dr. Su has been an extensive user of PAT sensors in the pilot plant at Purdue University and has a profound understanding and personal experience of the challenges in the sensor design.

3. **Intensified process integration for pharmaceutical continuous manufacturing**

Advantages of continuous manufacturing in the pharmaceutical industry have been demonstrated through extensive research studies in the last two decades, but mostly at the conceptual or theoretical level. Only few continuous manufacturing processes have been approved by the US Food and Drug Administration thus far, e.g., the Vertex Orkambi and Janssen Darunavir. To move from a conceptual continuous design to a practically operating pilot plant or manufacturing process, the efforts of regulators, academics, and industry are now focused more on process integration. For example, unit operations that are operated in batch mode, e.g., crystallization, filtration, drying, blending, granulation, tableting, etc., are to be integrated together and operated in continuous streamline mode, i.e., the end-to-end manufacturing as showcased in Novartis-MIT center for continuous manufacturing.

Advantages of continuous operation with the above process integration include the increased production capacity by running continuously and the reduced product variation by using steady-state operation. Nonetheless, it should be pointed out that pharmaceutical continuous manufacturing is not limited to this simple process integration, it also opens up another avenue of research towards a much more intensified process integration. For example, the wet granulation can be intensified by using a twin-screw extruder that integrates unit operations of feeding, blending, granulation. Another example is the spherical crystallization by which API molecule either crystallizes on porous excipient polymer materials or simultaneously crystallizes and aggregates to form an agglomerate. These spherical API particles could be directly tableted, shortening almost the whole secondary manufacturing process. Furthermore, other innovative manufacturing technologies are also emerging, e.g., the scaled-down drop-on-demand manufacturing for individualized medicines, which is prototyped and started-up at Purdue University at Professor G.V. Reklaitisâ€™s research group.

Therefore, it is envisioned that an intensified process integration for pharmaceutical continuous manufacturing, specifically a miniaturized and individualized manufacturing, could play an important role in the current trend of Industry 4.0 for smart manufacturing. Dr. Su has research experience covering the whole pharmaceutical secondary manufacturing process from crystallization to the end of tableting and was also involved in several discussions on the Mini Pharm project at Purdue University and is interested in setting up an intensified â€œsmart factoryâ€ in the future, specialized in biomolecular or protein drugs.

Teaching Interests:

1. **Undergraduate level**

Teaching at undergraduate level would require the experience and capability to interpret the rigid scientific terminologies in understandable language, which is usually a challenge for new lecturers. Being trained in Essential Teaching Skills and as a teaching assistant for two chemical engineering course modules in Loughborough University (7th in 2018 Times Good University Guide) in UK prepared me the first experience of this journey. Also, having laid a solid foundation in Chemical Engineering, especially through the advanced course modules in National University of Singapore (8th in 2017 QS world university ranking for Chemical Engineering), I would feel comfortable teaching across a broad range of subjects at undergraduate level. Furthermore, I strongly believe that mathematical tools are the best scientific language to verify the understanding of complex and dynamic chemical and biochemical processes. My background in Process Systems Engineering (PSE) has helped me gain a wide spectrum of knowledge on process modelling, simulation, control, and optimization. Hence, teaching for undergraduate course modules related to Process Dynamics and Control would be a preference for me, covering general topics of introduction to process control, chemical process modelling, Laplace transform, transfer functions, and control design, etc.

Reference books of â€œProcess Dynamics and Controlâ€ by Dale E. Seborg, Duncan A. Mellichamp, and Thomas F. Edgar or â€œProcess Control: Modelling, Design, and Simulationâ€ by B. Wayne Bequette are recommended.

Besides, it is also a good practise to have undergraduate students involved in ongoing research projects as helpers to senior PhD students to deepen their understanding in course modules and to have hands-on experience in research and engineering practise. As a faculty project advisor at Purdue University, I have the experience of supervising two undergraduate students in crystallization process control and NIR sensor design. I feel this experience help me a lot on how to communicate and learn with the new generations.

2. **Postgraduate level**

The greatest advantage of being a frontline researcher is our fresh first-hand experience on tackling the technical difficulties in research and better understanding of the expectations from the young Master or PhD students. For example, the experience of developing the state-of-the-art high resolution finite volume method for population balance model, solving complex integral partial derivative and algebraic equations (IPDAEs), multi-objective optimization using Non-Dominated Sorting Genetic Algorithm (NSGA-II), implementing the nonlinear model predictive control (NMPC), etc., would make me confident not only to help postgraduate students understand those methodologies, but also to encourage them to explore many open problems in chemical engineering.

As a lecturer for the course modules in the Doctoral Training Centre (DTC) in UK EPSRC Continuous Manufacturing and Crystallization (CMAC) research center, I have had the chances to experience the teaching of specific advanced topics on population balance modelling for crystallization, modelling and optimization of batch and continuous crystallization using MATLAB and gCRYSTAL software, etc. Hence, depending on the departmental research areas, course modules at postgraduate level may focus on specific topics, e.g., separation and purification using crystallization, crystallization process modelling and control, crystallization process optimization and control, or the advanced process control, etc.

Reference books of â€œHandbook of Industrial Crystallizationâ€ by Allan S. Myerson and â€œCrystallization: Basic Concepts and Industrial Applicationsâ€ edited by Wolfgang Beckmann are recommended. Besides, tutorial questions adapted from current research publications would be used to encourage postgraduate students to have comprehensive and critical literature-reading and problem-solving skills.

References

1. Hubert C, Lebrun P, Houari S, Ziemons E, Rozet E, Hubert P. Improvement of a stability-indicating method by Quality-by-Design versus Quality-by-Testing: a case of a learning process. *J. Pharm. Biomed. Anal.* 2014;88:401-409.

2. Igne B, Juan AD, Jaumot J, et al. Modeling strategies for pharmaceutical blend monitoring and end-point determination by near-infrared spectroscopy. *International Journal of Pharmaceutics*. 2014;473(1-2):219-231.

3. Nomikos P, MacGregor JF. Multi-way partial least squares in monitoring batch processes. *Chemometr. Intell. Lab. Syst.* 1995;30:97-108.

4. Su QL, Chiu MS. Monitoring pH-shift reactive crystallization of L-Glutamic acid using moving window MPCA. *Journal of Chemical Engineering of Japan*. 2016;49(7):680-688.

5. Su Q, Hermanto M, Braatz R, Chiu M. Just-in-Time-Learning based extended prediction self-adaptive control for batch processes. *Journal of Process Control*. 2016;43:1-9.

6. Mockus L, Peterson JJ, Lainez JM, Reklaitis GV. Batch-to-Batch variations: a key compoenet for modeling chemical manufacturing processes. *Organic Process Research & Design*. 2015;19:908-914.

7. Yu LX, Amidon G, Khan MA, et al. Understanding pharmaceutical quality by design. *The AAPS Journal*. 2014;16(4):771-783.

8. Juran JM. *Juran on quality by design: the new steps for planning quality into goods and services.* New York: The Free Press; 1992.

9. Potter C. PQLI application of science- and risk-based approaches (ICH Q8, Q9, and Q10) to exisiting products. *Journal of Pharmaceutical Innovation*. 2009;4(1):4-23.

10. Ierapetritou M, Muzzio F, Reklaitis G. Perspectives on the continuous manufacturing of powder-based pharmaceutical processes. *AIChE Journal*. 2016;62(6):1846-1862.

11. Pernenkil L. *Continuous blending of dry pharmaceutical powders*. Boston, MA 2008. PhD Thesis.

12. Lee SL, O'Connor TF, Yang X, et al. Modernizing pharmaceutical manufacturing: from batch to continuous production. *Journal of Pharmaceutical Innovation*. 2015;10(3):191-199.

13. Yu L. Continuous Manufacturing Has a Strong Impact on Drug Quality. *FDA Voice*. April 12, 2016. https://blogs.fda.gov/fdavoice/index.php/2016/04/continuous-manufacturing-has-a-strong-impact-on-drug-quality/. Accessed April 11, 2017.

14. Singh R, Godfrey A, Gregertsen B, et al. Systematic substrate adoption methodology (SAM) for future flexible, generic pharmaceutical production process. *Computer and Chemical Engineering*. 2013;58:344-368.

Diab S, Gerogiorgis DI. Process modelling, simulation and technoeconomic evaluation of crystallization

15. antisolvents for the continuous pharmaceutical manufacturing of rufinamide. *Computers and Chemical Engineering*. 2018;111:102-114.

16. Koswara A, Nagy ZK. On-Off feedback control of plug-flow crystallization: a case of Quality-by-Control in continuous manufacturing. *IEEE Life Sciences Letters*. 2017;3(1):1-4.

542289 High Performance Polymers for Water Purification and Energy Storage/Generation Applications: Rational Design Guided By Fundamental Structure/Property Relations

Abstract:

Research Interests:

My research interests are in the broad area of polymer science and engineering. More specifically, I am interested in designing functional polymeric materials (e.g., membranes and adsorbents) for water treatment and energy storage/generation applications, with a particular emphasis on improving the fundamental understanding of solute transport in dense and porous polymers. The research in my group will strive to marry the fields of chemical engineering, chemistry, and materials science to tackle some of the biggest technological problems facing society today. A primary focus of my future research program will be to establish general guidelines for rationally controlling polymer structure (both physical and chemical) to achieve a desired set of transport properties for solutes of practical interest. Such structure/property rules will be generated via a combined experimental and theoretical approach involving advanced polymer synthesis and characterization techniques, as well as theoretical modeling of solute transport through polymers. This fundamental knowledge will be leveraged to develop functional polymers that exhibit radical improvements in performance over existing state-of-the-art materials, potentially revolutionizing current industrial separations practices.

One grand challenge facing society today is water scarcity. Increasing global demand for clean water, driven by population growth, agricultural expansion, and climate change, among other reasons, coupled with depleting freshwater resources pose a major technological challenge to humanity. Moreover, the inherent connection between water availability and energy production further exacerbates this issue. Treatment of non-conventional water resources (e.g., seawater and wastewater) will be critical for meeting global water demands. Currently, membrane and adsorbent-based technologies are most widely implemented for water treatment applications due to their energy efficiency, small footprint, and operational simplicity. However, current state-of-the-art materials are plagued by a number of issues, including poor stability and inadequate selectivity for certain solutes, and are ill-equipped to handle treatment of non-conventional water sources (e.g., flowback water from fracking, brines, etc.) and waters containing more specialized contaminants (e.g., pharmaceuticals, organics). My research group will seek to address these issues by developing novel polymer membranes, as well as composite membranes, that are specifically designed to treat unconventional waters containing unique contaminants in an energy efficient and cost-effective manner. Design of such materials will be guided by fundamental knowledge of the relationship between polymer structure and solute transport properties.

Polymer membranes are also utilized as solid electrolytes in energy storage/generation applications such as fuel cells and batteries. Membranes for these applications must selectively permeate certain ionic species (e.g., lithium ions in lithium ion batteries or hydrogen ions in hydrogen fuel cells) while simultaneously hindering permeation of others. Although some of these technologies have been successfully commercialized, they still suffer from severe shortcomings, for example safety issues in lithium ion batteries due to use of flammable solvents, slow charge/discharge rates, etc. One avenue to mitigate (or eliminate) these issues and improve the overall performance of these technologies is to design better membranes, an endeavor that could be facilitated by a molecular understanding of the dominant factors governing solute transport in such materials and how polymer structure can be tuned to control solute transport properties. However, such fundamental understanding is scarce in the open literature. My research group will be well-equipped to fill in these basic scientific knowledge gaps by extending the synthesis, characterization, and modeling techniques developed for water-purification membranes to systems relevant for fuel cells and batteries (e.g., lithium ion batteries, redox flow batteries, aqueous batteries, etc.).

Postdoctoral Project: "*Synthesis of functionalized porous aromatic frameworks for rapid and selective removal of contaminants (e.g., boron) from aqueous solutions.*"

Under supervision of Jeffrey R. Long, Department of Chemistry, University of California, Berkeley

Ph.D. Dissertation: "*Ion sorption and transport in ion exchange membranes: importance of counter-ion condensation.*"

Under supervision of Benny D. Freeman and Donald R. Paul, McKetta Department of Chemical Engineering, The University of Texas at Austin

Research Experience:

I was first exposed to scientific research during the summer following my sophomore year in high school when I participated in a high school intern program at Columbia University Medical Center in an organic chemistry laboratory. Since then, my scientific curiosity has led me to a number of different research groups, exposing me to different research areas. During my undergraduate studies, I spent one summer in a biomedical engineering laboratory at Columbia University performing research in the area of nanobiotechnology. Later, I joined a materials science research group in the Center for Functional Nanomaterials at Brookhaven National Laboratory where I performed research on organic/inorganic hybrid solar cells and nanoscale metal oxide patterning. As an NSF graduate research fellow in the Department of Chemical Engineering at UT Austin, my project focused on improving the fundamental understanding of ion and water transport in ion containing polymers. The project was largely experimental in nature, involving the synthesis and characterization of polymer membranes, but it also involved a theoretical component. Notably, I developed a unifying theoretical framework for describing, and in some cases predicting with no adjustable parameters, ion partitioning and transport in ion containing polymers. I am currently a postdoctoral scholar in the Department of Chemistry at UC Berkeley, and my research project involves the synthesis and characterization of functionalized porous aromatic frameworks for use as high performance adsorbents for water treatment applications, specifically boron capture from water. This diverse research experience motivates me to establish and lead a highly collaborative and multidisciplinary research program that exists at the interface of chemical engineering, chemistry, and materials science.

Teaching Interests:

I developed a passion for teaching when I served as an undergraduate teaching assistant for a Linear Algebra course during my undergraduate studies at Stony Brook University. I found the experience to be exciting, challenging, and extremely rewarding, so I signed up to be a teaching assistant for a Multivariable Calculus course the next semester. During my Ph.D. studies, I was a graduate teaching assistant for two undergraduate level courses, Introduction to Chemical Engineering Analysis and Thermodynamics, and one graduate level course, Polymer Science. My responsibilities as a teaching assistant varied from designing homework and exam problems to lecturing large classes, and my passion for teaching has grown more with each experience. As a faculty member, I am interested in teaching core chemical engineering courses such as Thermodynamics, Transport Phenomena, and Reaction Kinetics, as well as more specialized courses such in polymer science and materials chemistry. I am also excited to develop specialized graduate level courses in areas related to my research (e.g., mass transfer in polymeric materials). As a chemical engineer by training, however, I would be delighted to teach any other courses depending upon departmental needs.

In addition to teaching, I have extensive experience mentoring a number of high school, undergraduate, and graduate students from various ethnic and socioeconomic backgrounds. As a faculty member, I will make it a priority to build a diverse research group, as I strongly believe that a diverse way of thinking is necessary to solve some of the greatest problems facing humanity today. I will also strive to form strong relationships with both local and national high schools and develop research programs for high school students, especially those from underrepresented backgrounds who otherwise may not be exposed to scientific research, since I strongly believe that early exposure to scientific research plays an instrumental role in pursuing a career in the STEM fields, as it did for me.

542392 Molecule Separation and Conversion Using Novel Porous Material

Abstract:

Research Interests: Metal-organic frameworks; gas adsorption and separation; membrane separation and reaction; energy storage

Teaching Interests: Thermodynamics, transport, porous materials and applications

Metal-organic frameworks (MOFs) are a series of novel porous materials which are composed of metal centers and organic linkers through coordination bonds. MOFs usually have 3D porous structures, large surface areas and pore volume, tunable pore size and chemical composition, which provide promising future in applications in molecule separation and conversion. A potential application example is CO_2 separation using MOFs. Unsaturated metal centers have been found to be the key for low partial pressure CO_2 adsorption. A robust hydrophobic MOF membrane showed high CO_2 permeance even under 80%RH, which is close to real flue gas condition. Proposed future work is focusing on 1) Oriented asymmetric MOF membrane for molecule separation such as CO_2 separation from natural gas and bioethanol separation; 2) MOF materials as catalyst precursors used in fuel cell and other electrochemical energy storage systems; 3) Biocompatible magnetic MOFs for molecule delivery, sensing and separation. The unlimited pairs of metal centers and organic linkers, chemical versatility, and easy to tailored pore and functionality open up great opportunities for MOFs in potential energy and environment applications.

541884 Engineering Anisotropy a New Design Strategy for Membrane Gas Separations

Abstract:

Research Interests: My research lies at the interface of engineering, transport phenomena, and rational material design. Specifically I am developing new strategies to design novel materials to control mass diffusion processes, with a special emphasis on membrane separations.

Teaching Interests: One of the pillars in Chemical Engineering core-curriculum education is transport phenomena. This subject is often characterized by undergraduate students as difficult and abstract. My teaching goals are focused on designing new active learning strategies that allow a more engaging and effective introduction and study of heat, mass, and momentum transport processes.

The design of membrane materials for gas separations has primarily relied on tuning the membrane chemical properties with the aim of tailoring the permeability and selectivity for the compounds of interest. From a physical point of view, manipulating the permeability and selectivity allows controlling the flux magnitude of the species in a prescribed way. Although this approach has been extensively used to design membrane and separation systems, the possibility of controlling flux direction to perform separations has remained largely unexplored. In this work, we show how metamaterial theory and effective medium approaches can be used to design novel anisotropic membranes that can be exploited for membrane-based gas separations. The proposed membranes consist of multilayer systems made of isotropic materials in which rationally designed anisotropic diffusion is controlled by the structural arrangement of the constituent materials rather than their chemical composition. Our novel anisotropic membranes operate by directing molecules of different species toward different places in the permeate side. Separation occurs as a result of the rerouting of mass flux in addition to differences in solubility and diffusivity. In terms of efficiency, we show that the selectivity of anisotropic membranes can be increased by orders of magnitude in comparison with typical isotropic materials while retaining similar permeabilities. Our work paves the way for the development of a new approach to design membrane materials and separation processes by taking advantage of an often overlooked intrinsic property of mass flux, its directionality.

542068 Membrane Technology and Bioengineering for Sustainable Products and Processes

Abstract:

Saurav Datta, PhD

Assistant Professor, Department of Biotechnology, Indian Institute of Technology Roorkee

Research Interests: **Membrane Technology and Bioengineering for Sustainable Products and Processes**

Recent years have witnessed worldwide major thrust in moving towards sustainable and smart economy by integrating sound scientific principles and breakthrough engineering practices. We embrace that opportunity and envision employing the fundamentals of Chemical Engineering, Chemistry and Biotechnology to develop sustainable products and processes. Our research activities encompass the following two core areas: (1) Biomimetic/bioinspired membranes for advanced bioprocessing, and (2) Biofuels and biobased chemicals formation.

1. **Biomimetic/bioinspired membranes for advanced bioprocessing**

- Engineering enzyme immobilized membrane bioreactor: Enzymes enable technologies with higher efficiency (high specificity of enzymes), lower energy consumption (ambient temperature operations) and environment friendliness (no use of toxic chemicals). In spite of that the industrial applications of enzymes are limited primarily due to the structural sensitivity of the free enzymes and challenges in downstream processing. We aim to engineer enzyme immobilized membrane reactors for efficient bioprocessing. One approach is to develop phospholipid bilayer functionalized, enzyme immobilized biomimetic membrane with improved functionality of the enzymes. We further aim to extend this study with enzyme cascades for value-added, bio-based chemicals formation.

- Membrane based bioseparations: Traditional membrane based bioseparations suffer from poor productivity and fouling. We aim to develop functionalized membrane based advanced bioseparations with superior performance. Membranes will be functionalized with polyelectrolytes/polymers/nanomaterials using different fucntionalization chemistries and will be utilized for various applications, such as separation of virus particles, isolation of milk components and purification of proteins.

2. **Biofuels and biobased chemicals formation**

- Improving the efficiency of bioethanol formation: Dilute acid pretreatment of lignocellulosic biomass produces inhibitors, such as HMF and furfural, which are detrimental to the downstream enzymatic hydrolysis and fermentation to produce bioethanol. We have isolated and identified a bacterium that is capable of efficient removal of these inhibitors. The bacterium survives with only HMF and/or furfural as the sole carbon sources. Besides, it selectively consumes the inhibitors over the sugars. We aim to deliver a bioprocess for efficient removal of the inhibitors and enhanced production of bioethanol. We also propose to study the degradation pathway of the inhibitors by the bacterium and plan to explore the opportunities of converting the inhibitors to valuable chemicals.

- Biological production of valuable organic acids: Organic acids, such as 3-hydroxypropionic acid (3HP) and succinic acid, are listed among the 12 most important value-added platform chemicals from renewable resources by the U.S. D.O.E. We plan to investigate biological production of these organic acids using renewable resources, such as crude glycerol. We also propose to develop an integrated separative bioreactor for simultaneous production and isolation of these bio-based chemicals using both suspended and immobilized microorganisms.

Teaching Interests:

1. Courses offered

(a) Core courses: (i) Chemical Reaction Engineering, (ii) Heat Transfer Operations, (iii) Mass Transfer and Separations, and (iv) Process Calculations

(b) Elective Courses: (i) Bioseparation Engineering, and (ii) Bioreactor Design and Analysis

2. Proposed Elective Courses

(i) Membrane Technologies for Energy, Environment and Bioengineering, and (ii) Introduction to Mathematical Modeling for Bioengineers

542100 Morphology Engineering of Carbon Molecular Sieve Membranes for Advanced Separations

Abstract:

Research Interests:

The high "energy penalty" associated with traditional separation techniques like distillation and amine scrubbing has motivated interest in advanced membrane based novel materials and processes. These membranes are useful for a wide range of applications such as post-combustion CO_2 capture, natural gas purification and olefin-paraffin separation in petrochemical streams. My research will focus on advanced **C**arbon **M**olecular **S**ieve (CMS) membranes in highly productive hollow fiber configurations. CMS membranes, formed *via* pyrolysis of specialty polymeric precursors under various atmospheres, can be tuned to address all of the above-mentioned separations. Engineering of a bimodal distribution of micropores and ultramicropores in CMS membranes, provides high penetrant flux and extraordinary angstrom-level molecular discriminating ability.

I will tackle the core fundamental issues to facilitate the development of these membranes as the preferred tools to revolutionize large scale energy intensive separations. To achieve this goal, advanced material characterization methods are needed to complement transport-based analyses. This ambitious undertaking will begin by pursuing the following three proposed thrusts, ideally in collaboration with experts in advanced complementary methods.

Proposal 1- The goal of the first project will be to not only clarify but also control the "*precursor to carbon*" formation pathway to create new CMS membranes. Besides extensive transport characterization and analysis, microscopic and spectroscopic techniques will be the main tools to achieve the goals of this thrust. A specific objective of this project involves precise control of the "defect ultramicropores" present in the CMS structure. Achieving this control can allow fine-tuning of the diffusivity of penetrant molecules without significantly reducing the overall sorption capacity of the membranes.

Proposal 2- To implement CMS membranes practically, configuration into asymmetric hollow fiber formats with high surface area to volume ratios is paramount. The presence of a "*hyperskin*" at the outermost selective CMS surface layer reduces productivity, without improving molecular selectivity. Understanding the mechanism behind the formation of this "*hyperskin*" and developing strategies to eliminate it, will be the primary aim of the second proposal. Techniques like **P**ositron **A**nnihilation **L**ifetime **S**pectroscopy (PALS) will be used in conjunction with transport studies and processing to elucidate and control the "*hyperskin*" properties.

Proposal 3- Unlike the hyperskin, avoiding large scale substructure collapse involves attention to the *macroporous support* that has no separation function, per se. Such collapse during the pyrolysis process in asymmetric CMS fibers, lowers overall membrane productivity. Existing solutions to overcome the support collapse fail at high pyrolysis temperatures that are needed for many applications. My third research proposal will address this higher pyrolysis temperature regime using alternate anti-collapse strategies. This project will enable creation of next generation asymmetric CMS membranes for diverse applications not accessible by current approaches.

Research Background

I have worked on both liquid and gas separation membranes during my doctoral and postdoctoral research. As part of my postdoctoral research with Prof. William J. Koros at Georgia Tech, I have focused on CMS hollow fiber membranes. At GT, I discovered and analyzed the presence of a "*hyperskin*" in CMS membranes. I also worked on approaches to mitigate physical aging in CMS membranes, which can cause undesirable loss in productivity over time. This work helped clarify the existence of non-slit larger ultramicropores in CMS membranes which may be used as tools to precisely engineer the membrane performance.

During my PhD research with Prof. Ilsoon Lee at Michigan State University, I surface modified commercial nanofiltration membranes using the layer-by-layer assembly of polyelectrolytes. My surface modified membranes showed higher permeability than commercially available membranes, but with equivalent or higher perchlorate ion removal. I next extended the use of these to more complicated systems like actual aggressive wastewater. These membranes were modified further with addition of clay-nanoplatelets to improve their fouling resistances.

Teaching Interests:

Based on my background in Chemical Engineering, I can teach most of the core courses like Kinetics, Thermodynamics, Mass Transfer, and Transport Phenomena. I would include a section on membrane materials and processes in the Mass Transfer curriculum, if given an opportunity to teach that course. If there is enough interest, I would also like to introduce an Advanced Separations or Membrane Materials and Processes course as graduate elective courses. During my PhD research, I served as the Teaching Assistant (TA) for Transport Phenomena and during my postdoctoral research, I worked as the TA for introductory sophomore class- Chemical Process Principles. I have also mentored 8 undergraduate researchers during my PhD and postdoctoral research and found this opportunity to be highly rewarding.

Selected Publications

- **Sanyal, O.;** Hicks, S.T.; Bhuwania, N.; Hays, S, Kamath, M.J.; Karwa, S.; Swaidan, R.; Koros, W.J. "Cause and Effects of Hyperskin Features on Carbon Molecular Sieve (CMS) Membranes," *Journal of Membrane Science* **551C**, 113-122, (2018).
- **Sanyal, O.;** Zhang, C.; Wenz, G. B.; Fu, S.; Bhuwania, N.; Xu, L.; Rungta, M.; Koros, W.J. "Next Generation Membranes-Using Tailored Carbon," *Carbon*, **127**, 688-698, (2018).
- **Sanyal, O.;** Liu, Z.; Yu, J; Meharg, B. M.; Hong J. S; Liao, W.; Lee, I. "Design of fouling-resistant clay-embedded polyelectrolyte multilayer membranes for wastewater effluent treatment," *Journal of Membrane Science*, **512**, 21-28, (2016).
- **Sanyal, O.;** Liu, Z.; Meharg, B. M.; Liao, W.; Lee, I. "Development of polyelectrolyte multilayer membranes to reduce the COD level of electrocoagulation treated high-strength wastewater," *Journal of Membrane Science*, **496**, 259-266, (2015).
- **Sanyal, O.;** Sommerfeld, A.N.; Lee, I. "Design of ultrathin nanostructured polyelectrolyte-based membranes with high perchlorate rejection and high permeability," *Separation and Purification Technology***145**, 113-119, (2015).
- **Sanyal, O**.;Lee, I. "Recent progress in the application of layer-by-layer assembly to the preparation of nanostructured ion-rejecting water purification membranes," *Journal of Nanoscience and Nanotechnology* **14**, 2178-2189, (2014).

541269 Advanced Porous Materials for Scalable Molecular Separation: IntegrationÂ of Material, and Process, and Engineering

Abstract:

Research Interests:

Molecular sieving membranes have created interest as high-performance separation systems for production of petro-based and renewable fuels and chemicals. In particular, Metal Organic frameworks (MOFs), and zeolites have created great potentials for use in high-performance membrane-based separation systems for applications such as olefin production or Xylene separation. However, despite the well-documented demonstrations of MOF and zeolite membranes capable for highly selective separation of molecules with effective size differences as small as 0.01nm, commercialization is hampered by the processability, and processing cost. Moreover, only a few MOF, such as ZIF-8, UIO-66, HKUST-1, among almost unlimited MOFs has been studied due to lack of universal scalable membrane fabrication techniques. I will focus on the investigating possibility of materials, rational design and engineering of energy-efficient molecular separations membrane. Integration of materials, process, and Engineering an will be a critical step towards improving the current energy-intensive chemical processes. will be a critical step towards improving the current energy-intensive chemical processes.

Teaching Interests:

1. Chemical and Biomolecular Separation Processes
The objective of this course is to familiarize the chemical engineering students with the fundamental principles of separation processes. Course topics will include chemical species conservation, macroscopic balances, convective diffusion and mass transfer, separation processes (absorption, distillation, extraction and adsorption), membranes, filtration, sedimentation, separation processes in microelectronics, bio-separations. Throughout the course, a wide variety of mass transfer principles will be introduced to broaden the discussions.

2. Transport Phenomenon

Momentum transport, together with heat and mass transport, constitute the subject of transport phenomena, which plays a central role in much of science and engineering. A rigorous introduction to the fundamentals of momentum transport - commonly known as fluid mechanics - will be provided through this course. The focus will be on isothermal incompressible Newtonian fluids, as their flow behavior is relevant to a broad range of applications and the understanding gained provides a strong foundation for study of more advanced topics. Emphasis will be placed on the use of scaling, non-dimensionalization, and asymptotic analysis to obtain approximate solutions to various problems and to gain physical insight into fluid mechanical phenomena. The techniques and ways of thinking learned in this course will be applicable to mathematical models far beyond those encountered in fluid mechanics.

2. Molecular Thermodynamics and Energy System â€" Thermodynamics

This course will provide a survey of formal aspects of thermodynamics and its use in practical situations. The underlying principles of thermodynamics and application of these principles to problems of reaction and phase equilibria in pure phases and solutions will be covered in this course. In addition to classical thermodynamics, some elementary statistical mechanics will be introduced in order to help provide insight into the molecular origins of bulk behavior.

542500 Synthetic Post-Translational Circuits for Cell-Mediated Therapy of Diseases Involving Immune Dysfunction

Abstract:

Research Interests:

My doctoral research was completed with Dr. Josh Leonard in the Department of Chemical and Biological Engineering at Northwestern University. During this time, I applied synthetic biology tools to study the structure-function relationships of native receptors as well as developed synthetic receptors that drive expression of a gene of interest in response to extracellular proteins. Using this system, which was the first synthetic receptor able to detect and respond to soluble extracellular ligands, we demonstrated secretion of IL-2, a proinflammatory cytokine, in response to VEGF, an anti-inflammatory cytokine, in T cells.

I am currently a postdoctoral researcher with Dr. Jim Collins in the Institute for Medical Engineering and Science at MIT where I developed a novel platform for synthetic post-translational circuits in mammalian cells. This research is the first synthetic circuit that uses post-translational modifications similar to but completely independent from native signaling pathways. Unlike other synthetic circuits, signals are transduced by the same phosphorylation-based mechanism as native signaling pathways, which enables the same fast and layered response that facilitates tighter control of cellular responses and more flexible outputs. When connected upstream to synthetic receptors, engineered signaling pathways can be configured to sense and respond to changes in extracellular ligand concentration. These circuits are completely orthogonal but can be easily integrated with native signaling pathways. They are also modular and tunable, which allows for the same full range of outputs as native signaling pathways.

My lab will specialize in developing and utilizing synthetic post-translational circuits to better understand and treat diseases such as cancer that are associated with immune dysfunction.

Research objective 1 â€“ Develop and utilize post-translational circuits to study in vivo dynamics within tumor cells and the tumor microenvironment as the tumor is established, grows, and metastasizes. The speed, sensitivity, and reversibility of synthetic post-translational circuits will enable real-time measurements of dynamics within the tumor, including the levels and types of cytokines in the tumor microenvironment as well as the activation of key signaling pathways within both cancer cells and tumor-associated immune cells.

Research objective 2 - Develop post-translational circuits that integrate input from the tumor microenvironment and transduce the signal into therapeutically beneficial responses. Outputs will include activation or inhibition of native signaling pathways and the secretion of proinflammatory cytokines to stimulate the immune system.

Research objective 3 - Develop novel platforms for targeted inducible control of mechanisms similar to post-translational modifications found in natural biological systems, such as methylation, acetylation, and ubiquitination. This will enable faster and finer control over cellular responses, such as the regulation of gene expression, RNA processing, and protein function.

My past experience in developing post-translational circuits makes me uniquely qualified for this innovative research, which has the potential for groundbreaking developments in both synthetic biology and cell-mediated therapy. By employing the same types of post-translational modifications that regulate cellular behavior in nature, we open the door to a larger array of methods to enable tighter and faster control, more closely resembling the many levels of regulation that exist in the mammalian cell. This work will help to elucidate the complicated mechanisms by which cells transduce information and drive the development of safer and more effective cell-mediated therapies for disease treatment.

Teaching Interests:

I am comfortable teaching a wide range of classes at any educational level including any of the core chemical engineering courses or biology-related electives. I am also excited about developing a new class on synthetic biology that will cover the basic techniques and technologies in synthetic biology then move on to real-world topics including applications, safety, and ethics. This class will emphasize written and oral communication.

542509 Engineering a Purple Non-Sulfur Bacterium to Expand Symbiotic Nitrogen Fixation

Abstract:

Research Interests:

When bio-available nitrogen is limited, plant growth and thus crop yields are reduced. Synthetic fertilizers have answered this need, nourishing crops that feed nearly 50% of the world's current population. While synthetic fertilizers have largely enabled the Green Revolution, their production and use have introduced new problems. Five percent of the natural gas and two percent of the energy produced in the world annually are consumed to make ammonia. When the fertilizer is applied to the soil, microbes convert it to nitrate. Plants can utilize nitrate, but it is easily washed into the water supply where it can exceed safe drinking water limits. Nitrate accumulation in waterways also contributes to hypoxic marine dead zones. Many prokaryotes use the enzyme nitrogenase to fix their own nitrogen and some form symbiotic associations with plants, greatly reducing the plant's need for soil nitrogen. Expanding symbiotic nitrogen fixation to a larger number of crop species is the motivation for my research.

Developing nitrogen-fixing chloroplasts could support a plant's nitrogen requirements. Expressing the *nif* cluster, the genes required to produce nitrogenase, from a diazotrophic cyanobacterium in a non-diazotrophic cyanobacterium could help uncover the requirements of nitrogen fixation in a photosynthetic host, including chloroplasts which evolved from a cyanobacterium. My graduate work focused on creating synthetic regulators for the predictable and specific expression of the *nif* cluster in the model non-diazotrophic *Synechocystis* sp. PCC 6803. When expressing genes in a new host, the native regulation that protected the proteins or minimized the required cellular resources is often missing. I developed genetic circuits to control transcription of the *nif* cluster in response to environmental conditions important for the function of nitrogenase, including oxygen, nitrogen, and light. To begin to mimic the complexity of gene expression in the native host, I created regulators that incorporated a feed forward loop and others that responded to two inputs with AND logic, in addition to single-input (inducible) promoters. I replaced the regulatory region between divergent genes in the *nif* cluster with oxygen-responsive promoters that increased nitrogenase activity by 50% in *Synechocystis*. This construct could also be used to probe the cluster for unidentified promoters.

My post-doctoral research has focused on engineering the metabolically versatile purple non-sulfur bacterium (PNSB) *Rhodopseudomonas palustris* CGA009 to utilize its ability to convert waste products, such as depolymerized lignin and carbon dioxide, into value-added compounds, including polyhydroxybutyrate and putrescine. I am currently developing synthetic regulators of gene expression since only one orthogonal, constitutive promoter has been used for gene expression in published literature, and it has not even been characterized. My work includes expanding an inducible promoter's dynamic range, testing a second orthogonal inducible system, generating an RBS library to expand and tune expression, and increasing the bacterium's sensitivity to selection pressure. I am also characterizing the range of lignin-derived compounds photosynthetic, nitrogen-fixing *R. palustris* can degrade and the metabolic pathways it uses for that degradation.

My future research interests focus on engineering *R. palustris* to form nitrogen-fixing nodules in a legume that already establishes a symbiotic relationship with one of the PNSB's close phylogenetic relatives. *R. palustris* can be found in the tropical rainforest rhizosphere and is known to promote plant growth. This PNSB is phylogenetically close to both *Bradyrhizobium japonicum*, which forms nitrogen-fixing root nodules in soybeans, and the photosynthetic stem-nodulating *Bradyrhizobium* sp. ORS278; yet, *R. palustris* has not been found in endosymbiotic relationship with any plant. Most research on the symbiotic relationship between legumes and nitrogen-fixing bacteria have focused on Nod factor-dependent plants or an existing symbiotic bacterium. Engineering a bacterium with many of the genes, proteins, and traits that are common among the stem-nodulating *Bradyrhizobia* but that is not an endosymbiont could uncover the crucial steps needed to achieve nitrogen-fixing endosymbiosis, first in the Nod factor-independent *Aeschynomene evenia*. Using the synthetic biology tools that I am currently developing, I will begin by expressing genes from two horizontally-acquired islands present in the photosynthetic stem-nodulating *Bradyrhizobia*, but not *R. palustris*. Another early project will be the introduction of mevalonate production, which is known to induce calcium spiking and the expression of symbiotic genes in vascular plants that form an endosymbiotic relationship with mycorrhizal fungi or rhizobial bacteria. This work could make significant contributions to the expansion of nitrogen-fixing symbiosis in crops.

Successful Proposals:

University of Nebraska Office of Research and Economic Development Revision Award, University of Nebraska Core Facility Grant Program for New Users Award, Nebraska Corn Board, NSF Graduate Research Fellowship, People, Prosperity and the Planet Student Design Competition for Sustainability

Postdoctoral Project:

"Harnessing the metabolically versatile *Rhodopseudomonas palustris* as a model bacterium"

Under the supervision of Dr. Rajib Saha, Department of Chemical and Biomolecular Engineering, University of Nebraska

PhD Dissertation:

"Genetic Circuits for Transcriptional Regulation in *Synechocystis* sp. PCC 6803"

Under the supervision of Dr. Tae Seok Moon, Department of Energy, Environmental & Chemical Engineering, Washington University in St. Louis

Selected Publications:

CM Immethun and TS Moon. in *Synthetic Biology of Cyanobacteria* Ch. 13. Synthetic Gene Regulation in Cyanobacteria (Springer/Nature Press, 2018).

CM Immethun, DM DeLorenzo, CM Focht, D Gupta, CB Johnson, TS Moon. Physical, Chemical and Metabolic State Transcriptional Regulators Expand the Synthetic Biology Toolbox for *Synechocystis* sp. PCC 6803, *Biotechnol Bioeng.* **114**:7 (2017).

N Wan, DM DeLorenzo, L He, L You, **CM Immethun**, G Wang, EEK Baidoo, W Hollinshead, JD Keasling, TS Moon, YJ Tang. Cyanobacterial Carbon Metabolism: Fluxome Plasticity and Oxygen Dependence, *Biotechnol Bioeng.* **114**:7 (2017).

CM Immethun, KM Ng, DM DeLorenzo, YC Lee, B Waldron-Feinstein, TS Moon. Oxygen-Responsive Genetic Circuits Constructed in *Synechocystis* sp. PCC 6803, *Biotechnol Bioeng.* **113**:2 (2016).

Teaching Interests:

Knowledge acquired in college covers only a small part of what an engineer will need to know throughout their career. It provides the foundation from which we start solving a problem; yet, the broad, complex and dynamic situations engineers face requires the confidence to think creatively and the ability to direct our own learning. Students' development of a firm knowledge base from which they have the confidence to solve the variety of problems they will face throughout their careers is my first educational objective. My students will tackle realistic problems designed to increase their comfort with resolving underdetermined systems, addressing the complex web of needs and constraints, thinking creatively, and communicating effectively. Practicing an organized and dynamically controlled thought process in new learning situations is the second educational objective for my students. By assessing the task at hand, evaluating the available resources, planning the approach, applying and monitoring the planned strategies, reflecting on the success of the planned strategies, and revising by adjusting the approach, securing additional resources and sometimes even re-assessing the task, students become self-directed learners. My courses utilize goal-directed practice with targeted feedback to help them embrace the learning process and realize its utility in solving the problems they will encounter. I returned to academia propelled by the desire to help students construct the foundation from which they can make important contributions to the challenges we face as a society. My instructional strategies and assessment methods are designed to support that goal and my educational objectives.

Selected Teaching Experience:

Teaching Assistant, Chemical Process Dynamics and Control - Received the EECE Graduate Student Teaching Award, based on student evaluations

Advisor, International Genetically Engineered Machine (iGEM) Competition - Washington University in St. Louis and University of Nebraska teams

Mentor, 11 Graduate Rotation and Undergraduate Students, Washington University in St. Louis - 7 are co-authors on my publications

Judge, International Genetically Engineered Machine (iGEM) Competition's Giant Jamboree

Teacher Workshop Instructor, BioBuilder® Educational Foundation

542154 Proteins Nanoparticles with Control of Shape, Size, and Valency for Therapeutics

Abstract:

Research Interests:

Protein therapeutics, especially antibodies, have revolutionized medicine in the last 30 years. These molecules perform exquisite functions that are not yet found in synthetic molecules—protein therapeutics feature molecular recognition and catalytic rate enhancement activities that specifically interact amongst a sea of molecules that are found in the body and in cells. However, the engineering of these proteins is currently limited to mutations to the amino acid sequence and chemical modification. By treating protein materials as nanoparticles, my future work will create new protein therapeutics that have improved pharmacokinetics and pharmacodynamics. In analogy to nanoparticles, the size, shape, surface chemistry, valency, and functionality of proteins will be chemically controlled to tune serum half-life, improve tissue- and cell-specific targeting, and control the function of binding and catalytic domains. To create these new protein materials, functional protein domains are scaffolded by fibrous protein domains, such that the properties of the functional and scaffold domains are varied independently. By decoupling the scaffold from function, protein nanoparticles are built from the bottom up with chemically tailorable properties. My work will seek to answer three questions:

1. How do material properties of a protein scaffold affect activity?

2. What is the effect of multi- and hyper-valency on antibody binding to cells?

3. Can antibody binding and enzyme activity be tuned by changing the local chemical environment in a protein material?

Successful Proposals

American Cancer Society Postdoctoral Fellowship

National Science Foundation Graduate Research Fellowship

Past Research

Proteins materials are made by crosslinking individual protein molecules. As a postdoctoral fellow with Professor Milan Mrksich at Northwestern University, I used a site-specific crosslinking method to create megamolecules, protein structure notable for their giant size, homogenous chemical composition, and tunable shapes. I used energy transfer to measure the end-to-end distance of megamolecules and found that linear megamolecules are relatively rigid and can be varied in size from 5–20 nm. These megamolecules served as a scaffold for antibody and enzyme domains, and the binding affinity of antibody megamolecules was increased by orders-of-magnitude by varying size and valency. This work impacts the development of next-generation therapeutic proteins such as bispecific antibodies.

Many proteins with interesting material properties, such as spider silk, are difficult to produce in bacteria due to toxicity to or degradation by the bacteria as they accumulate in the cell. As a graduate student with Professor Danielle Tullman-Ercek at the University of California, Berkeley, I improved production of these proteins by engineering a bacterial protein secretion system to accumulate protein outside of the cell. Proteins are unfolded during secretion, and I showed that proteins spontaneously refold and achieve post-translational modification after secretion. I also controlled expression of the secretion system gene cluster to increase the titer by 10-fold and enable macroscopic hydrogels to be made from secreted proteins. This work impacts the biotechnology industry, where therapeutic proteins are still produced and harvested from inside bacteria.

Teaching Interests:

I am interested in teaching courses in chemical kinetics, biochemical engineering, process design, thermodynamics, and introductory chemical engineering. Further, I am interested in creating an upper-division course in synthetic biology that will analyze academic and industrial efforts through case studies.

Teaching Experience and Philosophy

My goal as a teacher is for my students to build a physical intuition for molecular phenomena, and I do this through analogies and small-group problem solving. I have served as a graduate student instructor for courses in process design and biochemical engineering, for which I was awarded an outstanding instructor award. In addition, I seek additional teaching opportunities by giving guest lectures in courses on protein engineering and biosensors. I am also interested in teaching lay audiences through science outreach and I work with middle-school students in Chicago on science projects.

Mentorship Experience

My teaching extends to the laboratory, where I have mentored 11 students. As a mentor, my goal is to give students the tools to develop into independent researchers. My focuses are on designing good experiments and presenting their work. I run a journal club with students and colleagues where the focus is on the experimental design from high quality publications. To work on their writing skills, I give feedback on research reports, with particular focus on the introduction and motivation of their work. Finally, I support students to present their work at national conferences, where they have presented 16 posters and oral presentations.

Selected Publications

1. *Taylor, E.L., ***Metcalf, K.J.,** Carlotti, B., Lai, C.-T., Modica, J., Schatz, G., Mrksich, M., Goodson, T. "Long-Range Energy Transfer in Protein Megamolecules," *Submitted and available upon request*.

2. *Zhou, S., ***Metcalf, K.J.,** Bugga, P., Grant, J., Mrksich, M. "A photoactivable reaction for nanoscale patterning of multiple proteins," *Submitted and available upon request*.

3. **Metcalf, K.J.,** Slininger Lee, M., Jakobson, C.M., Tullman-Ercek, D. "An Estimate Is Worth About a Thousand Experiments: Using Order-of-Magnitude Estimates to Identify Cellular Engineering Targets," *Submitted and available upon request*.

4. **Metcalf, K.J.,** Bevington, J.L., Rosales, S., Burdette, L.A., Valdivia, E., Tullman-Ercek, D. "Proteins adopt functionally active conformations after type III secretion," ***Microbial Cell Factories***, 2016. DOI: 10.1186/s12934-016-0606-4

5. Azam, A., Li, C., **Metcalf, K.J.,** Tullman-Ercek, D. "Type III Secretion as a Generalizable Strategy for the Production of Full-Length Biopolymer-Forming Proteins," ***Biotechnology and Bioengineering***, 2015. DOI: 10.1002/bit.25656

6. **<u>Metcalf, K.J.,</u>** Finnerty, C., Azam, A., Valdivia, E., Tullman-Ercek, D. "Using Transcriptional Control to Increase Titers of Secreted Heterologous Proteins by the Type III Secretion System," ***Applied and Environmental Microbiology***, 2014. DOI: 10.1128/AEM.01330-14

542331 Biosensor Mediated Evolution of Biosynthetic Pathways for Biomanufacturing

Abstract:

Title: Biosensor mediated evolution of biosynthetic pathways for biomanufacturing

Author: Niju Narayanan, Postdoctoral Associate, Los Alamos National Laboratory, Los Alamos, New Mexico

Research Interests: Depleting petroleum reserves, energy security concerns and global climatic changes have driven our search for alternate technologies in order to shift to a more bio-based sustainable society. Biosynthetic pathways provide an inexpensive route from renewable materials for the production of molecules that are currently sourced from petroleum. the success of a biosynthetic pathway is highly context dependent - microbial host, pathway genes and their expression levels, promoters, enzyme activities, product and intermediate metabolite titers, cofactor availability, and often their delicate balance is offset owing to the complexities of biology. Advances in synthetic biology methods has enabled the construction and rapid generation of billions of cells with varying metabolic design. However, the evaluation of such designs by traditional metabolite measurement techniques currently available, lag by several orders of magnitude. Genetically encoded biosensors that act as screens for productive cells can solve these problems, identify optimal combination of genetic components, and enhance product titers and yields.

My research interest is in the development of biosensors and their application in the engineering of microbial metabolic pathways for human needs; including polymers, fuels and pharmaceuticals.

Research Experience: My research background has provided me with a wide range of experience in synthetic biology, high throughput technology, metabolic engineering, protein engineering, recombineering, molecular biology, biochemistry, fermentation technology, downstream processing, and analytical methods, some are described below:

Postdoctoral Study:

1. "Genetic biosensors for small molecules to engineer biosynthetic pathways in *Pseudomonas putida*" Advisors: Dr. Ramesh Jha and Dr. Taraka Dale, Los Alamos National Laboratory

This project focusses on using transcription factor based whole cell biosensors with fluorescence readouts for enhancing product titers and yields, evolving their promoters for transition to alternate production hosts, enzyme evolution, strain performance and selection of best producers, adaptive evolution for better growth of engineered host strains, utility of renewable non-conventional feedstocks etc.

2. Biosynthesis of Isopentenol in *Escherichia coli* via the Methyl Erythritol Phosphate Pathway

Advisor: Prof. Greg Stephanopoulos, Massachusetts Institute of Technology

The project was focused on improving isopentenol titers produced by the MEP pathway by soluble and functional expression of dephosphorylase and the pathway enzymes, maturation of the Fe-S cluster proteins, investigation of possible iron donors, NADP regeneration routes for cofactor supply, development of product and pathway metabolite analysis methods, culture media and isopentenol extraction procedures.

3. Pathways for synthesis of Advanced biofuels

Advisor: Prof. Greg Stephanopoulos, Massachusetts Institute of Technology

This project was focused on isovaleryl coA pathway for synthesis of C5 compounds, identification, cloning, characterization and variants of putative pivalyl coA mutases, investigation of prenol synthesis from isovaleryl coA pathway, development of analytical methods for pivalic acid and isovaleric acid determination.

PhD Dissertation: "Molecular and genetic strategies to enhance the functional expression of recombinant proteins in *Escherichia coli*"

Advisors: Prof. C. Perry Chou and Prof. Bill Anderson, University of Waterloo

The research work during graduate study was focused on the exploration of 'strategies for functional expression of recombinant proteins in *E. coli* using model proteins (lipase, penicillin acylase, yellow fluorescent protein, β-galactosidase, hCD83ext). They included posttranslational modifications, cytoplasmic and periplasmic expression, surface display, extracellular secretion, molecular chaperones, site directed and error-prone mutagenesis, stress regulatory pathways, host strains, and optimization of culture conditions.

Industrial collaboration project with Argos Therapeutics at University of Waterloo

"Development of 'bioprocess for production of recombinant human therapeutic protein hCD83ext"

Advisor: Prof. Murray Moo-Young and Prof. C. Perry Chou

The project included optimization of culture media and bioreactor conditions for high level expression of hCD83ext, Development and optimization of purification protocol for hCD83ext: GST affinity column, thrombin cleavage, ion-exchange chromatography, ultrafiltration, analytical methods for GST assay and endotoxin assay, construction of hCD83 mutants for improved protein stability

Funding/Awards:

NSERC Postdoctoral Fellowship, 2013

Murray Moo-Young Biotechnology Scholarship, 2008

NSERC Graduate Studentship, 2009

International Doctoral Student Award, 2006-2009

Teaching Interests: My background in Chemical Engineering, coupled with my research training in Biological Engineering has equipped me to teach a variety of courses in Chemical Engineering, Biotechnology, Bioengineering, Biochemistry, Protein Engineering, Metabolic Engineering and Synthetic Biology. However, I would be particularly interested in teaching 1. Biochemical Engineering and 2. Synthetic Biology. The coursework would be designed to encompass theoretical and computational approaches, practical laboratory classes, experimental model systems, and current research & classic literature.

Teaching Experience: I have had the opportunity to be a teaching assistant for various courses at the University of Waterloo and at Indian Institute of Technology Delhi. Some of them include Bioprocess Engineering, Advanced Biochemical Engineering, Engineering Biology, Introduction to Biotechnology, Chemical Engineering Concepts, and Physical Chemistry. This has given me teaching experience for 5-6 years at undergraduate level and around 8 months at the graduate level. Teaching helped me to hone my technical knowledge in the curriculum subjects and develop my interpersonal and communication skills and

I was able to enjoy a good rapport with my students and colleagues.

Supervisory experience: I have had the chance to mentor and work with 7 graduate students, 4 visiting scholars, 7 undergraduate co-op students and 3 technical staff during the course of my research career of graduate studies at UWaterloo, the Industrial collaboration project for therapeutic protein production at UWaterloo, and postdoctoral study at MIT and LANL. Most students who started without any background in biotechnology and bioengineering, developed the essential skills within a short period of time through research training. This is also demonstrated by their co-authorship on our publications.

Teaching Certificate Program: I have completed a teaching certificate program at MIT which was based on seven workshops aimed at development of teaching skills, exposure to research on teaching and learning, and structure of future learning. The workshop was followed by two microteaching sessions that are videotaped, evaluated and feedback provided.

Manuscripts in preparation

1. **Narayanan, N.**, Dale, T., and Jha, R. K. Adapted laboratory evolution and muconate-sensor based screening for robust growth of *Pseudomonas putida* and high product titers on corn stover hydrolysate.
2. **Narayanan, N.**, Henely, S., Dale, T., and Jha, R. K. Engineering a *b*-ketoadipate biosensor in *Pseudomonas putida* and evolution of aromatic catabolism pathway for biomanufacturing.
3. Jha, R. K., **Narayanan, N.**, and Dale, T. Genetically encoded sensor for alleviating product inhibition in an industrially relevant enzyme: Implications to biosynthesis of cis, cis muconic acid in *Pseudomonas putida* KT2440
4. **Narayanan, N**., Edgar, S., Woolston, B., and Stephanopoulos, G. Maturation of Fe/S clusters of Methyl erythritol phosphate pathway enzymes in *Escherichia coli*.
5. **Narayanan, N**., Woolston, B., and Stephanopoulos, G. NADPH regeneration in *Escherichia coli* improves the biosynthesis of isopentenol via Methyl erythritol phosphate pathway.
6. **Narayanan, N**., Duggar, D., Pereira, B., and Stephanopoulos, G. Identifying and engineering of putative pivalyl CoA mutase for the biosynthesis of C5 branched chain compounds.

Publications in refereed journals

1. **Narayanan, N.**, Khan, M., and Chou, C. P. **(**2011**)** Enhancing functional expression of heterologous *Burkholderia* lipase in *Escherichia coli*. Molecular Biotechnology. 47: 130-143
2. Srivastava, P., Singh, P., **Narayanan, N.** and Deb, J.K. **(**2011**)** Physiological and biochemical consequences of host–plasmid interaction – A case study with *Corynebacterium renale*, a multiple cryptic plasmid containing strain. Plasmid. 65: 110-117
3. Zhang, L., **Narayanan, N.**, Brand, S. R., Nicolette, C. A., Baroja, M., Arp, J., Wang, H., Moo- Young, M., and Chou, C. P. **(**2010**)** Structural identification of recombinant human CD83 mutant variant as a potent therapeutic protein. Protein Expression and Purification. 73: 140-146
4. **Narayanan, N.**, Khan, M. and Chou, C. P. **(**2010**)** Enhancing functional expression of heterologous lipase B in *Escherichia coli* by extracellular secretion. Journal of Industrial Microbiology and Biotechnology. 37: 349-361
5. **Narayanan, N.**, and Chou, C. P. (2009) Alleviation of proteolytic sensitivity to enhance recombinant lipase production in *Escherichia coli*. Applied and Environmental Microbiology.75: 5424-5427
6. **Narayanan, N.**, and Chou, C. P. (2008) Physiological improvement to enhance *Escherichia coli* cell-surface display via reducing extracytoplasmic stress. Biotechnology Progress 24: 293-301
7. **Narayanan, N.**, and Chou, C. P. (2008**)** Periplasmic chaperone FkpA reduces extracytoplasmic stress response and improves cell-surface display on *Escherichia coli*. Enzyme and Microbial Technology. 42:506-513
8. **Narayanan, N.**, Follonier, S., and Chou, C. P. **(**2008**)** In vivo monitoring and alleviation of extracytoplasmic stress to recombinant protein overproduction in the periplasm of *Escherichia coli*. Biochemical Engineering Journal. 42: 13-19
9. **Narayanan, N.**, Xu, Y., and Chou, C. P. **(**2006**)** High-level gene expression for recombinant penicillin acylase production using the *ara*B promoter system in *Escherichia coli*. Biotechnology Progress. 22: 1518-1523

10. **Narayanan, N.**, Hsieh, M. Y., Xu, Y., and Chou, C. P. (2006) Arabinose-induction of *lac*-derived promoter systems for penicillin acylase production in *Escherichia coli*. Biotechnology Progress. 22: 617-625
11. Xu, Y., Hsieha, M. Y., **Narayanan, N.**, Anderson, W. A., Scharer, J. M., Moo-Young, M., and Chou, C. P. (2006) Cytoplasmic overexpression, folding, and processing of penicillin acylase precursor in *Escherichia coli*. Biotechnology Progress. 21: 1357-1365
12. Xu, Y., Weng, C. L., Hsieh, M. Y., **Narayanan, N.**, Anderson, W. A., Scharer, J. M., Moo-Young, M., and Chou, C. P. (2005) Chaperone-mediated folding and maturation of the penicillin acylase precursor in the cytoplasm of *Escherichia coli*. Applied and Environmental Microbiology. 71: 6247- 6253
13. **Narayanan, N.**, Roychoudhury, P. K., and Srivastava, A. (2004) Isolation of *adh* mutant of *Lactobacillus rhamnosus*.sp for production of L(+) lactic acid. Electronic Journal of Biotechnology. 7: 72-84
14. **Narayanan, N.**, Roychoudhury, P. K. and Srivastava, A. (2004) L (+) lactic acid fermentation and its product polymerization. Electronic Journal of Biotechnology.7: 167-179

542378 Cell-Free Bioprocess Engineering for a Renewable Carbon Future

Abstract:

The potential of cell-free synthetic biology and bioprocess engineering to change the world for the better is tremendous. Previous efforts in this field to date have explored a variety of highly impactful directions, including elucidation of the biochemical basis of life, the ability to conduct highly sophisticated cell-free protein synthesis, and demonstration of complex synthetic gene circuits, to name just a few. With the modern set of biotechnological tools available, the opportunities for further exploiting this approach are continually expanding—from developing low- or negative-carbon energy and chemical systems, to accelerating agricultural engineering, to environmental remediation, to defense biosensing applications. Effective implementation of such cell-free synthetic biology tools and conversion platforms would result in significant global impact; the development of these systems is my central passion.

Research Interests:

For decades, the roles of proteins in living systems have been studied *in vitro*, forming the basis of our understanding of life at the smallest scales. Modern biotechnology, in harnessing this knowledge, offers the potential to improve the world in many ways, from biorenewable chemicals to agriculture to human health and beyond. Significant hurdles have limited the impact of this research enterprise in some fields however – microbes producing solvents rapidly succumb to the cytotoxic environment they create, agricultural modifications remain bound by the rate of plant growth, and the means available to efficiently debottleneck metabolic pathways are lacking. Cell-free methods—the use of cell lysate or purified enzymes to replicate complicated pathways—allow approaches with the potential to revolutionize these fields, due to their ability to operate in a wide range of conditions, serve as a rapid prototyping platform, and enable effective model-driven pathway optimization.

My previous work at Virginia Tech, and subsequently at Cell-Free Bioinnovations, focused on biological hydrogen production using a purified cell-free pathway. In this project, I combined cellulose and hemicellulose hydrolysis, novel phosphorylation steps, the pentose phosphate pathway, and a soluble hydrogenase to demonstrate the enzymatic co-utilization of C6 and C5 biomass sugars. I then improved the productivity of this pathway dramatically through the use of a kinetic model of the system—experimentally observing a 3-fold rate increase, as predicted, based on adjusting of the ratio of the enzymes in the system.

During my time at NREL I have started applying a similar approach to a very different problem – biological lignin conversion. Once fully proven out, I anticipate model-driven pathway optimization will be widely useful for a number of metabolic engineering problems. Furthermore, this approach is highly complementary to alternative, lysate-based rapid prototyping cell-free efforts. While those approaches offer high throughput, empirical methods especially useful during discovery efforts, my approach allows rational optimization, dramatically accelerating the improvement of the rate, titer, and yield of the products of these novel pathways.

Future research objectives include the use of cell-free as a conversion platform, for the utilization of toxic substrates and/or production of value-added chemicals and fuels that can only be microbially produced at uneconomically low titers—again, a limitation that can often be mitigated through the use of purified enzymatic systems. Additional directions of interest include prototyping plant pathways, to accelerate agricultural engineering, and environmental applications, where complex, specific biochemical transformations could be catalyzed without the introduction of self-replicating entities. Due to my status as a veteran, I remain cognizant of defense synthetic biology needs as well, another role in which cell-free approaches have tremendous potential. In several of these directions I anticipate a number of continued collaborations with NREL researchers into the future.

My ultimate ambition is to help lead the translation of cell-free biotechnology into world-changing technologies. Recently, I had the privilege of serving in a role that I believe sets me up well to be such a pioneer in the field – as a Fellow at the Department of Energy's Advanced Research Projects Agency (ARPA-E). This role provided an incredible opportunity to lead discussions with researchers at the top of their fields around out-of-the-box, high-impact energy technologies of the future, and to carefully consider the optimal role for biology and bioengineering within the field of renewable energy. I also reviewed hundreds of concept papers, interacted extensively with researchers through pre-proposal white papers, and came to understand well the responsibilities of working on the funding decision-making side of science. My efforts at ARPA-E also dramatically bolstered my network – these contacts that have become an integral part of my research ideation, planning, and execution.

Teaching Interests:

The elements of my teaching philosophy include encouraging participation and critical thinking and connecting the object of instruction with subjects about which the students are passionate. This inquiry-based learning style seeks to harness active student engagement in their educational experience. Related to this approach, I am excited to explore concepts such as research-integrated curricula as well. This approach is often used in design courses, but has significant potential in a chemical engineering laboratory or unit operations laboratory context as well. For example, exploring the concepts of reaction engineering, separations, and analytic approaches through the production of biodiesel or biogas from a waste stream, fermentative conversion of glucose into a potential nutraceutical, or pretreatment and hydrolysis of local agricultural residue. Non-standard chemical and biological engineering subjects I am interested in building courses around include metabolic engineering, enzyme engineering, cell-free systems, anaerobic digestion, process scale-up, and entrepreneurship.

I firmly believe that the example of my own passion for the subjects of biological and chemical engineering can be contagious – creatively presenting the material can make the classroom not only instructional, but inspirational. I was fortunate to witness such exemplary educational leadership multiple times at Virginia Tech. In addition to the exploration of novel research topics, these experiences included the integration of experiential learning (such as demonstrating methane release following agitation of the bottom of a pond), incorporation of specific industry-relevant examples (such as the use of polysaccharides as quick-clotting agents in modern bandages), and the use of cutting-edge web applications (for example, to assess the homology of proteins from different organisms). These experiences significantly informed my own career trajectory, research interests, and teaching philosophy, an effect I hope to replicate for my students.

At present, my teaching experience includes one semester as a teaching assistant for Virginia Tech's Unit Operations Laboratory, presenting four guest lectures in the Unit Operations classroom portion of that course, and, this Fall, co-instructing the Cell Biology & Biochemistry course at the Colorado School of Mines. The most rewarding teaching I conducted while a graduate student, however, was the mentorship of two REU students during the summer before I defended my Ph.D. Both contributed significantly to the success of my project, and both became co-authors on the resultant publication. I look forward eagerly to building upon these experiences further in my next position.

A unique role I see for myself in the teaching sphere is in my example as a professor who is a veteran and by mentoring fellow veterans. Former members of the military are quite common members of the student population, thanks to support such as the GI Bill – but faculty who have been there themselves are exceptionally rare. Student veterans are incredibly valuable – their passion, dedication, attention to detail, and, very often, ability to set the example and lead their peers, is in my opinion an underutilized resource of human capital. I hope to be lucky enough to encourage some of these veterans to build upon these skills and, if so inclined, contribute to the diversity of the research and development enterprise.

541922 Stochasticity, Complexity, and Multiscale Dynamics in Cancer Progression and Drug Response

Abstract:

Research Interests:

Biological systems differ from human engineered systems in three primary respects: 1) they are highly stochastic; 2) they are extraordinarily complex; and 3) they exhibit dynamics over a wide range of temporal and spatial scales. These characteristics are especially prevalent in cancer, which is a complex, adaptive disease characterized by genetic mutation, phenotypic plasticity, and cell-cell interactions among tumor cells and with the tumor microenvironment. Tumors are highly heterogeneous and exhibit complex responses to drugs and other therapeutic interventions. Patient outcomes to therapy can vary widely, but it is not uncommon for an initial favorable response to be followed by tumor recurrence and metastasis, which is very often fatal. The mechanisms that govern drug evasion and tumor reestablishment remain poorly understood.

It is becoming increasingly clear that to unravel the complexities of cancer a systems engineering approach is necessary. Cancer can be seen as a reverse engineering problem in that it is fundamentally a result of the breakdown of the cell's natural tumor suppressing machinery. Reestablishing or rewiring this apparatus requires a deep understanding of the mechanisms that underlie normal and cancer cell function. Computational models provide a means by which current understanding of biological mechanisms can be formalized mathematically. *In silico* predictions can be made, tested experimentally, and experimental results used to update and refine the model. This iterative approach of model building and refinement is a powerful tool for constructing a detailed and comprehensive understanding of a biological system.

Here, I present results of combined computational and experimental investigations of *in vitro* responses of isoclonal non-small cell lung cancer and melanoma cell populations to targeted drugs. Our results point to cancer cell populations employing a "bet hedging" strategy whereby cells diversify across multiple phenotypes in the absence of drug to increase the chances of population survival to a future drug onslaught. We hypothesize that the surviving subpopulation can persist for an extended period of time and is susceptible to genetic mutation, acting as a reservoir from which tumor recurrence may emerge. However, the subpopulation may also constitute a phenotypic bottleneck vulnerable to secondary treatment. Characterizing the molecular nature of the surviving subpopulation is thus of critical importance for improving patient outcomes. Work is currently underway towards this end, including the recent submission of a K22 grant application aimed at building a comprehensive, mechanistic model of the biochemical pathways underlying cell fate decisions in individual cancer cells.

Teaching Interests:

My expertise lies in chemical kinetics, thermodynamics, and numerical methods and I have experience in biochemistry, statistics, theoretical computer science, and software engineering. During my years as a postdoctoral researcher I have had the opportunity to mentor numerous undergraduate and graduate students and have given numerous lectures on systems biology. I enjoy teaching and make a concerted effort to engage students by providing analogies that convey complex concepts in an easy-to-understand manner. My philosophy of teaching is that nothing is ever as complicated as it seems and the best way to learn something is to do it yourself. Therefore, I aim to guide students in the right direction but ultimately leave it to them to solve a problem. It is a feeling of great satisfaction when something "clicks" for a student and they have that moment of deep understanding. I am prepared and enthusiastic about teaching full semester courses at both the undergraduate and graduate levels and look forward to mentoring graduate students and postdocs in their research endeavors. I am also strongly committed to minority outreach and increasing recruitment of underrepresented groups to higher education.

Select Publications:

Z.W. Jones, R. Leander, V. Quaranta, **L.A. Harris***, D.R. Tyson*, "A drift-diffusion checkpoint model predicts a highly variable and growth-factor-sensitive portion of the cell cycle G1 phase," *PLoS One* 13, e0192087 (2018). (**equal authors*)

L.A. Harris*, M.S. Nobile*, J.C. Pino*, A.L.R. Lubbock, D. Besozzi, G. Mauri, P. Cazzaniga and C.F. Lopez, "GPU-powered model analysis with PySB/cupSODA," *Bioinformatics* 33, 3492-3494 (2017). (**equal authors*)

L.A. Harris*, P.L. Frick*, S.P. Garbett, K.N. Hardeman, B.B. Paudel, C.F. Lopez, V. Quaranta and D.R. Tyson, "An unbiased metric of antiproliferative drug effect *in vitro*," *Nat. Methods* 13, 497-500 (2016). (**equal authors*)

L.A. Harris, J.S. Hogg, J.J. Tapia, J.A.P. Sekar, S. Gupta, I. Korsunsky, A. Arora, D. Barua, R.P. Sheehan and J.R. Faeder, "BioNetGen 2.2: Advances in rule-based modeling," *Bioinformatics* 32, 3366-3368 (2016).

J.S. Hogg*, **L.A. Harris***, L.J. Stover, N.S. Nair and J.R. Faeder, "Exact hybrid particle/population simulation of rule-based models of biochemical systems," *PLoS Comput. Biol.* 10, e1003544 (2014). (**equal authors*)

L.A. Harris and P. Clancy, "A 'partitioned leaping' approach for multiscale modeling of chemical reaction dynamics," *J. Chem. Phys.* 125, 144107 (2006).

L.A. Harris and A.A. Quong, "Molecular chemisorption as the theoretically preferred pathway for water adsorption on ideal rutile TiO2(110)," *Phys. Rev. Lett.* 93, 086105 (2004).

542136 Biomolecular Engineering and Magnetic Resonance for Structural Biology and Synthetic Biology

Abstract:

Research Interests:

Methods based on the physical principle of magnetic resonance, such as magnetic resonance imaging (MRI), utilize electromagnetic waves and magnetic field to noninvasively probe the location and structure of molecules. While MRI provides spatial information of molecules, its twin brother, nuclear magnetic resonance (NMR) spectroscopy, aims to probe the atomic-resolution details within a given molecule. Similarly, NMR is a unique noninvasive structural biology method that can study proteins under physiological hydration and temperature and in some cases, inside living cells. My scientific goals are to use biomolecular engineering to innovate NMR and MRI methods, allowing previously unattainable studies of molecules in intact cells and tissues, and on the other hand, to use NMR and MRI to characterize the performance of engineered molecules and cells.

As an example of this approach, I engineered biomolecular contrast agents that enabled acoustically modulated magnetic resonance imaging (AM-MRI), a new variant of MRI that uses ultrasound modulation [1-3]. While MRI can provide whole-body images at high resolution, there is a paucity of genetically encoded contrast agents, equivalent to green fluorescent protein (GFP) for optical imaging, that can enable MRI to visualize specific cellular processes such as gene expression. To address this technological gap, I engineered gas vesicles (GVs), a class of hollow gas-filled protein nanostructures evolved in photosynthetic microbes. The air inside GVs enabled their detection by MRI, and uniquely, such contrast was erasable by ultrasound treatment, which permitted selective imaging of these agents in the presence of background tissue contrast and solved a major issue that had plagued the use of existing MRI contrast agents.

While genetically encoded MRI contrast agents enabled the imaging of gene expression *in vivo*, my PhD research focused on the development of novel solid-state NMR (ssNMR) methods for studying atomic-resolution structure and function of membrane proteins in their native lipid bilayer environment. My research ranged from quantum mechanics theory of NMR pulse sequences [4-5], novel resonance assignment methods [6-8], to structure determination of a mercury transporter protein [9], which both demonstrated the frontier of the methodology and provided important insights into the bacterial mercury bioremediation pathway.

By working at the interface of two different disciplines – biomolecular engineering and magnetic resonance – we can develop innovative technologies and insights that will influence both basic biology research and the development of macromolecular and cellular therapeutics.

Selected Publications:

1. **Lu GJ**, Farhadi A, Szablowski JO, Lee-Gosselin A, Barnes SR, Lakshmanan A, Bourdeau RW, Shapiro MG (2018) Acoustically modulated magnetic resonance imaging of gas-filled protein nanostructures. *Nature Materials* 17:456–463. *News and views*. *Journal front cover*.
2. **Lu GJ** #, Farhadi A #, Mukherjee A, Shapiro MG (2018) Proteins, air and water: reporter genes for ultrasound and magnetic resonance imaging. *Curr. Opin. Chem. Biol.* 45:57-63. (# co-author)
3. Lakshmanan A #, **Lu GJ** #, Farhadi A #, Nety SP #, Kunth M, Lee-Gosselin A, Maresca D, Bourdeau RW, Yin M, Yan J, Witte C, Malounda D, Foster FS, Schröder L, Shapiro MG. (2017) Preparation of biogenic gas vesicle nanostructures for use as contrast agents for ultrasound and MRI. *Nature Protocols* 12(10):2050-2080. (# co-author)
4. **Lu GJ**, Opella SJ (2014) Mechanism of dilute-spin-exchange in solid-state NMR. *J. Chem. Phys.* 140(12):124201.
5. **Lu GJ**, Opella SJ (2013) Motion-adapted pulse sequences for oriented sample (OS) solid-state NMR of biopolymers. *J. Chem. Phys.* 139(8):084203.
6. **Lu GJ**, Park SH, Opella SJ (2012) Improved ^{1}H amide resonance line narrowing in oriented sample solid-state NMR of membrane proteins in phospholipid bilayers. *J. Magn. Reson.* 220:54-61.
7. **Lu GJ**, Son WS, Opella SJ (2011) A general assignment method for oriented sample (OS) solid-state NMR of proteins based on the correlation of resonances through heteronuclear dipolar couplings in samples aligned parallel and perpendicular to the magnetic field. *J. Magn. Reson.* 209(2):195-206.
8. **Lu GJ**, Opella SJ (2014) Resonance assignments of a membrane protein in phospholipid bilayers by combining multiple strategies of oriented sample solid-state NMR. *J. Biomol. NMR* 58(1): 69-81.
9. **Lu GJ**, Tian Y, Vora N, Marassi FM, Opella SJ (2013) The structure of the mercury transporter MerF in phospholipid bilayers: a large conformational rearrangement results from N-terminal truncation. *J. Am. Chem. Soc.* 135(25):9299-9302.
10. **Lu GJ**, Garen CR, Cherney MM, Cherney LT, Lee C, James MNG (2007) Expression, purification and preliminary X-ray analysis of the C-terminal domain of an arginine repressor protein from Mycobacterium tuberculosis. *Acta Crystallogr F* 63(Pt 11):936-939.

Awards/Honors:

Lu GJ (P.I.), Shapiro MG, Tirrell DA (mentors) NIH Pathway to Independence Award (K99/R00) Starting 08/2018.

Teaching Interests:

I like to implement an active learning environment in the classroom and have hands-on experience in teaching both laboratory course and lecture. In graduate school, I served as a teaching assistant in analytical chemistry laboratory, and presented weekly tutorial seminars for protein biochemistry course. At Caltech, I have given lectures in the Molecular Imaging course offered in the Chemical Engineering department. My Ph.D. background in chemistry and biochemistry offer a unique angle for me to teach in a Chemical Engineering department, while my exposure to quantitative physics and multidisciplinary research gives me confidence and enthusiasm to teach in various disciplines.

541622 Expanding the Biosynthetic Potential of Living Systems

Abstract:

Research Interests:

The ability to repurpose biomolecules has led to technology that forms the basis of a wide range of industries from biotech to food manufacturing to renewable fuels and plastics. Despite the relatively limited set of building blocks for constructing biomolecules, cells are able to achieve extraordinary functional complexity. Functional biomolecules such as carbohydrates, proteins, and nucleic acids are constructed from as few as four monomeric units. However, they possess unique chemical properties that emerge from their ability to have their sequence encoded precisely by genetic and biosynthetic machinery. In contrast, synthetic molecules can contain an almost unlimited range of functional groups and be composed of any number of unique building blocks. Yet chemical synthesis remains limited by our technical ability to construct chiral and sequence-specific macromolecules at scale, a feat more easily accomplished by biological systems.

My research aims to use fundamental approaches in synthetic biology, chemical biology, biosynthesis, and biomolecular engineering in reprogramming life with the overall goal of developing new solutions to address challenges in medicine, materials, and the environment. Towards this goal, I seek to extend the chemical functionalities of the amino acid, nucleic acid, and carbohydrate building blocks available in living systems for biosynthesis through discovery and engineering of new enzymes and pathways that can be genetically encoded. The newly *expanded metabolism* will drive the cellular production of biological macromolecules with unique chemical and physical properties that can be used for diverse application that include manufacturing of antibody-drug conjugates, renewable bioplastics and fuels, and biomaterials. In this poster I will present approaches, challenges, and my previous work in this area.

Teaching Interests:

Graduate school has presented me with opportunities to teach both undergraduates and graduate students. While a graduate student, I was head Graduate Student Instructor (GSI) for Biomolecular Engineering, a graduate-level chemical engineering course, and co-GSI for Technical Communication for Chemical Engineers. In Biomolecular Engineering I helped both lecture and design a class which showcased biological systems using fundamental principles. As an Assistant Professor, I aim to incorporate themes from my own research in order to demonstrate the relevance of the subject to current trends in science. For example, lessons taught as a research practicum, with unknown solutions to current problems, can serve as an efficient means to introduce critical thinking and scientific approaches. In addition to being able to teach core chemical engineering courses, I also feel uniquely positioned to introduce new courses geared towards the chemical and biological interface of our field. In particular, I am interested in designing a course aimed at drug discovery, design, and biomanufacturing.

Dissertation Work: "Discovery and application of a pathway for the biosynthesis of halo-, alkene-, and terminal alkyne amino acids" – Michelle Chang Lab – University of California, Berkeley

Successful Proposal: "Synthetic biology approaches to bio-orthogonal chemistry" – NSF (1710588)

Pre-doctoral Fellowship: HHMI Gilliam Fellow (2016-2019)

Presentation at AIChE 2018: "*Discovery of a Pathway for Halogenated, Terminal Alkene, and Terminal Alkyne Amino Acid Biosynthesis*" - (Session: Advances in Biocatalysts and Biosynthesis | Wednesday, Oct 31st at **8 am**)

541833 Single-Cell Analysis for Advancing Synthetic Biology

Abstract:

Research Interests:

The aim of synthetic biology is to design and build novel biological functions and systems. The promise of synthetic biology relies on the successful integration of engineering principles and biological systems. Single-cell heterogeneity, as a fundamental aspect in all biological systems, holds the key for further advancing synthetic biology. However, the impact of it on synthetic biology has not been extensively explored. Analyzing, understanding and utilizing single-cell heterogeneity will promote our ability of creating and altering the heterologous cellular community, which is the key for many applications in synthetic biology (e.g. bioproduction, microbiome engineering and immunotherapy etc.). My research will focus on (1) supplying novel parts and tools for synthetic biology through single-cell analysis and (2) advancing the understanding and engineering of heterogeneous cellular community through single-cell system biology. The synergistic effects of these two research directions will ultimately allow engineering heterologous cellular community with high precision and efficiency. In my future research, I plan to fully utilize the power of microfluidics, synthetic biology, metabolic engineering and bioinformatics to push the boundaries of bioengineering. In the short-term, I will expand our current synthetic biology toolbox through single-cell droplet microfluidic bioprospecting and bioengineering. In the long-term, my research will focus on advancing our ability to engineer complex cellular community using synthetic parts and design principles acquired through single-cell analysis. This research program will enhance our understanding of single-cell biology, enable the exploration of biological "dark matter" and advance our ability to design and build biological functions and systems to address the major challenges in the future.

Research experience:

My graduate research with Dr. Hal Alper at UT Austin focused on applying synthetic biology and metabolic engineering strategies to engineer nonconventional yeast *Yarrowia lipolytica*for high lipid production. Successfully rewiring the oleaginous yeast *Y. lipolytica*'s innate lipogenic capacity is critical for transforming it into a production platform for numerous fatty acids based value-added chemicals such as biofuels and nutrients. To achieve this goal, I first expanded *Y. lipolytica*'s genetic toolbox by developing hybrid promoters and centromere regulated plasmids to allow efficient manipulation of gene expression. Powered by these tools, I further employed combinatorial and evolutionary engineering strategies to increase the lipid production titer to over 40 g/L. I also performed comparative sequencing analysis to identify genomic and transcriptomic changes associated with the high lipogenic phenotype. These work demonstrated the synergistic effects of synthetic biology, genome scale analysis and metabolic engineering for reprograming the metabolic wiring of cells.

My postdoc work with Dr. Adam Abate at UCSF focused on droplet microfluidics based single-cell analysis. As a platform method for single-cell analysis, droplet microfluidic analysis offers an opportunity to exam the cellular heterogeneity and diversity in an unprecedented level of throughput and sensitivity. During my postdoc training, I have worked on the interface of droplet microfluidics and single-cell analysis: I

showcased the power of droplet based *in vitro* compartmentalization for engineering secretion phenotype by isolating *Y. lipolytica* strains with high riboflavin production; I applied single-cell RNA-seq (Drop-seq) to map cellular heterogeneity in lung fibrosis induced by bleomycin injury in mice and identified a profibrotic macrophage niche after tissue injury; I developed a novel workflow for highly parallel genome wide expression analysis of yeast colonies by combining microgel cultivation and Drop-seq. These work demonstrated the advantages of throughput, sensitivity and modularity of droplet microfluidics. It also showcased the needs for further developing and applying droplet based single cell analysis in bioengineering and biomedical researches.

Teaching Interests:

Teaching is more than just passing on information. It is also a process of nurturing critical, creative and divergent thinking, as well as acquiring ability to execute effectively and to work as a team collaboratively. I plan to use the proven system of lectures in combination of problem sets and group discussion for teaching technical expertise. I believe that granting research opportunities to undergraduates is a powerful tool for education, mentorship, and career development. As for the graduate level, I will focus on refining and deepening technical expertise, teaching technical writing, providing access to equipment and funding, and fostering career development. I will supply the overall picture and the access of new concepts and methods of the research field. I will also be available for discussing and exploring the potential focus of the research field in the future.

Serving as teaching assistant at UT Austin, I have taught biochemistry laboratory and general microbiology course. I have also successfully mentored several undergraduate and junior graduate students, who have either gone on to contribute to industry or received NSF GRFP awards. Based on my academia background and training, I am willing to teach undergraduate lecture courses such as Introduction of Chemical and Biomolecular engineering and Biochemical Engineering. I also plan to develop course focusing on introducing core concepts and methods in the fields of metabolic engineering, synthetic biology, single-cell analysis and microfluidics.

Selected publications:

1. **Liu L**, Dalal C, Heineike H, Abate A. "High throughput gene expression profiling of yeast colonies with microgel-culture Drop-seq." *In preparation*.
2. Aran D*, Looney A*, **Liu L***, (equal contribution) Hsu A, Fong V, Sheppard D, Abate A, Butte A, Bhattacharya M. "Reference-based annotation of single-cell transcriptomes identifies a profibrotic macrophage niche after tissue injury." bioRxiv. *284604 (Nature Immunology, in revision)*.
3. Wagner J*, **Liu L***, (equal contribution) Yuan S, Venkataraman M, Abate A, Alper H. "High-throughput screening of secretory phenotypes with double emulsion." Metabolic Engineering. 2018, 47, 346-356.
4. **Liu L**, Markham K, Blazeck J, Zhou N, Leon D, Otoupal P, Alper H. "Surveying the lipogenesis landscape in Yarrowia lipolytica through understanding the function of a Mga2p regulatory protein mutant." Metabolic Engineering. 2015, 31, 102-111.
5. **Liu L**, Pan A, Spofford C, Zhou N, Alper H. "An evolutionary metabolic engineering approach for enhancing lipogenesis in *Yarrowia lipolytica*." Metabolic Engineering. 2015, 29, 36-45.
6. **Liu L**, Otoupal P, Pan A, Alper H. "Increasing expression level and copy number of a *Yarrowia lipolytica*plasmid through regulated centromere function." FEMS Yeast Research. 2014, 14 (7), 1124-1127
7. Blazeck J*, Hill A*, **Liu L***, (equal contribution) Knight R, Miller J, Pan A, Otoupal P, Alper H. "Harnessing *Yarrowia lipolytica*lipogenesis to create a platform for lipid and biofuel production." Nature Communication. 2014, 5:3131.
8. Blazeck J*, **Liu L***, (equal contribution) Redden H, Alper H. "Tuning gene expression in Yarrowia lipolytica using a hybrid promoter approach." Applied and Environmental Microbiology. 2011, 77(42), 7905-7914.

544317 Application of Advanced Synthetic Biology Tools to Genetic Engineering and Bioprocessing

Abstract:

Research Interests:

I did my Ph.D. in Chemical Engineering with Prof. Ravi Kane in the Department of Chemical and Biological Engineering at Rensselaer Polytechnic Institute (RPI), where I developed optogenetic and synthetic biology tools for gene expression regulation. My research unites design principles of synthetic transcriptional systems and the mechanism of translation to invent new synthetic RNA systems. 1) I developed the first optogenetic system of controlling gene expression at translational level, based on the light-inducible reconstitution of a RNA binding domain and a translation initiation domain. 2) I have also invented a novel approach to control target mRNA translation using designable protein scaffolds (PUF domains) which enables either upregulating or downregulating the translation of a single mRNA. 3) To further extend this work, I utilized RNA binding domains to repress translation in a prokaryotic operon, enabling selectively regulating individual genes in a polycistronic mRNA, which cannot be achieved at the transcriptional level.

I am currently doing my postdoc work with Prof. Timothy Lu at Synthetic Biology Center at MIT, while co-mentored by Prof. Manolis Kellis at MIT Computer Science & Artificial Intelligence Laboratory (MIT CSAIL). My work focuses on the development of novel synthetic biology tools to disrupt disease diagnostics and drug manufacturing. I integrated synthetic biology and chemical engineering methods to establish a consolidated drug manufacturing platform of cell engineering, upstream and downstream bioprocessing for rapid, flexible and on-demand drug production. I developed a recombinase-based library screening method, which can be used to screen some difficult-to-screen libraries, and leverage machine learning to identify new 5'UTRs that can enhance protein expression in DNA vaccines and mRNA therapeutics. **A MIT invention based on this work was licensed to Senti Bioscience, Inc.**

Last year, **I was awarded $100,000, as a recipient of the inaugural Shark Tank Grant, by Broad Institute of MIT and Harvard** to develop a novel approach to detect rare mutations without prior knowledge of the locations for early cancer detection. In addition, I also applied cutting-edge NGS methods to mapping the single-cell epigenome, genome and transcriptome for multi-omics study of melanoma cells to discover new biomarkers for early diagnostics.

In the future, my lab will focus on the integration of synthetic biology, next-generation sequencing (NGS), single-cell technologies, genomics and artificial intelligence (AI) to develop AI-powered high-throughput profiling and screening strategies for drug target validation and lead optimization. In addition, I will adapt my flexible drug platform to cell-free gene expression systems, and develop new aptamer-based selective purification approach to produce high-quality biological drugs in a few hours at any location. I will also continue my liquid biopsy-based early cancer and Alzheimer's disease detection work in collaboration with Kellis and other bioinformatics labs. I aim to advance precision medicine with an integrated platform of early disease detection, drug target discovery and lead optimization, and point-of-care drug manufacturing.

Teaching Interests:

I enjoy teaching and mentoring. I had served as a teaching assistant for Transport Phenomenon and Bioprocessing and Biochemical Engineering Lab classes when I was a graduate student at RPI. I was a lab liaison for Innovation Teams (i-Team) at MIT for worked with three students in Sloan School of Management to explore the opportunities of commercializing the technologies that I co-developed in the lab. I am excited to teach core courses such as Transport Phenomena, Separation Processes and Chemical Reaction Engineering, or bio-related courses or labs including Bioprocessing Engineering, Biochemical Engineering and Chromatography. I am also passionate to develop new courses on synthetic biology and genomics, or other industrial related topics such biomanufacturing and next-generation sequencing.

Selected Publications:

1. **Cao J***, Novoa EM*, Zhang Z, Kellis M, and Lu TK. High-throughput engineering of 5'-untranslated regions by machine learning-powered library design and recombinase-based library screening. (Manuscript in preparation; Provisional patent filed in June 2018) (* Equal contribution)
2. **Cao J**, Perez-Pinera P, Lowenhaupt K, Wu M, Purcell O, Fuente-Nunez C, Lu TK, Versatile and on-demand biologics co-production in yeast. *Nature Communications*, 2018, *9*, 77.
3. **Cao J***, Fuente-Nunez C*, Ou R, Torres M, Pande S, Sinskey AJ, Lu TK, Yeast-based synthetic biology platform for antimicrobial peptide production. *ACS Synthetic Biology*, 2018, *7*, 896-902. (* Equal contribution)
4. **Cao J**, Arha M, Sudrik C, Mukherjee A, Wu X, Kane RS, A universal strategy for regulating mRNA translation in prokaryotic and eukaryotic cells. *Nucleic Acids Research*, 2015, *43*, 4353-4362.
5. **Cao J**, Arha M, Sudrik C, Schaffer DV, Kane RS, Bidirectional regulation of mRNA translation in mammalian cells by using PUF domains. *Angewandte Chemie International Edition*, 2014, *126*, 5000-5004. (Selected as a **Hot Paper** by the editors)
6. **Cao J**, Arha M, Sudrik C, Bugaj L, Schaffer DV, Kane RS, Light-inducible activation of mRNA translation in mammalian cells. *Chemical Communications*, 2013, *49*, 8338-8340.

Patents and Innovations:

1. "Tools for Next Generation *Komagataella* (*Pichia*) Engineering." Inventors: Lu TK, Perez-Pinera P, **Cao J**, Purcell O. US Patent Application No.: 62/360,731, International Patent Application No.: PCT/US2017/041509. Filed: July 11, 2017.
2. "pVAX-TP901, pVAX-phiC31, pVAX-BxB1, pVAX-R4, pCAG-phiBT1-NLS_IRES_Zeo, pCAG-Wbeta-NLS_IRES_Zeo, pFUGW-BxbI-attP-EYFP-hygro, HEK-LP" Inventors: Lu TK, **Cao J**, and Barbara J. MIT Case No. 19018. **This technology is licensed to Senti Bioscience, Inc.**
3. "High-Throughput Untranslated Region Engineering and Screening." Inventors: Lu TK, Kellis M, **Cao J**, Novoa EM, and Zhang Z. U.S. Provisional Application No.: 62/685,421. Filing Date: June 15, 2018.

542349 A Group Contribution Method for Heat Capacity Estimation of Hydrocarbons

Abstract:

Research Interests: Thermodynamics Properties and group contribution method

Teaching Interests: Power Engineering and Engineering Thermophysics

A Group Contribution Method for Heat Capacity Estimation of Hydrocarbons

Y Song, X Zhao,

MOE Key Laboratory of Thermo-Fluid Science and Engineering, Xi¡¯an Jiaotong University, Xi¡¯an, Shaanxi 710049, China

Heat capacity is an important physical quantity. Its data is the basis for various thermodynamic calculations and is also one of the most important thermodynamic data. Obtaining relatively accurate heat capacity data has a very important role in understanding the structure of matter, determining the phase transition of matter, and identifying the purity of materials. In general, the heat capacity data is usually determined by experiments, but with the development of modern production and technology, there are already millions of known compounds, of which tens of thousands are commonly used. In this case, the experimental determination of heat capacity data is very limited. For serious application, therefore, using the estimation method to calculate a large amount of heat capacity data is an important way to make up for the lack of experimental data.

In this paper, we have modified the equation proposed by Ruzicka and Domalski, and we use a second-order group additivity method for the estimation of the liquid heat capacity (CpL) of hydrocarbons. To assess the accuracy of the modified equation for estimating liquid heat capacity, reliable data from the NIST database was used to compare with estimated heat capacity values. The results can be seen that the hydrocarbons estimates are relatively close to the reference values in the NIST database. The relative error of the method from the melting point to the boiling point is almost below 3.5%. Even if the error does not exceed 10% near the critical temperature. This greatly improves the accuracy of the original method.

In conclusion, compared with the method of Ruzicka and Domalski, the modified equation proposed in this paper is applicable to the liquid heat capacity estimation of some hydrocarbons. It not only maintains the high prediction accuracy of the original method below the boiling point, but also improves the estimation accuracy from the boiling point to the critical temperature. It can be used to predict the liquid heat capacity near the critical temperature.

Key Words: *group contribution method, heat capacity estimation, modified equation.*

542354 Theoretical Calculation of Ethane Thermal Cracking Temperature

Abstract:

Research Interests: Thermodynamics Properties and Thermal cracking

Teaching Interests: Power Engineering and Engineering Thermophysics

Theoretical Calculation of Ethane Thermal Cracking Temperature

Y Song, X Zhao,

MOE Key Laboratory of Thermo-Fluid Science and Engineering, Xi¡¯an Jiaotong University, Xi¡¯an, Shaanxi 710049, China

Heat management problem of an engine has already become increasingly prominent along with the hypersonic flight vehicle¡¯s use gradually. Fuels provide power as propellant and serve as coolant at the same time. Endothermic hydrocarbon fuel is the most promising fuel to solve heat management problem, which can satisfy engine¡¯s cooling requirement through thermal cracking that absorbs heat. Alkanes which have the longer carbon chains would form new hydrocarbon fuel through cracking under high temperature. However, thermodynamic properties and combustion properties of hydrocarbon fuel cause huge changes between and after cracking. Therefore, research on cracking temperature and cracked products of alkanes is vital to engine¡¯ research on cooling characteristic and combustion properties. In order to preliminarily verify the theoretical calculation model, ethane is used as an example.

This paper obtains the global chemical reaction equations of ethane which undergo cracking reaction by means of free radicals reaction mechanism, some principles of cracking reaction of ethane and global models mechanism firstly. Then, the paper obtains theoretical cracking temperatures under different pressures incorporating Gibbs¡¯ Free Enthalpy. Comparing calculated cracking temperature with some literature values, for instance, we know theoretical cracking temperature of ethane is 1236K and literature value is 1200K under the pressure of 8 MPa. Within the error allowed, the results with theoretical calculation are correct.

In conclusion, the cracking temperature of ethane increases with the pressure increases, analyzing the laws between the temperature through theoretical calculation and pressure. Under lower pressure, cracking temperature increases rapidly with the change of pressure. Under higher pressure, cracking temperature¡¯ tendency of increase becomes flat. In addition, compared with literature data, the theoretical calculation model is correct

540802 Using X-Ray Science to Study Structure and Ultrafast Dynamics in Liquids

Abstract:

4th Year Postdoctoral Researcher

Research Interests:

My research interests are in the field of Chemical Physics with a specialization in liquids and aerosols. Aerosols enable us to reach metastable conditions which are difficult to achieve in bulk samples and consequently, materials can be studied far from equilibrium by aerosolizing them. I am interested in studying structure of materials and how they explain thermodynamic and kinetic properties of aqueous systems relevant to modern chemical engineering. A novel way to do this is by performing experiments at state-of-the-art ultrabright x-ray sources. E.g., X-ray scattering of micron-sized water droplets can reveal structural changes on supercooling; Ultrafast pump-probe experiments can help us observe dynamics on a femtosecond timescale. Such experiments are now possible due to X-ray Free Electron Laser (XFEL) sources which were not available a decade ago. I would also like to complement these studies with molecular modeling results by collaboration with theoreticians.

Successful proposals: I have written proposals to ultrabright x-ray facilities like SACLA in Japan, ESRF in France and APS in USA to perform experiments related wide angle and small angle x-ray scattering of supercooled bulk water.

Postdoctoral Project: Building an experimental set-up to study supercooled water and performing experiments at various ultrabright x-ray facilities to study its unique properties.

Principal Investigator: Dr. Anders Nilsson, Department of Physics, Stockholm University.

PhD Dissertation: Nucleation and Droplet Growth During Co-condensation of Nonane and D_2O in a Supersonic Nozzle

Under the guidance of Dr. Barbara Wyslouzil, Department of Chemical and Biomolecular engineering, The Ohio State University (OSU).

Research Experience:

My research career started when I was admitted for a M.S.+Ph.D. Program at the Chemical Engineering Department at OSU in September 2008. I studied nucleation and droplet growth of water and n-alkane nanodroplets and their mixtures. Supersonic nozzles are an effective way for dehydration of natural gas without the use of chemicals or excess energy. My research involved producing nanodroplets (nucleation) by condensation of vapors in a supersonic nozzle and following their growth. The vapors consisted of pure water, pure n-alkanes or mixtures of them. I used three experimental techniques to monitor this process- Pressure trace measurements (PTM), Fourier transform infrared (FTIR) spectroscopy and Small angle x-ray scattering (SAXS). I was also trained in the National school of Neutron and x-ray scattering school at Argonne national Lab and Oak Ridge National Lab to study the principals of scattering theory.

I joined Dr. Anders Nilsson for post-doctoral research in March 2015 where I along with a PhD student, built a set-up for supercooling microdroplets of water in vacuum to study structure and dynamics of supercooled water and answer questions about the anomalous behavior of this critical compound. The resulting set-up has been used at different X-ray synchrotron and XFEL sources like European Synchrotron Research Facility (ESRF), APS at Argonne National Lab (ANL) in USA, Linac Coherent Light Source (LCLS) at the SLAC National Accelerator Laboratory in USA, SACLA XFEL in Japan and PAL-XFEL in Korea. I have also led three of these beamtimes at SACLA, APS and ESRF. The experiments done with this set-up have resulted in high impact publications e.g. in Science (where I was the 2nd out of 12 authors) and Nature Communications.

I also led a project on correlating water tetrahedrality based on the peak split in the first peak of water's structure factor . I worked with theoreticians like Dr. Lee-Ping Wang and Dr. Jeremy Palmer on this project and became familiar with the different complexities of molecular modelling.

Teaching Interests:

I have a teaching philosophy that encourages students on asking questions and keeps the classroom environment more interactive. Since my teaching experience in Freshman engineering at OSU, I have interacted with students from different backgrounds. Based on my subsequent assignments as teaching supervisor/assistant, I have developed the following views on teaching:

Fundamental concepts: By the end of the course, the students should be able to grasp the fundamental concepts of the course. I designed tutorials that helped the students achieve this goal. A relevant example is scaling laws during transport phenomenon courses. The concept of thermal diffusivity in a micron-sized film and a thick wall of a house can be described by the same equation with differing time scales. Such universal applicability of any phenomenon makes students appreciate the subject and helps them retain it better.

Communication skills: According to Richard Feynman, whether you really understand a topic is based on whether you can explain that topic on a freshman level. I have realized that this is an important skill for anybody with higher education and requires good communication skills.

Teaching experience: I have been involved in the following courses as a Teaching Assistant at OSU:

Timeline	Course	Duties
September 2009-June 2010	Freshman Engineering FE 189	Trained students to write programs in MATLAB and utilize Autodesk Inventor for engineering graphics; graded homework and exams, and conducted office hours to answer students' questions.
Winter 2011, Winter 2012	ChBE 779 Air Pollution	Conducted tutorials and occasionally delivered

and Autumn 2012		lectures to a class of about 30-40 students for three academic terms; graded homework and exams, and conducted office hours

Experience as a supervisor: When I was studying for my PhD at The Ohio State University, I was a supervisor/mentor to an undergraduate student. I introduced him to concepts of nucleation theory and aerosol technology. I also trained him with hands-on work in the laboratory. I also supervised a group of 4 students where they used equipment in our lab for their Bachelors degree project. In total, I have supervised bachelor degree projects for 4 students and acted as a mentor for 1 student. During my employment as a post-doctoral researcher at Stockholm University, I was designated as an assistant supervisor to Doctoral student who joined in September 2015.

In the future, I would like to teach students about thermodynamics with specialization in metastable states and phase transitions.

Future Direction:

I would like to study dynamics and thermodynamics of supercooled water by building novel experimental set-ups because it is experimentally challenging. Even though water is one of the most important compounds for life, it is not yet understood completely. My plan is to modify the experimental set-up which I built during my postdoctoral research and achieve colder temperatures and higher pressures (as illustrated in figure 1) and investigate the "fragile to strong" transition of water and possibly the liquid-liquid critical point of water which is like the holy-grail in this topic. I would also work together with theoreticians and other experimentalists to interpret the results which I get from different experiments like x-ray scattering and spectroscopy.

Figure 1: Phase diagram of water. Figure adapted from Poole et al. The blue arrows show the different experimental conditions which I propose to explore. The hypothesized liquid-liquid critical point (LLCP) is shown as a ? and is yet to be explored experimentally.

<u>I. Flat sheet set-up to study deeply supercooled water at ambient pressure:</u>

My plan is to build a set-up to form flat sheets (1 μm or less) of water to probe temperatures below the Widom line (229 K at 1 atm). With such small sample volume, I can probe even colder temperatures by delaying homogenous freezing. I have made preliminary attempts to make these sheets work and the results look promising.

<u>II. Aqueous-organic ideal mixtures:</u>

I would also like to study the effect of pressure on water. One way to study the effect of pressure on water is by addition of a second component. Although addition of some solutes delays the crystallization, they may also dilute the anomalous behavior. However, the right kind of solute can delay crystallization and still preserve its anomalies. This ideal mixture can then be delivered in the form of supercooled microdroplets or flat sheets.

Selected Publications:

1. Kyung Hwan Kim*, Alexander Spaeh*, **Harshad Pathak**, Fivos Perakis, Daniel Mariedahl, Katrin Amann-Winkel, Jonas A Sellberg, Jae Hyuk Lee, Sangsoo Kim, Jaehyun Park, Ki Hyun Nam, Tetsuo Katayama and Anders Nilsson. *Maxima in the thermodynamic response and correlation functions of deeply supercooled water*, Science 2017, 358, 1589-1593.

2. **Harshad Pathak**, Jeremy Palmer, Daniel Schlesinger, Kjartan Thor Wikfeldt, Jonas A Sellberg, Lars GM Pettersson, and Anders Nilsson. *The structural validity of various thermodynamical models of supercooled water*,

Journal of Chemical Physics 2016, 145, 134507.

3. Andrew Amaya, **Harshad Pathak**, Viraj P Modak, Hartawan Laksmono, N Duane Loh, Jonas A Sellberg, Raymond G Sierra, Trevor A McQueen, Matt J Hayes, Garth J Williams, Marc Messerschmidt, Sebastien Boutet, Michael J Bogan, Anders Nilsson, Claudiu A Stan, and Barbara E. Wyslouzil. *How cubic can ice be?,* Journal of Physical Chemistry Letters, 2017, 8(14), 3216-3222.

4. **Harshad Pathak,** Judith Woelk, Reinhard Strey and Barbara Wyslouzil. *Co-condensation of Nonane and D_2O in a supersonic nozzle,* Journal of Chemical Physics 2014, 140, 034304.

5. **Harshad Pathak**, Gerald Wilemski, and Barbara Wyslouzil. *The structure of D_2O-nonane nanodroplets*, Journal of Chemical Physics, 2014, 140, 224318

6. **Harshad Pathak**, Kelley Mullick, Barbara Wyslouzil, and Shinobu Tanimura. *Nonisothermal Droplet Growth in the Free Molecular Regime,* Aerosol Science and Technology 2013, 47, 1310-1324.

541743 Molecular Simulations of Biological Self-Assembly

Abstract:

Research Interests: The overarching goal in my research program is the design and engineering of biologically related materials, ranging from protein self-assemblies to bio-nano sensing devices. Along the way of achieving this goal, my main contribution is understanding the physics behind the biological phenomena including but not limited to protein folding/non-folding/misfolding, DNA hybridization, aqueous solvation and bio-nano interfaces. The methods that I use to execute the research mostly relies on molecular simulations, computational modeling, and programming. A key aspect of my research is the continuous contribution to the improvement of computational methods.

This poster will include my experience in atomistic simulations of biologically related materials including protein folding/unfolding/non-folding/misfolding, aqueous solvation and bio-nano interfaces.

Teaching Interests: I am extensively experienced in and ready to teach all core subjects of Chemical Engineering at the undergraduate level, including Physical Chemistry, Classical and Statistical Thermodynamics, Phase and Reaction Equilibria, Chemical Reaction Engineering, Transport Phenomena, Applied Mathematics and Numerical Methods. Additionally, I have an advanced knowledge on Molecular Simulation Techniques, Physical Chemistry, Statistical Thermodynamics, Numerical Methods, Polymer Physics that will help me design and teach advanced classes of the mentioned subjects. I am also keen on designing online-friendly teaching materials and demonstrative tutorials as well as online repositories of the teaching material.

542870 Multi-Dimensional Single Cell Analysis with a Chemistry, Materials, and Nanotechnology Toolset

Abstract:

The single cell is the unit cell of biology, and modern **single cell analysis can now reveal the heterogeneity of biology**. However, just as single cell resolution unveils heterogeneity invisible to bulk assays, **multidimensional analysis illuminates the dark spaces in biological networks** – the interplay between proteins, transcripts, and other biomolecules that is invisible to any single measurement method. Capturing this level of resolution in single cells requires effective coordination in nanoliter volumes across a panel of chemistry, molecular biology, and nanotechnology tools. My research interests are to **develop nanomaterials and molecular probes to isolate single cells, and sieve and strain them for multi-dimensional single cell analysis.** In particular, my research vision is to use my expertise in cell perturbation to **add the aspect of intracellular delivery to single cell analysis:** to deliver constructs into cells that can process and output information and add more dimensions to cell analysis.

Postdoctoral work: "Integrated measurements of transcripts and proteins in single cells"; "Microfluidic platforms for multi-dimensional single cell analysis"

James Heath, California Institute of Technology and now Institute for Systems Biology.

Graduate dissertation: "Engineering cell access using nanostraws"

Nicholas Melosh, Stanford University

Research Interests:

I've focused on two facets of nanobiotechnology in my career, cell manipulation and cell analysis. As a postdoc in the Heath lab, I created a platform to analyze proteins and transcripts in the same cell to be published shortly. The guiding principle of the project is to maximize the leverage of established techniques: we used bead-based single cell transcriptomics and barcoded protein immunoassays on the same single cells, and we encoded the position of each cell into DNA barcodes to reconstruct a spatial transcriptomic map of cells to match the proteomics. This project was instrumental for me to both confront the challenges of single cell measurements including proteomics and transcriptomics, and also to witness the impact of single cell analysis through the Heath lab's translational projects at UCLA's medical school and consortia like the Parker Institute for Cancer Immunotherapy.

In graduate school, I developed a nanoscale platform for intracellular delivery with Nick Melosh. We used high aspect ratio nanostructures – nanostraws – to create fluidic conduits to cells, toeing the fine line between effective intracellular delivery and cell disruption. The project itself required a balance of nanomaterials engineering, cell and molecular biology, and surface science, and I gained an appreciation for the biophysical barriers to cell access and the nanoscale levers to pull to manipulate cells. We used the nanostraws to deliver time-resolved signals, cell-impermeable probes, and more importantly, to assist collaborators in studying biology. It is this intersection – intracellular delivery as applied to multi-dimensional analysis – where I see my research group making the most impact, by delivering probes of cellular information into cells to couple with transcriptome sequencing.

In the future, I'm interested in coupling single cell transcriptomics to other single cell observations. As with my integrated proteomics-transcriptomics postdoc work, the primary tool will be to use DNA barcodes. My research program revolves around developing new methods to selectively deliver barcodes, or attach logic to barcodes to couple their sequencing to biological parameters. This includes materials level approaches using nanoparticles or microfluidics, biological assays such as competitive binding probes and DNA base-pairing logic, and computation-level strategies, especially to add information into DNA barcodes. With tools capable of linking transcriptomics to other measurements such as protein levels or fluorescent intensity, I can turn my focus to biological questions for heterogeneous systems, especially in my background field of cancer.

Both my postdoc and graduate work have been well funded. My current Ruth Kirschstein postdoc fellowship (NIH F32) received a first percentile 12.0 impact score, and my PhD was funded by both NSF and NDSEG graduate fellowships. Outside of my primary research projects, I've had numerous formative research experiences, from surface deposition at Argonne National Labs, to DNA information storage in Irene Chen's lab at Harvard, to my twin undergraduate research positions at MIT, in phosphoproteomics and biologically template materials, with Forest White and Angela Belcher respectively.

Teaching Interests:

In addition to being a teaching assistant in graduate school (MATSCI 380 – Nanobiotechnology, 2013 and 2015), I've been focused on mentorship and I'm looking forward to a teaching program for postdocs that I've enrolled in for the winter (STEP at University of Washington). I benefited from excellent mentors as an undergrad, and since then I've mentored 6 undergraduates, as well as other graduate students. I've published or submitted papers with all but one of the undergraduates, and one of them submitted a first author publication under my guidance. I would be happy to teach courses of almost any flavor, as my own undergraduate coursework at MIT spanned my two majors of mathematics and materials science, as well as two minors of biomedical engineering and chemistry. I'm most excited to teach courses similar to the nanobiotechnology course that I TA'ed, literature based survey courses that reconstruct a sense of history behind a field through the progression of papers published.

<u>Selected Publications</u>:

Google scholar: 5 first author papers, h-index: 12

1. **Alexander M Xu**, Sally A Kim, Derek S Wang, Amin Aalipour, Nicholas A Melosh. Temporally resolved direct delivery of second messengers into cells using nanostraws. Lab on a Chip, 2016.
2. **Alexander M Xu**, Amin Aalipour, Sergio Leal-Ortiz, Armen H Mekhdjian, Xi Xie, Alexander R Dunn, Craig C Garner, Nicholas A Melosh. Quantification of nanowire penetration into living cells. Nature Communications, 2014.
3. Jules J VanDersarl*, **Alexander M Xu***, Nicholas A Melosh. Nanostraws for direct fluidic intracellular access. Nano Letters, 2011.
4. **Alexander M Xu**, Paul H Huang. Receptor tyrosine kinase coactivation networks in cancer. Cancer Research, 2010.
5. **Alexander M Xu**, Qianhe Liu, Kaitlyn L. Takata, Sarah Jeoung, Yapeng Su, Igor Antoshechkin, Sisi Chen, Matthew Thomson, James R Heath. Integrated measurement of cytosolic proteins and transcripts in single cells. In peer review.

543894 Life-Cycle and Techno-Economic Assessment of Microalgal Biorefinery for Biological CO_2 Sequestration

Abstract:

Research Interests:

In recent years, global warming driven climate change is threatening the existence and questioning the survivability of entire humanity. The major cause is due to the increase in concentrations of greenhouse gases in the atmosphere. Use of fossil fuels and other anthropogenic causes further elevate the issue that demands immediate measures to be taken. In such scenarios, looking for bio-based options is a good strategy as it is renewable, sustainable and does not pose a threat to economic security. My research interest lies in finding solutions to transform and unleash the wealth that is present in the waste and aim for a sustainable future using techniques and principles of biochemical engineering. I have experience in microalgal cultivation using flue gas in indoor and outdoor systems, CO_2 sequestration, waste water treatment, development of biorefinery, life-cycle analysis, techno-economic assessment, modelling and simulation of chemical process using PC-SAFT and gSAFT in Aspen Plus and gPROMS.

PhD Research:

My doctoral dissertation involves microalgal cultivation in open raceway reactors and closed photobioreactors, CO_2 sequestration from flue gas using microalgae, process integration with wastewater treatment, biomass generation for biorefinery application such as fish feed, biofuel etc. and performing process techno-economic and life cycle analysis.

Research experience and Awards:

- **Fulbright-Kalam Climate Fellow**- I worked with Prof. Warren Seider from the Department of Chemical and Biomolecular Engineering at the University of Pennsylvania. We have developed a CO2 intensified process for the extraction of algal oil and its conversion to biodiesel using microbubbles. The techno-economics and life cycle analysis of the process was investigated using aspen Plus Process Analyzer.
- **Newton-Bhabha Fellow**- I worked in the laboratory of Prof. Will Zimmerman at the University of Sheffield, UK in the Department of Chemical and Biological Engineering. The work involved use of CO_2 microbubbles generated by fluidic oscillator to strip ammonia off from the landfill leachate and using the processed leachate as inexpensive growth medium for microalgal cultivation and lipid accumulation.
- **Benjamin L Van duuren Award**

Teaching Interests: During my doctoral study at IIT Kharagpur, I have supervised 3 undergraduate and 2 post-graduate students, each for their one year long dissertation work. I have also evaluated detailed lab reports of undergraduate students to assess and provide suggestions to improve their writing abilities.

Future Directions: As a faculty candidate I would like to apply the knowledge that I acquired in Process Modelling and Simulation of biofuel production processes through my experience in cultivation of algae, harvesting, extraction of bio fuels from algae, downstream processing to various fundamental and industrial problems in Chemical Engineering.

Publications:

Yadav G and Sen R (2017). A microalgal green refinery concept for biosequestration of flue gas carbon dioxide vis-à-vis wastewater remediation and bioenergy production: Recent technological advances. Journal of CO2 Utilization, 17: 188-206.

Subramanian G, **Yadav G** and Sen R (2016). Rational use of mixotrophic microalgae for reducing carbon footprint of microalgal cultivation strategies in photobioreactors: A technological perspective. RSC Advances, 6:72897-72904.

Yadav G, Karemore A, Dash SK and Sen R. (2015). Performance evaluation of a green process for microalgal CO2 sequestration in closed photobioreactor using flue gas generated in-situ. Bioresource Technology. 191: 399-406.

Yadav G, Singh A, Bhattacharya P, Yuvraj J and Banerjee R (2013). Comparative analysis of solid-state bioprocessing and enzymatic treatment of finger millet for mobilization of bound phenolics. Bioprocess and Biosystems Engineering. 36 (11): 1563-1569.

Singh A, Kuila A, **Yadav G** and Banerjee R (2011). Studies on the process optimization of antioxidant rich polyphenol from okara. Food Technology and Biotechnology, 49 (3) 322â€"328.

Yadav G, Seider WD, Soh L and Zimmerman J (2018). Process Intensification of Algae Oil Extraction to Biodiesel. (In) 13th International symposium on Process Systems Engineering (PSE 2018). (Ed) Mario R. Eden, Marianthi G. Ierapetritou, Gavin P. Towler. Pp. 1699-1704. Elsevier, USA.

Yadav G and Sen R (2018). Sustainability of Microalgal Biorefinery: Scope, Challenges, and Opportunities. (In) Sustainable Energy Technology and Policies: A Transformational Journey. (Ed) Sudipta De, Santanu Bandyopadhyay, Mohsen Assadi, Deb A Mukherjee. Pp 335-351. Springer, Singapore.

Yadav G, Kumar, R and Sen R (2015). Fermentation techniques in bioenergy production. (In) Marine bioenergy: Trends and developments. (Ed) Se-Kwon Kim and Choul-Gyun Lee. Pp 111-134. CRC Press, USA.

Karemore A, R Dineshkumar, **Yadav G**, Subramanium G and Sen R (2015). Photobioreactors for improved algal biomass production: Analysis and Design considerations. (In) Algal Biorefinery: an integrated approach (Ed) Debabrata Das. Capital Publishing Company. Springer International Publishing.

537674 Triggerable Tissue Depth of Externally-Triggerable Drug Delivery Systems for on-Demand Nerve Block

Abstract:

Externally triggerable drug delivery systems have provided promise in personalized drug delivery where the time and site of drug release can be controlled by the caregiver or patient, non-invasively and effectively. However, major limitations stand in the path of translation of these systems, one of which is the applicability of activating such drug delivery systems in deep tissues, at the target site. Here, developed ex vivo and validated in vivo, we studied the tissue depth-dependence of externally-activated drug release from drug delivery systems. We developed an approach predict the payload release at varying tissue depths, and the results were validated by experimental results ex vivo and in vivo nerve block models. Light and ultrasound were chosen as the external triggers as many drug delivery systems have been developed based on these energy sources and they possess desirable properties as external triggers: ultrasound has been considered to penetrate tissue effectively, and near infrared light penetrates deeper than UV or visible light. We found that the key factor to achieve payload release at the desired tissue depth was not only the tissue penetrating properties of the energy source, but also the sensitivity of the drug delivery system towards the external energy.

Our results demonstrated that two important parameters should be considered for the activation of drug delivery systems in deep tissue: 1) the ability of the activation energy to penetrate tissue, 2) the sensitivity of the system to the activation source. Although ultrasound penetrated through tissue more effectivly than near-infrared light, the higher sensensitivity of the light-triggerable system allowed more effective cargo release at tissue depths between 0 and 2 cm. In vivo nerve block results demonstrated that our theoretical model provided a reasonable estimate of the activated release at different tissue depths.

538075 Approaches for Creating Smart Insulin Delivery Systems

Abstract:

Approaches for Creating Smart Insulin Delivery Systems

Lisa R Volpatti, Morgan A. Matranga, Abel C. Cortinas, Robert S. Langer, Daniel G. Anderson

Koch Institute of Integrative Cancer Research & Dept. of Chemical Engineering, MIT, Cambridge, MA 02139

Motivation: Diabetes mellitus is a class of diseases in which the body either does not produce (Type 1) or is insensitive to (Type 2) insulin, resulting in poorly controlled blood glucose levels. To combat chronic hyperglycemia and prevent severe complications such as blindness, cardiovascular disease, and kidney failure, Type 1 and many Type 2 diabetic patients are dependent on multiple daily doses of exogenous insulin. Patient compliance is a major challenge, however, due to the need to measure blood glucose levels up to six times a day, inject the appropriate amount of insulin, and carefully monitor diet. To reduce this large patient burden, an attractive diabetes therapy is a glucose-responsive insulin delivery system that would self-regulate a patient's blood glucose levels.

Results: Here we will discuss three types of chemically glucose-responsive insulin delivery systems that employ the enzyme glucose oxidase as a glucose sensor. Glucose oxidase converts glucose to gluconic acid thus reducing the pH of the microenvironment when glucose levels are high. This change in pH acts as a trigger to release insulin on demand. The first system (Fig 1a) uses a pH responsive polymer, acetalated-dextran, to form nanoparticles that physically encapsulate both insulin and glucose oxidase. The particles rapidly degrade in the presence of acid affording this system the advantage of being a fast acting therapeutic. The second system (Fig 1b) is comprised of alginate microgels that encapsulate these nanoparticles to create a depot of insulin for sustained glucose-responsive release *in vivo* for up to two weeks. The third system (Fig. 1c) is based on the electrostatic complexation of insulin to positively charged polymers, such as polyethylenimine and poly(beta-amino esters). When the pH is reduced below insulin's isoelectric point, the complex will dissociate, releasing insulin only in response to elevated levels of glucose. The synthesis, formulation, *in vitro* characterization, and *in vivo* results in both a healthy and diabetic mouse model will briefly be discussed for each of these systems, highlighting their pros and cons.

Conclusions: Here we show various approaches to creating a self-regulated insulin delivery system that achieves tight glycemic control *in vivo*. We conclude with perspectives on the future of glucose-responsive insulin delivery and discuss ideas for further work.

536351 Biomolecular Engineering of Acousto-Magnetic Protein Nanostructures for Non-Invasive Imaging of Cellular Function

Abstract:

Genetically encoded optical reporters such as green fluorescent protein (GFP) have revolutionized biomedical research by enabling observations of biological processes in engineered cells and transgenic animals. However, such optical reporters are fundamentally limited by the ~ 1 mm penetration depth of light in opaque tissues. As cellular therapies advance towards rodent models and ultimately humans, this limitation becomes increasingly severe. Therefore, we aim to use biomolecular engineering to develop new classes of genetically encoded agents capable of communicating with deeply penetrant forms of energy, such as magnetic field and sound waves.

Here we introduce gas vesicles (GVs), a class of gas-filled protein nanostructures discovered in certain photosynthetic microbes as their flotation devices. GVs possess a hollow gas interior, ~ 200 nm in size, enclosed by a 2 nm protein shell that is permeable to gas but excludes liquid water (**Fig. a-b**). We hypothesized that the gaseous interior of GVs would have a different magnetic susceptibility from the surrounding aqueous tissue and could thus produce susceptibility-based MRI contrast (**Fig. c**). Indeed, we observed robust T2, T2*, and quantitative susceptibility contrast of GVs at nanomolar concentrations (**Fig. d**). A unique feature of GVs is that they can be irreversibly collapsed under acoustic pressure above genetically determined thresholds, leading to the rapid dissolution of their gaseous contents. We demonstrated in several scenarios including brain tissues of living mice (**Fig. e**) that the MRI contrast produced by GVs can be erased remotely *in situ* by ultrasound, an orthogonal non-invasive imaging modality. Such acoustically modulated reporters would overcome a major challenge in MRI posed by background contrast from endogenous sources by allowing reporters to be identified specifically based on their acoustic responses. Next, we tested whether GVs can be used as MRI reporter genes. As a proof-of-concept, the expression of GVs was placed under the control of the chemical IPTG. When *E. coli* cells expressing this circuit were exposed to IPTG, they showed MRI contrast, which was absent from controls (**Fig. f**). Furthermore, multiplexed imaging to differentially visualize several species of GVs could be achieved either by scanning ultrasound pressures from low to high to selectively collapse certain GVs (**Fig. g**) or from distinct MRI fingerprints of GVs that differ in size and shape (**Fig. h**). Finally, we ask whether GVs can be engineered as dynamic sensors of biological signals. We found that clustering of GVs induced potentially by a biological signal could produce a marked change in T2*-weighted images, as experimentally shown by streptavidin-biotin cross-linking (**Fig. i**). These *in vitro* experiments established a set of principles allowing us to use GVs as acoustically modulated MRI reporters *in vivo* (Lu et al, *Nature Materials*, 2018).

By coupling the complementary physics of MRI and ultrasound, this genetically encoded and engineerable nanostructure gives rise to a distinct modality for molecular imaging with unique advantages and capabilities, especially in basic research of biological processes in deep-lying tissues and the development and tracking of therapeutic microbes.

References: Lu GJ, Farhadi A, Szablowski JO, Lee-Gosselin A, Barnes SR, Lakshmanan A, Bourdeau RW, Shapiro MG (2018) Acoustically modulated magnetic resonance imaging of gas-filled protein nanostructures. *Nature Materials*. http://doi.org/10.1038/s41563-018-0023-7

532384 A New Antifouling Strategy with Active Surface Topography

Abstract:

A new antifouling strategy with active surface topography

Huan Gu[1, 2], Sang Won Lee[1,2], Dacheng Ren[1,2,3,4*]

[1]Department of Biomedical and Chemical Engineering, [2]Syracuse Biomaterials Institute, [3]Department of Civil and Environmental Engineering, [4]Department of Biology, Syracuse University, Syracuse, NY 13244, United States

Bacteria can attach and form multicellular structures, known as biofilms, on virtually any abiotic or biotic surface. Due to the protection of extracellular polymer substrates (EPS) and slow growth, biofilm cells are up to 1,000 times more resistant to antimicrobials compared to their planktonic counterparts. Biofilms related high-level drug tolerance causes serious problems of persistent biocorrosion and biofouling in industry, as well as chronic infections in humans. The significance of biofilm-associated problems has stimulated intensive efforts on developing materials and surface modification techniques to prevent bacterial adhesion. However, the studies to date are largely limited to static features, which can be overcome by microbes over time and does not provide long-term prevention. To better control biofilm formation, we engineered active surface topographies that not only can effectively prevent initial bacterial adhesion but also achieve on-demand removal of mature biofilms. In this approach, rationally designed antifouling surface topographies are actuated by external stimuli with programmable beating frequency and force to repel bacteria and disrupt mature biofilms. In a proof-of-concept study, systematically designed micron-scale poly(dimethylsiloxane) (PDMS) surface topographies filled with biocompatible superparamagnetic Fe_3O_4 nanoparticles were actuated with an external magnetic field (5 mT). The antifouling effect of this surface system was first validated against 48 h *Pseudomonas aeruginosa* PAO1 biofilms. By actuating the square-shaped surface topographies with a height of 10 μm, side width of 2 μm, and inter-pattern distance of 5 μm for 3 min, more than 99.9% (3 logs) biofilm cells were removed. Because this strategy can be applied to different materials, it is a new generation of antifouling technology with promising biomedical applications.

532970 Developing Platform Biomaterials: From Messenger RNA Delivery to User-Friendly Synthetic Hydrogels

Abstract:

Leveraging principles from engineering and fundamental chemical synthesis, our research encompasses two primary areas: RNA delivery and user-friendly synthetic hydrogel development. Our interest in these areas is unified by a common question: how can we use rational design principles from organic chemistry to create platform biomaterials with improved function? In our talk, we will highlight both of our primary research interests, with an overarching goal of demonstrating how synthetic chemistry approaches can afford biomaterials for various applications. Below, our two primary subsections are described in brief.

<u>RNA Delivery</u>

Given their instability in serum and their limited ability to passively transfect cellular membranes, RNAs require delivery vectors to realize their full clinical potential.[1,2] Messenger RNAs (mRNAs), for example, could be used for biomedical applications including genomic engineering, cancer immunotherapy, and protein replacement strategies.[3] But when mRNAs are administered without a delivery vector into the blood stream, the immunological response, renal clearance, and rapid degradation prevent meaningful levels of accumulation (and subsequent translation) of the mRNA within target cell populations.[4] Lipid nanoparticles (LNPs) have emerged as an attractive non-viral delivery platform for improving the safety and potency of RNA therapeutics.[5,6] Nevertheless, their complex molecular composition makes it difficult to understand which chemical parameters ultimately impact the function of the LNP, thereby making it challenging to improve the efficacy and safety profiles of next-generation RNA delivery vectors. To address these questions, we have *i.* synthesized and purified rationally designed lipid materials of precise chemical structure, *ii.* formulated these materials into mRNA LNPs using microfluidic approaches, and *iii.* evaluated LNP potency and safety in vitro and in vivo as a function of the lipidâ€™s chemical structure. Our lipids, which are actively being pursued as lead materials with pharmaceutical partners, ultimately represent robust RNA delivery vectors with high potency, low toxicity, and tunable biodistribution profiles independent of molecular targeting ligands and sequence modifications.

<u>User-Friendly Synthetic Hydrogels</u>

Given their tunable mechanical properties and resemblance to soft tissue, hydrogels are advantageous in a wide range of biomedical applications including 3-dimensional cell culture, controlled release, and tissue engineering.[7,8] Recent advances in chemistry and materials science have led to the proliferation of synthetic hydrogels with reproducible mechanical properties for biomaterials application.[9] However, these systems can be difficult to implement broadly within academic and professional laboratories due to challenges associated with the synthesis, formulation, and scalability of these hydrogels and their precursors.[10] Here, we aim to address some of these limitations by presenting a class of mechanically- and kinetically-tunable hydrogels whose gelation occurs at physiologically relevant temperatures without the need for initiators, specialized laboratory equipment, or complex monomer synthesis. Specifically, our hydrogel self-assembles upon bench top mixing of commercially available small molecules with a decagram-scalable polyethylene glycol derivative. The design, synthesis, and mechanical properties of our hydrogel will be discussed alongside its application for 3-dimensional cell culture for scalable biomaterials evaluation. In doing so, we not only hope to present upon an alternative hydrogel platform for biomedical application, but also to highlight the importance of rational chemical design for the development of next generation biomaterials.

<u>References:</u>

1. Whitehead KA, Langer R, Anderson DG *Nat. Rev. Drug Discov.* **2** (2009) 516
2. Fenton OS, Olafson KN, Pillai PS, Mitchell MJ, Langer R *Advanced Materials*, Accepted.
3. Sahin U, Kariko K, Tureci, C *Nat. Rev. Drug Discov.* **13** (2014) 759-780
4. Kauffman KJ, Webber MJ, Anderson DG *J. Control. Release* (2016) 227-234
5. Fenton OS, Kauffman KJ, McClellan RL, Appel EA, Dorkin JR, Tibbitt MW, Heartlein MW, DeRosa F, Langer R, Anderson DG *Advanced Materials* **28** (2016) 2939-2943
6. Fenton OS, Kauffman KJ, Kaczmarek JC, McClellan RL, Jhunjhunwala S, Tibbitt MW, Zeng MD, Appel EA, Dorkin JR, Mir FF, Yang JH, Oberli MA, Heartlein MW, DeRosa F, Langer R, Anderson DG *Advanced Materials* **29** (2017) 1606944
7. Hoffman AS *Adv Drug Delivery Rev* **54** (2002) 3-12
8. Peppas NA, Hilt JZ, Khademhosseini A, Langer R **18** (2006) 1345-1360
9. Tibbitt MW, Anseth KS *Biotechnol Bioeng* **103** (2009) 655-663
10. DeForest CA, Anseth KS *Annual Rev. of Chem. and Biomol. Eng.* **3** (2012) 421-444

531262 Enzymatically Powered Surface-Associated Self-Motile Protocells

Abstract:

Protocell engineering is an active area of research in which one tries to re-create fundamental cell processes, such as motility, secretion, and division, in synthetic structures.[1, 2] Numerous examples of building synthetic motile systems in solution have been developed. A notable self-motile microstructure is the microswimmer, in which bisegmental rods made of platinum (Pt) and gold (Au) swim in hydrogen peroxide (H_2O_2) solutions.[3-5] The Pt end spontaneously oxidizes H_2O_2 and the Au end reduces water, leading to a directional flow which induces autonomous directional motion of the rod toward the Pt end.[5] Similar electrophoretic motions have been seen in both bimetallic structures, such as nanowires[6] and Janus nanospheres.[7, 8] Other platforms for recreating motility in solution using protocellular capsules have involved engineering a built-in asymmetry.[3-5, 9-11] Wilson and co-workers have made stomatocyte-like structures using asymmetric polymersome design, in which enzymatic activity of entrapped catalytic motifs lead to the directional expulsion of products and hence propulsion.[12] Recently, Battaglia and co-workers made synthetic asymmetric swimming nanoparticles that displayed directional motion into the glucose-rich brain by entrapping the enzyme glucose oxidase (which generates H_2O_2 from glucose) and catalase (which converts H_2O_2 to water and oxygen) within polymersomes.[13] These systems all involve the motion of a particle in solution and an inherent asymmetry in either the geometry of the particle or the distribution of its enzymatic elements.

In this study, we demonstrate the motility of an adherent particle by harnessing an asymmetric fluctuation of enzyme activity in a protocell. This type is more akin to eukaryotic cell motion, in which traction is mediated by adhesion and in which motility requires the breakage and formation of adherent contacts.[14] This work is different than seen with microswimmers in solution, because our vesicles do not have an inherent asymmetry and are adherent to a surface. We hypothesize that fluctuations in the force generation of enzymatic reactions on the surface of the protocell is sufficient to break adhesive contacts between polymer vesicles and the surface, and drive their autonomous motion, mimicking formation and breakage of adhesive contacts displayed by real mamalian cells. We incorporate the enzyme catalase into the lumen of biotinylated polymersomes that are adhesive to a solid surface that has been functionalized with a low density of avidin. Given that each turnover event in a typical enzyme reaction generates approiximately 10 pN of force[10, 15, 16] and the force to break an biotin-avidin bond is approximately 170 pN in an observation time of 3 millisecond,[17] a differential turnover of only @ 17 substrate molecules would be necessary to break a biotin-avidin contact and induce motion. Based on the large number of catalase molecules in the vesicle and its high turnover rate,[11] such a fluctuation seems achievable. We show that enzymatic reactions indeed induce motion similar to a random-walk and that the motility depends on the concentration of the substrate in the solution. Our results demonstrate that enzymatic reactions can be harnessed to generate motility of surface-adherent protocells, opening up numerous possibility for making motile protocells of different activity and specificity.

[1] W. S. Jang, S. C. Park, E. H. Reed, K. P. Dooley, S. F. Wheeler, D. Lee, D. A. Hammer, *Soft Matter* **2016,** *12*, 1014-1020.

[2] C. Martino, S. H. Kim, L. Horsfall, A. Abbaspourrad, S. J. Rosser, J. Cooper, D. A. Weitz, *Angewandte Chemie-International Edition* **2012,** *51*, 6416-6420.

[3] W. F. Paxton, C. C. Kistler, A. Olmeda, A. Sen, S. K. St Angelo, Y. Y. Cao, T. E. Mallouk, P. E. Lammert, V. H. Crespi, *Journal of the American Chemical Society* **2004,** *126*, 13424-13431.

[4] W. F. Paxton, A. Sen, T. E. Mallouk, *Chemistry-a European Journal* **2005,** *11*, 6462- 6470.

[5] W. F. Paxton, P. T. Baker, T. R. Kline, Y. Wang, T. E. Mallouk, A. Sen, A., *Journal of the American Chemical Society* **2006,** *128*, 14881-14888.

[6] N. Mano, A. Heller, *Journal of the American Chemical Society* **2005,** *127*, 11574- 11575. [7] T. C. Lee, M. Alarcon-Correa, C. Miksch, K. Hahn, J. G. Gibbs, P. Fischer, *Nano Letters* **2014,** *14*, 2407-2412.

[8] P. M. Wheat, N. A. Marine, J. L. Moran, J. D. Posner, *Langmuir* **2010,** *26*, 13052- 13055.

[9] K. K. Dey, X. Zhao, B. M. Tansi, W. J. Mendez-Ortiz, U. M. Cordova-Figueroa, R. Golestanian, A. Sen, A., *Nano Letters* **2015,** *15*, 8311-8315.

[10] H. S. Muddana, S. Sengupta, T.E. Mallouk, A. Sen, P. J. Butler, *Journal of the American Chemical Society* **2010,** *132*, 2110-2111.

[11] C. Riedel, R. Gabizon, C. A. M. Wilson, K. Hamadani, K. Tsekouras, S. Marqusee, S. Presse, C. Bustamante, C., *Nature* **2015,** *517*, 227-U288.

[12] D. A. Wilson, R. J. M. Nolte, J. C.M. van Hest, *Nature Chemistry* **2012,** *4*, 268-274.

[13] A. Joseph, C. Contini, D. Cecchin, S. Nyberg, L. Ruiz-Perez, J. Gaitzsch, G. Fullstone, X. H. Tian, J. Azizi, J. Preston, G. Volpe, G. Battaglia, *Science Advances* **2017,** *3*.

[14] D. A. Lauffenburger, J. J. Linderman, *Receptors: Models for binding, trafficking and signaling*. Oxford Press: New York, **1993**.

[15] P. J. Butler, K. K. Dey, A. Sen, *Cellular and Molecular Bioengineering* **2015,** *8*, 106-118

[16] S. Sengupta, K. K. Dey, H. S. Muddana, T. Tabouillot, M. E. Ibele, P. J. Butler, A. Sen, *Journal of the American Chemical Society* **2013,** *135*, 1406-1414.

[17] R. Merkel, P. Nassoy, A. Leung, K. Ritchie, E. Evans, *Nature* **1999,** *397*, 50-53.

529587 Rational Fabrication of Polymer-Graphene Based Scaffolds/Devices Using 3D Bioprinting and Microfluidics to Control Stem Cell Differentiation and Fate Commitment

Abstract:

The focus of my research endeavors is to elucidate the mechanisms underlying the relationship between advanced biomaterials and cellular interactions. Particularly I am pursuing research efforts in the development of functional biomaterials targeting 3D bioprinting for tissue/organ regeneration, drug/gene delivery for cancer therapy/theranostics, and bioinspired polymeric films/devices for biomedical applications. Each of these efforts have resulted in valuable contributions to the field and posed broader impact in translational research; however, here I will mainly discuss our past, present and future research efforts regarding the 3D bioprinted scaffolds for stem cell-based therapies and neural regeneration.

The clinical use of stem cell-based therapies for regeneration purposes is promising; however, it is limited due to the current difficulties associated with 1) non-scalable and reversible differentiation protocols; 2) design of implantable scaffolds with desired properties mimicking the complex 3D extracellular matrix microenvironment; and 3) precise control on the final fate of the implanted cell population.[1-3] In order to address these limitations, my research is concentrated on biomaterial design for stem cell-laden implantable scaffolds-based cellular therapies with a particular interest in stem cell differentiation and fate commitment for neural regeneration.

In our previous studies, we uniquely integrated various physical and chemical cues (using longitudinal micropatterns, surface gradient of chemicals/growth factors and neurotropic factors loaded micro/nanoparticles) in porous polymeric films providing controlled release of active components, proper microenvironment for cells and directed nerve growth.[4] In order to circumvent the limitations of 2D cell culture platforms, we exhibited significant control on mesenchymal stem cells (MSCs) to Schwann cell-like phenotypes (SCs) transdifferentiation via chemical stimuli by changing the 3D microstructural/mechanical properties of gelatin-based conduits.[5] Moreover, to eliminate the need for expensive chemicals and growth factors for chemical stimuli-based stem cell differentiation, we also demonstrated successful transdifferentiation of MSCs into SCs via sole electrical stimuli on conductive graphene-based 2D substrates.[6] This study selected as cover art for Advanced Healthcare Materials Journal, received Star Award from the Society for Biomaterials and resulted in provisional patent filing. In order to provide a mechanistic understanding of transdifferentiation process, we have recently conducted a detailed proteomics analysis and identified significantly regulated proteins and related cellular pathways upon MSCs to SCs transdifferentiation via chemical stimuli.[7]

Based on this background, we currently focus on designing 3D bioprinted, graphene microcircuits integrated, cell-laden, conductive, biodegradable and implantable conduits/scaffolds with tunable microstructural/mechanical properties. The main idea is that the synergetic effect of electrical and microstructural cues applied via sophisticated conduit design have the potential to eliminate the multiple procedures for ***in vitro*** transdifferentiation and enable precise control in situ-***in vivo*** transdifferentiation and fate commitment accompanied by directed migration and axonal regeneration. We also conduct a detailed mechanistic investigation to understand the regulated genes, proteins and cellular pathways upon transdifferentiation using the synergetic effect. For this study, I have recently submitted NIH Pathway to Independence Award (Parent K99/R00) proposal, which is currently pending. Another ongoing effort is to use 3D graphene foams integrated to the biodegradable polymeric films possessing surface micropatterns and porous microstructure as scaffolds for stem cell differentiation via electrical stimuli for neural regeneration. In this study, the purpose is to use electrical stimuli within a conductive 3D microstructure to control stem cell differentiation. In alignment with these studies, another aspect of my current research is focused on developing biodegradable polymer based, 3D microstructured, flexible electronic devices/films using a novel microfluidic approach coupled with 3D origami as an easy, fast, cost effective and green fabrication method. Our efforts for this study have resulted in submission of an invention disclosure and together with 3D bioprinting, it will be the main foundation of my future research endeavors. These flexible electronic devices hold promising potential to be used in various biomedical, tissue engineering and robotics applications. Thus, we investigate the biointerface of these devices for the purpose of microfluidics and organ/lab-on-a-chip applications for neural regeneration, which will be the main future directions.

Overall, the ultimate goal will be to make realistic 3D tissue models that mimic the actual cellular arrangement of tissues/organs using different strategies and mechanistically elucidating material-cell and cell-cell interactions.

References:

1. **Uz, M.;** Das Suprem, R.; Ding, S.; Sakaguchi Donald, S.; Claussen Jonathan, C.; Mallapragada Surya, K., Advances in Controlling Differentiation of Adult Stem Cells for Peripheral Nerve Regeneration. Advanced Healthcare Materials 0 (0), 1701046.
2. **Uz, M,** and Mallapragada, S.K.; Smart Materials for Nerve Regeneration and Neural Tissue Engineering. Smart Materials for Tissue Engineering: Applications, Qun Wang (Ed.), Royal Society of Chemistry, 2017, pp.382-408. ***(Print ISBN: 978-1-78262-484-4)***
3. J. Sandquist, **M. Uz,** A.D. Sharma, B.B. Patel, S.K. Mallapragada, D.S. Sakaguchi, Stem Cells, Bioengineering, and 3-D Scaffolds for Nervous System Repair and Regeneration: G.L. Zhang and L.D. Kaplan (Eds.), Neural Engineering: From Advanced Biomaterials to 3D Fabrication Techniques, Springer International Publishing 2016, pp. 25-81.
4. **Uz, M.,** Sharma, A., Adhikari, P., Sakaguchi, D., and Mallapragada, S.K., "Development of Multifunctional Films for Peripheral Nerve Regeneration Conduits, ***Acta Biomat.*** (2017). Available online. https://doi.org/10.1016/j.actbio.2016.09.039
5. **Uz, M.,** Buyukoz, M., Sharma, A., Sakaguchi, D., Altinkaya, S., and Mallapragada, S.K., "Gelatin-Based 3D Conduits for Transdifferentiation of Mesenchymal Stem Cells into Schwann-cell Like Phenotypes", ***Acta Biomat.***, 53, 293-306 (2017).
6. Das, S.,[#] **Uz, M.,**[#] Ding, S., Lentner, M., Hondred, J., Cargill, A., Sakaguchi, D.S., Mallapragada, S., and Claussen, J., "Electrical Differentiation of Mesenchymal Stem Cells into Schwann Cell-Like Phenotypes using Inkjet Printed Graphene Circuits", ***Healthcare Mater.***, 6(7), 1601087 (2017). Journal cover image. doi: 10.1002/adhm.201601087. ***([#] These authors put equal contribution)***
7. Anup D. Sharma, Jayme Horning, **Metin Uz,** Pawel Ciborowski, Surya K. Mallapragada, Howard E. Gendelman, and Donald S. Sakaguchi; "Proteomics Analysis of Mesenchymal to Schwann Cell Transdifferentiation.", ***Journal of Proteomics, June 2017, In Press,*** https://doi.org/10.1016/j.jprot.2017.06.011

525068 Using Biological Heterogeneity to Understand Disease: From Single Cells to Personalized Medicine

Abstract:

Heterogeneity is emerging as a defining feature of biology. This biological heterogeneity can be seen at multiple characteristic scales – from individual patients to single cells – and at multiple regulatory levels – genomic, transcriptomic, proteomic, and metabolomic. Originally perceived as “biological noise”, this heterogeneity is now being leveraged to understand fundamental principles of biology and design more complete and effective therapies to treat diseases.

My work uses computational methods to use biological heterogeneity at multiple scales to understand disease progression, understand dynamic regulation of disease processes, and predict more effective disease therapies and personalized therapies. Specifically, I use a combination of “big data” analyses from individual patients, single-cell RNA-seq and proteomic analyses, and personalized genome-scale metabolic modeling to understand dynamic metabolic regulation in and predict therapies for multiple diseases including non-alcoholic fatty liver disease and breast cancer in individual patients and in patient populations.

AUTHOR INDEX

AUTHOR INDEX

AUTHOR INDEX

AUTHOR INDEX

AIChE
120 Wall Street, FL 23
New York, NY 10005-4020

ISBN 978-1-5108-7607-1

CURRAN ASSOCIATES INC.
proceedings.com